T0135058

Studies in Computational Intelligence

Volume 710

Series editor

Janusz Kacprzyk, Polish Academy of Sciences, Warsaw, Poland
e-mail: kacprzyk@ibspan.waw.pl

About this Series

The series "Studies in Computational Intelligence" (SCI) publishes new developments and advances in the various areas of computational intelligence—quickly and with a high quality. The intent is to cover the theory, applications, and design methods of computational intelligence, as embedded in the fields of engineering, computer science, physics and life sciences, as well as the methodologies behind them. The series contains monographs, lecture notes and edited volumes in computational intelligence spanning the areas of neural networks, connectionist systems, genetic algorithms, evolutionary computation, artificial intelligence, cellular automata, self-organizing systems, soft computing, fuzzy systems, and hybrid intelligent systems. Of particular value to both the contributors and the readership are the short publication timeframe and the worldwide distribution, which enable both wide and rapid dissemination of research output.

More information about this series at http://www.springer.com/series/7092

Dariusz Król · Ngoc Thanh Nguyen
Kiyoaki Shirai
Editors

Advanced Topics in Intelligent Information and Database Systems

 Springer

Editors
Dariusz Król
Faculty of Computer Science
and Management
Wrocław University of Science
and Technology
Wrocław
Poland

Kiyoaki Shirai
School of Advanced Science
and Technology
Japan Advanced Institute of Science
and Technology
Nomi, Ishikawa
Japan

Ngoc Thanh Nguyen
Faculty of Computer Science
and Management
Wrocław University of Science
and Technology
Wrocław
Poland

and

Division of Knowledge and System
Engineering for ICT
Faculty of Information Technology
Ton Duc Thang University
Ho Chi Minh City
Vietnam

ISSN 1860-949X   ISSN 1860-9503   (electronic)
Studies in Computational Intelligence
ISBN 978-3-319-85965-1   ISBN 978-3-319-56660-3   (eBook)
DOI 10.1007/978-3-319-56660-3

Printed on acid-free paper

This Springer imprint is published by Springer Nature
The registered company is Springer International Publishing AG
The registered company address is: Gewerbestrasse 11, 6330 Cham, Switzerland

Preface

The concept and scope of information and database systems have been diversified recently and re-attracted widespread attention from academics and researchers at all level of experience. Many years of common practice have indicated that a large set of new techniques might be very helpful to solve some of the challenging real-world problems.

What has really happened? Firstly, the information and database systems have actually incorporated intelligence into their applications. Now, these systems may perform sophisticated, multidisciplinary tasks which are not possible by traditional computing paradigm. Furthermore, intelligent systems can imitate and automate some smart behaviours of thinking being. They are capable of learning, varying their state or action in response to past experience. Secondly, a dramatic increase in our ability to collect data from various devices and applications becomes a big problem. Internet yields every second a huge and constant flood of data. The digital data is doubling in size every two years. Therefore, new developments to exploit these innovations are strongly expected. On the other hand, as technologies become more complex, their links to science become stronger. The modern information and database systems need to address all these issues and updates still requiring further progress in the area.

This timely book published in the flagship Springer series "Studies in Computational Intelligence" presents a theory and practice of the ongoing research in intelligent information and database systems. The focus of this volume is on a broad range of methodological approaches and empirical reference points including algorithmics, artificial and computational intelligence, collaborative systems, decision management and support systems, natural language processing, image and text processing, Internet technologies, and information and software engineering. The carefully selected contributions to this volume were initially accepted for presentation as posters during the 9th Asian Conference on Intelligent Information and Database Systems (ACIIDS 2017) held on 3–5 April 2017 in Kanazawa, Japan.

The level of contributions corresponds to that of advanced scientific works, although several of them could be addressed also to non-expert readers.

The volume brings together 47 chapters divided into six main parts:

- Part I. From Machine Learning to Data Mining.
- Part II. Big Data and Collaborative Decision Support Systems,
- Part III. Computer Vision Analysis, Detection, Tracking and Recognition,
- Part IV. Data-Intensive Text Processing,
- Part V. Innovations in Web and Internet Technologies, and
- Part VI. New Methods and Applications in Information and Software Engineering.

The initial Part I explores different classification algorithms, optimization methods, and data mining techniques. Part II deals with the challenge of managing big and temporal data participating in collaborative, decision-making intelligent systems. Part III examines the latest developments in the field of computer vision and image processing, including collision detection, plant identification, defect classification, parking space prediction, foreground detection, and tracking of bone reparation process. In Part IV, the biggest in this book, we encompass a wide spectrum of approaches to automatic translation, latent semantic analysis, multi-label text classification, content analysis, photo-documentation, sentiment analysis, multi-sentence compression, plagiarism checking, and text summarization. Part V contains topics about indoor positioning, schema validation for open data, job description language, the effectiveness of knowledge-driven Web application, and building responsive data tables. A variety of methods and applications in information and software engineering are presented in the last part of this book (Part VI). It includes the range of subject matter devoted to data-driven forecasting model, compliance checking between the specification and implementation, test data generation, using agile methods, automated test cases generation from UML diagrams, optimal path calculation based on the distance, and finally, solving the state space explosion in modular model checking of component-based software.

In concluding, we would like to thank all the authors contributed to this book. We are also very grateful to the Program Committee members of ACIIDS 2017 who rigorously reviewed the papers with remarkable expertise and always constructive feedback provided to the authors, even under the pressure of extremely tight deadlines. Without often critical but substantive assistance, this volume would have been much less than it is. In addition, we extend our thanks to the editor of this series, Prof. Janusz Kacprzyk, and the executive editor from Springer, Dr. Thomas Ditzinger, for their continuous support and cooperation.

It is worth emphasizing that much theoretical and empirical work remains to be done. It is encouraging to find that more research on intelligent information and database systems is still required. All things considered, the prospects for the next ACIIDS conferences look good all around.

We hope the readers will find this book interesting, useful, and informative, and it will give them a valuable inspiration for original and innovative research.

Wrocław, Poland   Dariusz Król
Wrocław, Poland   Ngoc Thanh Nguyen
Nomi, Japan   Kiyoaki Shirai
January 2017

Contents

Part VI   New Methods and Applications in Software Engineering

Part I
From Machine Learning to Data Mining

Analyzing Accident Prone Regions by Clustering

Shuvashish Paul, Ashik Mostafa Alvi, Mahmudul Alam Nirjhor,
Shohanur Rahman, Adeeba Kashfee Orcho
and Rashedur M. Rahman

Abstract Traffic accidents and injuries related to them have unfortunately become a daily incident for the people of Bangladesh and this is particularly true for people living in Dhaka City. This paper aims to identify the most hazardous regions for such incidents within the Dhaka Metropolitan Region as well as assess their influences. This research effort collects accident related data from the Accident Research Institute (ARI) at Bangladesh University of Engineering and Technology (BUET), Dhaka. This paper utilizes the k-means clustering and expectation maximization method to cluster related incidents together.

Keywords Traffic accidents · Clustering · Accident prone region

S. Paul · A.M. Alvi · M.A. Nirjhor · S. Rahman · A.K. Orcho · R.M. Rahman (✉)
Department of Electrical and Computer Engineering, North South University,
Plot – 15, Block – B, Bashundhara 1229, Dhaka, Bangladesh
e-mail: rashedur.rahman@northsouth.edu

S. Paul
e-mail: reisende@outlook.com

A.M. Alvi
e-mail: alvibdj@gmail.com

M.A. Nirjhor
e-mail: 1nirjhor@gmail.com

S. Rahman
e-mail: shohan.nsu.cse@gmail.com

A.K. Orcho
e-mail: adeeba.kashfee@gmail.com

© Springer International Publishing AG 2017
D. Król et al. (eds.), *Advanced Topics in Intelligent Information
and Database Systems*, Studies in Computational Intelligence 710,
DOI 10.1007/978-3-319-56660-3_1

1 Introduction

Road accidents, formally known as traffic collisions refer to the incident where one vehicle collides with another vehicle, pedestrians, animals or any other object in the road. They can end up causing disability, disfigurement, loss of property and even loss of life in the most extreme cases. Several factors are known to influence such incidents—namely vehicle design, as well as several environmental, roadway and motorist characteristics. It is estimated that as many as 54 million people were affected globally by traffic collisions in the year of 2013 alone [1]. This ended up resulting in as many as 1.4 million fatalities [2]. These trends are continuing to surface in Bangladesh as well.

Since Bangladesh is a developing country, the penetration rate for car ownership is rather low. In fact, only about 3 people in every 1000 actually had access to a personal vehicle of their own in 2010 [3]. However, this is steadily rising as people's purchasing power increases. This means that the rate of traffic accidents too is on the rise in Bangladesh [4]. Accidental injury research was created to tackle this problem and has been largely successful in identifying such trends on a global scale. Such efforts however are not common in Bangladesh and a large portion of incidents are actually never even properly reported. For every two traffic collisions resulting in a fatality in the United States, there are as many as 160 such deaths in Bangladesh [5]. It is no surprise that Bangladesh reportedly has one of the highest road fatality rates in the world [6]. To realize national trends and factors fueling this problem, Bangladesh University of Engineering and Technology (BUET) has established the Accident Research Institute (ARI) which aggregates accidental injury as well as severity data through a direct collaboration with the Dhaka Metropolitan Police (DMP) [7]. We have collected some pre-aggregated data from the ARI and analyzed it to discover relevant trends as well as Dhaka's deadliest accident spots on a clustered level.

In brief, the major achievements of our work are: (i) Find and visualize the most accident prone areas within the Dhaka Metropolitan Region, (ii) Associate their impacts with local clusters built from the dataset, (iii) Build a cluster map signifying the most dangerous regions within Dhaka.

The remaining of the paper is arranged in 5 sections. Section 2 briefly discusses about the related work in this area. Section 3 showcases our research methodology in detail. Section 4 presents our research findings and Sect. 5 presents our final thoughts on the topic.

2 Related Works

Research related to the topic of traffic collisions and accidental injury is historically unfortunately rather uncommon in Bangladesh. In a recent study, it was found that nearly 22% of all reported traffic collisions in Bangladesh took place within Dhaka

Metropolitan City [8]. It was also found that large proportions of the main street network contained "blackspots," areas of high relevance that were the largest contributors to traffic accidents.

Another research effort pointed out these constraints, e.g., institutional weakness, lack of professional expertise, political support and proper policies to combat such problems in Bangladesh [9]. One researcher used Multinomial Logit Models (MNL) to analyze similar data but finally switched to Ordered Probit Regression since MNL could not provide ordering of severity levels [6]. They modeled injury severity using a 4 point Likert scale when preprocessing the accidental injury data.

Several researchers have recommended the use of k-means and its variant k-modes alongside Latent Class Clustering (LCC) in analyzing road accident data [10]. They also reported that LCC is computationally infeasible when the data contains a large number of categorical attributes. Other researchers also note that hierarchical clustering (such as Ward's method or single linkage method) can also be beneficial when analyzing road safety and incident reports [11]. DBSCAN, which is a density based clustering algorithm, is also highlighted when dealing with spatial data (such as longitude and latitude) as being particularly good as they are not affected by outliers. Fuzzy clustering approach has previously been used in determining blackspots that lead to accidents [12]. The authors utilized Fuzzy C-Means (FCM) clustering which is an extension of the k-means algorithm to the fuzzy framework. It utilizes membership degrees instead of binary attributes that says "this point is a member of the cluster," or "this point is not a member of the cluster."

3 Methodology

The dataset(s) we acquired from the Accident Research Institute (ARI) at Bangladesh University of Engineering and Technology (BUET) were unfortunately only available as a Portable Document Format (PDF). It was also pre-aggregated by intersections/major city areas.

3.1 The Dataset

As the dataset was only available as PDF and as none of the statistical and data mining software suites allow PDF inputs, we had to have it manually converted back to the Comma Separated Values (CSV) format. Our dataset was pre-aggregated by major intersections within Dhaka City and included a range of X/Y coordinates defined atop an arbitrary axis that takes Dhaka's GPO as the origin point and lays out other regions accordingly for a period of 10 years (2002–2012). An example value of those two attributes would be 304850–304879—which describes the range of the value of the X-coordinate for the area canonically known

Table 1 The dataset showcasing used attributes

Ro...	Tacc	Fatal	Ninjury	Pacc	Pfatal	Pninjiury	Peracc	Xavg	Yavg
1	16	12	3	11	10	1	68.800	304,865	72,275
2	19	13	5	9	9	0	47.400	304,385	72,875
3	16	16	0	14	14	0	87.500	304,805	72,305

as Jatrabari. A small snapshot (of 3 records) showcasing the attributes of dataset (pre-analysis) can be seen in Table 1.

Table 1 showcases all of the attributes that we are taking into consideration when clustering. We will briefly describe them: (i) TACC stands for Total Accidents; it is the number of total incidents in that aggregation (data) point, (ii) Fatal stands for the amount of accidents that prove fatal for one or more involved people, (iii) NINJURY stands for normal injury, it describes the amount of non-fatal injuries sustained, (iv) PACC stands for pedestrian accident, it is the number of incidents where pedestrians are involved, (v) PFatal stands for pedestrian fatalities, the number of incidents that result in one or more pedestrian fatalities, (vi) PNINJURY stands for pedestrian normal injury, the number of incidents that result in non-fatal injury to pedestrians, (vii) PERPACC stands for "% of pedestrian accidents," it simply signifies how many (out of total accidents for that location) incidents involved a pedestrian in it (viii) XAvg and YAvg roughly describe the aggregation point in a X–Y coordinate system *after* data preprocessing. That step is explained in detail in the next section. It is worth noting that several attributes were not considered as a part of the research (such as name of the general aggregation point and what type of road it is), we have chosen to ignore them from our analysis.

3.2 Data Preprocessing

Unfortunately, nominal clustering atop a range of values would not assist us in generating locally significant clusters to show the most hazardous areas within Dhaka. We had to get this converted into one value so we could represent this as a numeric attribute. Initially, we had chosen to simply take the starting part of the coordinate ("left of the hyphen") and discard the rest. However, this proved to introduce large amounts of error since Dhaka is a relatively small place area wise.

We then decided to split each of these attributes into two portions—one that was in the left of the hyphen, and the other that was in the right. Afterwards, the rounded (up) mathematical average of the two numbers was chosen to represent each location in the X–Y plane.

$$x(new) = \text{ceil}\left(\frac{x(left) + x(right)}{2}\right)$$

Several duplicate attributes had to be renamed to make the data importable in the data mining platforms that we put to use. The original dataset for example had two attributes with the same name of "Fatal." As a part of preprocessing, we also filled in missing values using the global mean for that attribute or column. This data was then converted into the Attribute-Relation File Format through the use of a Java program that is available as a part of the Weka tool suite [13] under the namespace of "*weka.core.converters*". The final dataset contains an aggregated report of 303 accidents out of which 189 (nearly 62%) proved fatal for one or more of the people involved in the accident. Preliminary analysis of the dataset is given: (i) Total Fatalities: 189 (62% of Total Accidents), (ii) Pedestrian Involvement 149 (49% of Total Accidents), (iii) Pedestrian Fatalities: 126 (42% of Total Accidents) (iv) Non-fatal/Normal Injury Count: 99 (33% of Total Accidents). The most common intersection count was Tee which appeared 24 times.

We also analyze the histograms for different attributes. For example, we can see that the majority of the values of TACC, which stands for Total Accidents, fall within the range of 0–24. We then apply data mining methods to form the clusters and attain aggregated results. These methods are: (i) K-means Clustering (with Euclidean distance) (ii) Expectation Maximization. In the next section, we describe *k*-means clustering in detail.

3.2.1   K-Means Clustering

K-means clustering is popular among all other clustering techniques used in data mining. The algorithm can be divided up into three steps: step (1) Given a list of records or examples, assign each record to the cluster that has the "nearest" distance. In this instance, we are using the Euclidean distance; step (2) Calculate the new mean of each cluster and create new centroids using them; step (3) Repeat from step-1 until the centroids no longer change

Usually, centroids are either randomly assigned from random records (which is the Forgy method), or a random cluster is assigned to each record and the update process is begun [14]. We are using the *k*-means implementations of Weka and Rapidminer in this paper. The main attribute(s) that we are clustering in this case are the X and Y coordinates of the X–Y plane system. They describe positional information within the grid, and we aggregate around this to derive the fatality and injury rates.

The attributes labelled SN (Serial Number), Name (Canonical Name of the Area), Type (Intersection Type), Xarea (X range of the Area) and Yarea (Y range of the Area) were ignored when generating the cluster (since they are nominal, and *k*-means cannot deal with nominal values directly.)

Now, before we could run the clustering algorithms, we had to figure out suitable values of *k* since both *k*-means and Expected Maximization (EM) need the user to tell them the number of clusters that we would like to form. We did some tests that measured the Sum of Squared Error (SSE) for $k = 2$, $k = 3$ and $k = 4$. Those results can be seen in the Table 2.

Table 2 SSE comparison for different values of K

KSSE	Distribution
17.70	C0: 48, C1: 6
15.83	C0: 47, C1: 5, C2: 2
12.7346	C0: 41, C1: 5, C2: 1, C3: 7

The distribution field in the above table describes how many records ended up in which cluster. C0: 47 for example states that 47 records were in cluster 0. SSE calculation was done using Weka.

Keeping in mind that Dhaka is a small city and partitioning it up in too many clusters will probably make the bigger picture impossible to see, we decided to try working with $k = 3$ and $k = 4$. K = 5 was ignored because it resulted in the creation of clusters with only one element in it. We ultimately decided to use $k = 3$ as it had an acceptably lower SSE value and did not result in the creation of single member clusters. We did the modelling in both Weka and Rapidminer (to be able to compare between multiple runs and implementations).

The Centroid Table(s) Table 3 containing the clustered values can be seen below.

Cluster 0 ends up with 87% of all records (47 aggregation points), Cluster 1 has only 9% (5 aggregation points) while Cluster 2 has 4% (2 aggregation points). Similarly, in iteration 2 using Rapidminer, the following Table 4 is discovered.

We end up with a similar cluster membership breakdown where Cluster 0 has 38 aggregation points, Cluster 1 has only 1 aggregation point and Cluster 2 has 15 aggregation points. These tables show the values of each centroid that is at the center of each cluster generated. For example, using Weka, the average value of total accidents for Cluster 0 is 4.10. For cluster 1, this is 18.6, the highest among all the clusters. Cluster 2 is in the middle with 8.5. It is important to note that k-means has a chance of differing from implementation to implementation as the results entirely depend on how the initial centroids are chosen. The average of sum of squared errors in three clusters is 15.82 for k-means clustering algorithm.

3.2.2   Expectation Maximization Clustering

Expectation Maximization, commonly known as EM refers to a recurrent process of visualizing the maximum likelihood estimates of parameters that belong to some statistical model. More information on EM could be found elsewhere [15, 16]. We will be using the EM implementation of the (Weka) in this paper. We chose to create 3 clusters with maximum iterations set to 100 and minimum standard deviation set to 1.0E-6. Table 5 depicts the result derived for EM algorithm. The values in the brackets represent the standard deviation across the mean value as derived by the EM algorithm while the values outside the brackets are the avg. mean across that cluster. For example, the average count of Total Accidents in Cluster 0 is 5.8688 ± 2.23, while in Cluster 1 it is 17.1636 ± 4.18 and in Cluster 2 it is 1.4823 ± 1.15. Other values are distributed similarly. The cluster membership

Table 3 Centroid values for each cluster's attributes

Cluster	Attribute	Toolkit	Value
Cluster_0	Total Accidents	Weka	4.1064
Cluster_0	Fatal Accidents	Weka	2.3404
Cluster_0	Normal Injury	Weka	1.5957
Cluster_0	Ped. Accident.	Weka	1.6383
Cluster_0	Ped. Fatality	Weka	1.3404
Cluster_0	Ped. Normal Injury	Weka	0.383
Cluster_0	% of Ped. Acc.	Weka	37.3085
Cluster_0	X-coord	Weka	304493.5106
Cluster_0	Y-coord	Weka	72881.7021
Cluster_1	Total Accidents	Weka	18.6
Cluster_1	Fatal Accidents	Weka	13.6
Cluster_1	Normal Injury	Weka	3.6
Cluster_1	Ped. Accident.	Weka	11.8
Cluster_1	Ped. Fatality	Weka	11.2
Cluster_1	Ped. Normal Injury	Weka	0.8
Cluster_1	% of Ped. Acc.	Weka	63.78
Cluster_1	X-coord	Weka	304607
Cluster_1	Y-coord	Weka	72975
Cluster_2	Total Accidents	Weka	8.5
Cluster_2	Fatal Accidents	Weka	5.5
Cluster_2	Normal Injury	Weka	3
Cluster_2	Ped. Accident.	Weka	6.5
Cluster_2	Ped. Fatality	Weka	3.5
Cluster_2	Ped. Normal Injury	Weka	3
Cluster_2	% of Ped. Acc.	Weka	75.7
Cluster_2	X-coord	Weka	304415
Cluster_2	Y-coord	Weka	72777.5

Table 4 Rapidminer clustering around significant attributes

Attribute	Cluster_0	Cluster_1	Cluster_2
TACC	5.211	8	6.467
Fatal	3.105	6	4.333
NINJURY	1.632	6	2.067
PACC	2.474	4	3.400
PFatal	2.026	4	3
PNINJURY	0.526	0	0.533
PERPACC	37.784	50	49.200
Xacg	304520	304595	304447
Yavg	72626.711	75515	73369.333

Table 5 Clustered attribute values according to EM

Attribute	Tool	Cluster_0	Cluster_1	Cluster_2
Total accidents	Weka	5.8688 (2.23)	17.1636 (4.18)	1.4823 (1.15)
Fatal accidents	Weka	3.34 (1.94)	12.50 (3.41)	0.86 (0.94)
Normal injury	Weka	2.20 (1.97)	3.50 (1.80)	0.72 (1.21)
Ped. accidents	Weka	2.53 (1.08)	11.16 (3.33)	0.41 (0.49)
Ped. fatalities	Weka	1.95 (0.96)	10.16 (3.81)	0.41 (0.50)
Ped. normal injuries	Weka	0.72 (0.74)	1.17 (1.07)	0 (0)
% of Ped. accidents	Weka	46.03 (16.63)	66.48 (15.81)	25.38 (36.66)
X-coord	Weka	304501.47 (156.97)	304525.02 (253.14)	304492.85 (240.13)
Y-coord	Weka	72872.81 (584.56)	72991.41 (628.94)	72874.23 (435.07)

distribution included 26 (48%) aggregation points within Cluster 0, 6 (11%) aggregation points within Cluster 1 and 22 (41%) aggregation points within Cluster 2. The log likelihood, which is a statistical attribute for measuring the quality of a model, was −25.27 for this estimation. Lower values are preferable because it indicates maximization of the EM likelihood function. Table 5 depicts the results.

4 Our Findings

If we take the standard deviation values from EM into account, it appears that both k-means and EM seem to produce similar results. To illustrate this, we manually attach labels to the dataset aggregation points as follows—total accident counts greater than 15 is labelled 'highly dangerous,' counts between 5 and 14 is labelled 'moderately dangerous,' and counts less than 5 are labelled 'moderately safe.' We then compare the results of the two clustering algorithms to find the best one. Those results are illustrated in the validation table, i.e., Table 6. It compares the two clustering algorithms based on their accuracy.

As we can see from the Table 6, EM results in better accuracy. We believe this is due to suboptimal centroid formation on the k-means implementations (since it is randomized on every run)—EM does not have to deal with this problem. Generally speaking, k-means can be thought of as a variant of EM with an assumption that it will result in the creation of spherical clusters. It does this by "hard assigning" data points to specific clusters when it converges, while EM only "soft assigns," as in it is open to entertaining the possibility of a certain point belonging to any centroid.

After going through the results generated, we can come up with the following findings. To make this easier to visualize, we have plotted the clusters atop a map of the Dhaka City in Fig. 1.

Table 6 Cluster validation between methods

Method	Correctly classified records	Accuracy (%)
Expectation Maximization (EM)	48	88.89
K-means	41	75.93

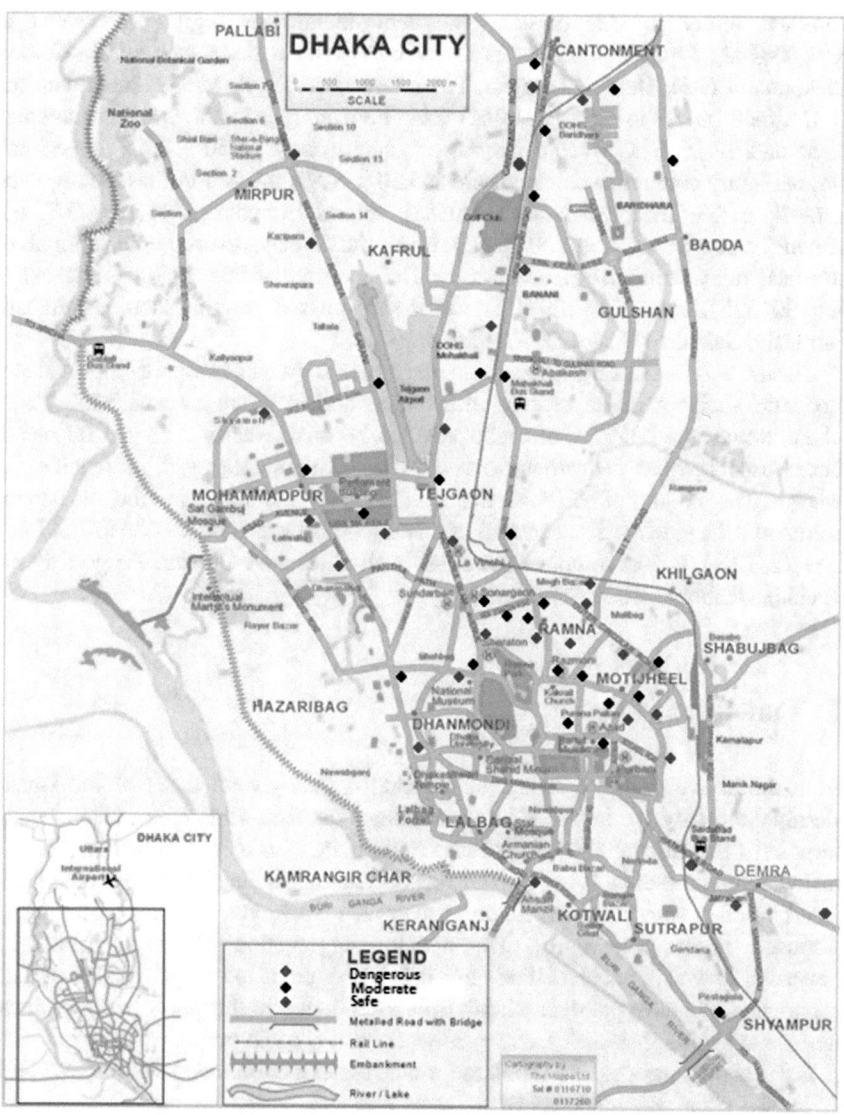

Fig. 1 Cluster map

Cluster 0 proves to be the safest with only around 4 or 5 total accidents reported in the entire dataset when using k-means. EM however finds another set with a much lower accident count—around 1.48 ± 1.15 within Cluster 2. Fatal accidents among those are even lower (around 0.86 ± 0.94), normal injury rate is around 0.72 ± 1.21. On average, pedestrians are involved in 0.41 ± 0.49 incidents out of which 0.41 ± 050 incidents result in fatalities. No pedestrians suffer a normal or non-fatal injury in this cluster. The centroid of this region is located at $X = 304492.85$ and $Y = 72874.23$ and encompasses regions such as Zia Colony Cantonment Gate, Osmani Uddyan, Tongi Diversion Road, Mohakhali and so on.

Cluster 1 proves to be the deadliest using both analytic methods with an average of around 17.16 ± 4.18 accidents out of which nearly 12.50 ± 3.41 prove fatal. Normal injury count is the highest here at 3.50 ± 1.80, and pedestrians are involved in nearly 67% of these incidents. The pedestrian accident count is 11.16 ± 3.33 out of which nearly 10.16 ± 3.81 prove to be fatal. Pedestrians suffer normal or non-fatal injuries in 1.14 ± 1.07 cases. The centroid of this region is located at $X = 304525.02$ and $Y = 72991.41$ and it encompasses regions such as Jatrabari, Farmgate, Saidabad, Jasim Uddin Crossing, etc.

Cluster 2 (while using k-means) and Cluster 0 (while using EM) provides a moderate accident prone cluster where total accidents are around 5.87 ± 2.23 where around 3.34 ± 1.94 of them prove to be fatal. Nearly 2.53 ± 1.08 out of these cases involve pedestrians out of which 1.95 ± 0.96 end in fatalities in average. The amount of accidents that result in non-fatal injuries to the pedestrians is around 0.72 ± 0.74. The centroid of this region is located at $X = 304501.47$ and $Y = 72872.81$ and encompasses regions such as Shapla Chattar, Progoti Sarani (Badda), Shahbag, Bijoy Sarani, etc.

5 Conclusion

From the above findings, it can be concluded that several areas of the Dhaka Metropolitan City are indeed a lot more dangerous than other areas. The primary purpose of this research paper is to locate those areas so measures can be taken to deal with the problems. We are glad to report that our findings match up with those found by other researches. Kamruzzaman et al.'s work on a similar subject matter produced similar results [6]. Traditionally, aggregation has been used with geospatial as well as accidental research data in this arena to recognize areas of high importance. We have put data-mining techniques to use in this paper to do the same and can conclude that such techniques work well for this purpose.

Data scarcity has posed a problem for our research effort since day one. While it is known that Bangladesh is a very accident prone country [1, 2], the reporting/recording of information relating to these incidents is often entirely manual and forgotten. We want to give suggestions to higher authorities for better management of accident related data. More extensive research initiative could be taken by other researchers on this computerized rich data set.

References

1. Global Burden of Disease Study 2013, Collaborators: Global, Regional, and National Incidence, Prevalence, and Years Lived with Disability for 301 Acute and Chronic Diseases and Injuries in 188 Countries, 1990–2013: A Systematic Analysis for the Global Burden of Disease Study 2013, 20 Apr 2016. doi:10.1016/S0140-6736(15)60692-4
2. Global, Regional, and National Age–Sex Specific All-Cause and Cause-Specific Mortality for 240 Causes of Death, 1990–2013: A systematic analysis for the global burden of disease study 2013. The Lancet **385**(9963), 117–171 (2015). doi:10.1016/s0140-6736(14)61682-2
3. The World Bank. Motor Vehicles (per 1,000 People): The World Bank. 2013. Archived from the Original http://data.worldbank.org/indicator/IS.VEH.NVEH.P3 (2014). Accessed 20 Apr 2016
4. Momin, S.M.: Traffic accidents: are the drivers solely to blame? The Independent. http://www.theindependentbd.com/printversion/details/10945 (2015). Accessed 20 Apr 2016
5. Walsh, D.: Can an app really make Bangladesh's notoriously deadly roads safer? The Guardian. http://www.theguardian.com/world/2015/mar/13/bangladesh-deady-road-accidents-criticalink-app-emergency-services (2015). Accessed 20 Apr 2016
6. Kamruzzaman, M., Haque, M.M., Ahmed, B., Yasmin, T.: Analysis of traffic injury severity in a mega city of a developing country. Paper presented at the 4th Road Safety International Conference held in Sydney, Australia, 4–5 Mar 2013
7. Road Safety Facts, 1st edn. Accident Research Institute (ARI), Dhaka (2013). Web, 20 Apr 2016
8. Hoque, M.S., Mahmud, S.M., Kawsar, C., Siddiqui, A.: Road Safety in Bangladesh and Some Recent Advances, 1st edn. Accident Research Centre (ARC), Dhaka (2012). Web, 20 Apr 2016
9. Mahmud, S.M., Hoque, M.S.: Proceedings of the 4th Annual Paper Meet and 1st Civil Engineering Congress, 22–24 Dec 2011, Dhaka, Bangladesh. https://www.k4health.org/sites/default/files/30bfullpaperroadsafetyresearch.pdf. Accessed 20 Apr 2016
10. Kumar, S., Toshniwal, D.: A data mining framework to analyze road accident data. J. Big Data. doi:10.1186/s40537-015-0035-y. Accessed 20 Apr 2016
11. Doğru, N., Subaşi, A.: Comparison of clustering techniques for traffic accident detection. Turk. J. Electr. Eng. Comput. Sci. **23**, 2124–2137 (2015). doi:10.3906/elk-1304-234
12. Sazi, Y.: Fuzzy clustering approach for accident black spot centers determination. In: Fuzzy Logic—Emerging Technologies and Applications (2012). doi:10.5772/35521
13. Waikato University: Converting CSV to ARFF. WEKA. https://weka.wikispaces.com/ConvertingCSVtoARFF. Accessed 20 Apr 2016
14. Hamerly, G., Elkan, C.: Alternatives to the K-means algorithm that find better clusterings. In: Proceedings of the Eleventh International Conference on Information and Knowledge Management—CIKM '02 (2002). doi:10.1145/584792.584890
15. Chen, Y., Gupta, M.R.: EM demystified: an expectation-maximization tutorial. University of Washington, 20 Apr 2016
16. Hartley, H.O.: Maximum likelihood estimation from incomplete data. Biometrics **14**(2), 174 (1958). doi:10.2307/2527783

Analyzing Life Insurance Data with Different Classification Techniques for Customers' Behavior Analysis

Md. Saidur Rahman, Kazi Zawad Arefin, Saqif Masud, Shahida Sultana and Rashedur M. Rahman

Abstract Analyzing data of life insurance companies gives an important insight on how the customers are reacting to the offered insurance policies by the companies. This information can be used to predict the behavior of future policy holders. Life insurance companies maintain a large database on their customers and policy related information. Data mining technique applied with proper preprocessing of data prove to be very efficient in extracting hidden information from data stored by life insurance companies. There are many data mining algorithms that can be applied to this huge set of data. The main focus of our work is to apply different classification techniques on the data provided by a life insurance company of Bangladesh. Attribute selection techniques are applied to properly classify the data. Classification techniques proved to be very useful in classifying customers according to their attributes. A comparative analysis of the performance of the classifiers is also reported in this research.

Keywords Data mining · Data balancing · Life insurance · Machine learning · Custer behavior analysis

Md. Saidur Rahman · K.Z. Arefin · S. Masud · S. Sultana · R.M. Rahman (✉)
Department of Electrical and Computer Engineering, North South University,
Plot – 15, Block – B, Bashundhara, Dhaka 1229, Bangladesh
e-mail: rashedur.rahman@northsouth.edu

Md. Saidur Rahman
e-mail: saidurrahman@northsouth.edu

K.Z. Arefin
e-mail: kazi.arefin@northsouth.edu

S. Masud
e-mail: saqif.masud@northsouth.edu

S. Sultana
e-mail: shahida.sultana@northsouth.edu

© Springer International Publishing AG 2017
D. Król et al. (eds.), *Advanced Topics in Intelligent Information
and Database Systems*, Studies in Computational Intelligence 710,
DOI 10.1007/978-3-319-56660-3_2

15

1 Introduction

Life insurance company deals with a huge amount of data acquired from their policy holders. The data is generally related with customers' personal information, health information and current financial status. Companies also record information about the policy the customers are buying and premium payment. Most often these attributes indicate whether a customer can be regular or irregular in his installment payment. This information is very crucial for an organization like life insurance companies because they want to design their policy in such a way that it can attract most of the customers who are willing to pay installments.

A life insurance company is a financial organization which offers insurance policies to its customers. An insurance policy is a legal agreement between an insurance provider and an insurance policy holder. The insurance company promises to pay a designated amount of money to the insurance holder according to the insurance policy. The amount of money and conditions in which the policy holder is paid depends on the insurance policy or agreement offered by the insurance company. Policy holders often have to pay a monthly or yearly premium against their policy. Here comes the matter of regularity of a customer. The insurance company is interested about the regular customers because it increases their chance of profit as a company. Again customers who are regular are benefited from the insurance policy they have bought because it ensures their chances of successful insurance claim.

To classify the customers as regular or irregular different data mining techniques prove to be useful. Mainly classification algorithms are used to classify different types or classes of data from a dataset. Our target is to classify the customers based on their given attributes so that we can predict the class label for future customers. Classification techniques such as JRIP, Naïve Bayes, IBK, PART etc. are used and their performance is compared to find the best suited classifier for the acquired dataset. Data mining techniques are generally used to find interesting patterns in the dataset to provide useful information for the future. For our dataset the data is imbalanced on a particular class label. So we have to balance our data using balancing techniques such as Random Over Sampling (ROS) and Random Under Sampling (RUS).

2 Related Work

Kirlidog and Asuk's [1] worked on fraud detection approach on health insurance company data. They applied anomaly detection, clustering techniques and classification to detect anomaly in data. First they used statistical approach on the data collected for calculating percentage of rejected claims. Next they divided their anomaly detection into two criteria. First they considered excessive claim by different types of health centers. Next they considered individual health centers and calculated excessive claim for them. The combined result gave them a predictive model for anomaly detection

Xiayun and Danyue [2] also designed an algorithm called RB algorithm to detect outlier in the evaluation of client moral risk. The algorithm along with the application of density factor detected the chances of client to have high moral risks. This RB algorithm first selected an initial resolution for a data set and the density was updated if the cluster size increased with the changing resolution of the algorithm. They also considered seven paramount attributes that effected greatly in client moral risk. They worked on a data of Chinese Health Insurance Company.

Yan and Xie [3] worked with data mining techniques on the Chinese insurance companies. They proposed decision tree based classification and application of data mining in CRM (Client Relation Management) model. They also proposed to apply it in risk management. They also mentioned about the need of insurance companies to maintain a large data center or warehouse to efficiently store the information.

Goonetilleke and Caldera [4] used data mining techniques on life insurance company data to evaluate the possibility of attrition of a customer. They first selected effective attributes using CFS (co-relation based feature selection) method. Then they used different classification algorithms and ranked the effective attributes. As the initial data set was imbalanced they used cost sensitive learning approach. They considered different stages of an insurance policy and tried to evaluate the attrition probability of customer depending on the current stage of his insurance policy. They also evaluated the classification techniques on different classification evaluation metrics. Finally they developed a cost matrix to evaluate the attrition cost of a customer.

Thakur and Sing [5] used decision tree based classification technique for developing a prediction system on customer data. They had a training data of customers who wanted vehicle insurance from online. Based on the customers' attributes they classified new customer for their interest in online insurance. They evaluated their classifier based on its accuracy and error rate. Their main target was to classify based on the age of customer and educational status and the type of vehicle they own. They built a system that provided all the necessary information for vehicle insurance online.

3 Dataset and Tools

We have collected the data from Prime Islami Life Insurance Company Ltd., of Bangladesh which has the data from almost every division of Bangladesh. The timeline of data was from 2011 to 2014 and it has data of about 282,282 policy holders. We want to determine that if a customer is regular or irregular in installment payment. So we introduced a new attribute named 'Regularity'. This attribute consists of two values which are regular/irregular. We have also assumed that the customers who are regular can complete the installments of policy and the irregular ones will fail to do it. To properly assess a customer as capable of full installment payment we required data that would range over a larger timeline which is rare to acquire so we assumed the regularity factor. We determined the regularity based on the following facts which are the starting date of a policy, the last payment dates of every policy and if a customer has paid his dues before these dates. One major factor is that the dataset was greatly imbalanced considering the 'Regularity' attribute. We had to maintain the ratio throughout data preprocessing.

We used "WEKA" for implementing different algorithms of balancing and classification techniques and used MySQL server for storing and preprocessing the dataset.

4 Methodology

Our approach contains mainly three steps—"Data Preprocessing", "Attribute Evaluation" and "Classification Technique Implementation". At first in "Data Preprocessing", we prune the less important attributes from the dataset and then prune the incomplete records. Sampling is used as the original dataset is huge. But we need to maintain the ratio of 'class' attribute in sample dataset as of that in original dataset. Then, in "Attribute Evaluation", we figure out which attributes are worthy to consider for classification. In "Classification Technique Implementation", we explicitly balance the dataset and then apply different classification techniques.

To find out the most effective classifier, we go through a number of processes. We can breakdown our tasks or processes in following parts

1. Data Preprocessing
2. Attribute Evaluation/Ranking
3. Classification Technique Implementation

    3.1 Class Label Balancing
    3.2 Classification

To preprocess the data it is required that we understand the collected data properly. As insurance policies are difficult to understand in general we have to study about the methods of the storage of data by an insurance company. We prune the data so that the attributes that could have maximum effect on specifying a customer as regular or irregular could be obtained. In the attribute evaluation/ranking section we apply certain attribute evaluator techniques and compare them so that the best set of attributes could be chosen for classification. We again apply class label balancing techniques so that we could balance the data that is previously unbalanced with respect to class label. We use a number of classification techniques such as JRIP, SMO etc. and compare them so that we could determine the best classification model. Details of these parts are described in the following subsections.

4.1 Data Preprocessing

Data preprocessing part is a two steps process—"Pruning" and "Sampling".

4.1.1 Pruning

The initial data set contains more than 100 attributes. Most of these attributes are found to be irrelevant by us. Based on our study and research we select 10 attributes (POLICY TERM, AGE, SEX, OCCUPATION, URBAN-RURAL, MARITAL STATUS, SUM-ASSURED, DIVISION, PREMIUM PAYMENT MODE and REGULARITY) that we figure out effective in determining the regularity of a customer. We also have to discard many of the records because they contain null values. Brief descriptions of these attributes are as follows.

(a) Policy Term (PT): This refers to the time span through which the policy holder will pay his or her premiums.
(b) Age (A): It points the age of the policy holder when s/he starts the policy.
(c) SEX (SX): It marks the gender of the policy holder.
(d) OCCUPATION (O): The occupational status (i.e., teacher, electrician or housewife) of the policy holder, when s/he starts the policy, is described here.
(e) URBAN-RURAL (UR): This refers to the policy holders dwelling status whether s/he is a city dweller or lives in village area.
(f) MARITAL STATUS (MS): It marks the marital status (married, unmarried or divorced, single) of the policy holder.
(g) SUM-ASSURED (SA): The pre-decided amount which the insurer promises to pay the nominee in case of the policyholder's death.
(h) DIVISION (D): It marks the division (Dhaka, Sylhet etc.) of Bangladesh, the policy holder registered his or her policy from.
(i) PREMIUM PAYMENT MODE (P): It marks the breakdown of the payment of the policy holder's total premium (i.e., yearly or monthly)
(j) REGULARITY (R): It marks the policy holder as "regular" or "irregular" based on whether s/he is paying the premiums on time.

4.1.2 Sampling

We initially have a data set of about 282,282 policy holders and we prune the data to about 10,000 policy holders for faster processing of data. We also maintain initial ratio of the data so that the integrity of the data is not lost while pruning. The reason behind doing so is that the use of classification techniques over full dataset is highly time-consuming and also it does not make any notable accuracy.

4.2 Attribute Evaluation

The attributes selected during pruning as relevant for classification often prove to be ineffective in action. To analyze effectively, we may have to discard some attributes which are selected during pruning. If we take all the 10 attributes for classification the classification would be unnecessarily sparse and difficult to interpret. The model also tends to be data driven which is it tries to memorize its training set and fits the testing set accordingly.

We apply different attribute evaluator techniques to determine the most effective attributes for classification. The first two techniques which we apply are 'information gain attribute evaluation' and 'gain ratio attribute evaluation techniques'. From 'information gain attributes evaluation' we found P, PT, SA, A, O and from 'gain ratio attribute evaluation' we found P, PT, SA, A, MS attributes are relevant.

We use "Greedy Step Wise" algorithm as the search method for Correlation based Feature Selection (CFS). In our approach, we get "Premium Payment Mode" (P) as the most effective attribute for our dataset.

We also use 'Classifier attribute evaluation' technique which determines the effective attributes for the designated classifier.

For each classifier that we use later we chose the "worthy" attributes by applying these techniques. Here, the "worthy" attributes are those which are common in the findings from the 'information gain attribute evaluation' and 'gain ratio attribute evaluation techniques' and 'Classifier attribute evaluation' for that specific classifier. This is the common practice for finding the relevant attributes for classification.

4.2.1 Classification Technique Implementation: Class Label Balancing

In Fig. 1, we see that the original ratio of regular and irregular percentage is maintained in the sample data. Maximum (almost 76%) of our data belongs to 'irregular' class which indicates there is an imbalance in the data set. The problem occurs when we try to apply classification techniques on this imbalanced data set as the result of the classification becomes biased towards the class of larger percentage. It means that most of the prediction is labeled as the major class which in this case is the 'irregular' class. To balance our data set we decide to apply both under-sampling and over-sampling techniques. We apply 'RUS' (Random Under Sampling) for under-sampling and SMOTE [6] (Synthetic Minority Over Sampling Technique) for over-sampling.

RUS algorithm tries to minimize the number of majority class randomly by a ratio. In our analysis we set the distribution spread value as 1 so that the ratio of majority and minority becomes 1. SMOTE algorithm tries to over sample the minority class using k-nearest neighbor calculation without sample replacement. It means SMOTE creates synthetic values by calculating random k-nearest neighbors from the designated minority class and increases the number of tuples in the data set. We set the parameters of the SMOTE algorithm as follows, K = 5; It indicates

Fig. 1 Regularity ratio in the main dataset and sample dataset

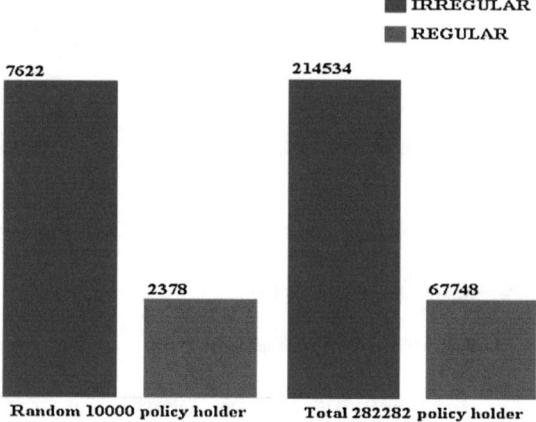

the number k-nearest neighbors considered T = 2378; The original number of records in data set labeled as regular N = 220.5%; amount of SMOTE percentage which means the minority class is oversampled by 220.5% from the original data. Total 5244 data of regular class is synthetically created and added to our data set.

4.2.2 Classification Technique Implementation: Classification

After pruning, balancing and attribute selection of data, we apply different classification techniques to figure out which techniques are most effective and yield more accurate results. We apply two methods to balance our dataset. One of them is the RUS and another one is SMOTE which balances the data by under sampling and over sampling respectively. We applied RIPPER, Naïve Bayes, IBK, SMO, Multilayer Perceptron and PART on our data. Each algorithm is tested on both balanced data acquired from applying the RUS and SMOTE methods. We have compared the ROC curve among the two balanced data for each algorithm. This process is done to figure out which algorithm is effective. Below some classification techniques are briefly discussed.

'Ripper' is a classification algorithm. It is a rule based direct classifier. It is best for imbalanced class distribution as it uses a validation set for preventing over-fitting of model. It uses general to specific rule growing approach with Foil's information gain to measure the performance of the rule. It works well with large data set as it does not consider all of the training data for rule production. 'Naïve Bayes' is another classification approach which uses the Bayesian probability considering every attribute independent from each other and calculates their probability when probability of class is given. It chooses the value of class label that maximizes the output of probability of attributes used for classification. This algorithm considers the attributes to be independent. ROC curve for the implementation of JRIP and Naïve Bayes are shown in Fig. 2.

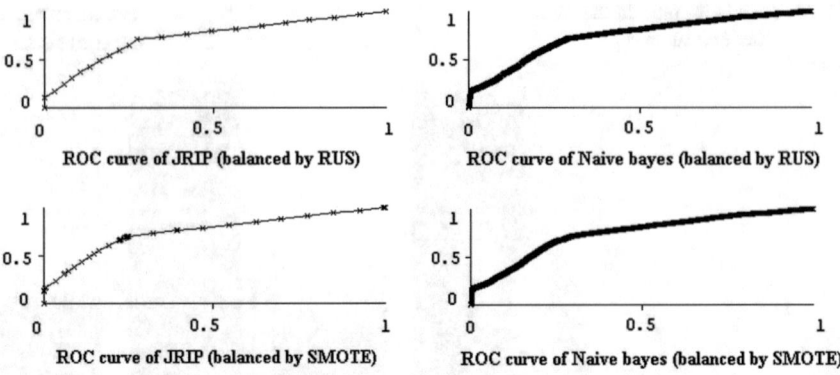

Fig. 2 ROC curve of JRIP and Naïve Bayes after applying RUS & SMOTE

'IBK' (K-nearest Neighbor Classifier) is a classifier that uses the theory of distance based weighting. It tries to find the nearest neighbors depending on the cross validation. It depends solely on how many nearest neighbors are considered during calculation. 'SMO' (Sequential Minimal Optimization) is a 'SVM' (Support Vector Machine) algorithm used for classification. It solves quadratic problems by dividing them into smaller parts and uses analytical approach to solve them. ROC curves for the implementation of IBK and SMO are shown in Fig. 3. As for parameter in IBK, we have used k = 5, seed = 1, and Euclidean distance is used for distance calculation.

'MP' (Multilayer Perceptron) is an artificial neural network that uses a graphical approach to adapt to a selective problem. It is constructed as a graph of nodes with connected edges with weights which are updated during learning. The update occurs during 'back propagation' a method used by the MP to minimize the loss function. 'PART' is classification technique that uses the divide and conquers

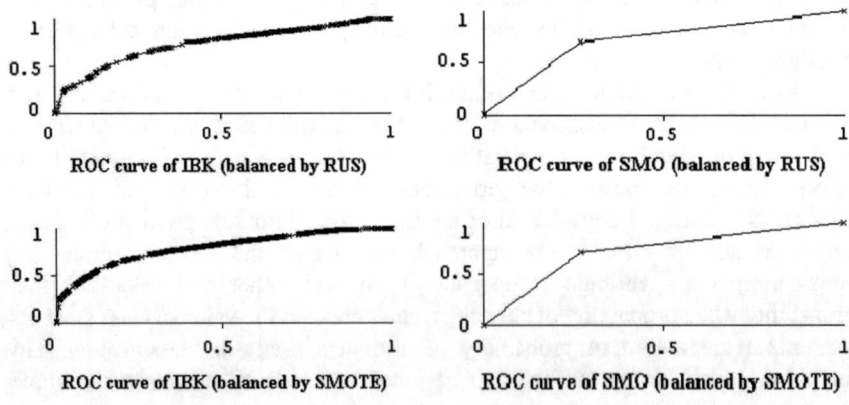

Fig. 3 ROC curve of IBK and SMO after applying RUS & SMOTE

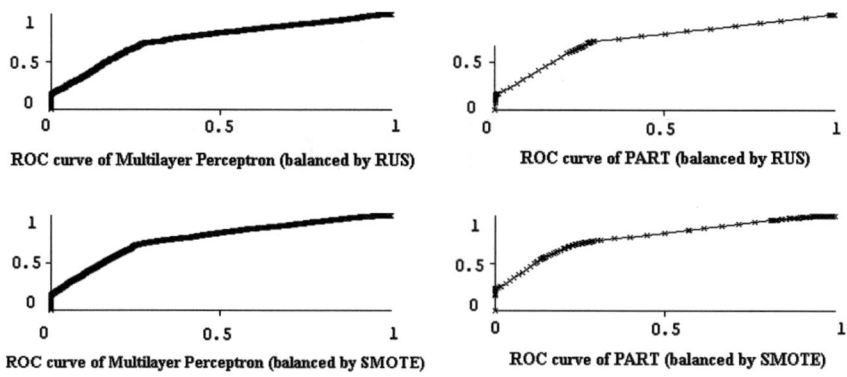

Fig. 4 ROC curve of multilayer perceptron and PART after applying RUS & SMOTE

strategy. It generates a partial tree on the present instances and creates rule on the generated tree. ROC curves for the implementation of MP and PART are shown in Fig. 4.

To compare the results we consider FP rate, precision, F-measure, AUC for ROC, correctly classified instances which are well known factors for evaluating a classifier. FP rate considers the false positive rate of a classifier from its confusion matrix. It designates that how many times the classifier predicts a false class as true class. Precision indicates general precision of the classifier which is the ratio of total records and number of correct classifications. F-measure is the harmonic mean of recall and precision with a weighted value. AUC stands for area under the curve. It calculates the area under ROC curve to show the performance of a classifier. Correctly classified instances are just the percentage of correctly classified instances from total classified instances.

5 Evaluation

The training set and testing set which is needed for classification techniques are generated by cross validation technique. We use tenfold cross validation technique as it is better than other techniques to validate the dataset (Figs. 5 and 6).

FP rate and correctly classified instances are two very important factors in determining the performance of a classifier. Lower FP-rate and higher correctly classified instances represent better efficiency of a classifier. From our experiment we see that IBK has the highest FP-rate (0.319) on RUS but is lower in correctly classified instances (68%) so it is one of the poorest performing classifier. RIPPER is efficient in correctly classified instances (73%) and is also lower on FP-rate (0.27) for SMOTE, and for RUS it has almost the same ratio. SMO works moderately on both RUS and SMOTE. Naive Bayes has a high FP-rate on RUS (0.31) and is also less efficient on correctly classified instances (68%). MP has an average value of

Fig. 5 Comparison of the performance of different classifier based on FP Rate

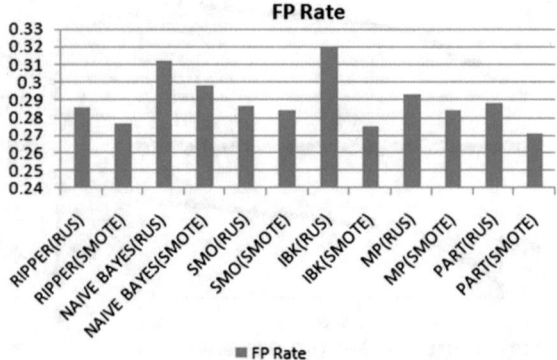

Fig. 6 Comparison of the performance of different classifier based on correctly classified instances

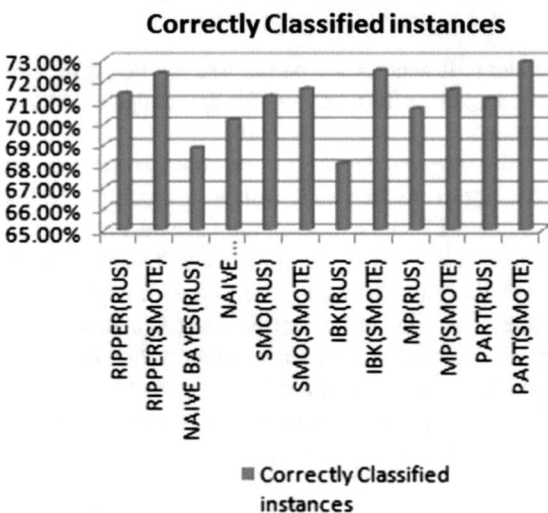

FP-rate and correctly classified instances compared to others. From the results it is evident that balancing the data with SMOTE and using PART classification on the balanced data gives the maximum correctly classified instances (72.86%) and lowest FP-rate (0.271). It is also high on precision (0.732) and on F-measure (0.728) and AUC of ROC (0.773). The performance of PART algorithm is significant in this dataset compared to others.

6   Conclusions

Our goal is to find a classifier that could effectively classify a non-regular customer from a regular customer of an insurance company. To do this initially we face some problems in preprocessing stage. To solve this we use first attribute selection methods to select the proper attributes that can have maximum effect on the classification. It proves to be very effective in action. We also use balancing algorithms on our data to balance the data. Without applying balancing techniques the classification is mostly favored by the general class. But after balancing the results that we get are quite good. As the balancing is done maintaining the initial ratio, the result is equally applicable on original data set.

7 Future Work

Data mining techniques are very useful to apply on life insurance companies data. The regularity of a customer for installment payment depends on certain important factors that the company stores which are obviously user specific and very sensitive. The company that provided us the data could not provide user specific information such as the actual income of the policy holder, health condition of the policy holder etc. which can be integrated in the attributes effecting classification of a customer. We intend to collect these user sensitive information which we believe will effect strongly in building a more specific and effective classifier in future.

References

1. Kirlidog, M., Asuk, C.: A fraud detection approach with data mining in health insurance. Proc.-Soc. Behav. Sci. **62**, 989–994 (2012)
2. Xiaoyun, W., Danyue, L.: Hybrid outlier mining algorithm based evaluation of client moral risk in insurance company. In: 2010 The 2nd IEEE International Conference on Information Management and Engineering (ICIME), pp. 585–589. IEEE (2010)
3. Yan, Y., Xie, H.: Research on the application of data mining technology in insurance informatization. In: Ninth International Conference on Hybrid Intelligent Systems, 2009. HIS'09, vol. 3, pp. 202–205 (2009)
4. Goonetilleke, T.O., Caldera, H.A.: Mining life insurance data for customer attrition analysis. J. Ind. Intell. Inf. **1**(1)
5. Thakur, S.S., Sing, J.K.: Mining customer's data for vehicle insurance prediction system using k-means clustering—an application. Int. J. Comput. Appl. Eng. Sci. **3**(4), 148 (2013)
6. Chawla, N.V., Bowyer, K.W., Hall, L.O., Kegelmeyer, W.P.: SMOTE: synthetic minority over-sampling technique. J. Artif. Intell. Res. 321–357 (2002)

Classification of Product Rating Using Data Mining Techniques

Pinku Deb Nath, Sowvik Kanti Das, Fabiha Nazmi Islam, Kifayat Tahmid, Raufir Ahmed Shanto and Rashedur M. Rahman

Abstract Data mining is the procedure to find patterns and necessary details from huge amount of data collected from various sources for a period of time. The target of our research is to classify the rating of individual products in an online shopping website based on price, discount, number of items left, sellers, count of likes and seller followers. The online shopping website from where we collected the data for prediction is kaymu.com.bd which is an online store in Bangladesh. The product rating that we are going to predict gives the correct rating of each product that not only depends on a single user's rating but the overall rating considering views of every other user. This helps user to decide what product to buy and how good it actually is.

Keywords Data mining · Product rating · Online shopping · Machine learning

P.D. Nath · S.K. Das · F.N. Islam · K. Tahmid · R.A. Shanto · R.M. Rahman (✉)
Department of Electrical and Computer Engineering, North South University,
Plot – 15, Block – B, Bashundhara, Dhaka 1229, Bangladesh
e-mail: rashedur.rahman@northsouth.edu

P.D. Nath
e-mail: pinku.nath@northsouth.edu

S.K. Das
e-mail: sowvik.das@northsouth.edu

F.N. Islam
e-mail: fabiha.islam@northsouth.edu

K. Tahmid
e-mail: Kifayat.navid@northsouth.edu

R.A. Shanto
e-mail: Raufir.ahmed@northsouth.edu

© Springer International Publishing AG 2017
D. Król et al. (eds.), *Advanced Topics in Intelligent Information and Database Systems*, Studies in Computational Intelligence 710,
DOI 10.1007/978-3-319-56660-3_3

1 Introduction

Shopping is a common interest of people. At present the Internet provided us the facilities of online shopping that has made our life easier. Now anyone can shop from home, saving all the effort to go to showrooms and buy things. But since people do not need to go shopping themselves how would they know the things they are buying are of good quality and whether they serve well. Our research comes up with assistance to customers providing the reliable product ratings out of 5, when no E-Commerce center in Bangladesh has done a thorough analysis of correct rating of products. The research has basically been done on kaymu.com.bd using web crawler to extract and create dataset necessary to find the product rating. In most online shopping websites, we see the rating of product, but this is often not satisfactory since if only one customer gives the product a rating of 5, it remains as 5 which might be different from other people who might have not rated the product but given bad reviews. Therefore, the given rating of product is somewhere vague and might not be correct. Our research performs an extensive analysis using price, discount, number of items left, count of likes and seller followers. The data mining methods and other techniques that are used here in our research are: Import.io for data collection, J48, Naïve Bayes, Multilayer perceptron in Weka.

2 Related Works

Many e-commerce related researches have been already done to improve the marketing strategy and profit margin of many organizations. In [1] the authors discussed various techniques of clustering, association rule mining and other data mining procedures applied on user internet usage to understand the behavior and motive of the customers. The authors extracted patterns based on the purchase and browsing information of the past customers to predict the behavior of the future customers. Similar work has been shown on a research by Patels and Chauhan [2] where they have used web mining to find plausible patterns in the usage of the customers to improve the user interface of the websites and provide customized services. Pattern discovery techniques were also used here. The authors analyzed on perspective of business shopping organizations but we predicted to help customers to get the right product.

Relatively new methods like Neural Networks have been suggested for web mining in the research paper by Crone and Soopramanien [3] to estimate the probabilities of class labels. In this research, several variables related to user information, product and past internet shopping data are used to classify consumers into "online shoppers", "browsers" and "non-internet shoppers" using the data collected from surveys in UK. The authors in [4] predicted the success of future films with selected/information provided on IMDB before release. They faced few problems initially with the format of source data which they converted to suitable formats using data mining tools and developed java application to process and extract data to overcome this problem. The prediction techniques are not as same as ours. We provide an easier way to collect data from an e-commerce website like kaymu.com.bd and process it but outcome is similar to find out the rating. In addition to finding patterns in the data, it is also important to visualize the patterns and effectiveness of the generated model. There are many visualization tools such as scatter plots, grand tour, data cube etc. [5]. The types of data required for the application of web mining in e-commerce are customer information, commodity information and server information [6]. Customer information refers to the personal data of the customers, commodity information refers to the product features such as price, amount left etc. and server information refers to the cookies, logs generated by a user session [6]. There are three sub-sections of web data mining and they are web content-mining, web structure-mining and web usage-mining [6]. Web content-mining is the process finding patterns, models or knowledge from the contents of a web page, web structure-mining is the process of recognizing the underlying correlations among the web pages and other online objects and web usage-mining is the process of mining browsing patterns from the usage information of the customers [6]. In our paper, we have explored the domain of web content-mining using import.io. Collection of web content information is primarily done using a crawler. There are some bespoke software tools and services to crawl webpages and collect data such as Kimono Labs, import.io and Crawlbot [7].

In addition, there are many scientific approaches for collecting data such as Mining Data Records in Web Pages (MDR) algorithm proposed in [8] which discovers sets in web pages with some common features by comparing the attributes of the child node in the tree generated by following hyperlinks etc. [7]. Since there are many e-commerce sites in many regions of the world and selling products of high diversity, the data analysis should be independent from both various languages and specific product domains [7]. The quick discovery of patterns and reacting to those patterns are of paramount importance to the online retailers. For example, in one study it was found that if a product stays online for a long time then the probability that it will be sold becomes very low [9].

3   Dataset and Tools

The online shopping website that is used in our research for data extraction to classify the product rating is kaymu.com.bd and import.io to scrap the html pages. We have used "WEKA" for implementing different algorithms' of pruning and preprocessing the dataset and applying various classification techniques.

4 Methodology

Our approach is divided into three categories: "Data Collection", "Preprocessing" and "Classification".

4.1 Data Collection

Here we have used import.io. Import.io is a web application where there are various modules to collect data in various fashions from the webpages. We use the crawler module to collect data. Crawler was used to crawl the entire website, specifically the product pages within the website.

From the product page it is decided that to collect 10 information given in the Fig. 1 (relationship is shown in Table 1). We consider them important for the classification of rating. For doing this the attribute names are set manually from the crawler interface of import.oi. There are two ways to train the crawler so that it can retrieve data from all the product pages of the website. One way is to click the item from the webpage which can be accessed from the crawler interface. Another way is

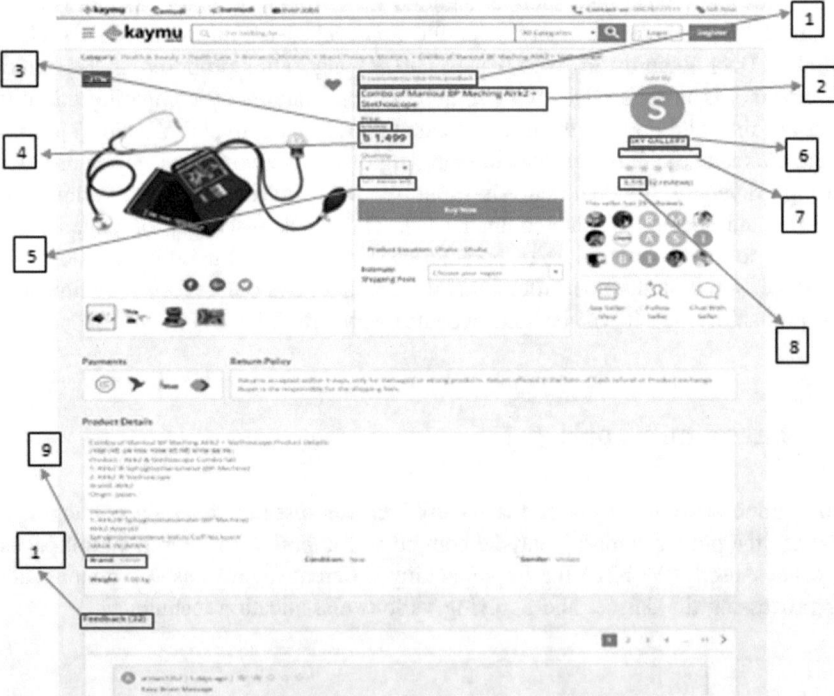

Fig. 1 Kaymu.com.bd product page

Table 1 Attribute relation and description

Attribute number from Fig. 1	Attribute name	Attribute type	Attribute description
1	product_like	Numerical	Number of likes of the product given by customers
2	product_name	Nominal	Name of the product
3	price_before	Numerical	Price of the product without discount
6	seller	Nominal	Name of the seller/company who is selling the product
7	seller_selling_for	Nominal	The number of years, month or days the seller is selling products
8	rating	Numerical	Rating out of five (average of all the ratings given by the customers). In this example 3.7/5 we took the numerical value 3.7
9	brand	Nominal	The name of the brand of the product
10	product_feedback	Numerical	Number of people who have given a comment on this product

to manually override the Xpath (reads through the source code of the webpage). The attributes where Xpath was used are given below:

rating: .//*[@itemprop="rating"]/@content

reviews: .//*[@itemprop="votes"]/@content
product_feedback: .//*[@class="row-small-12 mtl pbn"][contains(.,"Feedback")]
product_likes: .//*[@class="s-bold gray-medium wishlist likes"]/@data-wishlist-count
products_left: .//*[@class="quantity gray-medium s-bold"]/@data-qty

It is even possible to collect more attributes but it seems that the 10 attributes are enough to classify the rating. Once all the attributes are finished training from one product page we have to test it on four more product pages to see if we have made our training right. After finishing the training phase, we start our crawler. The starting page of the crawler: http://www.kaymu.com.bd/. Data extracted form pages with template (product page): http://www.kaymu.com.bd/{words-num}.html$. The collected data was downloaded from import.io as csv (comma separated format).

4.2 Data Preprocessing

The collected data which is in csv format is converted to arff format which is weka's default format. First we detect the outliers and extreme values using weka filters and then we discretize some of the numerical attribute for better classification results. The discretization of the attribute seems to worsen the performance of the classification so we have decided later to omit the discretization. Among the attribute it is observed that the seller_follower has 8%, product_feedback has 8%, product_likes has 51%, and rating has 8% missing values. Two of the nominal attributes like Sellers and Brand are removed as they seem to impact the results of the classification in a negative way. We calculate the discount from the price_before and price_now attribute using excel and give name the new attribute discount because removing the price_before attribute gives better classification results. The attribute seller_selling_for has a number of varying nominal values like "1 days", "2 days", "3 days" etc. As a result, it takes a good amount of time to build our classification model so we decide to merge these attributes into three categories:

New: less than 1 month
Oldnew: less than 1 year and greater than 1 month
Old: more than 1 year

The new merged seller_selling_for attribute has not made a significant impact on the classification results but it seems to reduce the time to build classification models. The final distribution of the tenfold cross validation training data is given in Fig. 2.

4.3 Classification

Generally, there are two types of classifiers and they are discriminative and generative classifiers. Discriminative classifiers build a function from an input set to class label and generative classifiers build a model of a joint probability and predict the class label of an input instance using Bayes rules [10]. In Weka, there are many ways to configure the classifiers which affect the model constructed by the classifiers. For example, we can choose the minimum number of records to consider at each node of the tree. This will affect the size of the tree, for example, the decision tree with higher minimum number of records per node will have less nodes compared to the decision tree with fewer records per node. If we set the "prune tree" option to be true, then the generated tree will be even smaller. However, greater accuracy of the decision tree is achieved when the minimum number of records per node is low and the tree is unpruned. We have used tenfold cross validation which lowers the possibility of over fitting the training set into the model.

We have used oneR classifier to determine the most significant attribute. OneR classifies based on one of the attributes that gives the minimum number of error. In our case product_feedback gave the minimum number of errors. The model generated by oneR is 91% accurate with 5245/5704 instances correctly classified on training set.

We can also determine the baseline accuracy of the dataset using the zeroR classifier. In the zeroR classifier, the class label which occurs frequently is determined and then the classifier labels all the records with the frequently occurring class label. In our dataset, the baseline accuracy was 83%. Hence, if the accuracy of any model is greater than 83% then we can say that the model is useful. A model is considered good if the Area under the curve (AUC) of the Receiver Operating Curve (ROC) is above 0.5. If the AUC value close to 0.5 then the model randomly assigns class labels to the records. If the AUC value is less than 0.5 then the model is more likely to give wrong labels to records.

We experimented with three classifiers to observe which classifier created the most accurate model for our dataset. The classifiers used are J48 Decision Tree, Naïve Bayes and Multilayer Perception Neural Network classifiers.

4.3.1 J48 Decision Tree

The J48 Decision Tree is based on the updated version of the C4.5 Decision Tree algorithm. The J48 Decision Tree is a discriminative classifier. We have used tenfold cross validation for building the model. We have experimented and saw that when the minimum number of records per node is low and the tree is unpruned then the accuracy is high. Since we have used cross validation, the chances of over fitting was low. Hence, we have used minimum 2 records per node and left the tree unpruned.

The AUC value of the J48 model is above 0.98 for all the class labels except for the class label 2, which is satisfactory. The correctly classified instances are 98.1241% and incorrectly classified instances are 1.8759% which is also appreciable. Figure 2 shows the ROC plot of the J48 classifier.

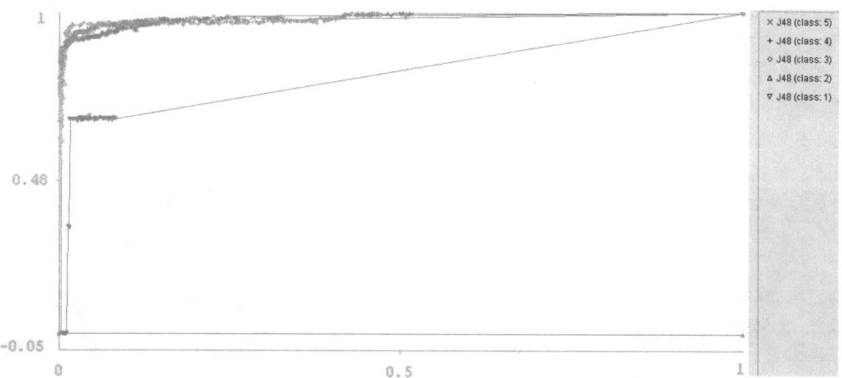

Fig. 2 J48 ROC Plot

4.3.2 Naive Bayes

Naïve Bayes is a simple probabilistic classifier which uses Bayes theorem to predict the probabilities of class labels. In Naïve Bayes, the attributes are considered independent of each other and they contribute to the class label of a record independently [11]. Maybe there is some inter-dependence in the attributes of the data-set for which reason Naïve Bayes performed poorly as a classifier.

4.3.3 Multilayer Perception

Multilayer Perception is an implementation of Neural Network which has two non-linear activation functions each of which maps weighted inputs to the output of each neuron. The activation functions are:

$$y(v_i) = \tanh(v_i) \quad \text{and} \quad y(v_i) = \left(1 + e^{-v_i}\right)^{-1}$$

Here, the first function is the hyperbolic tangent and its value ranges from −1 to 1, y(i) is the i th neuron output and v(i) is the weighted sum of input synapses and the second function is the logistic function.

5   Evaluation

From the application of the three classifiers, we have observed that the model generated by the J48 Decision Tree algorithm has performed the best in classifying the rating for a product with a satisfactory AUC in ROC curve for the class labels 3, 4 and 5. We have also observed that the Naïve Bayes created a bad model for the classification of the rating for a product. The Neural Network based Multilayer Perception has created a model that has a satisfactory performance in the classification of the ratings. The model generated by the Multilayer Perception has taken around 10 s to construct the model while the models for the J48 and Naïve Bayes are generated within 3 s. The performance of the classifiers can be observed using the bar charts in Figs. 3, 4 and 5 which represent the Average Accuracy, Average F-Measure and Average AUC of the J48 Decision Tree, Naïve Bayes and

Fig. 3 Performance statistics chart of J48 decision tree classifier

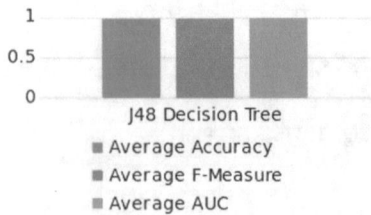

Fig. 4 Performance statistics chart of multilayer perception classifier

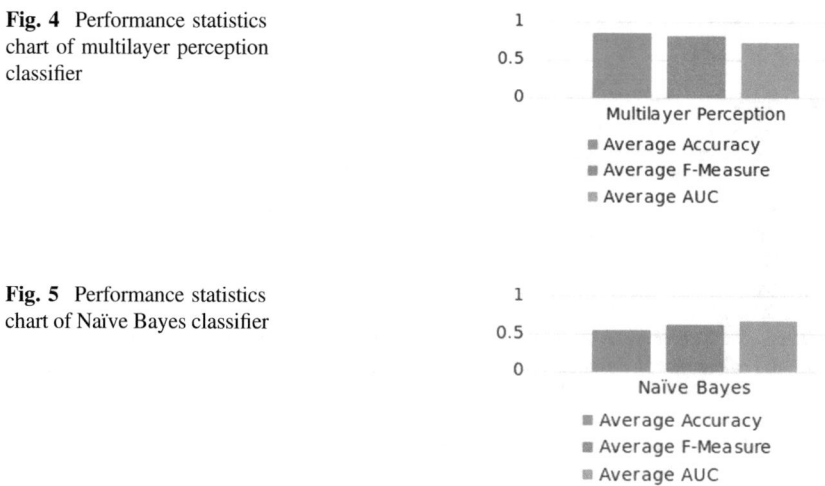

Multilayer Perception
- Average Accuracy
- Average F-Measure
- Average AUC

Fig. 5 Performance statistics chart of Naïve Bayes classifier

Naïve Bayes
- Average Accuracy
- Average F-Measure
- Average AUC

Multilayer Perception classifiers respectively. The unique characteristic of our data is that most of the products have rating 4 while some of the records have rating 3 and 5. Only three instances have rating 1 and none of the records have a rating of 2. Hence, generated models have performed well in classifying products that should receive rating 4, 5 or 3 because of the abundance of such products for training and testing datasets. We faced many problems and discrepancies while collecting data such as missing values etc. In addition, there were products that received many likes but few people rated that product. There were also products that had many good reviews but their rating did not reflect the positive sentiments of the customers. There might also be the situations where a product received a rating from a customer but the customer did not buy the product. The presence of such issues can reduce the effectiveness of the models generated by the data mining tools.

6 Conclusion and Future Work

The online shopping website from where we collected the data for prediction is kaymu.com.bd which is an online store in Bangladesh. The product rating that we are going to predict gives the correct rating of each product that not only depends on a single user's rating but the overall rating considering views of every other user. This helps user to decide what product to buy and how good it actually is. Possible future works include collaborating with the e-commerce owners and working on customer usage data that can be collected from the server end of the e-commerce websites. We can also suggest e-commerce website to ask their customers to fill a survey some days after they have bought a product because these survey data will more accurately reflect the sentiments of the customers.

References

1. Rastegari, H., Sap, M.N.M.: Data mining and e-commerce: methods, applications, and challenges. Jurnal Teknologi Maklumat 116–128 (2008)
2. Patel et al.: Web mining in e-commerce: pattern discovery, issues, and application. Int. J. P2P Netw. Trends Technol. **1**(3), 40–45 (2011)
3. Crone, S.F., Soopramanien, D.: Predicting customer online shopping adoption—an evaluation of data mining and market modelling approaches. In: Data Mining Conference (DMIN), pp. 20–23 (2005)
4. Saraee, M., et al.: A data mining approach to analysis and prediction of movie ratings. In: Data Mining V, pp 344–352. WII Press, UK (2004)
5. Zhang, F.: The application of visualization technology on e-commerce data mining. In: Second International Symposium on Intelligent Information Technology Application (2008)
6. Zhao, W., Lin, H.: WEB data mining applications in e-commerce. In: 9th International Conference on Computer Science and Education, Canada, pp. 557–559 (2014)
7. Horch, A., et al.: Mining e-commerce data from e-shop websites. IEEE Trustcom (2015)
8. Liu, B., et al.: Mining data records in web pages. In: SIGKDD USA, 2003, pp 601–606
9. Dlamini, M.G., et al.: Extracting interesting patterns from e-commerce databases to ensure customer loyalty. In: Proceedings of 2015 IEEE 12th International Conf. Networking, Sensing and Control, Taiwan, pp. 382–387
10. Andrew, Ng.Y., Michael, J.L.: On discriminative vs. generative classifiers: a comparison of logistic regression and Naïve Bayes. http://ai.stanford.edu/~ang/papers/ (2002)

MASS: A Semi-supervised Multi-label Classification Algorithm with Specific Features

Thi-Ngan Pham, Van-Quang Nguyen, Duc-Trong Dinh, Tri-Thanh Nguyen and Quang-Thuy Ha

Abstract Multi-label Classification (MLC), which recently has attracted numerous attentions, aims at building classification models for objects assigned with multiple class labels simultaneously. Existing approaches for MLC mainly focus on improving supervised learning which needs a relatively large amount of labeled data for training. In this work, we propose a semi-supervised MLC algorithm to exploit unlabeled data for enhancing the performance. In the training process, our algorithm exploits the specific features per prominent class label chosen by a greedy approach as an extension of LIFT algorithm, and unlabeled data consumption mechanism from TESC. In classification, the 1-Nearest-Neighbor (1NN) is applied to select appropriate class labels for a new data instance. Our experimental results on a data set of hotel (for tourism) reviews indicate that a reasonable amount of unlabeled data helps to increase the F1 score. Interestingly, with a small amount of labeled data, our algorithm can reach comparative performance to a larger amount of labeled data.

Keywords Semi-supervised clustering · Multi-label classification (MLC) · Specific feature · Semi-supervised multi-label classification

T.-N. Pham · V.-Q. Nguyen · D.-T. Dinh · T.-T. Nguyen · Q.-T. Ha (✉)
Vietnam National University (VNU), University of Engineering and Technology (UET),
Hanoi, Vietnam
e-mail: thuyhq@vnu.edu.vn

T.-N. Pham
e-mail: nganpt.di12@vnu.edu.vn

V.-Q. Nguyen
e-mail: quangnv_570@vnu.edu.vn

D.-T. Dinh
e-mail: trongdd_58@vnu.edu.vn

T.-T. Nguyen
e-mail: ntthanh@vnu.edu.vn

T.-N. Pham
The Vietnamese People's Police Academy, Hanoi, Vietnam

© Springer International Publishing AG 2017
D. Król et al. (eds.), *Advanced Topics in Intelligent Information and Database Systems*, Studies in Computational Intelligence 710,
DOI 10.1007/978-3-319-56660-3_4

1 Introduction

In the domain where an example can simultaneously belong to multiple classes, MLC aims at identifying a subset of predefined class labels for a given unlabeled instance. The multi-label classification has received increasingly attention and been applied to several domains, including web categorization, tag recommendation, gene function prediction, medical diagnosis and video indexing [1–6].

The most well-known approach to MLC is to use a different classifier for each label. The final labels of each instance are then obtained by using an aggregation scheme where the predictions of the individual classifier are combined. This approach has the advantage of its simplicity and disadvantage of ignoring the correlation among labels, thus, in certain situation, it can show performance degradation. In commonly existing approaches, all the class labels are discriminated based on the same feature representation. In other words, they use the identical feature set for different class label functions in computation. Since each label relies on its own specific characteristics, this approach might be not optimal. In several approaches, a document collection is divided into two groups based on positive and negative instances [7–9], then specific features are built in different ways. For example, Zhang and Lei [7] proposed an intuitively effective *multi-label learning with Label specIfic FeaTures* algorithm (named LIFT), which builds the specific features of each label by applying clustering analysis on its positive and negative instances, and then carry out training and testing by exploiting the clustering results. Similarly, Zhang et al. [8] proposed to employ spectral clustering to figure out the closely located local structures between positive and negative instances, and exploit the clustering results in classification. Huaqiao et al. [9] built the label-specific features by computing and selecting high density features on the positive and negative instance set for each class. Finally, each class label is classified based on its specific features.

Clustering, a basically unsupervised learning technique, groups a data set into clusters such that data in the same clusters are more similar (i.e., regarding to a distance measure) to each other than those in another cluster [10]. Clustering can be used to aid the text classification in term of discovering the kinds of structure in training examples. Due to the fact that obtaining labeled data is costly and time consuming, the combination of both labeled and unlabeled data in semi-supervised classification framework provides a more effective and cheaper approach to increase the performance.

Recently, semi-supervised clustering is an approach to semi-supervised classification [11–15]. Self-Organizing mapping (SOM) clustering is used to identify the label of the unlabeled data in non-ambiguous nodes by using the label of their nodes, then data in the clusters are used to train a multi-layer perception classifier [12]. This method significantly improves the classification performance on all the experimental datasets. Demirez et al. [14] presented a semi-supervised clustering algorithm which finds a set of clusters by minimizing a linear combination of cluster dispersion measured by mean square error and cluster impurity measure.

Zhang et al. [15] introduced a novel semi-supervised learning method, called *TExt classification using Semi-supervised Clustering* (TESC). In clustering process, TESC uses labeled texts to capture silhouettes of text clusters, next the unlabeled texts are added the corresponding clusters to adjust the centroid. These clusters are used for classification phase. Given a new unlabeled text, the label of the nearest text cluster is used to assign to the unlabeled text.

Kong et al. [16] proposed a model of transductive multi-label classification by using label set propagation, called TRAM. Firstly, TRAM formulates the transductive multi-label learning as an optimization problem to exploit unlabeled data. Secondly, TRAM develops an efficient algorithm which has a closed-form solution for this optimization problem and assigning label sets to unlabeled instances. The key in this method is to use the test data for optimization. In addition, the test examples are also used as unlabeled examples.

In this paper, we propose a novel *semi-supervised algorithm for Multi-label clASSification* (called MASS), which can exploit both *unlabeled data* and *specific features* to enhance the performance. By determining the prominent label in specific collection, the dataset is then divided into three different subsets; and the semi-supervised clustering is applied in each subset to extract features specific to each label or label set. The method of extracting specific features is an extension of LIFT algorithm proposed by Zhang and Lei [7]. In addition, MASS has some key breakthrough in using semi-supervised clustering to exploit both labeled and unlabeled data together at the same time as mentioned in TESC of Zhang et al. [15].

The rest of this paper is organized as follows. Section 2 introduces our newly proposed algorithm for MLC text classification. This section will give more details about the process of constructing specific features and using semi-supervised clustering in building MLC. Section 3 evaluates the proposed algorithm using experiments. Conclusions are shown in the last section.

2 The Proposed Algorithm

2.1 Problem Formulation

Supervised Multi-label Classification

Let $\overline{D}^L$ be the input labeled document collection with a set L of q labels, i.e., $L = \{l_1, l_2, \ldots, l_q\}$, where each document in $\overline{D}^L$ is assigned a non empty subset of labels $label(d \in \overline{D}^L) \subseteq L$ The task of MLC is to construct the classification function $f: \overline{D}^L \to 2^L$, so that, given a new unlabeled document d^u, the function identifies a set of relevant labels $f(d^u) \subseteq L$.

Semi-supervised Multi-label Classification

Let $\overline{D} = \{\overline{D}^L, \overline{D}^U\}$ be a document collection, where $\overline{D}$ and $\overline{D}^U$ are the collections of labeled and unlabeled documents, correspondingly. The task of semi-supervised MLC is to construct the classification function $f: \overline{D} \to 2^L$. The goal in the training step is to find a partition C from $\overline{D}$, such that $C = \{C_1, \ldots, C_m\}$, where $C_i = \{d_1^{(i)}, \ldots, d_{|C_i|}^{(i)}\} (1 \leq i \leq m)$, $\bigcup_{1 \leq i \leq m} C_i = \overline{D}$, and $C_i \cap C_j = \varnothing (1 \leq i \neq j \leq m)$. For all documents in C_i, they are given the same non-empty label set (called cluster-label) l_{C_i}.

In traditional unsupervised clustering method, the number of cluster is often predefined and manually chosen. However, in our model, the number of clusters m is automatically identified based on the label set in combination with the labeled and unlabeled data set.

After we have obtained the partition C, given a new unlabeled document $d^u \in D^U$, f employs the 1-nearest neighbor to get the nearest cluster $C_j = \arg\min_{C_p} dis(d^u, c_p)$, and c_p is the centroid of the text cluster C_p and $dis(.)$ is the distance between data points, then the cluster label of C_j is assigned to d^u, i.e., $l(d^u) = l_{C_j}$. Our contribution is to consume labelled and unlabeled data to find the partition C to form classification model f, which could predict class label set of unlabeled texts D^U.

2.2   Brief Summary of LIFT and TESC

LIFT Algorithm

LIFT was proposed for enhancing the performance of supervised multi-label classification using label-specific features. With assumption that label-specific features, i.e. the most specific characteristics, could improve the classification. Concretely, LIFT, at the first step, aims at figuring out features with label-specific characteristics, so as to provide appropriately discriminative information to facilitate its learning as well as classification. For each class label $l_k \in L$, the set of positive and negative training instances are founded as the set of the training instances with and without label l_k, respectively. After that, clustering analysis is performed on its positive and negative sets to extract the features specific to l_k. In the second step, q binary classifiers, one for each class label l_k using l_k-specific features, are used to check whether a new instance has the label l_k. The approach in LIFT is supervised method in which the input is labeled dataset for training process and the output is a classification model including the family of q classifiers corresponding to q labels. Given an unseen examples, its associated label set is predicted by going through q classifiers to get prediction for each label.

TESC Algorithm

TESC was proposed for single label classification where each instance can be associated with only a single class label. In this work, the task of constructing classification model is based on a semi-supervised clustering. The basic assumption is that the data samples come from multiple components. Therefore, in the training step, TESC uses clustering to identify components from both labeled and unlabeled texts. The labeled documents are clustered to find the silhouettes of documents, then, the unlabeled documents are added to adjust the clusters. The label of cluster is assigned to the newly added unlabeled documents.

Let $D = \{D^L, D^U\}$ be a document collection, where D^L and D^U are the collections of (single-label) labeled and unlabeled documents, correspondingly. Let L be the label set on D^L including q labels, i.e. $L = \{l_1, l_2, \ldots, l_q\}$ and C be the partition on D after process of semi-supervised clustering (i.e., the training phase)

$$C \leftarrow TESC(D, L).$$

After this process, the resulted cluster set C is regarded as the model of the classification function. In the classification step, given a new document the label of the nearest cluster is used as the predicted label of the new document, i.e., given an unseen example, the label of its nearest cluster $c_j \in C$ is used to assign to it.

2.3 Proposed Algorithm

In our approach, we construct the specific features for each label and label set based on the idea proposed in LIFT with several improvements. In LIFT, the authors build the features specific to each label in the same manner. In our model, the first step is to find the prominent labels in a cluster following to the greedy approach, i.e., select the best choice at the moment (or local optimization) with an assumption that this would lead to a globally optimal solution. Since, with the labels with few occurrences, it is not good enough to form a cluster, we proposed to select the maximum occurrence label (the prominent label) as clue to build clusters.

Next step in LIFT is to extract features specific to each label by k-means clustering technique on its positive and negative samples. Our model makes some important changes in this stage. We divide a document collection into three different document subsets: (1) documents with expansion of the only prominent label λ, (2) documents with a set of label including λ, and (3) documents without λ. After that, we perform semi-clustering analysis on these three subsets to get a partition on collection of unlabeled and labeled documents. The semi-supervised clustering technique in TESC is applied in our model to consume unlabeled documents, i.e., an unlabeled document is added to its nearest cluster, and its label set is the same as the cluster label. Finally, the partition on the dataset of both labeled and unlabeled documents is used as classification model. No additional classification algorithm is

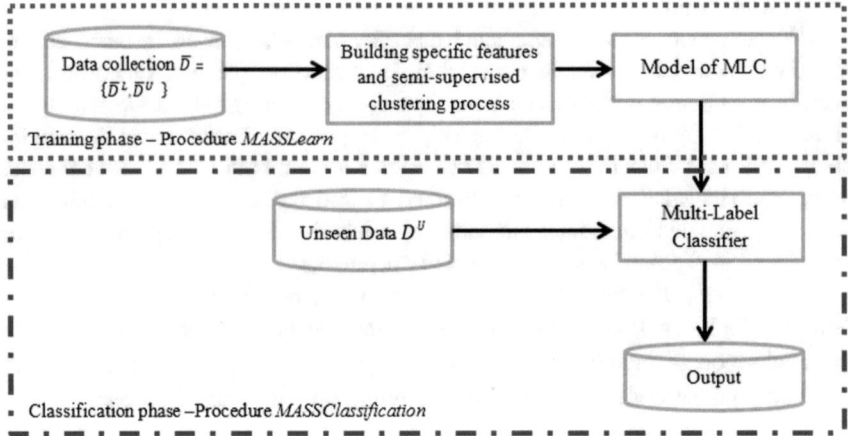

Fig. 1 The proposed semi-supervised MCL model

used in our approach. This is different from the LIFT which uses q (i.e., the cardinality of the label set) binary classifiers with label-specific features in classification phase.

The proposed algorithm comprises of two phases as described in Fig. 1: one is the training phase, which uses clustering to identify the components (i.e., clusters) from both labeled and unlabeled texts based on the prominent label; The other phase is classification, which identifies the nearest text cluster to label the unlabeled text D^U.

In training phase, we use the semi-supervise clustering method in [15] to take advantages of TESC algorithm to get partition on the text collection $\overline{D}$. We name training procedure *MASSLearn*(.), of which the pseudo-code is shown in Fig. 2.

In order to find the partition C (i.e., the model of our classification algorithm), we first initialize $C = \{\}$; then call *MASSLearn*$(\overline{D}, \{\}, L, C)$. The resulted set of text clusters C is regarded as components and used to predict labels of unlabeled texts in classification phase as shown in Fig. 3.

In classification process, the input includes unlabeled texts need labeling. The output is the collection of labels corresponding to each text in unlabeled texts. We calculate the distances from unlabeled text to the centroids of all clusters to find out the nearest centroid. Then the label set of the nearest cluster will assign to the unlabeled text.

Procedure $MASSLearn$ ($\overline{D}, L_1, L_2, C$)

Inputs:

$\overline{D}$: collection of labeled and unlabeled texts, $\overline{D} = \{\overline{D}^L, \overline{D}^U\}$, where $\overline{D}^L$ and $\overline{D}^U$ are the set of labeled and unlabeled documents, correspondingly

L_1: set of available default labels for all the instances in $\overline{D}^L$

L_2: set of possible labels which could be labeled for some instances in $\overline{D}^L$

Output:

C: a collection of labeled text clusters

1. Let λ be the label in L_2 with the maximum occurrences in $\overline{D}^L$
2. Create a new label set L^* including three assistant labels: $\lambda_1 \leftarrow L_1 \cup \{\lambda\}$, $\lambda_2 \leftarrow L_1 \cup \{\lambda\} \cup \{other\ labels\}$, and $\lambda_3 \leftarrow \{label\ set\ without\ \lambda\}$

   /* Annotate the data set $\overline{D}^L$ as single label data according to three macro assistant labels λ_1, λ_2, and λ_3 to use TESC algorithm*/

   $\overline{D}' \leftarrow Annotate(\overline{D}, \{\lambda_1, \lambda_2, \lambda_3\}); \quad C^* \leftarrow TESC(\overline{D}', L^*)$

3. Divide $\overline{D}$ into three subsets of labeled and unlabeled texts regarding to the partition C^*:

   a. $\overline{D}_1$: subset of labeled and unlabeled texts in which labeled texts are assigned the label λ_1

   b. $\overline{D}_2$: subset of labeled and unlabeled texts in which labeled texts are assigned the label λ_2

   c. $\overline{D}_3$: subset of labeled and unlabeled texts in which labeled texts are assigned the label λ_3

4. On $\overline{D}_1$:

   $C \leftarrow C \cup TESC(\overline{D}_1, L_1 \cup \{\lambda\})$

5. On $\overline{D}_2$:

   a. If all instances in $\overline{D}_2$ have the same label set L3

   $C \leftarrow C \cup TESC(\overline{D}_2, L_1 \cup \{\lambda\} \cup L3)$

   b. Else

   // Instances in $\overline{D}_2$ are not identical in set of labels.

   // Move λ from L2 to L1

   $MASSLearn(\overline{D}_2, L_1 \cup \{\lambda\}, L2 \setminus \{\lambda\}, C)$

6. On $\overline{D}_3$:

   a. If all instances in $\overline{D}_3$ are assigned by the set of labels L4

   $C \leftarrow C \cup TESC(\overline{D}_3, L4)$

   b. Else

   // Instances in $\overline{D}_3$ are not identical in set of labels.

   // Delete λ from L2

   $MASSLearn(\overline{D}_3, L1, L2 \setminus \{\lambda\}, C)$

Fig. 2 Pseudo-code of clustering process

```
Procedure MASSClassification
Input:
    C: collection of labeled text clusters C = {C_1, ..., C_m}
    D^U: collection of unlabeled texts
Output:
    L^U: a collection of labels corresponding to each text in D^U

    1.  For each d^u ∈ D^U
    2.      C_temp ← C_0 // C_0 is the first cluster in C
    3.      l_{d^u} ← l_{C_temp} //l_{C_temp} is the label set of cluster C_temp
    4.      For each C_j ∈ C
    5.          Dis(d^u, C_j) ← ||d^u - C_j|| //we use Euclidean distance here
    6.          If Dis(d^u, C_temp) > Dis(d^u, C_j)
    7.              C_temp ← C_j
    8.          End if
    9.      End for
    10.     l_{d^u} ← l_{C_temp}
    11.     Add l_{C_temp} to L^U
    12. End for
```

Fig. 3 Pseudo code of classification procedure

3   Experiments and Results

3.1   The Datasets

We built three datasets of labeled, unlabeled, and testing data from thousands of reviews retrieved from several famous Vietnamese websites on tourism and hotels. After some preprocessing steps on the datasets, i.e., main text content extraction, word segmentation, and stop word removal, we got about 1800 reviews. 1500 reviews were manually tagged to create the labelled set of 1250 reviews, and the testing set of 250 reviews. The rest of 300 reviews were left intact to create unlabeled set. We considered reviews on five aspects: (a) location and price, (b) service, (c) facilities, (d) Room Standard, and (e) Food.

3.2   Experimental Results

We took several experiments with different configurations to evaluate the effect of the proposed algorithm. In order to analyze the contribution of the labeled data, we also generated some subsets of size of 500, 750, 1000, 1250 reviews. The contribution of unlabeled data is also evaluated in each category with different size of 0, 50, 100, 200, 300 reviews.

We also built one baseline using supervised algorithm of SVM for MLC, i.e., five binary SVM classifiers, one for each class label. The baseline worked best on the training set of 750 reviews, hence, we used this result for later comparison.

In our model, we used the label-based measures for evaluation [4]. For each class label y_j, TP_j, FP_j, TN_j and FN_j, which are the number *of true positive, false positive, true negative* and *false negative* test samples, were recorded. Let $B(TP_j, FP_j, TN_j, FN_j)$ be some specific binary classification measures (e.g., $B \in \{P, R, F1\}$, where $P = TP_j/(TP_j + FP_j)$, $R = TP_j/(TP_j + FN_j)$, and $F1 = \frac{P*R}{2(P+R)}$. The micro-averaging measures are calculated as follows:

$$B_{micro} = B\left(\sum_{j=1}^{q} TP_j, \sum_{j=1}^{q} FP_j, \sum_{j=1}^{q} TN_j, \sum_{j=1}^{q} FN_j \right)$$

where q is the total number of labels. For these metrics, the bigger value, the better classification performance.

The results of the experiments are reported in the Table 1. We observed that the proposed solution's results are very promising in all experiments in comparison

Table 1 The results of experiments

Training dataset size	Unlabeled dataset size	Precision$_{micro}$ (%)	Recall$_{micro}$ (%)	F1$_{micro}$ (%)
Baseline		68.50	60.00	63.90
500	0	77.40	81.10	79.20
	50	81.40	77.70	79.50
	100	80.60	78.70	79.70
	200	83.00	82.50	**82.70**
	300	79.60	80.40	80.00
750	0	77.70	81.50	79.60
	50	82.40	81.30	81.80
	100	82.10	82.30	**82.20**
	200	80.70	82.50	81.60
	300	79.00	82.30	80.60
1000	0	80.10	79.60	79.80
	50	80.70	81.00	80.90
	100	81.30	83.30	82.30
	200	81.00	84.40	82.60
	300	82.40	83.90	**83.20**
1250	0	79.40	82.70	81.00
	50	80.70	80.70	80.70
	100	80.90	79.90	80.40
	200	81.60	83.30	**82.40**
	300	78.50	82.70	80.50

with the baseline. It indicates that the proposed algorithm may take reasonable contribution to the multi-label classification approaches. We found that in each category of training dataset, the system outperforms in experiments of using different unlabeled sets than the case of using no unlabeled texts. This is the reason that various selections of unlabeled texts involved in MASS can improve the performance of text classification. These experiments also show the role of labeled data in proposed model characterizing the silhouettes of the text clusters. Although the increase in size of labeled dataset also makes some contribution to the performance of system in general, the best result in each category seems to be stable with different number of unlabeled texts. By dividing the dataset into three different sub-datasets, MASS also overcomes the limitation in computational complexity.

4 Conclusions

In this paper, we proposed MASS—an approach for semi-supervised MLC to exploit label-specific features. Using two basic assumptions including the effect of label-specific features in learning process and the multiple components in each label which can be identified by clustering, our proposed model brings major contribution in building label-specific features for multi-label learning with an approach of semi-supervised clustering technique. The experimental results show the promising trends in MASS for the MLC. Our work is currently seen as the initial step, more improvements, e.g. the method to effectively select unlabeled instances, or post-processing to prune the resulted clusters to remove outliers, should be done to evaluate the proposed approach.

Acknowledgements This work was supported in part by VNU Grant QG-15-22.

References

1. Elisseeff, A., Weston, J.: A kernel method for multi-labelled classification. NIPS **2001**, 681–687 (2001)
2. Rousu, J., Saunders, C., Szedmák, S., Shawe-Taylor, J.: Kernel-based learning of hierarchical multilabel classification models. J. Mach. Learn. Res. **7**, 1601–1626 (2006)
3. Silla Jr., C.N., Freitas, A.A.: A survey of hierarchical classification across different application domains. Data Min. Knowl. Discov. (DATAMINE) **22**(1–2), 31–72 (2011)
4. Tsoumakas, G., Katakis, I., Vlahavas, I.P.: Mining Multi-label Data. Data Min. Knowl. Discov. Handb. **2010**, 667–685 (2010)
5. Trohidis, K., Tsoumakas, G., Kalliris, G., Vlahavas, I.P.: Multi-label classification of music into emotions. ISMIR **2008**, 325–330 (2008)
6. Zhang, M.-L., Zhou, Z.-H.: ML-KNN: a lazy learning approach to multi-label learning. Pattern Recognit. (PR) **40**(7), 2038–2048 (2007)
7. Zhang, M.-L., Lei, W.: LIFT: multi-label learning with label-specific features. IEEE Trans. Pattern Anal. Mach. Intell. **37**(1), 107–120 (2015)

8. Zhang, J.-J., Fang, M., Li, X.: Multi-label learning with discriminative features for each label. Neurocomputing **154**, 305–316 (2015)
9. Huaqiao, Q., Zhang, S., Liu, H., Zhao, J.: A multi-label classification algorithm based on label-specific features. Wuhan Univ. J. Natl. Sci. **16**(6), 520–524 (2011)
10. Basu, S.: Semi-supervised clustering: probabilistic models, algorithms and experiments. University of Texas at Austin (2005)
11. Tian, D.: Semi-supervised learning for refining image annotation based on random walk model. Knowl. Based Syst. 72–80 (2014)
12. Dara, R., Kermer, S., Stacey, D.: Clustering unlabeled data with SOMs improves classification of labeled real-world data. In: Proceedings of the 2002 International Joint Conference on Neural Networks, pp. 2237–2242 (2002)
13. Luo, X., Liu, F., Yang, S., Wang, X., Zhou, Z.: Joint sparse regularization based sparse semi-supervised extreme learning machine (S3ELM) for classification. Knowl. Based Syst. **73**, 149–160 (2015)
14. Demirez, A., Bennett, K., Embrechts, M.: Semi-supervised clustering using genetic algorithms. In: Proceedings of Artificial Neural Networks in Engineering (ANNIE-99), pp. 809–814 (1999)
15. Zhang, W., Tang, X., Yoshida, T.: TESC: An approach to text classification using semi-supervised clustering. Knowl. Based Syst. **75**, 152–160 (2015)
16. Kong, X., Ng, M.K., Zhou, Z.-H.: Transductive multilabel learning via label set propagation. IEEE Trans. Knowl. Data Eng. **25**(3), 704–719 (2013)

Parallel Self-organizing Map Using Shared Virtual Memory Buffers

**Noor Elaiza Bt Abd Khalid, Muhammad Firdaus B. Mustapha,
Azlan B. Ismail and Mazani B. Manaf**

Abstract Parallel implementation of Self-organizing Map (SOM) has been studied since last decade. Graphic Processing Unit (GPU) is one of most promising architecture for executing SOM in parallel. However, there are performances issues are highlighted when imposing larger mapping and dataset size onto parallel SOM that executed on the GPU. Alternatively, heterogeneous systems that soldered GPU together with Central Processing Unit (CPU) are introduced in order to improve communication between CPU and GPU. Shared Virtual Memory (SVM) is one of features in OpenCL 2.0 which allows the host and the device to share a common virtual address range. Thus this research proposes to introduce a parallel SOM architecture that suitable for both GPU and heterogeneous system with the aim to compare the performance in term of computation time. The architecture comprises of three kernels that executed on two different platforms (1) discrete GPU platform and (2) heterogeneous system platform that tested using SVM buffers. The experimental results show the parallel SOM running on heterogeneous platform has significant improvement in computation time.

Keywords Parallel self-organizing map · GPU computing · OpenCL

N.E.B.A. Khalid (✉) · M.F.B. Mustapha (✉) · A.B. Ismail · M.B. Manaf
Faculty of Computer and Mathematical Sciences, Universiti Teknologi MARA,
Shah Alam, Malaysia
e-mail: elaiza@tmsk.uitm.edu.my

M.F.B. Mustapha
e-mail: firdaus19@gmail.com

A.B. Ismail
e-mail: azlanismail08@gmail.com

M.B. Manaf
e-mail: mazani@tmsk.uitm.edu.my

© Springer International Publishing AG 2017
D. Król et al. (eds.), *Advanced Topics in Intelligent Information
and Database Systems*, Studies in Computational Intelligence 710,
DOI 10.1007/978-3-319-56660-3_5

49

1   Introduction

Graphic Processing Unit (GPU) is a many core processor consisting hundreds or even thousands of compute cores that has been used to process the applications of scientific computing and scientific simulations or also called General Purpose Graphic Processing Unit (GPGPU) [1]. GPU computing has become popular since the introduction of GPU programming framework such as Compute Unified Device Architecture (CUDA) in 2007 and Open Computing Language (OpenCL) in 2009 [2]. Many researchers are trying to take advantages of GPU computing to execute Self-organizing Map (SOM) algorithm in parallel manner. Some researchers agree that GPU variant shows the significant speed up for large data compared to Central Processing Unit (CPU) variant [3, 4]. Both comparisons are proven that GPU computing achieves better performance in terms of computation time. Moreover, GPU computation time will reduce when the increment of input dimension and SOM network size compared to execute on CPU [5]. On the other words, the GPU computation is suitable to handle large dataset with high dimension. However, [6] address the larger mapping size and feature dimensions, the slower the computation time for both CPU and GPU. The major issue is addressed by researchers in executing parallel SOM on GPU is memory utilization increase when processing large mapping size [5–7]. The high memory utilization leads to high rate of memory transfers which will burden the processing time.

For the meantime, some researchers attempt to decompose several steps of the algorithm with the aim to execute SOM in parallel. There are different configurations of task decomposition on SOM algorithm has been applied and observably many researches works are found in the literature perform decomposition on calculate distance and find the Best Matching Unit (BMU) steps. For instance, [6, 8] decompose calculate Euclidean distance and BMU searching process. Some researchers try to decompose calculate distance, find BMU, and update the neurons' weights [5, 7, 9, 10]. Meanwhile, [4, 11] decompose initialize neuron weights. From the literature shows that the major steps have been decomposed include; initialize weights, calculate distance, find the BMU, and update the weights. Figure 1 illustrates steps in SOM algorithm that are decomposed by researchers specifically using GPU computing platform.

Recently, heterogeneous systems using GPU has become attractive computing model given the available scale of data-parallel performance and GPU programming framework such as OpenCL. The heterogeneous systems that combine CPU and GPU on a single chip are capable to share the same memory which leads to improve communication between each other [12]. Thus, the aim of this paper is to explore the parallel SOM architecture that suitable for executing discrete GPU and heterogeneous system architecture. The architecture comprises of three kernels that executed on two different platforms; (1) discrete GPU platform and (2) heterogeneous system platform that tested using SVM buffers; coarse-grained and fine-grained buffers. The performance of both platforms are measured based on computing time in seconds.

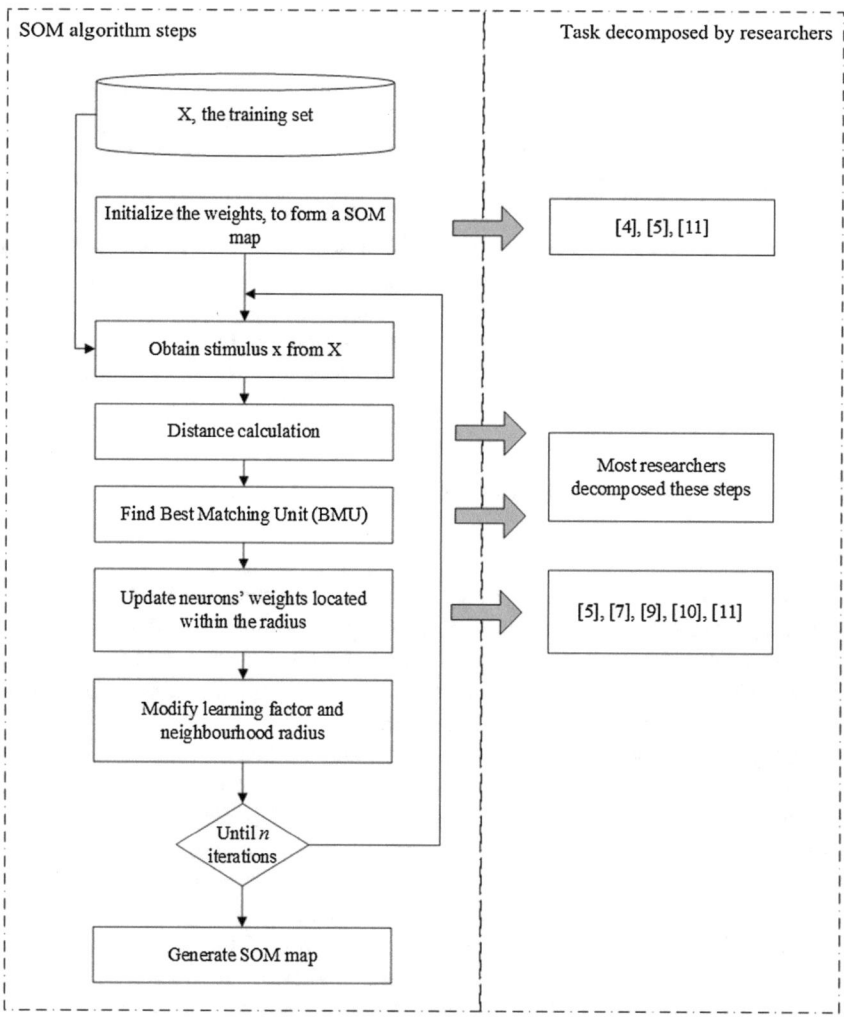

Fig. 1 Task decomposition of SOM algorithm

2 Evolution of Parallelism in Hardware

Nowadays, CPU approximately touches its limit whereas increasing the frequency of CPU will consume large power. As an alternative, modern graphic cards or GPU take role of powerful computation hardware. The performance gap between CPU and GPU becomes wider as GPU achieves seven times for gigaflops and bandwidth metrics compared to CPU [13].

There are another types of accelerator core that gained interest over last decade such as Field-Programmable Gate Arrays (FPGAs) and the Cell Broadband Engine

(Cell BE) [13]. However, the GPUs are most popular among these accelerator cores because large numbers of desktop and laptops computers have a dedicated GPU compared to FPGAs and the Cell BE are only found in specially ordered setup. Additionally, the future of the Cell BE is currently uncertain and FPGAs too hard to program for general-purpose computing [13].

The most recent technology, heterogeneous systems, that incorporated CPUs and GPUs together on a single integrated circuit (IC) chip, is quickly becoming the design paradigm for today's platform because of their impressive parallel processing capabilities [14]. The introduction of heterogeneous programming models such as OpenCL 2.0 in July 2013 is to improve the communication between CPU and GPU. This framework treats the GPUs as a first-class computing device which allows the GPUs to manage their own resources, as well as access some of the CPU's resources.

2.1   GPU Programming Framework

OpenCL 1.2

OpenCL is a framework of parallel programming that can be used for programming a heterogeneous collection of central processing units (CPUs), GPUs and other discrete computing devices are organized into a single platform [12]. An OpenCL device or GPU is divided into one or more compute units (CUs) where each CU has one or more processing elements (PEs).

An OpenCL program is executed on a host and the host is connected to one or more GPUs. The host code portion of an OpenCL program runs on a host processor according to the models native to the host platform. The OpenCL program host code submits various commands to a command queue, to be executed by processing elements within the device. The command can be of different types, such as for execution, memory management, or synchronization.

Meanwhile, the device code or kernel is executed on GPU. Kernels are sets of instructions that are executed in parallel. Each kernel program is stored in a separate file with the extension of. cl. However, the main problem in performance for OpenCL 1.2 applications is data transfers between the host code and device code [14]. Moreover, the memory management in OpenCL 1.2 still relied on the programmer to take care of data movement between the CPU and the GPU.

OpenCL 2.0

OpenCL 2.0 is the next release of OpenCL framework which introduced new features that concentrate on managing the heterogeneous system. This feature is to overcome the data transfers between CPU and GPUs. OpenCL 2.0 introduced Shared Virtual Memory (SVM) which allows the host and the device to share a common virtual address range [12]. This reduces overhead by eliminating deep copies during host-to-device and device-to-host data transfers. Deep copies involve completely duplicating objects in memory [14]. SVM implementations can be described the following below:

- Coarse-grained: includes synchronization during mapping and unmapping of memory objects, along with during kernel launch and completion. Accordingly the memory object updates after the completion of kernel and the unmapping of memory.
- Fine-grained: the synchronization occurs during the implementation of SVM buffers. Therefore the memory objects are updated coherently for both CPU and GPU.

2.2 Proposed Architecture

This paper proposes to parallelize all of the three steps using separate kernels code. The first kernel is to calculate the distance between neurons and a current input vector. The second kernel is to find BMU for each input vector. The BMUs values are then used by the third kernel to update the map appropriately. The parallel SOM will be tested on three different buffers; (1) non-SVM buffers, (2) coarse-grained buffer SVM, and (3) fine-grained buffer SVM. Figure 2 shows the proposed architecture of parallel SOM.

Initially, the input data are retrieved and stored into an array and follows by initialization of SOM parameter such as learning factor and weights. These tasks are performed at host side. In order to execute the kernels three functions are created for providing setting, initializing parameters, and calling the kernels. For example, the calculate distance function uses to call Calculate Distance kernel and it is done the same way with the other two kernels.

All of the kernels are implemented on the device side. The Distance Calculation kernel is to calculate the distance between neurons and current input vector and store the distance values into an array. It is represented by amount of work-items that is equal to the number of neurons in the SOM map. As such, each work-item of the kernel is responsible for finding the distance between a single neuron and the current input vector. This research applies Manhattan distance calculation.

Meanwhile, the Find BMU kernel applies reduction method with the aim of finding BMU in parallel. The kernel utilizes work unit the same number of neurons on SOM map. This process includes two stages where the first stage is to divide the work unit per compute unit (CU). The work unit per CU then is divided by the size of local work group in order to acquire the amount of work units for each processing element must deal with. Each work-item in the work-groups will find the minimum distance among the distance values covered by the work groups. The minimum distance value identified by the kernel is stored in a local array. The second stage is to find minimum distance for each CU from the minimum values in the local array. The minimum values of each CU then stored into global array and the host will determine the winning neurons.

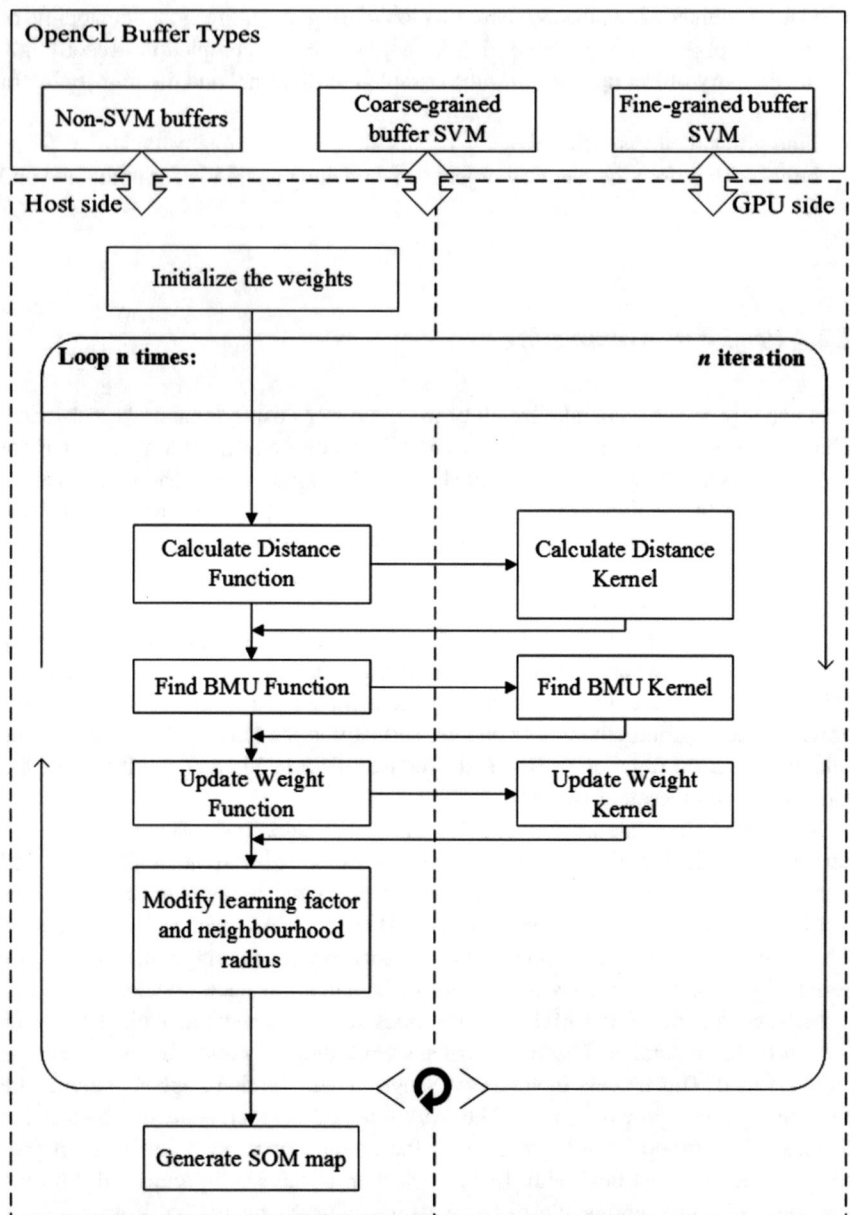

Fig. 2 The proposed architecture of parallel SOM

On the other hand, the third kernel involves updating the weight of neurons based on learning rate and neighborhood function. The learning rate describes how much a neuron's vector is changed during an update according to how far away the

Table 1 The implementation setting of buffers

Type of OpenCL buffers	Parameter declaration	Allocating the parameters
Non-SVM	`cl::Buffer` `input_buff;`	`input_buff = cl::Buffer` `(device_context, CL_MEM_READ_WRITE |` `CL_MEM_USE_HOST_PTR, sizeof(float)*` `input_size * input_length, input,` `&err);`
Coarse-grained SVM	`float*` `input_buff;`	`input_buff = (float*)clSVMAlloc` `(oclobjects.context,` `CL_MEM_READ_WRITE, input_size` `*input__length *sizeof(float,0);`
Fine-grained SVM	`float*` `input_buff;`	`input_buff = (float*)clSVMAlloc` `(oclobjects.context,` `CL_MEM_READ_WRITE|` `CL_MEM_SVM_FINE_GRAIN_BUFF,` `input_size * input_length * sizeof` `(float,0);`

neuron is from the BMU on the map. The BMU and its close neighbors will be changed the most, while the neurons on the outer edges of the neighborhood are changed the least. Right after executing the three kernels, modify learning factor and neighborhood radius take place. All of the steps include in the loop block will repeat until n iteration or epoch before the SOM map is generated.

2.3 OpenCL Buffers Types

With the interest to evaluate the performance of SVM buffers, the proposed architecture is tested with three different type of OpenCL buffer. The following Table 1 shows the implementation of three buffers.

The non-SVM buffer is following the OpenCL 1.x specification, meanwhile the SVM buffers that comprise of coarse-grained and fine-grained buffers are using OpenCL 2.0 specification. The fine-grained SVM applies CL_MEM_SVM_FI-NE_GRAIN_BUFF in order to activate fine-grained compared to coarse-grained SVM. Additionally, the synchronization point of coarse-grained SVM occurs during mapping or unmapping of memory objects and kernel launch and completion. While the synchronization point of fine-grained SVM happens during the executing of the memory objects.

3   Computation Experiment

3.1   Experimental Setup

In this study, Bank Marketing dataset from UCI Machine Learning Repository is applied for the computation experiments. Three sizes of dataset has selected; 5000, 10000 and 15000. Table 2 depicts the description of the dataset and experimental design for this paper. The experiments are conducted in order to examine the performance of parallel SOM using three different buffers; non-SVM (NSVM), coarse-grained SVM (CG), and fine-grained SVM (FG). The evaluations are based on computation time that divided into three: total kernel time, total setup time, and total time. The experiments were conducted on a laptop equipped with Intel i7-6700HQ processor, 16 GB of RAM and built in Intel® HD Graphics 530. This processor belongs to the Skylake family which supports the OpenCL 2.0. It is equipped with 4 CPU cores and 24 number of execution units placed at GPUs.

3.2   Computation Results and Analysis

The results of the computation experiments are presented in Fig. 3. Performance of three different buffers are included into the figure using the following label; NSVM, FG, and CG. Each buffer has tested on three different sizes of dataset and four SOM mapping size. The NSVM is executed on OpenCL 1.x platform meanwhile CG and FG are performed on OpenCL 2.0.

From the results, FG outperforms NSVM and its sibling for every dataset size. The parallel SOM triggered by FG well utilize SVM features in OpenCL 2.0 due to CPU and GPU efficiently share a common virtual address space where it is removing the need to explicitly copy buffers back and forth between the two devices [14]. Meanwhile the CG performs the worst among of three buffers where it suffers from consuming the most total setup time. The CG buffers which also utilize SVM feature but apply `clEnqueueSVMMap` and `clEnqueueSVMUnmap` for the synchronization point likely to burden the processing. Overall, all of the buffers types share the same trends as the bar chart raise higher when executing larger dataset size and mapping size.

Table 2 The experimental setting

Dataset parameters		SOM parameters		Performance measurement
No. of samples	No. of parameters	Iterations	Mapping size	Time, s
5000	3	30	10 × 10	Total kernel time
10000			20 × 20	Total setup time
15000			30 × 30	Total time
			40 × 40	

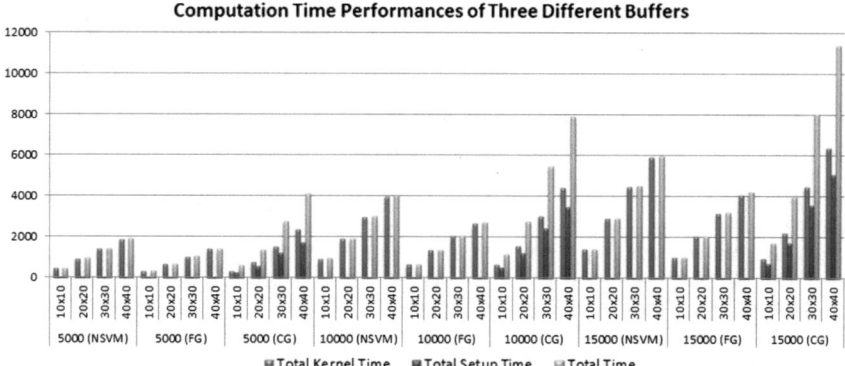

Fig. 3 Computation time performances of three different buffers

4 Conclusion

This paper proposes a parallel SOM architecture that comprise of three kernels. There is a possible way to parallelize the SOM algorithm through decomposing three major steps; calculate distance, find the BMU, and updating the weights. On the other note, GPU computing offers a great solution for SOM parallelism where the large amount of calculation could be catered by massive parallelism. The aim of this architecture is to accelerate SOM training. From the results show that the proposed parallel SOM architecture is capable to execute parallel SOM especially for FG buffers.

Acknowledgements This work was funded by Ministry of Higher Education (MOHE) of Malaysia, under the Fundamental Research Grant Scheme (FRGS), grant no. FRGS/81/2015 and Academic Staff Bumiputera Training Scheme (SLAB). The authors also would like to thank the Universiti Teknologi MARA for supporting this study.

References

1. Perelygin, K., Lam, S., Wu, X.: Graphics Processing Units and Open Computing Language for parallel computing. Comput. Electr. Eng. **40**(1), 241–251 (2014)
2. Kirk, D.B., Hwu, W.W.: Programming Massively Parallel Processors. Elsevier (2013)
3. Wittek, P., Darányi, S.: Accelerating text mining workloads in a MapReduce-based distributed GPU environment. J. Parallel Distrib. Comput. **73**(2), 198–206 (2013)
4. Lachmair, J., Merényi, E., Porrmann, M., Rückert, U.: A reconfigurable neuroprocessor for self-organizing feature maps. Neurocomputing **112**, 189–199 (2013)
5. Gajdos, P., Platos, J.: GPU based parallelism for self-organizing map. In: Advances in Intelligent Systems and Computing, IHCI 2011, vol. 179, pp. 3–12 (2013)
6. Hasan, S., Shamsuddin, S.M., Lopes, N.: Machine learning big data framework and analytics for big data problems. Int. J. Adv. Soft Comput. Appl. **6**(2), 1–17 (2014)

7. McConnell, S., Sturgeon, R., Henry, G., Mayne, A., Hurley, R.: Scalability of self-organizing maps on a GPU cluster using OpenCL and CUDA. J. Phys. Conf. Ser. **341**, 12018 (2012)

8. Moraes, F.C., Botelho, S.C., Filho, N.D., Gaya, J.F.O.: Parallel high dimensional self organizing maps using CUDA. In: 2012 Brazilian Robotics Symposium Latin American Robotics Symposium, pp. 302–306 (Oct. 2012)

9. Khan, S.Q., Ismail, M.A.: Design and implementation of parallel SOM model on GPGPU. In: 2013 5th International Conference Computer Science Information Technology, pp. 233–237 (Mar. 2013)

10. Wang, H., Zhang, N., Créput, J.-C.: A Massive Parallel Cellular GPU Implementation of Neural Network to Large Scale Euclidean TSP. In: Castro, F., Gelbukh, A., González, M. (eds.) Advances in Soft Computing and Its Applications: 12th Mexican International Conference on Artificial Intelligence, MICAI 2013, Mexico City, Mexico, 24–30 November 2013, Proceedings, Part II, pp. 118–129. Springer, Berlin, Heidelberg (2013)

11. Faro, A., Giordano, D., Palazzo, S.: Integrating unsupervised and supervised clustering methods on a GPU platform for fast image segmentation. In: 2012 3rd International Conference Image Processing Theory, Tools Applications IPTA 2012, pp. 85–90 (2012)

12. Khronos OpenCL: OpenCL Specification (2014)

13. Brodtkorb, A.R., Hagen, T.R., Sætra, M.L.: Graphics processing unit (GPU) programming strategies and trends in GPU computing. J. Parallel Distrib. Comput. **73**(1), 4–13 (2013)

14. Mukherjee, S., Sun, Y., Blinzer, P., Ziabari, A.K., Kaeli, D.: A comprehensive performance analysis of HSA and OpenCL 2.0. In: 2016 IEEE International Symposium Performance Analysis System Software (April, 2016)

Parametric Optimization of the Selected Classifiers in Binary Classification

Daniel Kostrzewa and Robert Brzeski

Abstract The conception of classification is one of the major aspects in data processing. Conducted research present comparison of chosen classifiers' results of classification for a few data sets. All data were chosen from these available on UCI Machine Learning Repository web site. During realization of research, the optimization process of the results of classification was made on the modifiable parameters for particular classifiers. In this work, gathered result of classification was presented as well as conclusion and possibility of future work.

Keywords Classifier · Classification · WEKA · Accuracy · IBk · Logistic Base · Naive Bayes · Voted Perceptron · Zero R · SMO · SGD · UCI · Parameter · Optimization

1 Introduction

One of the major operation in data processing is classification. Just a possession of a set of data—values which reflect a process, situation, problem, is quite often totally insufficient. Only information or knowledge, obtained in the processing of data, starts to have a practical, useful value. One of the elements or steps of the data processing is a classification. The classification process involves assignment of the given data vector to a proper class, based on individual elements of the data vector. In the current research, attention has been focused on the binary classification. Related work in this subject was presented in Sect. 2. The very process of gathering the appropriate number of data describing the problem is not an issue of current research. To carry out presented research, the generally available data sets have been used. In this case, a few sets of data available on the UCI Machine Learning

D. Kostrzewa (✉) · R. Brzeski
Silesian University of Technology, Gliwice, Poland
e-mail: daniel.kostrzewa@polsl.pl

R. Brzeski
e-mail: robert.brzeski@polsl.pl

© Springer International Publishing AG 2017
D. Król et al. (eds.), *Advanced Topics in Intelligent Information and Database Systems*, Studies in Computational Intelligence 710, DOI 10.1007/978-3-319-56660-3_6

Repository [1–4] was chosen. A more detailed description of these data sets is provided in Sect. 3. For those data sets, subsequently were carried out a preliminary classifications, using almost thirty most popular and useful classifiers. Based on the best results, several classifiers were selected and then the optimization process of classification was conducted, by changing selected parameters for individual classifiers.

To carry out the classification process, the software provided in Weka [5] system (Waikato Environment for Knowledge Analysis) was used, in the form of Java classes. Each class contains code that implements one of the classifiers. These classes were used in the original software, that makes easier the implementation of the research. Facilitation involves the multiple executions of classification process, in accordance to a predetermined plan of using certain classification parameters and then saving the results in a suitable form. These results are presented and commented in Sect. 4. Summary of the research, and the ability to develop them is provided in Sect. 5.

The current study presented in this article is sort of continuation of the research presented in [6, 7]. There, the process of dimensionality reduction was made through feature selection, executed by IWO algorithm [8–10]. And the process of classification was done only by 1NN classifier. Authors would like to extend not only the process of dimensionality reduction, but also execute the process of classification, by many other classifiers and then to use this whole process, to different data sets. That is why the authors of this paper divided planed research into a few parts and in this article is presented one of them.

2 Related Work

One of the classical problem in data mining [11] is classification [12]. This process, called supervised learning, is executed on data set, previously properly collected and processed. As a result of binary classification, the data vector is classified into one of two subsets. To simplify, the first subset can be described as 'yes'—that fulfill certain criteria, a second as 'no'—do not fulfill the criteria. Whether the binary classification is appropriate, depends due to the fact, what type of set of data is consider and what is the need for having an adequate number of grouped subsets, to obtain the desired knowledge. Therefore, the current research has been done on the data sets, in which just such a grouping into two subsets is adequate.

Since, in the current study, the authors focused on the binary classification only, then might be worth to start with the Zero R [13] classifier. Due to simplicity of its operation, is often consider as reference point, for the classification results obtained by other classifiers. And the most common classifiers are: IBk—Instance-based learning algorithm [14], Voted Perceptron [15], SMO [16–18], Naïve Bayes [19], K* [20], Simple Logistic [21] and many more [22].

There are quite a lot of different classifiers, and many of them are still the subject to a process of development, in order to obtaining a higher classification accuracy [12, 23–26], achieving greater speed or less memory consumption.

The results obtained in the classification process can be compared by several criteria [12, 23–25], which are selected according to the nature of the classification purposes. The problem of classification is still intensively developed. In many situations obtained knowledge helps, or even becomes, the basis for decisions [11].

There are some study about performance of classification, like statistical comparisons of different classifiers over multiple data sets [27], classification made by few chosen classifiers over two data sets [28], comparative analysis of two classification algorithms (J48 and Multilayer Perceptron) on several data sets [29], ways of evaluating and comparing classifiers [12, 23–25], but authors have not been able to find the one, like presented in this article.

3 Overview of the Classifiers, Description of Research

Current research as the assumption includes only binary classification. Therefore, the data sets have been selected precisely under this terms—classification capabilities of each data vector into exactly one of two sets of vectors. In total, four separate, independent data sets were selected, different in size—in the number of data vectors, in the length of the vector—the number of attributes, in data character. Two of the data sets are differentiated (the vectors represent individual data subsets in number of 50%) two other are not.

In particular, all sets of data have a large number of attributes in the vector. This will allow in the future, to compare the results, to the one of the planned study— classification of the data sets, after the conducted process of dimensionality reduction.

All of the data sets, used in current research were prepared for the NIPS 2003 [30] variable and feature selection benchmark by Isabelle Guyon[1] [1–4, 31].

The first selected data set is DOROTHEA—a drug discovery data set. Chemical compounds represented by structural molecular features must be classified as active (binding to thrombin) or inactive [1, 31].

The original data set was provided by DuPont Pharmaceuticals Research Laboratories for KDD Cup 2001 [1, 32]. This data were modified for the purpose of the feature selection challenge. In particular, a number of distractor feature called 'probes', having no predictive power were added. The order of the features and patterns were randomized [1].

The second data set is DEXTER—a text classification problem in a bag-of-word representation. This is a two-class classification problem with sparse continuous input variables [2, 31].

---

[1]955 Creston Road, Berkeley, CA 94708, USA (isabelle@clopinet.com).

The original data set is a subset of the well-known Reuters text categorization benchmark [2, 31]. The used data were formatted by Thorsten Joachims and then a number of distractor feature called 'probes', having no predictive power were added. The order of the features and patterns were randomized [2, 31].

The third data set is ARCENE, where the task is to distinguish cancer versus normal patterns, from mass-spectrometric data. This is a two-class classification problem with continuous input variables [3, 31].

The original data were obtained from two sources: The National Cancer Institute (NCI) and the Eastern Virginia Medical School (EVMS). All the data consist of mass-spectra obtained with the SELDI technique [3, 31]. The original features indicate the abundance of proteins in human sera having a given mass value. Based on those features one must separate cancer patients from healthy patients. To the original data a number of distractor feature called 'probes', having no predictive power were added. The order of the features and patterns were randomized [3, 31].

The fourth data set is GISETTE—a handwritten digit recognition problem. The problem is to separate the highly confusable digits '4' and '9' [4, 31]. The data set was constructed from the MNIST data, that is made available by Yann LeCun and Corinna Cortes at http://yann.lecun.com/exdb/mnist/ [4, 31].

The digits have been size-normalized and centered in a fixed-size image of dimension 28 × 28. The original data were modified, in particular, pixels were samples at random in the middle top part of the feature, containing the information necessary to disambiguate 4 from 9 and higher order features were created as products of these pixels to plunge the problem in a higher dimensional feature space. Also a number of distractor features called 'probes' having no predictive power were added. The order of the features and patterns were randomized [4, 31] (Table 1).

In all data sets the classification process was realized, using several selected classifiers, which can be described as one of the most popular. The results are shown in Table 2. It contains the factor Accuracy (Acc) [23–25] (1) as one of the most typical representative, in determining the quality of classification. It does not mean that this is the best way to define quality in every situation, but this factor is most universal. The best result for a given set of data is bold marked, and result close to the best are italics. On this basis, for further study, the classifiers giving pre-best results were selected. These selected classifiers in Table 2 are marked in bold.

$$Acc = \frac{TP + TN}{TP + TN + FP + FN} \tag{1}$$

Lack of the data in the table means, that the classification process was stopped, due to excessive execution time. Classification typically last from less than a second, to several minutes. That is why we decided, that the maximum time is 2 h and after this time the process as the entirely ineffective, will be interrupted.

Table 1 The quantity of data vectors in the data set

Name of data set	Number of positive examples	Number of negative examples	Total number of examples	Number of attributes
DOROTHEA Training set	78	722	800	100 000
DOROTHEA Validation set	34	316	350	100 000
DEXTER Training set	150	150	300	20 000
DEXTER Validation set	150	150	300	20 000
ARCENE Training set	44	56	100	10 000
ARCENE Validation set	44	56	100	10 000
GISETTE Training set	3 000	3 000	6 000	5 000
GISETTE Validation set	500	500	1 000	5 000

Table 2 Initial result—accuracy factor

Classifier\data set	Dorothea	Dexter	Arcene	Gisette
Ada Boost M1	*0.937*	0.81	0.65	0.866
Bayes Net	–	0.9	0.72	0.905
Decision Stump	*0.937*	0.667	0.66	0.831
Filtered classifier	0.929	0.823	0.67	0.959
Hoeffding Tree	0.903	0.587	*0.74*	0.844
IBk	0.906	0.527	*0.83*	*0.962*
Iterative classifier optimizer	0.929	0.897	0.71	0.904
J48	0.911	0.777	0.66	0.942
JRip	0.929	0.853	0.64	0.936
K*	–	0.5	0.56	–
LMT	–	*0.91*	0.64	–
Logistic	–	0.84	0.66	*0.975*
Logistic Base	*0.937*	*0.91*	0.64	*0.976*
Logit Boost	0.923	0.897	0.71	0.904
Naive Bayes	0.934	0.9	*0.74*	0.905
Naive Bayes multinomial	0.911	*0.917*	0.69	0.92
Naive Bayes multinomial text	0.903	0.5	0.56	0.5
Part	0.877	0.797	0.66	0.95
Random tree	0.874	0.68	0.7	0.847

(continued)

Table 2 (continued)

Classifier\data set	Dorothea	Dexter	Arcene	Gisette
Randomizable filtered classifier	0.849	0.557	0.65	0.592
Rep Tree	*0.937*	0.803	0.65	0.944
SGD	0.934	**0.93**	*0.79*	**0.977**
SGD Text	0.903	0.5	0.56	0.5
Simple Logistic	*0.937*	*0.91*	0.64	*0.976*
SMO	0.931	*0.91*	**0.84**	*0.976*
Voted Perceptron	**0.949**	0.89	*0.78*	*0.964*
Zero R	0.903	0.5	0.56	0.5
Min except for **Zero R**	0.849	0.5	0.56	0.5
Max	0.949	0.93	0.84	0.977
Average except for **Zero R**	0.917	0.777	0.682	0.877

Result obtained by Zero R classifier can be considered as reference point for other classifiers. For better comparison, at the end of Table 2, are additionally: the worst result obtained for particular data set, the best one and average calculated from the results.

4 Tests—Optimization of Classifiers

As appropriate research, the entire set of tests, on the same data sets were implemented, and executed by classifiers giving pre-best results. Additionally the different parameters were used by the classification algorithms. Due to size limitations of this article, it is not possible to present all of the partial results of the classification process.

That is why the authors took only final best gained result. As a summary in Table 3 the comparison of accuracy factor is presented: before (Default parameters—**D**) and after optimization process (Optimized parameters—**O**). The Table 3 also present percent of improvement (%), together with parameters for which it was obtained. Lack of parameter's values in Table 3 means, the obtained result did not depend on this parameter. Value 'min' is a minimum from appropriate column, except classifier Zero R and Logistic Base (for which there are no parametric optimization). In similar way is chosen value 'max' as maximum and calculated average of these values.

During parametric optimization for individual classifiers following parameters, in given range, were tested:

IBk [22] parameters:

- I—weight neighbours by the inverse of their distance (used when K > 1). (Default = off). In the research was checked off and on.

Table 3 Comparison of accuracy factor before and after optimization process

Classifier\data set	Dorothea			Dexter			Arcene			Gisette		
	D	O	%	D	O	%	D	O	%	D	O	%
IBk (parameters)	0.906	0.906	0	0.527	0.543	3.164	0.83	0.84	1.205	0.962	0.966	0.416
	K = 1			K = 17			K = 4 I or F = on			K = 3		
Logistic Base	0.937		–	0.91		–	0.64		–	0.976		–
Naïve Bayes	0.934	0.946	1.223	0.9	1	11.11	0.74	0.74	0	0.905	0.905	0
	D = on			K = on			turned off			turned off or D = on		
SGD	0.934	0.94	0.612	0.93	0.933	0.358	0.79	**0.89**	12.66	0.977	**0.982**	0.512
	L = 10			L = 0.00001 F = 1 N = on			R = 1.0 F = 1 N = on			E = 2000 F = 1 N = off		
SMO	0.931	0.931	0	0.91	0.927	1.832	0.84	0.85	1.190	0.976	0.981	0.512
	N = 0 or N = 2 L ≤ 0.1			N = 2 L ≤ 0.01			N = 0 C = 0.01			N = 1 L ≤ 0.01		
Voted Perceptron	0.949	**0.949**	0	0.89	0.927	4.120	0.78	0.88	12.82	0.964	**0.982**	1.867
	E = 1 I = 1 or 2 or 3 E = 0.02 I = 14 or 15			E = 2 I ≥ 6			E = 5 I >=10			E = 2 I = 14 or 15		
Zero R	0.903		–	0.5		–	0.56		–	0.5		–
Min	0.906	0.906	0	0.527	0.543	0.358	0.74	0.74	0	0.905	0.905	0
Max	0.949	0.949	1.223	0.93	1	11.11	0.84	0.89	12.82	0.977	0.982	1.867
Average	0.931	0.934	0.367	0.831	0.866	4.117	0.796	0.84	5.575	0.957	0.963	0.661

- F—weight neighbours by 1—their distance (used when K > 1). (Default = off). In the research was checked off and on.
- K—number of nearest neighbours (k) used in classification. (Default = 1). In the research were tested values from 1 up to 20.

The nearest neighbours coefficient was tested cross with I and F parameters.

Logistic Base [22] classifier does not have any parameters, and obtained result as well as from **Zero R** can be consider as reference point.

Naive Bayes [22] parameters:

- K—use kernel density estimator rather than normal distribution for numeric attributes.
- D—use supervised discretization to process numeric attributes.

In our research were tested all 3 possibilities (off, K = on, D = on).

SGD [22] parameters:

- F—set the loss function to minimize, where: 0 = hinge loss (SVM), 1 = log loss (logistic regression). (Default = 0). In our research were tested F = 0 and F = 1.
- L—the learning rate. (Default = 0.01). In our research were tested values from L = 10 to 10^{-6}.
- R—the lambda regularization constant. (Default = 10^{-4}). In our research were tested values from R = 10 to 10^{-6}.
- E—the number of epochs to perform (batch learning only, default = 500). In our research were tested values from E = 10 to 10 000.
- N—do not normalize the data. (Default = off). In our research was checked off and on.
- M—do not replace missing values. (Default = off). In our research was checked off and on.

In our research were tested L and R and E cross with F and N and M parameter.

SMO [22] parameters:

- C—the complexity constant C. (Default = 1). In our research were tested values from C = 0.01 to C = 100.0.
- N—whether to 0 = normalize/1 = standardize/2 = neither. (Default = 0). In our research were tested all 3 values.
- L—the tolerance parameter. (Default = 10^{-3}). In our research were tested values L = 10^{-6} to L = 1.0.
- V—the number of folds for the internal cross-validation. (Default = −1—use training data). In our research were tested values −1 and 10.

In our research were tested C and L cross with N and V parameter.

Voted Perceptron [22] parameters:

- I—the number of iterations to be performed. (Default = 1). In our research were tested values from 1 to 15.
- E—the exponent for the polynomial kernel. (Default = 1). In our research were tested values from −10 to 10.

In our research were tested I cross with E parameter.

The optimization process was carried out for a predetermined set of parameters.

The very process of optimization, is such a complex process, that theoretically could be realized infinitely. For the purposes of the current research it was carried out, for the presented, quite large set of parameters. Tests were performed in a crisscross pattern.

5 Conclusions

As can be seen in Table 3, the parametric optimization of classifiers, not always improve the quality of classification. Gained improvement is between 0 up to 12.8%. This shows that the default values are chosen not accidentally. It can be also seen, that even for a given set of data—in presented research the Dorothea one, any result after optimization, is not better, than the best one before optimization. For other data set it was possible to get better outcomes than the best one before optimization.

It can also be noted, that for different data sets, the best results are obtained by different classifiers. It comes not only because the nature of the collected data, but also because the different number of data vectors, and the percentage distribution of data vectors to the appropriate result subset. And because of data characteristic and gained knowledge, it is obvious that changing of different parameters, cause improvement of result for deferent data sets.

By analyzing results in Table 3 together with data in Table 1, the correlation between number of training vectors (Table 4—value: training) and improvement of classification (max of %; average %) was noted. The correlation [33] was calculated and presented in Table 4. As we can see there is a correlation with maximum improvement, obtained during optimization process, as well as with average improvement. The lower number of training vectors causes the higher possibility of making the improvement. Even stronger correlation can be noticed, with the average value of accuracy factor (what could be expected). It possible that for

Table 4 Pearson product-moment correlation coefficient

	Dorothea	Dexter	Arcene	Gisette	Correlation coefficient
Training	800	300	100	6000	–
Max of %	1.223	11.11	12.82	1.867	**−0.622**
Average %	0.367	4.117	5.575	0.661	**−0.609**
Average O	0.934	0.866	0.84	0.963	**0.790**

Dexter and Arcene data set, the number of training vectors is a little too small. So it is possible, that particularly for data sets with small number of training vectors, the default parameters for classifiers are not the best solution and especially in that situation it is worth to do the additional work. Of course we are aware that we have a very small number of sample test, for calculated correlation and in the future it should be proved for appropriate number of different data sets.

No difference were observed between the results obtained for the collection of balanced and unbalanced data sets.

Parametric optimization and selection of appropriate classifier is of course not the only possibility to improve the results. A very important aspect is also suitable preparation of the data, for which the process of classification is performed. The next step will be study of this nature, with reduction of dimensionality of used data set (feature selection; dimensionality reduction—both linear, for example PCA, and nonlinear, for example: Isomap, LLE, GPLVM).

As a summary can be recalled, that the present study is the continuation of the research presented in [6] and at the same time the introduction to the future one. The study will allow the comparisons with the result of the planned research, on reducing the dimensionality of the data sets. And especially in the context of continuation of the study and comparison of results, the current purpose was to maximize the gained results, to check in the future, how the preparation of the data affects the whole process of optimizations. This article presents both, the results obtained by a number of different classifiers, as well as those on the optimization process for pre-best classifiers.

Acknowledgements This work was supported by BKM16/RAu2/507 grant from the Institute of Informatics, Silesian University of Technology.

References

1. UCI Machine Learning Repository Dorothea data set. Irvine, CA: University of California, School of Information and Computer Science. http://archive.ics.uci.edu/ml/datasets/Dorothea
2. UCI Machine Learning Repository Dorothea data set. Irvine, CA: University of California, School of Information and Computer Science. http://archive.ics.uci.edu/ml/datasets/Dexter
3. UCI Machine Learning Repository Dorothea data set. Irvine, CA: University of California, School of Information and Computer Science. http://archive.ics.uci.edu/ml/datasets/Arcene
4. UCI Machine Learning Repository Dorothea data set. Irvine, CA: University of California, School of Information and Computer Science. http://archive.ics.uci.edu/ml/datasets/Gisette
5. Weka 3. http://www.cs.waikato.ac.nz/~ml/weka/
6. Josiński, H., Kostrzewa, D., Michalczuk, A., Świtoński, A.: The exIWO metaheuristic for solving continuous and discrete optimization problems. Sci. World J. **2014**, Article ID 831691 (2014). Hindawi Publishing Corporation
7. Josiński, H., Świtoński, A., Jędrasiak, K., Kostrzewa, D.: Human identification based on gait motion capture data. In: Lecture Notes in Engineering and Computer Science, vol. 2195, pp. 507–510. Hong Kong (2012)
8. Mehrabian, A.R., Lucas, C.: A novel numerical optimization algorithm inspired from weed colonization. Ecological Informatics **1**(4), 355–366 (2006). Elsevier

9. Pahlavani, P., Delavar, M.R., Frank, A.U.: Using a modified invasive weed optimization algorithm for a personalized urban multi-criteria path optimization problem. Int. J. Appl. Earth Obs. Geoinf. **18**, 313–328 (2012). Elsevier

10. Kostrzewa, D., Josiński, H.: The exIWO metaheuristic—a recapitulation of the research on the join ordering problem. In: Communications in Computer and Information Science, vol. 424, pp. 10–19. Springer (2014)

11. Haiyang, Z.: A Short Introduction to Data Mining and Its Applications. IEEE (2011)

12. Agrawal, R., Imielinski, T., Swami, A.N.: Database mining: a performance perspective. IEEE Trans. Knowl. Data Eng. **5**(6), 914–925 (1993)

13. Zero R classifier. http://weka.sourceforge.net/doc.dev/weka/classifiers/rules/ZeroR.html

14. Aha, D., Kibler, D.: Instance-based learning algorithms. Mach. Learn. **6**, 37–66 (1991)

15. Freund, Y., Schapire, R.E.: Large margin classification using the perceptron algorithm. In: 11th Annual Conference on Computational Learning Theory, New York, NY, pp. 209–217 (1998)

16. Platt, J.: Fast training of support vector machines using sequential minimal optimization. In: Schoelkopf, B., Burges, C., Smola, A. (eds.): Advances in Kernel Methods—Support Vector Learning (1998)

17. Keerthi, S.S., Shevade, S.K., Bhattacharyya, C., Murthy, K.R.K.: Improvements to Platt's SMO algorithm for SVM classifier design. Neural Comput. **13**(3), 637–649 (2001)

18. Hastie, T., Tibshirani, R.: Classification by pairwise coupling. Advances in Neural Information Processing Systems (1998)

19. John, G.H., Langley, P.: Estimating continuous distributions in bayesian classifiers. In: Eleventh Conference on Uncertainty in Artificial Intelligence, San Mateo, pp. 338–345 (1995)

20. Cleary, J.G., Trigg, L.E.: K*: an instance-based learner using an entropic distance measure. In: 12th International Conference on Machine Learning, pp. 108–114 (1995)

21. Sumner, M., Frank, E., Hall, M.: Speeding up logistic model tree induction. In: 9th European Conference on Principles and Practice of Knowledge Discovery in Databases, pp. 675–683 (2005)

22. Weka 3: Documentation. http://weka.sourceforge.net/doc.dev/allclasses-noframe.html

23. Costa, E.P., Lorena, A.C., Carvalho, A.C.P.L.F., Freitas, A.A.: A review of performance evaluation measures for hierarchical classifiers. In: Evaluation Methods for Machine Learning II, AAAI-2007 Workshop, pp. 182–196. AAAI Press (2007)

24. Powers, D.M.W.: Evaluation: from precision, recall and F-score to ROC, informedness, markedness & correlation. J. Mach. Learn. Technol. **2**, 37–63 (2011)

25. Provost, F., Fawcett, T., Kohavi, R.: The case against accuracy estimation for comparing classifiers. In: Proceedings of the ICML-98. Morgan Kaufmann, San Francisco, pp. 445–453 (1998)

26. Smith, M.R., Martinez, T.: Improving classification accuracy by identifying and removing instances that should be misclassified. In: Proceedings of the IEEE International Joint Conference on Neural Networks, pp. 2690–2697 (2011)

27. Demšar, J.: Statistical comparisons of classifiers over multiple data sets. J. Mach. Learn. Res. **7**, 1–30 (2006)

28. Kumar, Y., Sahoo, G.: Analysis of Bayes, Neural Network and Tree Classifier of Classification Technique in Data Mining using WEKA (2012)

29. Arora, R., Suman: Comparative analysis of classification algorithms on different datasets using WEKA. Int. J. Comput. Appl. **54**(13) (2012)

30. Guyon, I., et al.: Competitive baseline methods set new standards for the NIPS 2003 feature selection benchmark. Pattern Recognit. Lett. **28**, 1438–1444 (2007)

31. Guyon, I., Gunn, S.R., Ben-Hur, A., Dror, G.: Result analysis of the NIPS 2003 feature selection challenge (2004)

32. KDD Cup 2001. http://www.cs.wisc.edu/~dpage/kddcup2001/

33. Pearson, K.: Notes on regression and inheritance in the case of two parents. Proc. R. Soc. Lond. **58**, 240–242 (1895)

Prediction of Academic Performance During Adolescence Based on Socioeconomic, Psychological and Academic Factors

A.T.M. Shakil Ahamed, Navid Tanzeem Mahmood and Rashedur M. Rahman

Abstract Educational effectivity is paramount towards enhancing modernization. In our study we have taken into account various socioeconomic, psychological and academic factors to properly understand a student's life during adolescence and their effect on academic performance. In order to build a predictive model, we have pre-processed the data using dimensionality reduction, data balancing, discretization, and normalization and then classified the data using different machine learning techniques like Artificial Neural Net, K-Nearest Neighbors and Support Vector Machine. Lastly we have discovered patterns throughout the dataset in relation with academic performance.

Keywords Data mining · Predictive model · Machine learning

1 Introduction

Our study falls into the field of Educational Data Mining: a rather new area where academic performance is investigated and steps are figured out to enhance it. The aim of this study is set towards Higher Secondary Certificate (HSC) (12th grade equivalent) candidates; all data has been collected from recent HSC examinees: university freshers.

The goal of our study is to provide a prediction of HSC result from socioeconomic, psychological and academic factors and to discover pattern in attributes which allow students to visualize how the attributes influencing their educational

A.T.M. Shakil Ahamed · N.T. Mahmood · R.M. Rahman (✉)
Department of Electrical and Computer Engineering, North South University, Plot-15, Block-B, Bashundhara, Dhaka, Bangladesh
e-mail: rashedur@northsouth.edu; rashedur.rahman@northsouth.edu

A.T.M. Shakil Ahamed
e-mail: ahnafshakil@gmail.com

N.T. Mahmood
e-mail: navid.mahmood@hotmail.com

© Springer International Publishing AG 2017
D. Król et al. (eds.), *Advanced Topics in Intelligent Information and Database Systems*, Studies in Computational Intelligence 710, DOI 10.1007/978-3-319-56660-3_7

performance which help students to enhance the outcome of HSC exam. Guardians of students and responsible authorities could use such information generated from our model and help the students to enhance the HSC performance.

2 Related Work

To group the student according to their skill level in [1] the authors have used model based clustering, other clustering and K-mean technique. In paper [2] the authors have used quiz score, midterm marks, final marks, lab marks and CGPA as attribute to predict semester final grade of individual courses. They have used ANN, C4.5 and Naïve Bayes for the classification and the have managed to reach accuracy around 82%. In order to predict academic performance [3] class attendance, seminar, class test score and assignment marks have been used as attributes. Authors have used C4.5, CART and ID3 algorithms to design their predictive model.

3 Designing Survey Questionnaire and Data Set

Designing a predictive model greatly depends on the quality of the data set. We are very careful while designing the survey questionnaire, as the survey is the source of attributes for the predictive model. In order to get the relevant attributes, we have interviewed experts and explored different factors related to academic performance. We have made sure that every attribute is proven to be related to academic performance through published research. We have surveyed 423 students from different location and economic status for maintaining the versatility in the data set. Our data set consists of 33 attributes grouped into 3 main categories namely socioeconomic, psychological and academic as shown in Table 1.

4 Overview of Designing Predictive Model

The design of the predictive model is shown in Fig. 1. After getting raw data we have realized that data set needs to be preprocessed before applying any learning algorithm to augment the predictive accuracy. We have used data balancing technique to balance the data set and also used Principal Component Analysis (PCA) for dimensionality reduction then Optimal equal width binning for discretization and finally used normalization technique as pre-processing. Then we have applied 3 different classifiers namely Artificial Neural Network (ANN), K-Nearest Neighbors (K-NN) and Support Vector Machine (SVM) and after comparative study of the classifiers we have derived the best model for the prediction.

Table 1 All the attributes used in designing the predictive model

Categories	Factors
Socioeconomic and demographic factors	• Sex • Age • Mother's education • Father's education • Mother's job • Father's job • Average monthly family income • Place of education • Private tuition
Psychological factors	• Family size • Age difference with most immediate sibling • Parents' involvement in education • Parent's marital status • Absence/Death of parent(s) • Involvement in physical activity • Motivating teacher • Motivating parents • Academic ambitions • Time spent on social media • Romantic relation • Hanging out with friends • Alcohol/Drugs consumption • Smoking • Health status of participant • Health status of family members
Academic factors	• Early childhood education • Weekly study time • Previous failures • Number of absence(s) • Extracurricular activities • Accessibility of the internet • SSC (Secondary School Certificate) result (10th grade equivalent) • HSC result

5 Data Pre-processing

5.1 Data Balancing

Our data set is not balanced as shown in Fig. 2. The problem with of the imbalanced data can be visualized while mammographic pictures are used as data set. The picture contains 98% of normal pixels and 2% of abnormal pixels. So simply by guessing the pixel as normal one can be right for the 98% of the time [4]. We have used SMOTEBoost technique [5] which consists of Synthetic Minority oversampling technique and a boosting procedure to provide equal entropy to the both majority and minority class to enrich the quality of the accuracy of the classifiers.

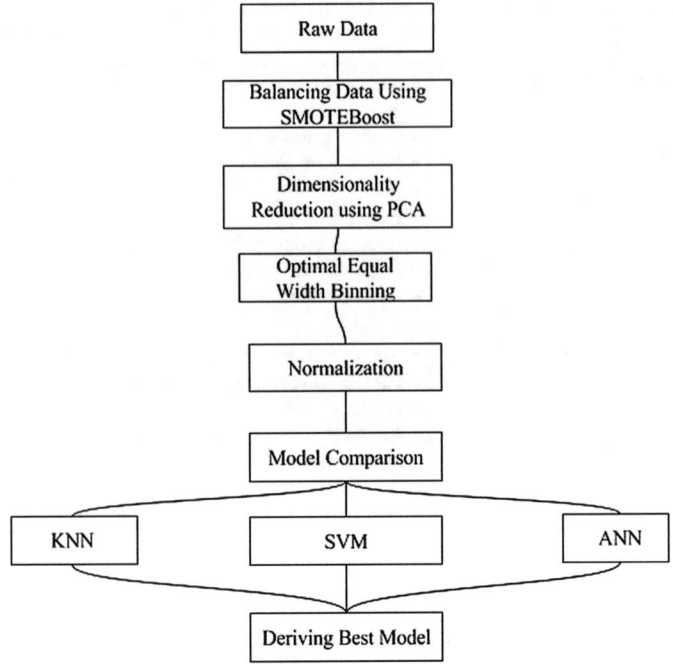

Fig. 1 An overview of our system

Fig. 2 Instances of target
class (HSC result)

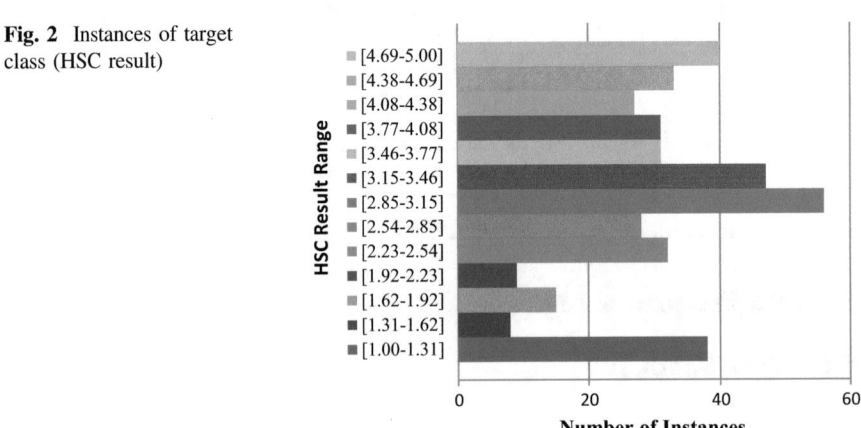

5.2 Dimensionality Reduction

The data set consists of 33 attributes; most of them are dispersed from each other
and some of them are redundantly related to the target class. Removing redundancy
bolsters classifiers by excluding the curse of the dimensionality reduction [6]. We

have used principal component analysis (PCA) to reduce the dimensionality of our data set. PCA applies an orthogonal transformation; so all resultant attributes are correlated to only the target class and not each other [7].

5.3 Discretization

Our data set consists of attribute such as SSC and HSC result which are continuous valued attributes. To discretize the data set we have implemented Optimal Equal Width binning [8] that searches dynamically for the optimal width and number of bins for the target class. In our data set we got 8 bins with the width of 0.5. As the HSC result is in the range of 1–5 the bins we got from the binning are: [−∞–1.5], [1.5–2], [2–2.5], [2.5–3], [3–3.5], [3.5–4], [4–4.5] and [4.5–∞].

5.4 Normalization

Our data set consists of attributes with different scales and units, for instance, "Weekly study time" has the unit of hour, on the other hand, average family income has the unit of Taka and "Family involvement" is measured in a scale of 5, conversely "Health status" is measured in the scale of 10. To ensure the equal contribution of the attribute to predict the target class we have used min-max [9] normalization technique to rescale the data set in order to produce better training set for the classifiers using the Eq. (1).

$$x' = \frac{x_i - x_{min}}{x_{max} - x_{min}} \tag{1}$$

This technique is able to preserve all the relationship among attributes and rescale them for equal contribution of the attribute to predict target class.

6 Learning Algorithms

After applying the pre-processing techniques we have used learning algorithms to predict the performance of HSC examination. We have spilt the data set in 70:30 for training and testing. In order to ensure consistency in class distribution throughout the data set we have approached stratified sampling to maintain the equal distribution in the training and testing set in order to discard the possibility of having adverse impact on predictive accuracy due to exhibiting inconsistency in the split parts of the data set.

6.1 K-Nearest Neighbors (K-NN)

K-NN learns from each instance from other known outputs of the data set. All attributes are transformed into vector space where each of the N dimensions of an attribute is represented by vectors [10]. K-NN uses Euclidian distance to measure the similarity metric amongst neighbors using the Eq. (2).

$$d_E(x, y) = \sum_{i=0}^{N} \sqrt{(x_i^2 - y_i^2)} \tag{2}$$

It works the following way in our system:

(a) Finding appropriate k-value to get the similar neighbors for classifying.
(b) Calculating distance between training sample and testing sample.
(c) Sorting the samples based on the distance.
(d) Applying majority vote across the k-nearest neighbors produced the predicted HSC result.

6.2 Support Vector Machine (SVM)

By constructing hyperplanes in a multidimensional feature space it breaks down class levels into different cases. Iteratively, SVM applies a training algorithm to produce an optimal hyperplane and also minimizes the error function in each iteration [11]. Error function is described in the Eq. (3).

$$\frac{1}{2} w^t w + c \sum_{i=1}^{N} \xi_i \tag{3}$$

Subject to: $y_i \left(w^T \phi(x_i) + b \right) \geq 1 - \xi_i$ and $\xi_i \geq 0, i = 1, \ldots\ldots, N$

where,

- C = capacity function
- w = Vectors of coefficients
- b = Constant
- ξ_i = Parameter that handles non-separable data
- i = Number of training cases (N)
- $y \in \pm 1$ = Class labels
- x_i = Independent variable
- Kernel ϕ transforms data to the feature space from the input.

As C increases, the error gets penalized more; the value of C can be set to avoid the problem of overfitting. Dot kernel has been applied for our system as shown below in Eq. (4).

$$K(X_{i,} X_{j,}) = \begin{cases} X_i \cdot X_j & Linear \\ (\gamma X_i \cdot X_j + C)^4 & Polynomial \\ \exp(-\gamma |X_i - X_j|^2) & RBF \\ \tanh(\gamma |X_i \cdot X_j + C) & Sigmoid \end{cases} \tag{4}$$

6.3 Artificial Neural Network

Artificial neurons perform its actions almost like their biological counterparts: they process their mechanism by transmitting information among neurons that helps each other to 'fire' based on given inputs. To 'fire' the activation function shown in Eq. (5) must be satisfied [12].

$$a_i = \sum_{J=1}^{N} W_{ji} x_j + \theta_i \tag{5}$$

where, x_j is the output from a neuron or an external input, W_{ji} is the weight and θ_i is the threshold.

We have used 'feed forward' architecture for designing our system; the outputs of a layer of neurons get fed into the inputs of the neurons from the next layer. The 'hidden' layer(s) exist in between input and output layers. Number of hidden layers of our system can be calculated using the Eq. (6).

$$(Number\ of\ Attributes + Number\ of\ Classes)/2 + 1$$
$$= (32 + 1)/2 + 1 = 17(approx.) hidden\ layers. \tag{6}$$

7 Result Analysis

7.1 K-Nearest Neighbors

While no pre-processing is used the KNN provides the accuracy of 30% as shown in Table 2 but after introducing pre-processing technique KNN it starts achieving better accuracy. After using PCA and SMOTEBoost it increases its accuracy to 50.35%. In addition when we have used Optimal Equal Width binning the accuracy has reached around 63% and after the normalization the accuracy has notably augmented to 70%. It is because KNN uses Euclidian algorithm to measure the

Table 2 Performance comparison of the different predictive models

Models	Accuracy (%)	Precision (%)	Recall (%)	F-measure (%)
ANN	70.05	55.96	59.21	57.54
ANN (PCA + SMOTEBoost)	71.43	79.87	65.68	72.08
ANN (PCA + SMOTEBoost + Binning)	77.01	81.77	87.50	84.54
ANN (PCA + SMOTEBoost + Binning + Normalized)	77.78	75.68	68.95	72.16
K-NN	30.35	16.32	19.25	17.66
K-NN (PCA + SMOTEBoost)	50.35	17.02	22.25	54.29
K-NN (PCA+ SMOTEBoost + Binning)	63.74	63.92	54.77	58.99
K-NN (PCA + SMOTEBoost + Binning + Normalized)	70.00	82.37	61.48	70.41
SVM	70.88	71.99	55.52	62.69
SVM (PCA + SMOTEBoost)	78.50	69.22	56.02	61.92
SVM (PCA + SMOTEBoost + Binning)	76.17	86.85	62.75	72.86
SVM (PCA + SMOTEBoost + Binning + Normalized)	75.93	85.43	57.36	68.64

distance between neighbors; by rescaling the attribute and normalizing the range bolster K-NN to achieve such high score.

7.2 Artificial Neural Network (ANN)

ANN has provided 70% of the accuracy without using any pre-processing technique, after SMOTEBoost and PCA it does not provide any promising change in accuracy. Using discretization with Optimal Equal Width Binning it enhances its accuracy by 7% and normalization does not have any significant impact on accuracy of ANN.

7.3 Support Vector Machine (SVM)

SVM provides accuracy of 70.88% without using any pre-processing technique. After balancing the data set using SMOTEBoost the accuracy has increased by 8%. As SMOTEBoost provides equal entropy to the majority and minority class, it bolsters SVM to provide better accuracy. Binning and Normalization lowers the accuracy of the SVM.

7.4 Key Findings

We have generated key findings using decision trees as shown in below. One of the observations is "higher family involvement notably increases the performance producing better result with same time spent on studying weekly". Family involvement has measured in the scale of 1–5 where 1 is least involvement and 5 is the highest involvement. As we can see in the rule below that with increasing family involvement performance gets better for students with having lower study time.

```
SSC result = 4.250
| Family Involvement = 1: (HSC result) 2.750
| Family Involvement = 3
| | Weekly Study Time = 1 hour: (HSC result) 3.750
| | Weekly Study Time = 2 hours: (HSC result) 4.000
| Family Involvement = 4: (HSC result) 4.250
| Family Involvement = 5
| | Weekly Study Time = 1 hour: (HSC result) 4.000
| | Weekly Study Time = 2 hours: (HSC result) 4.500
| | Weekly Study Time = 3 hours: (HSC result) 4.750
```

Students with the previous failure get better results with the increasing family involvement in their study as shown in the tree below.

```
Previous failures = 3 (Number of previous failures)
| Family Involvement = 2: (HSC result) 2.750
| Family Involvement = 3: (HSC result) 3.250
| Family Involvement = 4: (HSC result) 3.500
```

Another observation is "while parents get separated then having a romantic relation among students hampers the academic attainments". Conversely, another observation exhibits romantic relationship can cause enhancement in academic performance while parents are together.

```
Parent Status = Living Apart
| Romantic Relation = no: (HSC result) 2.750
| Romantic Relation = yes: (HSC result) 1
Parent Status = Living Together
| Romantic Relation = no: (HSC result) 2.250
| Romantic Relation = yes: (HSC result) 3.250
```

8   Conclusion

Highest accuracy of the system is 78.5% while SMOTEBoost is used along with PCA and the second highest accuracy is the 77.78% by ANN along with the PCA, SMOTEBoost, binning and normalization. Pre-processing has significant impact on classifiers in most of the time. Predictive model and the key finding through the visualization of the data set provide students and their parents an important instrument to get better academic performance in HSC.

References

1. Ayers, E., Nugent, R., Dean, N.: A comparison of student skill knowledge estimates. In International Conference on Educational Data Mining, Cordoba, Spain, pp. 1–10 (2009)
2. Jishan, S.T., Rashu, R.I., Mahmood, A., Billah, F., Rahman, R.M.: Application of optimum binning technique in data mining approaches to predict students' final grade in a course. Adv. Intell. Syst. Comput. 159–170 (2015). doi:10.1007/978-3-319-13153-5_16
3. Yadav, S.K., Bharadwaj, B., Pal, S.: Data mining applications: a comparative study for predicting student's performance (2012). arXiv preprint arXiv:1202.4815
4. Woods, K., Doss, C., Bowyer, K., Solka, J., Priebe, C., Kegelmeyer, P.: Comparative evaluation of pattern recognition techniques for detection of microcalcifications in mammography. Int. J. Pattern Recognit. Artif. Intell. 7(6), 1417–1436 (1993)
5. Chawla, N.V., Bowyer, K.W., Hall, L.O., Kegelmeyer, W.P.: SMOTE: synthetic minority over-sampling technique. J. Artif. Intell. Res. 16, 321–357 (2002)
6. Jimenez, L.O., Landgrebe, D.A.: Supervised classification in high-dimensional space: geometrical, statistical, and asymptotical properties of multivariate data. IEEE Trans. Syst. Man Cybern. 28(1), 39–45 (1997)
7. Jolliffe I.T.: Principal component analysis. Series: Springer Series in Statistics, XXIX, 487 p. 28 illus, 2nd edn. Springer, NY (2002). ISBN: 978-0-387-95442-4
8. Kayah, F.: Discretizing Continuous Features for Naive Bayes and C4. 5 Classifiers. University of Maryland Publications, College Park, MD, USA (2008)
9. Jayalakshmi, T., Santhakumaran, A.: Statistical normalization and back propagation for classification. IJCTE 3(1), 89–93 (2011). doi:10.7763/ijcte.2011.v3.288
10. Mullin, M., Sukthankar, R.: Complete cross-validation for nearest neighbor classifiers. Accessed from http://www.cs.cmu.edu/~rahuls/pub/icml2000-rahuls.pdf (2002)
11. Min, J., Lee, Y.: Bankruptcy prediction using support vector machine with optimal choice of kernel function parameters. Expert Syst. Appl. 28(4), 603–614 (2005). doi:10.1016/j.eswa.2004.12.008
12. Kar, A.: Stock prediction using artificial neural networks. Department of Computer Science and Engineering, IIT, Kanpur (n.d.)

A Super-Vector Deep Learning Coprocessor with High Performance-Power Ratio

Jingfei Jiang, Zhiqiang Liu, Jinwei Xu and Rongdong Hu

Abstract The maturity of deep learning theory and the development of computers have made deep learning algorithms powerful tools for mining the underlying features of big data. There is an increasing demand of high-accuracy and real-time object detection for intelligent communication and control tasks in embedded systems. More energy efficient deep learning accelerators are required because of the limited battery and resources in embedded systems. We propose a Super-Vector Coprocessor architecture called SVP-DL. SVP-DL can process various matrix operations used in deep learning algorithms by calculating multidimensional vectors using specified vector and scalar instructions, which enabling flexible combinations of matrix operations and data organization. We verified SVP-DL on a self-developed field-programmable gate array (FPGA) platform. The typical deep belief network and the sparse coding network are programmed on the coprocessor. Experiments results showed that SVP-DL architecture on FPGA can achieve 1.7 to 2.1 times the performance under a low frequency compared with that on a PC platform. SVP-DL on FPGA can also achieve about 9 times the performance-power efficiency of a PC.

Keywords Deep learning · Coprocessor · Vector · FPGA

1 Introduction

Deep learning technology has been one of the most inspiring and popular areas in the machine learning community, challenging complex tasks like recognition, mining and synthesis. The majority of Deep Learning Algorithms (DLA) belong to compute-intensive types of algorithms gaining good performance for large-scale data mining and analysis, which has inspired enormous investments from famous companies, such as Google, Facebook, Microsoft, IBM and Baidu; however, it remains

J. Jiang (✉) · Z. Liu · J. Xu · R. Hu
Faculty of Science and Technology on Parallel and Distributed Processing Laboratory,
National University of Defense Technology, Changsha 410072, China
e-mail: jingfeijiang@nudt.edu.cn

© Springer International Publishing AG 2017
D. Król et al. (eds.), *Advanced Topics in Intelligent Information
and Database Systems*, Studies in Computational Intelligence 710,
DOI 10.1007/978-3-319-56660-3_8

limited to high-powered clusters so far. There is an increasing demand of high-accuracy and real-time object detection in embedded systems, such as UAV, robot, auto-piloted cars, etc., for supporting intelligent communication and control tasks. Well-designed architectures are required to obtain better performance-power ratio that would be used in those new fields like real-time systems and mobile platforms. Therefore, customized architectures that accelerate deep neural network models have been built up in recent years. For example, Kim et al. [1] realized significant speedup for a 256×1024 RBM using Altera Stratix III FPGA. A high-throughput accelerator for machine learning [2] achieves extremely high performance-energy ratio, 452 Giga-Operations/s at 485 mW, because of its extraordinary design and advanced silicon process technology. Zhang et al. [3] proposed a FPGA-based accelerator for deep convolutional neural networks, achieving a significant performance of 61.62 GFLOPS under working frequency of 100 MHz in a VC707 FPGA board.

The main operations in DLA can be grouped into two categories. The first category is the convolution that expresses convolutional neural networks, which use small kernels to convolute matrices by certain strides. Their operation mode and memory access mode are specific, which are valuable to accelerate. Numerous studies have obtained good results on these networks [2, 3]. The second category is the large matrix operation that expresses full connection neural networks like Deep Belief Network (DBN), Sparse Coding (SC), autoencoder, et al., involving large matrix operations including matrix multiplication/addition, dot production, matrix summary/average, and some special complex functions. The complexity of this type of neural networks does not only involve large matrix operations but also the complicated connections among various matrix operations. Operation scale and type may vary depending on the network scale and the super-parameters. Thus, the accelerator platform should provide good flexibility to satisfy the variations of operation connections, operation type and implementation precision. This kind of works has remained limited thus far to specific architectures with minimal variability. A super vector coprocessor architecture for DLA (SVP-DL) is proposed in this paper that can process various matrix operations that involved in DLA by calculating multidimensional vectors.

2 Execution Framework of DLA

A deep learning model is a multi-layer structure with a building block to model each layer. For typical applications of object classification, a classifier layer is frequently added at the top of the structure, forming the whole model framework. Figure 1a shows the model framework of DLA. The model uses large scale inputs to update its parameters, thus learning some "features" of the input data. There are many typical models to construct the building block, such as Restricted Boltzmann Machine (RBM), SC, autoencoder and pooling methods. Labels are used to supervise the learning in some stages to fine-tune the whole model. The framework can use various building blocks based on different theories, but its execution flows and operation characteristics are similar.

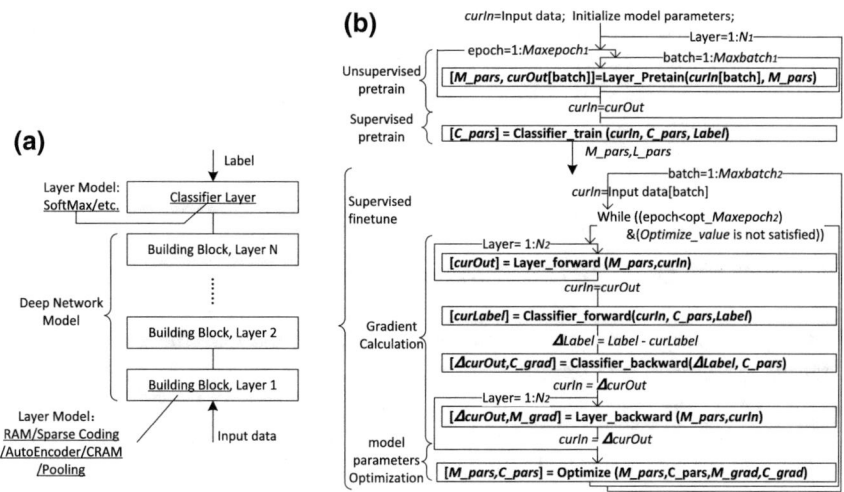

Fig. 1 **a** Framework of DLA. **b** Execution flow of DLA

Figure 1b summarizes the execution flow of typical DLA. The bold expressions are main learning processes, which divided into two stages: pretraining and fine-tuning. During pretraning, each layer learns the current input data (*curIn*) followed by optimization methods that update the layer model parameters (*M_pars*) in batches for *Maxepoch*$_1$ times, which generates the output data (*curOut*) for the next layer at the last epoch. The classifier layer uses its input data to generate actual label, which is compared with the actual label (*Label*) to update its model parameters (Classifier Parameters, *C_pars*). During finetuning, the model initially forward propagates the input data from layer 1 to the classifier, using current model parameters and classifier parameters, which generates current label (*curLabel*). Then the model back propagates the residual (*ΔLabel* and *ΔcurOut*) from the classier layer to layer 1, which calculates the gradient of the model functions (*C_grad* and *M_grad*). Using the gradients, the model then calculates the direction and step of the descent gradients to finetune the model parameters.

Model pretraining intends to minimize some cost function. For example, RBM [4] defines an energy function, which uses the contrastive divergence method to minimize, shown in Eq. 1. Where, w is the connection weight between layer input v and output h; a and b are biases of v and h respectively.

$$h^0 = logistic(v * w + a); v^{rec} = logistic(h^0 * w' + b)$$
$$h^{rec} = logistic(v^{rec} * w + a); \Delta w = -g = vh^0 - v^{rec}h^{rec} \qquad (1)$$

Other models, such as SC and autoencoder [5], acquire the gradients by calculating the partial derivative of its cost function J, as shown in Eq. 2. Where, f is the feature matrix,w is the connection weight and v is the layer input. Then, some optimization methods are used to calculate the direction and step of the descending gradients. The second step of Eq. 2 is similar to the model parameters optimization step of finetuning.

$$g = \frac{\partial J(f, w)}{\partial w} = (\frac{1}{m})(-2f'v + 2f'fw) + \lambda w. / \sqrt{w^2 + \varepsilon}$$
$$\Delta w = Optimize(g) = step(g) * direction(g) \tag{2}$$

The pretraining procedure and the gradients calculation are the main compute-intensive parts of the execution flow, which may run $Maxepoch_1$ times or times depends on the *optimize_value*. The optimization process, which only includes optimal search and model parameters updating, has more diversified operations but is not as time-consuming as others. The hardware of DLA must accelerate those compute-intensive process and can also execute all operations involved in those algorithms at the same time. The example of Eqs. 1 and 2 shows that the process includes matrix multiplications, matrix transpose multiplications, complex functions and vector multiplications.

3   Architecture of SVP-DL

SVP-DL architecture is illustrated in Fig. 2. It exhibits a heterogeneous architecture that constitutes a scalar processor and a vector processor. In the scalar processor, a block RAM (BRAM) stores the instructions, a register file that includes 32 general registers and some algorithm units. In the instruction decoding stage, instructions are recognized as scalars or vectors. Scalar instructions continue to execute in the scalar processor, whereas vector instructions are sent to the vector processor. Scalar data or addresses used in the vector processor can be stored in scalar registers.

In the vector processor, p computation units form an array that can process a vector in parallel. Each computation unit consists of a fixed-point arithmetic module, a logic module and several complex function modules. An accumulator tree is used to process elements accumulation. A vector register file (VRF) is composed of p separated $BRAMs$. The computation array can read two vectors of p elements each from the *VRF* and write a vector to the *VRF* in one cycle. The load/store unit processes the scalar and vector access between *VRF* and the memory interface. A multiplexer/demultiplexer tree is used to select available elements between computation units and *VRF*. The controller of vector instructions can control pipelined processing of multiple vectors in the computation array and the load/store unit pipeline. Thus, SVP-DL can process the matrix (called "super-vector") operations directly by hardware. The bit-width of the computation units/*VRF*, the parallelism p and the depth of *VRF* are parameterized in order to trade-off the precision and cost.

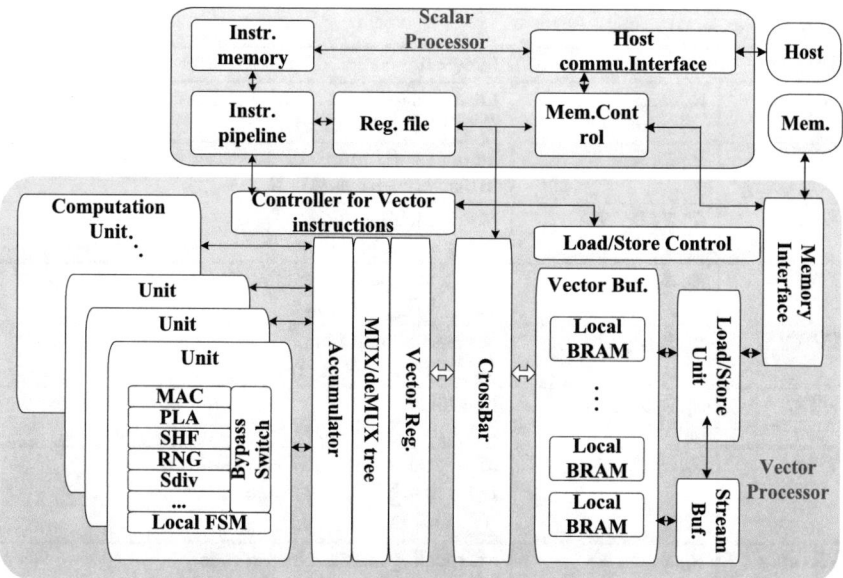

Fig. 2 Architecture of SVP-DL

4 Instructions Set Architecture

4.1 Instruction Format and Function

The instruction format of SVP-DL are encoded as an 80-bits binary with 5 fields: *opcode*, R_i, R_j, *imm*/R_k, *Size*. The operands are destination register, two source registers, immediate data, number of vectors respectively. In vector instructions, the scalar registers store the addresses of *VRF* and the memory. Table 1 describe the main vector instructions. Matrix operator A is addressed by value in R_j, B is addressed by the value in R_k, and C is addressed by value in R_i. If B or C is a scalar, it is stored in the scalar register R_i.

4.2 Implementation of the Main SuperVector Instructions

VLOAD/VSTORE. SVP-DL stores a matrix in the *VRF* as *Size* $*$ p array, as Fig. 3 shows. A large-scale matrix stored in the memory must be segmented into the data block of *Size* $*$ p to match the data organization in *VRF*. If matrix A with s rows and p columns is stored sequentially in the memory by rows. Thus, the *VLOAD*/*VSTORE* instruction can transfer the matrix between *VRF* and the memory in burst, which maximizes the use of the memory bandwidth. Good data segmentation and memory organization are very important to improve the performance of *VLOAD*/*VSTORE*.

Table 1 Vector instructions

Assembly		Function
VLOAD	$R_i, R_j, size$	Load a matrix from memory addressed in R_j to VRF addressed in R_i
VSTROE	$R_i, R_j, size$	Store a matrix from VRF addressed in R_j to memory addressed in R_i
VMAC	$R_i, R_j, R_k, size$	Matrix multiplication, $A_{size*p} * B_{p*p} = C_{size*p}$
VVMC	$R_i, R_j, R_k, size$	Matrix transposition multiplication, $A_{size*p} * B'_{p*p} = C_{size*p}$
VVVC	$R_i, R_j, R_k, size$	Vectors multiplication, $\sum_{i=1}^{size}(A(i :,)' * B(i, :)) = C_{p*p}$
VACC	$R_i, R_j, R_k, size$	Row addition, $B(1 : p) + \sum_{i=1}^{size}(A(i, :) = C_{1*p}$
VSUM	$R_i, R_j, R_k, size$	Matrix summary, $B(1 : p) + \sum_{i=1}^{size}(A(i, :) = D_{1*p}$, $\sum_{j=1}^{p} D = C_{1*1}$
VPooling	$R_i, R_j, R_k, size$	Maxpooling matrix A by row order, window size is a parameter
VADD	$R_i, R_j, R_k, size$	Matrix addition, $A_{size*p} + B_{size*p} = C_{size*p}$
VSUB	$R_i, R_j, R_k, size$	Matrix subtraction, $A_{size*p} - B_{size*p} = C_{size*p}$
VMUL	$R_i, R_j, R_k, size$	Matrix dot multiplication, $A_{size*p}. * B_{1*1} = C_{size*p}$
VDIV	$R_i, R_j, R_k, size$	Matrix dot division, $A_{size*p}./B_{1*1} = C_{size*p}$
VSMUL	$R_i, R_j, imm, size$	Matrix dot multiplication, $A_{size*p}. * imm = C_{size*p}$
VSADD	$R_i, R_j, imm, size$	Matrix dot addition, $A_{size*p}. + imm = C_{size*p}$
VSIG	$R_i, R_j, size$	Sigmoid functions, $\frac{1}{1+e^{-A_{size*p}}} = C_{size*p}$
VEXP	$R_i, R_j, size$	$e^{A_{size*p}} = C_{size*p}$
VLOG	$R_i, R_j, size$	$log(A_{size*p}) = C_{size*p}$
VSTAS	$R_i, R_j, size$	Random function, $(A_{size*p} > rand(sizeofA)) = C_{size*p}$

VMAC. Figure 4 shows the instruction implementation and an example. All matrixes are stored in *VRF* in row order. The *VRF* address in R_j changes every p cycles. A row of A_{size*p} is read out every p cycles. The multiplexer tree selects one elements of the row every cycle then broadcasts it to all the computation units. Address in R_k changes every cycle. A row of B_{p*p} is read out and distributed to the computation units. The array calculates p partial products in the multiply-accumulate pipeline. Address in R_i changes every p cycles, during which a row of matrix C is written to *VRF*.

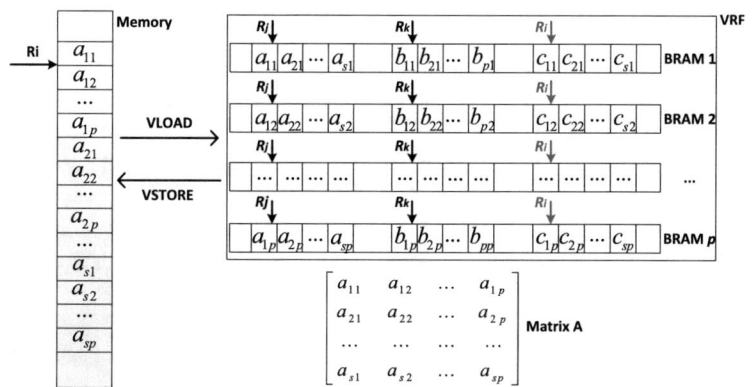

Fig. 3 Data organization for VLOAD/VSTORE

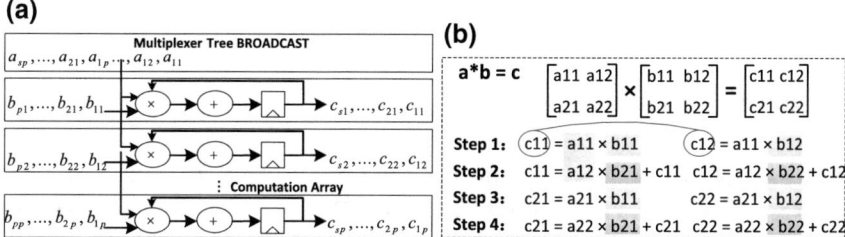

Fig. 4 **a** VMAC implementation. **b** An example of VMAC

VVMC. Equations 1 and 2 show that there are many operation pairs that initially use a matrix, and then use the transpose of that matrix. It is inefficient that a matrix which stored by rows is accessed by column in the next transpose operations. *VVMC* can still process matrix transposition multiplication using the row-order stored matrix. Figure 5 shows the implementation and an example. The matrixes are stored in the same manner as *VMAC*. The control is slightly different. The address in R_j changes every p cycles, which distributes the elements in a row to the computation units. The address in R_k changes every cycle and the address of R_i changes every p cycles. The multiply-accumulation pipeline for *VVMC* in computation units can produce an element of matrix C every cycle. A demultiplexer tree selects the right *VRF* ports to write back.

VVVC. Calculating two vectors also has numerous requirement, which are similar to the last expression in Eq. 1 and many gradients optimization processes. The control method and the implementation of *VVVC* is similar with *VMAC*, whereas the sequence for selecting the elements is different.

Others. VSIG, VEXP, VLOG, and VSTAS are four instructions that implement the primary complex functions. The linear approximation methods are adapted to implement the functions [6]. The remaining vector instructions are simple element-by-

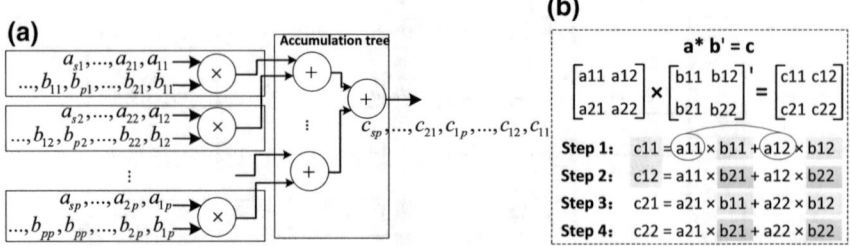

Fig. 5 **a** VVMC implementation. **b** An example of VVMC

element matrix operations, which use the same data path and computation units as the instructions described above. Only the instruction control and some multiplexer logic are different.

5 Algorithms Mapping

DBN for MNIST classification is the first typical example to show the application mapping method on SVP-DL, which has three RBMs and a classifier layer, with the model size of 1024-576-576-256-10. The dataset contains 32×32 pixel images of the digits. The entire model is pretrained and then finetuned. The conjugate gradient method is used to optimize the cost functions.

Pretraining sequence of Eq. 1 and layer backward sequence of Fig. 1 were mainly programmed using vector instructions. For each RBM layer, v, h, w and other intermediate matrices are all segmented into the blocks with size $p * batchsize$ and stored by row. The data in the blocks are stored by row and the blocks are stored by column. Thus, the procedures $v * w = h$ and $h * w = v$ can both be programmed using several vector instructions without changing storage order. We choose a small parallelism p of 64 to illustrate the miniaturization of SVP-DL. The *VRF* depth is approximately 6K. The bit-width of the main datapath and *BRAM* for DBN is 16 bits to keep precision according to software evaluation.

Table 2 illustrates an assembly program fragment calculates $v * w = h$. The $'@'$ strings represent addresses. Other strings are character constant. Table 3 shows DBN mapping result of three types of code. In program 1, we use minimal *VRF* without significantly affecting the performance. In program 2, we make full usage of *VRF* to improve performance. All pipeline delays and architecture parameters are known and the running cycles of the program can be calculated.

For SC of MNIST classification, the procedures of gradient and cost calculation using cross-optimization [7] method are mapped on SVP-DL. One layer SC was mapped with the model size of 784-256-10. The bit-width of the main datapath and *BRAM* for SC is 24 bits to keep the precision. Table 4 shows the result of the SC assembly program. Program 1 calculates the gradient and the corresponding cost

Table 2 Assembly program fragment of DBN

	Code		Indication
	VLOAD	R4, R5, $p*$ rownumber	; load a column blocks of w
	mov	R3, @v_vrf	; start address of v
	mov	R4, @w_vrf	; start address of w
	mov	R5, @h_vrf	; start address of h
	mov	R7, @h2_vrf	; use to store a block of temp h
	mov	R2, 0	; v loop counter
	VMAC	R7, R3, R4, batchsize	; first block
	beq	R2, esp	
loop:	VMAC	R5, R3, R4, batchsize	
	VADD	R7, R5, R7, batchsize	
esp:	iadd	R3, R3, batchsize	; increase v address by batchsize
	iadd	R4, R4, p	; increase w address by p
	iadd	R2, R2, 1	
	icmp	R6, R2, rownumber	
	bneq	R6, loop	
	mov	R8, @hbias_vrf	
	add	R8, R8, R1	
	VSADD	R7, R7, R8, batchsize	; add the bias
	VSIG	R7, R7, batchsize	; sigmiod h at R7 and write back
	mov	R3, @h1sum_vrf	; address of sum h
	add	R3, R3, R1	
	VACC	R3, R7, 1, batchsize	; calculate h average storing to R3

Table 3 Assembly programs result of DBN

Schedule	Program 1: pretraining using small VRF	Program 2: pretraining using large VRF	Program 3: finetuning
Instruction number	305	245	1320
VRF usage(depth)	3.3 K	5.6 K	6 K
Cycles	4.678×10^9	4.53×10^9	1.4×10^{12}

when f is fixed. Program 2 calculates the gradient and the corresponding cost when w is fixed. The running cycles of the program are counted based on the pipeline delays and architecture parameters, as shown in Table 4.

Table 4   Assembly programs result of SC

Schedule	Program 1: w gradient and cost	Program 2: f gradient and cost
Instruction number	260	361
VRF usage(depth)	296	268
Cycles	2017899	2445416

6   FPGA Verification of SVP-DL

SVP-DL is written in Verilog. A FPGA platform is designed to verify the prototype, which includes a Xilinx Virtex VII XC7VX485T FPGA with 4 GB DDR3 memory, a Xilinx Zynq FPGA and several serial links. Data and control message are transferred between the host and XC7VX485T through Zynq system. The Core Generator of Xilinx generated the DDR3 controller. A first in, first out (FIFO) is added between SVP-DL logic and the DDR3 controller to separate the asynchronous clock fields. We first synthesized SVP-DL without the DDR3 controller. When the bit-width is configured as 16 bits, the logic can achieve a frequency of approximately 248 MHz. Table 5 shows resource usage at the left side. The total chip resource is listed in the middle. We then constrained the DBN frequency to 200 MHz according to the generated DDR3 controller. Table 5 shows the synthesized results of resource usage at the right side. The FIFO for the clock fields and DDR3 controller occupy a significant capacity of *BRAM*. The other resources do not increase considerably. When the bit-width of SVP-DL is configured as 24 bits for SC, the logic usage (slices LUTs) increases about 7%. Whereas, the DSP number doubles because a 24-bits multiplier uses two DSP cores instead of one that a 16-bits multiplier uses.

Using the number of cycles in Table 3 and the frequency of 200 Mhz, DBN performance is calculated in Table 6. We ran the same DBN programmed in Matlab on a PC with AMD A8 5500 CPU 3.2 GHz and 4 GB memory. The performance is recorded in Table 6. It indicates that the small FPGA platform can achieve a little better performance with a low frequency compared with the PC platform. SVP-DL for SC has a wider bit-width that constricts its frequency to 100 Mhz. Using the number of cycles in Table 4, the performance is calculated in Table 7. FPGA platform for SC can achieve about twice the performance with a lower frequency compared with the PC platform.

Table 5   Resource usage of SVP-DL

	Without controller			With controller	
	Resource	Ratio%	Total	Resource	Ratio%
Slice register	56166	9%	607200	67635	11%
Slice LUTs	115407	37%	303600	120486	39%
DSP	132	4%	2800	132	4%
RAM	338	37%	1030	881	85%

Table 6 Performance of SVP-DL for DBN

	Program 1	Program 2	Program 3	DBN
Perf.(s)@200MHz	23.4	22.7	6998	7022
Perf.(s) on PC	51.4	51.4	12306	12358
SpeedUp	2.2	2.26	1.75	1.76

Table 7 Performance of SVP-DL for SC

	Program 1	Program 2
Perf.(ms)@100MHz	24.44	20.18
Perf.(ms) on PC	52.98	38.08
SpeedUp	2.17	1.89

Table 8 Performance/power ratio of SVP-DL

	FPGA platform	PC
DBN performance (s)	7022	12358
SC performance (ms)	20.18	38.08
Power (W)	19	95
DBN efficiency ratio	8.79	1
SC efficiency ratio	9.45	1

The power of the two platforms is measured. The efficiency is expressed as $\frac{1}{power*time}$. We set the PC efficiency as 1, whereas the FPGA platform exhibits about 9 times efficiency, as shown in Table 8. It indicates that SVP-DL on FPGA can use less power to achieve higher performance, which is favorable for deep learning applications requiring platform miniaturization.

7 Conclusion

SVP-DL is flexible because of its programmability and efficient because of its configurable parallelism to process matrix operations directly. The mapping results and performance of the most typical DLA demonstrate that SVP-DL and the platform hardware matched well. Our work presented a high-performance, low-power acceleration solution for those deep learning applications that require small and low-power platform. It can be used as a reconfigured IP core that is implemented on various FPGA platforms.

References

1. Kim, S., McMahon, P., Olukotun, K.: A large-scale architecture for restricted boltzmann machines. In: 18th IEEE Annual International Symposium on Field-Programmable Custom Computing Machines (FCCM), 2010, pp. 201–208. IEEE (2010)
2. Chen, T., Du, Z., Sun, N., Wang, J., Wu, C., Chen, Y., Temam, O.: Diannao: a small-footprint high-throughput accelerator for ubiquitous machine-learning. In: Proceedings of the 19th International Conference on Architectural Support for Programming Languages and Operating Systems, pp. 269–284. ACM (2014)
3. Zhang, C., Li, P., Sun, G., Guan, Y., Xiao, B., Cong, J.: Optimizing fpga-based accelerator design for deep convolutional neural networks: an analytical approach based on roofline model. In: Proceedings of FPGA (2015)
4. Hinton, G., Osindero, S., Teh, Y.: A fast learning algorithm for deep belief nets. Neural Comput. **18**(7), 1527–1554 (2006)
5. Lee, H., Ekanadham, C., Ng, A.: Sparse deep belief net model for visual area v2. Adv. Neural Inf. Process. Syst. **20**, 873–880 (2008)
6. Jiang, J., Hu, R., Mikel, L., Dou, Y.: Accuracy evaluation of deep belief networks with fixed-point arithmetic. COMPUTER MODELLING and NEW TECHNOLOGIES **6**, 7–14 (2014)
7. Schölkopf, B., Platt, J., Hofmann, T.: Efficient Sparse Coding Algorithms. MIT Press (2007)

Part II
Big Data and Collaborative Decision Support Systems

Bi-temporal Database Model for Legal Merger Transactions and Late-Arriving Information Problem: The Case of Polish Merger Market

Aleksander Buczek and Jacek Mercik

Abstract Most economical and financial software applications should take into consideration the temporal aspect of these processes. Bi-temporal databases guarantees that the results may be always reproduced exactly as they were at some previous point in time. The concept of bi-temporal data is reviewed when designing model of the merger transaction database. Special attention is paid to the problem of late-arriving information. The Polish merger market is presented as an example.

Keywords Bi-temporal database · Legal merger process · Polish merger market · Log-Cauchy distribution · Late-arriving information

1 Introduction

Nature of different economical processes is rather time-varying. Thus, most economical and financial software applications should take into consideration the temporal aspect of these processes. As every such application has a database in the back-end, this need heads us towards the concept of bi-temporal data. Databases which implement this concept are capable of representing reality in more accurate way than conventional databases as the latter ones capture only a single snapshot of the modelled reality [9].

A. Buczek (✉)
Wroclaw University of Science and Technology, Wroclaw, Poland
e-mail: aleksander.buczek@pwr.edu.pl

J. Mercik
WSB University in Wroclaw, Wroclaw, Poland
e-mail: jacek.mercik@wsb.wroclaw.pl

© Springer International Publishing AG 2017
D. Król et al. (eds.), *Advanced Topics in Intelligent Information and Database Systems*, Studies in Computational Intelligence 710,
DOI 10.1007/978-3-319-56660-3_9

The concept of bi-temporal data is built around two orthogonal time dimensions [3, 8, 9]:

- **Valid time** which is the time when the given fact is current in the modelled reality;
- **Transaction time** which is the time when the given fact is current in the database. It is the duration between information insertion to (logical) deletion. This type of time increases monotonically forward and is bounded between the time when the database was created and the current time. Moreover, transaction time cannot be changed and there is no way to amend errors which occurred in the past.

The implementation of both time dimensions allows showing from which point in time the data is correct in the database and exactly corresponds to the reality. This guarantees that the results may be always reproduced exactly as they were at some previous point in time and will take into consideration the problem of late-arriving information [7].

Other time dimensions can be still incorporated into database structure. Doucet [7] defines them as a **user-defined times** which are invariant as the time is passing. Moreover, Abelló and Carme [3] understand them as an uninterpretable attribute domain of time directly managed by the database user and not by the database system.

The presence of both time dimensions is beneficial feature of many software applications from:

- **Corporate sector**. They handle many crucial business processes like accounting, human resources, marketing or inventory management [8, 9]. Especially, highly regulated financial industry requires such database design to maintain their transaction books and credit/deposit portfolios in order to perform advanced reporting tasks;
- **Public administration sector**. These applications gather full auditable public data about citizens, companies or registered cars to mention few in a given country [10];
- **Healthcare sector**. These applications are capable of maintaining full patients' disease and therapy history [9].

The advantages of bi-temporal data are achieved at some costs. According to Kumar et al. [9] a significant increase in size as transaction time proceeds is a main disadvantage of bi-temporal database which also affects negatively its update and query times. Additionally, Jensen and Snodgrass [8] claim that query optimization is more complex than in the case of conventional databases. Moreover, Rozmus [10] states that it takes more time to build an application based on bi-temporal data concept as the development team needs to model more complex relationships.

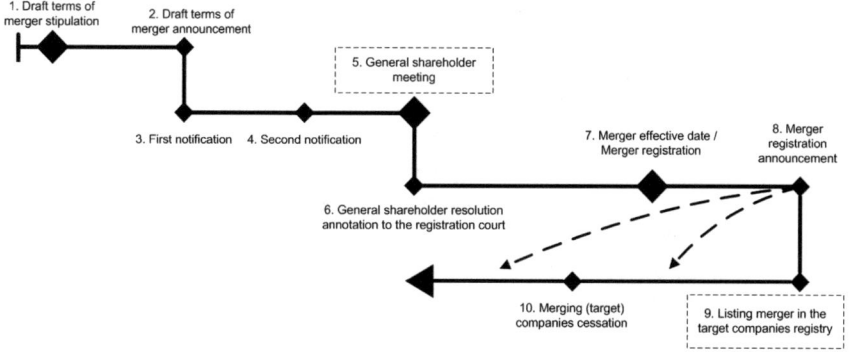

Fig. 1 Legal merger process. *Source* Buczek and Mercik [6]

2 Legal Merger Process

Legal merger consolidation can be executed as a merger by takeover or merger by formation of a new company [1]. This should be understood as transfer of all assets of a company or partnership to another company in exchange for the shares that the buying company issues to the shareholders or partners of the target company or partnership. In turn, the latter one assumes the formation of a company to which the assets of all merging companies or partnerships devolve in exchange for shares of the new company.

Both procedures differ in the way the assets are exchanged [6] for the shares but they still have many common elements which are presented in the order of appearance in Fig. 1.

3 Bi-temporal Database Model

Legal merger process is an economic phenomenon which is stretched over time. According to Buczek and Mercik [6], legal merger process in Poland lasts for around 4.5 months (median is equal to 140 days) on average and three-fourths of the transactions close between 4 and 6.5 months. During this period, some characteristics (for example initial capital level) of the counterparts involved in the process are often changing causing that the correct analysis is only possible when historical data is accessible in the database. Acquiring company which is taking over other entities in a given point in time, after some time may be again involved in another transaction and act on acquiring side or may change its role and become a company being acquired. What is more, transaction initiation and closure is publicly known only at the time when obligatory announcements presented in the Fig. 1 are posted. This cause that the time lags appear between:

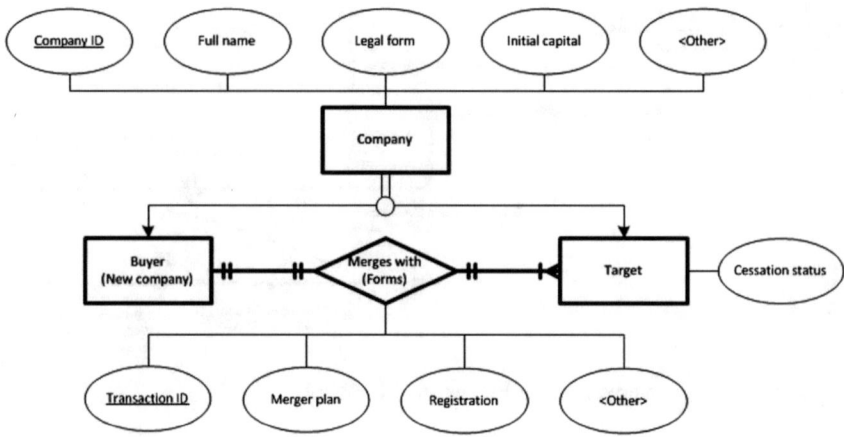

Fig. 2 Non-temporal entity-relationship diagram. *Source* Authors' own study. *Note* Crown's foot notation was used

- the stipulation of a merger plan and its announcement;
- the registration entry and its announcement.

This corresponds to the well-known late-arriving information problem. This fact together with the time-varying nature of the legal merger process imply that the concept of bi-temporal data is an appropriate design model of the merger trans-action database.

Jensen and Snodgrass [8] as well as Rozmus [10] suggest that non-temporal conceptual design of the database at a single point in time helps to understand better the analysed process. The presence of temporal aspects at this stage of the devel-opment introduces unnecessary complexity which is not needed when defining high-level logic of the process. Following their recommendation, legal merger process has been illustrated with the use of Entity-Relationship (ER) diagram (Fig. 2).

Entity-Relationship diagram takes into consideration two ways in which the companies may merge. In case of a merger by takeover always one company (buyer) buys one or more other companies (targets). From the other hand, when merger by formation of a new company occurs, then at least two or more companies form only one new company. It is worth noting, that buyer/new company and target entities are sub-entity of the master entity called company. Both sub-types of entities share nearly the same attributes. Target entity has one additional attribute called cessation status. It corresponds to the fact that in Poland the target companies should be deleted from National Court Register (KRS)[1] after the merger transaction.

Having conceptual design in place, it is very easy to create a conventional relational database model. Unfortunately, this kind of model will not take into

---

[1]National Court Register (*pol. Krajowy Rejestr Sądowy—KRS*).

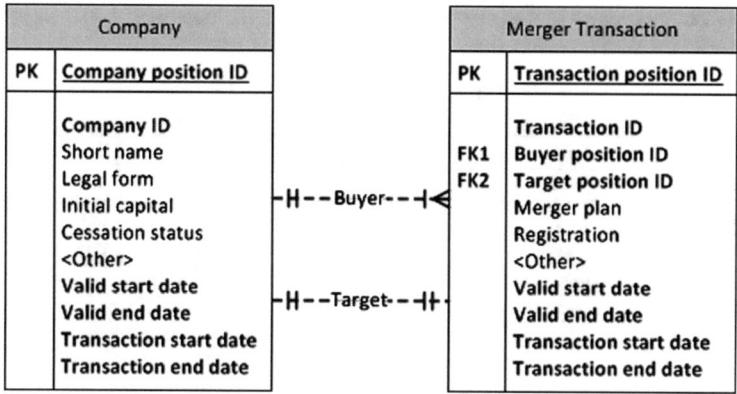

Fig. 3 Bi-temporal relational model. *Source* Authors' own study

consideration temporal nature of merger transactions. However many authors like Abelló and Carme [3], Jensen and Snodgrass [8] or Rozmus [10] state that it is possible to add required time dimensions to the relational model. When designing merger transaction database, this activity will lead to the bi-temporal model presented in the Fig. 3.

After extending database model with valid and transaction time identifiers, the primary keys associated with the company entity and merging relationship have lost their uniqueness. That was the reason why there were replaced with so called position primary keys [10] which are unique for every temporal state of the chosen company or transaction instance.

At this stage the design of bi-temporal database model is complete. The next step of the database development is connected with establishing the logic of the data insertion. For Poland, the Journal of the Ministry of Justice [2] is the best available source of information about legal mergers in Poland. It allowed collecting the information about **3870 merger transactions** which have taken place in the period between **1st January 2002 and 31st December 2013**.

In order to establish the bi-temporal data structure of the collected information, the following assumptions have been made:

- draft terms of merger stipulation formally initiates consolidation procedure in the real world;
- registration formally close the procedure in the real world;
- a single legal merger transaction is inserted in the database on the day when it is publicly announced;
- it may be updated in the database before the registration is announced;
- it must be updated on the day when the registration is announced.

Table 1 presents the dates mapping between bi-temporal database model and the legal merger process which is the result of the mentioned assumptions.

Table 1 Time dimensions mapping between bi-temporal database model and legal merger process

Date type	Merger plan		Registration	
	Stipulation	Announcement	Entry	Announcement
Valid start	Yes	No	Yes	No
Valid end	No	No	No[a]	No
Transaction start	No	Yes	No	Yes
Transaction end	No	No	No	Yes

Source Authors' own study
Note [a]Valid end date is equal to entry date shifted back by one day

3.1 Aggregated Transaction Statistics and Late-Arriving Information

Aggregated merger and acquisition market statistics[2] are vastly used in commercial research done by investment banks and consultancy companies to assess the current macroeconomic situation of a given country or region.

Completed volume statistic based on the registration entry is often biased with the problem of late-arriving information because transaction closure is always publicly announced with some delay. When the bi-temporal database model is introduced to record the merger transaction lifecycle, then it is possible to track the variability of the aggregated merger market statistics. As an example, the Fig. 4 presents completed volume aggregate of transactions closed in October 2008.

It is clearly visible that this aggregate for a given period (valid time dimension) depends on the day in which it is calculated (transaction time dimension). For Polish Merger Market October 2008 completed volume was equal to only 12 transactions on 1st November 2008. It nearly reached its true level of 53 transactions more than 20 days later, on 22nd November 2008. The last transaction closure was recorded on 14th October 2010 and then the true level of the aggregate was finally known.

3.2 Modelling Time Delay for Late-Arriving Information

The time delay magnitude of the merger registration entry appearance in the bi-temporal database can be modelled and analysed with the use of statistical techniques. Basically, this time delay can be defined as the difference between transaction start date and valid end date when the latter is not null.

---

[2]There are four main aggregated statistics of merger and acquisition market: announced, completed, withdrawn and backlog volumes. Their definitions can be found in Buczek [5].

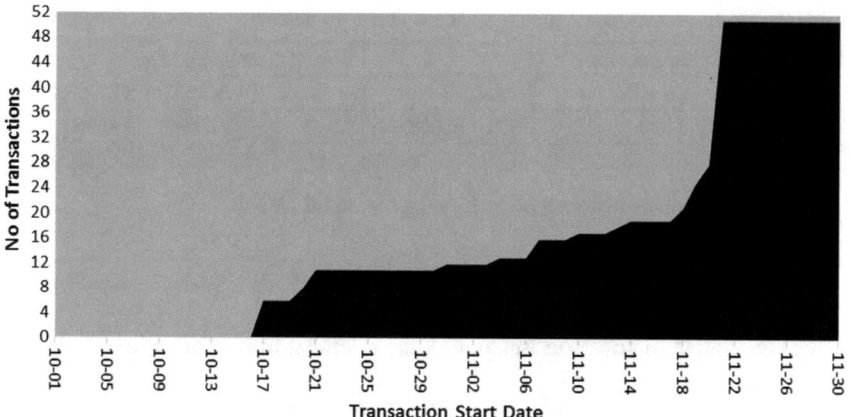

Fig. 4 Completed volume in October 2008. *Source* Authors' own study

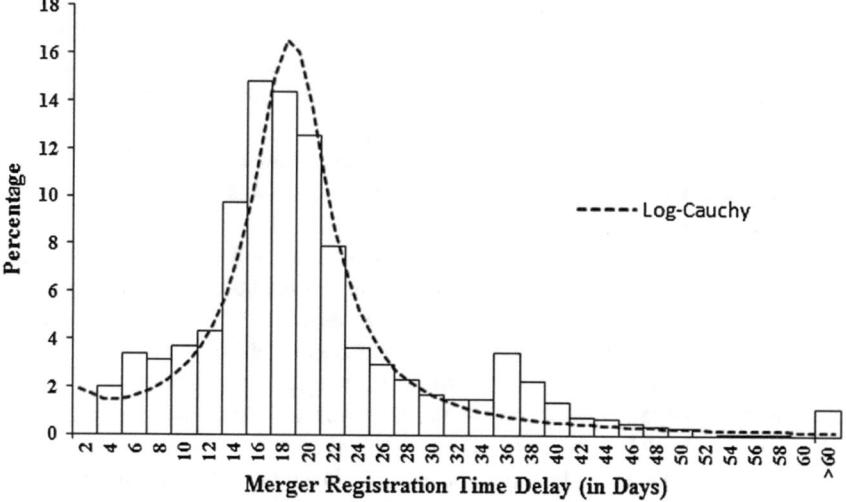

Fig. 5 Histogram of merger registration time delay with fitted density curve. *Source* Authors' own study

Histogram from the Fig. 5 was constructed on the sample containing 3571 observations. It reveals that the time delay of merger registration has a bell-shaped distribution with longer right tail indicating the positive skewness. Moreover, its distribution is leptokurtic as it has an acute peak near the mean and a fat right tail.

All these characteristics give a good indication that Log-Cauchy distribution could be a proper choice to model the delay of merger registration data arrival into bi-temporal database.

Table 2 Results of distribution fitting. *Source* Authors' own study

Distribution	Parameters		Kolmogorov-Smirnov test		
	Location (μ)	Scale (θ)	Statistics	P-value	Simulations
Log-Cauchy	2.873 (0.005)	0.219 (0.005)	3.309	0.036	10000

Log-Cauchy probability density function is equal to [4]:

$$f(x) = \sigma/x\pi/\left((\ln x - \mu)^2 + \sigma^2\right) \text{for } x > 0. \tag{1}$$

Its cumulative distribution function is then defined in the following way:

$$F(x) = 1/2 + \arctan((\ln x - \mu)/\sigma)\text{for } x > 0. \tag{2}$$

We estimate the parameters of the probability density function from the Eq. (1) with the use of Maximum Likelihood Method. Newton-Raphson Optimization with Line Search will be used to determine the parameter subset values for which the likelihood function reaches its maximum if it exists. As all the calculations are performed with the use of SAS software and its native procedure SEVERITY, more information about the parameter estimation and optimization methods can be found in the documentation [12].

Table 2 contains the results of the distributions parameters estimation. Standard errors are reported in the brackets. All parameters are significant—their standard errors are relatively small compared to the estimates.

Quality of fit is assessed with the use of Kolmogorov-Smirnov (KS) test [12]. Since the parameters of the considered Log-Cauchy distribution are estimated from the data, p-values need to be obtained from Monte Carlo simulations. We decided to use the algorithm proposed by Ross [11]. The number of trials was set to 10,000. At 1% significance level there is no evidence to reject the null hypothesis stating that the data sample is drawn from the reference Log-Cauchy distribution.

Decent quality fit is also visible when empirical cumulative distribution function is compared with the theoretical one (please refer to the Fig. 6). Slight discrepancies in the right tail appear beyond 90th percentile.

Distribution analysis of the merger registration time delay helps to determine the cut-off date for which aggregated transaction statistics presented in the previous section is very close (and sometimes equal) to its true level. According to Log-Cauchy distribution there is 95% probability that the delay will not be bigger than 70 days. This is a conservative assessment when compared to the respective delay of 39 days calculated on the empirical sample (please refer to the Table 3). In conclusion, the official reporting date of given aggregated merger transaction statistic for a given period should happen not earlier than after approximately 2 months after the end date of this period.

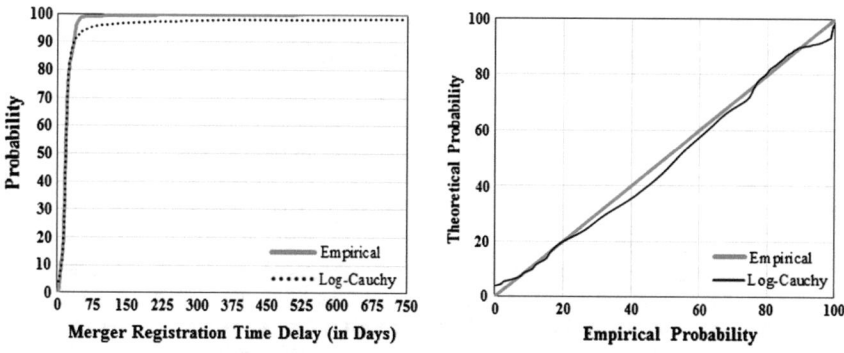

Fig. 6 Cumulative distribution function and probability-probability graphs. *Source* Authors' own study

Table 3 Quantiles and related measures. *Source* Authors' own study

Distribution	Min	5th Percentile	1st Quartile	Median	3rd Quartile	95th Percentile	Max
Log-Cauchy	–	4.445	14.219	17.697	22.025	70.448	–
Empirical	3	6	14	18	22	39	748

4 Conclusions

Bi-temporal databases guarantee that the results of the most economical and financial applications may be always reproduced exactly as they were at some previous point in time by taking into consideration the temporal aspect of these processes. The concept of bi-temporal data is capable of representing reality in much more accurate way than conventional databases especially when a case concerns modelling of the merger transactions. Well-known late-arriving information problem (because transaction closure is always publicly announced with some delay) together with the time-varying nature of the legal merger process imply that the concept of bi-temporal data is an appropriate design model of the merger transaction database.

The advantages of bi-temporal data are achieved at some costs. A significant increase in size as transaction time proceeds is a main disadvantage of bi-temporal database which also affects negatively its update and query times. The primary keys associated with the company entity and merging relationship have lost their uniqueness and they should be replaced with so called position primary keys which are unique for every temporal state of the chosen company or transaction instance.

When the bi-temporal database model is introduced to record the merger transaction lifecycle, then it is possible to track the variability of the aggregated merger market statistics. It is clearly visible that this aggregate for a given period (valid time dimension) depends on the day in which it is calculated (transaction time dimension). The results of the conducted statistical analysis suggests that official reporting date of aggregated merger transaction statistics for a given period should happen not earlier than approximately 2 months after the end date of this calculation period.

References

1. Act of 15 September 2000: The Code of Commercial Partnerships and Companies (in Polish)
2. Act of 22 December 1995 on issuing the Journal of the Ministry of Justice (in Polish)
3. Abelló, A., Carme, M.: A Bitemporal Storage Structure for a Corporate Data Warehouse. In: Kambayashi, Y., Mohania, M., Wöß, W. (eds.) DaWak 2003. LNCS, vol. 2737, pp. 109–118. Springer, Heidelberg (2003)
4. Bondesson, L.: On the Lévy measure of the lognormal and logcauchy distributions. Methodol. Comput. Appl. Probab. **4**, 243–256 (2002)
5. Buczek, A.: Aggregate size measures of merger market: empirical evidence from Poland, 2002–2013. Wroclaw School Bank. Res. J. **16**(3), 73–90 (2016)
6. Buczek, A., Mercik, J.: On conformance of legal merger duration with Burr type III and XII distribution. Wroclaw School Bank. Res. J. **15**(5), 597–608 (2015)
7. Doucet, R.: A strategy for storing bitemporal data in the dimensional model. Bus. Intell. J. **17**(2), 41–55 (2012)
8. Jensen, C.S., Snodgrass, R.T.: Temporal data management. IEEE Trans. Knowl. Data Eng. **11**(1), 36–44 (1999)
9. Kumar, A., Tsotras, V.J., Faloutsos, C.: Designing access methods for bitemporal data-bases. IEEE Trans. Knowl. Data Eng. **10**(1), 1–20 (1998)
10. Rozmus, S.: Database design with full history of data changes—the bitemporal database model and write operations. Biuletyn WAT **65**(1), 89–109 (2016) (in Polish)
11. Ross, S.M.: Simulation. Elsevier, Academic Press, San Diego (2006)
12. SAS Institute Inc.: SAS/ETS 9.3 User's Guide. SAS Institute Inc., Cary (2011)

Breathing Movement Analysis for Adjustment of Radiotherapy Planning

M. Penhaker, M. Novakova, J. Knybel, J. Kubicek, J. Grepl, V. Kasik and T. Zapletal

Abstract Aim of this work is the developing of algorithm, which extract and process the breathing movement data from CyberKnife log files for prospective conventional radiotherapy planning. The described algorithm enable verification and accuracy of the settings radiated tumor rims, which are used in conventional radiotherapy. This results should contribute to improving of the treatment of irradiated tumors with possible subsequent statistical evaluation.

Keywords Radiotherapy · Breath movement · CyberKnife · Unfavorable movements · Dose accuracy · Therapy

1 Introduction

Currently much research medicine focused on the most efficient treatment of tumors. In connection with it was put into operation CyberKnife radiotherapy machine, which now rank among the best that the irradiation of tumor tissue was

M. Penhaker (✉) · M. Novakova · J. Knybel · J. Kubicek · J. Grepl · V. Kasik
VSB – Technical University of Ostrava, FEECS, K450, Ostrava, Czech Republic
e-mail: marek.penhaker@vsb.cz

M. Novakova
e-mail: marketa.novakova@vsb.cz

J. Knybel
e-mail: Jakub.knybel@fno.cz

J. Kubicek
e-mail: jan.kubicek@vsb.cz

V. Kasik
e-mail: vladimir.kasik@vsb.cz

J. Knybel · T. Zapletal
Faculty Hospital, Ostrava, Czech Republic
e-mail: tomas.zaletal@fno.cz

© Springer International Publishing AG 2017
D. Król et al. (eds.), *Advanced Topics in Intelligent Information
and Database Systems*, Studies in Computational Intelligence 710,
DOI 10.1007/978-3-319-56660-3_10

developed. Among the most accurate technique to a tenth of a centimeter. However, even though errors occur and unnecessary contact with healthy tissue. And therefore constantly looking for new techniques that would be able to prevent such damage. CyberKnife instrument can also contribute in many ways when compared with the treatment that has been used as conventional radiotherapy and may show us whether the set parameters are sufficient, even when moving bearings caused breathing and other adverse effects. These movements are physiological components of the organism, which cannot be suppressed or eliminated [1, 2].

For by monitoring the patient's respiration waveform which arises due Cyber-Knife device during irradiation of a patient, detect deviations from tidal breathing patient to avoid unnecessary exposure of healthy tissue and this therapy was thus much more precise. CyberKnife device can compensate for an unwanted movements at millimeter level. In conventional radiotherapy so but not to the tumor and is attributed to the large safety rim. It was compared with data resulting from this program should be able to compare whether they are currently set up hems with conventional radiotherapy correct, or the percentage of cases that do not respond due to physiological movements of the patient's body [3, 4].

1.1   Reference Data

Before starting treatment, it is necessary to obtain reference data. Before shooting this data is acquired several radiographs. These images are scanned at an angle of 45° left and right of the patient's X-ray tubes, which are part of the device CyberKnife and are located near the ceiling of the facility. It is necessary to obtain a correlation model, which is determined by X-ray images in different phases of the respiratory movements, the system then has available information, wherein the bearing is in which phase of respiration. With this correlation model, the patient can be monitored during therapy because of armor, who wears three LEDs. They are captured by a special camera. After X-rays are taken at certain intervals to check whether the bearing position corresponds to the original model. Thus is obtained the breathing curve of the patient [5–7].

2   Problem Definition

Patient motion during radiation are stored in the. Log. This file is being prepared in the course of treatment, when the patient is wearing a special vest with LEDs. The movements of these diodes are scanned by special camera and motion information is written as an indication of patient movement in all 3 directions of movement in time.

Log always starts with a blank line, followed by information about the time when treatment was initiated. This time is given in seconds since midnight. On the third line there is information about the patients themselves. The data are in four columns. The first column is about time, therefore the length of irradiation. The remaining three columns of information about the movement of deposits during treatment. Is in the 3D system, it is therefore the X, Y, Z, the corresponding motions in direction X = Superior-Inferior (from head to feet), Y = leftright or laterolateral (left and right) and Z = anteriorposterir (from back to belly). Coordinates in one log may vary abruptly, because of treatment interruption and repositioning of the patient. The frequency of the resulting record is 25 Hz, i.e. 25 frames per second. The recording time is determined by the length of the actual treatment and the number of records for each patient is determined by the number fraction.

It is necessary to keep patient information. The log regards social security number and name. The main result should be the maximum and minimum values of deflections movement irradiated and monitored bearings. It is necessary to filter out the curve overshoot and adjust for desired outcomes. These data statistically to express their incidence and prevalence. Determine when treatment was discontinued. This can be detected thanks to longer delays in the first column of the log. It is necessary to keep the information about the number of interruptions during treatment to determine total treatment time, including delays in discontinuation without it. Visually is then displayed motion bearings in space [8–10].

When using the conventional method is a safety rim is set in the order of several centimeters, as opposed to device CyberKnife this rim can be minimized to a few millimeters. The rim is important to cover the vicinity of the tumor, which can be moved by moving the patient. In conventional radiotherapy is the rim more, because it is not possible to monitor and correct the patient's movement and must therefore cover the entire irradiated tumor size at all stages of the respiratory cycle. Use of the proposed program will help process the data from the respiratory curves, which are obtained during treatment. It should be developed, to improve current treatment of irradiated tumors.

The algorithm proposed for processing respiratory waveform will be used for control. Combining monitoring of respiratory curve in patients treated at the unit CyberKnife and conventional radiotherapy can be proved whether the security hem with conventional therapy.

Data processed and visualized in the program can and will be used for subsequent irradiation postanalysis quality and design of optimal methods of radiation and radiological treatment approaches.

3 Breath Movement Signal Analysis

The main objective of this work is to develop algorithm that would be able to modify data and display the requested information, including information on the treatment, the patient information. The program should be able to proceed the data from the respiratory curve overshoot using a suitable filter, display amplitude of respiration curve and calculate statistics on discontinuation of therapy, duration of each segment, and the number of deep breaths, which is necessary to set an appropriate function. Very important fact is also the sum of all departments, which indicates the duration of the treatment. Other requirements include displaying only the amplitude of the curve, i.e. their extremes, which correspond to the maximum inspiratory and expiratory time low. The final requirement for the output of the software is the ability to export the modified values in the file that would be possible to continue to process mathematical software.

In the case of the creation of this algorithm, you can use it for planning radiotherapy alone for patients and a treatment could then become again a little more accurate. Serve for processing already ex post treatment and statistical processing of patient motion and physiological movements in his body during treatment, which affect the actual course of irradiation. Will be used to control when the conventional radiation therapy that has been proven whether the currently set values of safety rim sufficient. All data will be possible to continue to process and gain better insights [11–18].

3.1 Source Data

Log file type is produced CyberKnife device. It contains information about a patient's course of treatment. Data on patient movement is written in three-dimensional system for the duration of radiation in real time under the name of the file. In the event log file type, it is necessary to make adjustments to allow data from it to correctly handle and apply. Log file type can be seen in the following figure (Fig. 1).

The first line in the file is empty. The second line indicate the date and time of patient treatment. The third line includes patient data. Patient's surname and birth number is hidden intentionally in order to preserve the anonymity of the patient's personal data. On the fifth row start to read data during treatment.

When you view the x-axis without modification can see a large number of confusing data, straight sections, which means interruptions. It is therefore easily detectable eye of times the treatment was discontinued. These sections are not alone when viewing axis desirable. Curve is necessary to exterminate them, but to

```
48708.659 DATE_ANCHOR Wed 03 Oct 2012 13:31:48
48708.659 INFO --> LoggerHandler::init: initialized id 6 with file ▮▮▮▮_▮▮▮▮▮▮_03Oct12_133148/Modeler.
48708.659 INFO --> Modeler log started Wed Oct 03 13:31:48 Central Europe Daylight Time 2012
        30149.248:-4.227 12.332 0.958
        30149.287:-4.213 12.339 0.925
        30149.325:-4.140 12.378 0.747
        30149.363:-4.057 12.422 0.547
        30149.401:-4.029 12.437 0.480
        30149.440:-4.043 12.430 0.514
        30149.478:-4.043 12.430 0.514
        30149.517:-3.997 12.454 0.403
        30149.556:-3.946 12.481 0.280
        30149.594:-3.923 12.494 0.225
```

Fig. 1 The original log file type of device CyberKnife

preserve the value of extremes. It may be noted periods corresponding sleep stages of treatment, when the patient is almost at rest and respiratory amplitude are almost identical, these findings can be seen mainly in the second half of the treatment.

3.2 Conversion of Angle

Data are recorded during treatment relative to the robot, which performs irradiation. For this reason, data spin about −45° relative to the patient. It is therefore necessary before rendering and other processing data converted by +45° (Fig. 2).

$$x_{(45°)} = (x* \cos 45°) - (y* \sin 45) = (x*0.7071) - (y*0.7071)$$
$$y_{(45°)} = (y* \cos 45°) - (y* \sin 45) = (x*0.7071) - (y*0.7071)$$

$$(1)$$

Each line of the log header including the time stamp. Time is measured in seconds since midnight, so it is necessary to convert these confusing numbers so they can be displayed as a time. The time is needed to view such information in the data table program, which displays the time of the header file that is stored in the data structure and is not recalculated, since there is no time to be recalculated.

During the treatment can occur to interruptions process, which are recorded as a crack in time sequence. These interruptions are most often due to readjustment of the patient has occurred, for example, motion to move the robot and the patient needs to get back into position to deposit illuminating correctly.

To determine the number of interruptions is running string, which explores times in the first column of the array and compared one after another. When it detects the difference in time greater than 0.1 s are entered information that the interruption has occurred. Indication of interval data to which it belongs, is listed in the blank

Fig. 2 Description axes to the robot and to the patient

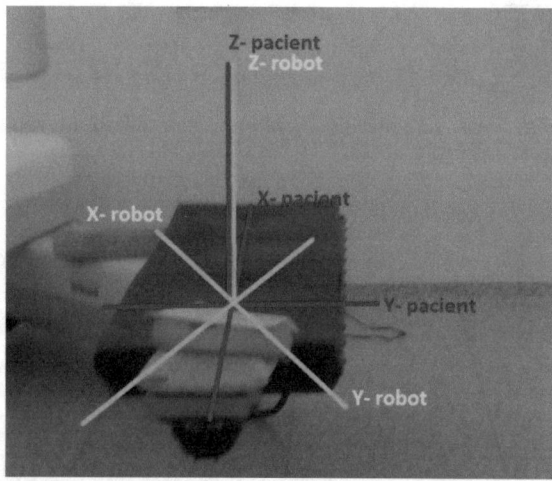

columns per row, it is subsequently traceable in the process the data is located, and also the number of times and the treatment was interrupted for whatever reasons. A line number at which interruption occurred while the recorded lap time, i.e. the length of time before the interruption.

3.3 Filtering Data

Data is written to the log of the frequency of 25 Hz. Companies such data in such a small time deviation cause considerable noise. For better display of curves it is essential to filter out data from all that is required. However, it is necessary to hold all the highs and lows of their value. The following figure shows the loaded breathing curve of the original data before filtering. They are clear overlaps and noise that would be for the following functions for finding maxima were considerable difficulties, and therefore it is necessary to eliminate these shortcomings (Fig. 3).

Processing of these signals was elected FIR filter. The values specified for the value of m is created ideal low-pass filter, which in the lower frequency signal outlet and upper frequencies suppresses signal. Filter Design (FIR 2 (30 f, m)) with the length of the pulse characteristics 30, was selected parameter as a compromise. The higher the value, the better the filter filters, but it is even more difficult to calculate. The filter is configured by the constants f and m. Then, the run command "filtfilt", the filtering without signal delay. This prevents unnecessary calculations and the process will accelerate. After starting the filter, they retained value extremes and timeline. To remove all unnecessary fluctuations in the signal (Fig. 4).

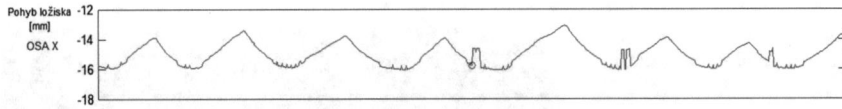

Fig. 3 Raw data from the log file containing information about the movement of breathing of the patient during treatment

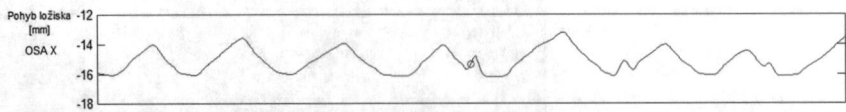

Fig. 4 Using FIR2 filter settings f = [0, 0.15, 0.15, 1] on the log file containing information about the movement of breathing of the patient during treatment

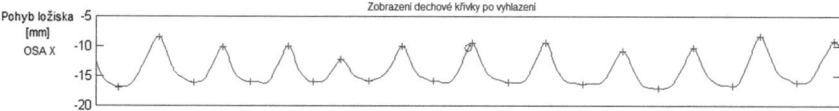

Fig. 5 Finding and marking extremes on breathing curve

3.4 Extremes

Extremes in a dataset representing the function of inhalation and exhalation, thus peaks curve. Their read correctly is important for some other functions, such as the number of deep breaths, deep breaths number, complete the maximum or minimum in the sector.

Extremes are sought through mathematical functions with the second derivative. This function compares point by point on each axis and searches extremes, if the scanned value is not extreme, is entered into the line as zero, otherwise, if the extremum thus found, its value is entered. In this case, therefore extrema written to one line and therefore cannot recognize if it is a maximum or minimum respectively extremes are not divided at the maximum and minimum. The program extremes shown as red crosses axes (Fig. 5).

Due to the large amount of data frequently a situation that is as extreme deemed value that only slightly advanced, but not the apex inhale or exhale, so the program functions (MaxMin), which compares consecutive searched extremes and seeks all that differ from each other by a value of 30. It is thus ensured that successive extrema better match the actual breathes. It is also high probability that of the peak will be followed by a minimum and maximum after him. The coefficient 30 is set so that the error is minimized depending on the files to be tested.

The following figure sought extremes without MaxMin used functions, i.e. without setting the minimum difference between two adjacent extremes. It is obvious that this function does not work correctly as it will help us find consecutive inhale and exhale.

In the data extremes are found in a small percentage of data inaccuracies that arise signal filtering and search entry highs and lows. The configuration is such that the errors are minimized. Errors were counted during final testing.

The resulting error is less than 5% and for the use of this program is the magnitude of this error is acceptable.

4 Representation of Algorithm Outputs

To visualize the proposed algorithm was used environment in MATLAB GUI editor. Beneath frames and tables are displayed itself breathing curves for each axis separately. The x-axis is additionally appear dimmed and ghosted interruption, at

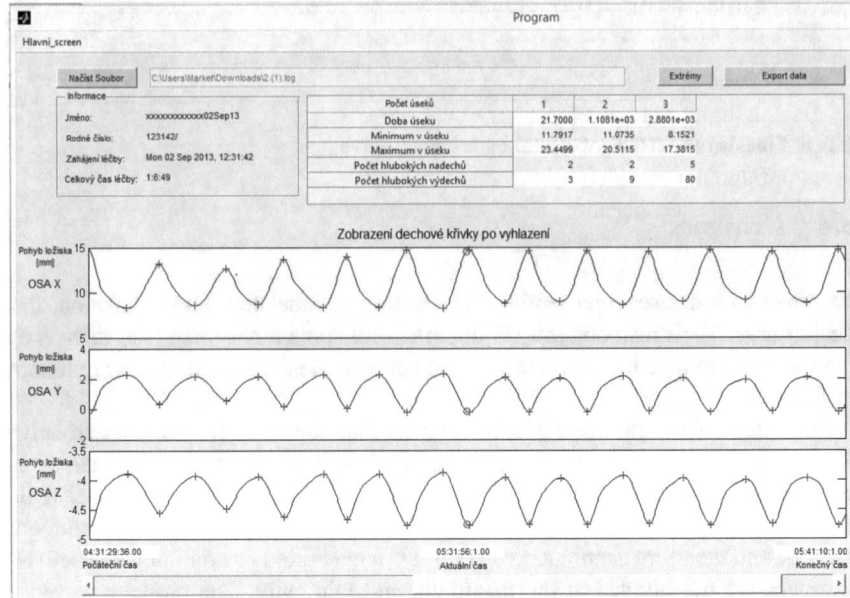

Fig. 6 User interface for visualization of algorithms for processing breathing curves from data log files CyberKnife system

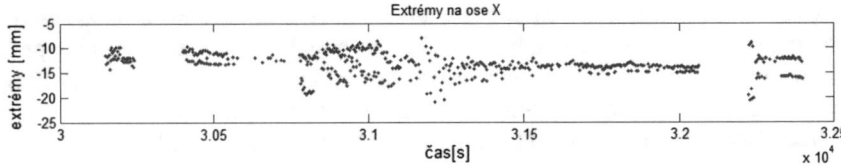

Fig. 7 Trend visualization of extreme breaths on the processed data

the point where you stopped the treatment and readjustment of the position of the patient. Red dots on the axes are extremes. Ranges on the axes of each graph are automatically adjusted by the program needs to be arranged graphs (Fig. 6).

Extremes are displayed in all three axes and the main program for each curve as red crosses. Only extremes for each axis without the original curve, the dot plot. Thus viewed extremes are much clearer than if they were shown the original curve (Fig. 7).

5   Testing of the Algorithm

An important part of the testing was to determine how big bug occurs when displaying highs and lows limits. Due to the large amount of data when the number of extremes in the order of several hundred to several thousand, it was not possible to go through the entire curve, that was calculated, how many errors occurs in each file. Were selected because 300 values in each file. The first 100 values at the beginning of treatment, the second 100 values from the center of treatment and the last 100 values from the end of treatment. Number of errors on all 300 values were averaged and converted to an error in percentage terms. This error in neither case did not exceed 5%. All values are only approaching this value. Given the large number of data this error is acceptable to use the program.

Another important point was to test whether the data is loaded and displayed. This has been proven by in the left frame to monitor the patient and displays graphs breathing curves in all three axes. Because the display feature followed immediately successive additional functions, it is therefore necessary to work load at 100% files. All remaining functions in the file is loaded as those that calculate and display the number of sections, the number of deep breaths, time segments, total treatment time, and other extremes.

During testing on all available files file was not found, for which the program had a single mistake. Not found error that would prevent proper operation of the program. Set functions thus meet the requirements of the assignment. The following table demonstrates the functionality at the ten test files. I bearing have been known in the file name. Necessary information has been reading the data, the total time of patient treatment, the number of found extremes, number of sections, a number of treatment interruption, whether the program is able to export data, whether it worked button display the extremes of the controlled values, errors in these controlled areas and conversion to error in percentage. On average, the bug was found in the extremes of 2.75%. It is a value calculated from the following table.

The table shows in the first column of the registered name bearing. This figure is nowhere mentioned in the log, I was known from the filename of the program for testing and demonstrating the functionality of the program for different areas of bearing position. The second column refers to retrieve the file (Table 1).

If the file is loaded the table is yes or no, if not to load the file. Total time is displayed in the program after the file is loaded, its values are given in the table as the next column. Its distinct values show that there were different types of files. The number of sections is again what is shown in the program. Demonstrates the functionality of sub-functions program, which seeks discontinuation. Presented value once again demonstrates the functionality of the program. The following column has an export value of yes/no to the functioning/malfunctioning button to export the data and create a .xlsx file. The fact that the file is created correctly, demonstrates additional column that shows the number of extremes. This value is only deductible in the program from an open file .xlsx, a complement to the table proves correct functionality. Column to view the extremes again has a value of

Table 1 Testing results processing algorithm breathing curve of log files

Bearings	Loading	Total time of therapy	Number of segments	Export	Number of extremes	Visualization of extremes	Verified extremes	Errors in extremes	Error in %
Hepar	Yes	1:53:18	3	Yes	2515	Yes	300	6	2
Hepar	Yes	1:39:45	4	Yes	3316	Yes	300	5	1.6
Hepar	Yes	0:58:37	2	Yes	881	Yes	300	3	1
Lungo	Yes	0:31:45	4	Yes	534	Yes	300	12	4
Lungo	Yes	0:59:39	5	Yes	1516	Yes	300	13	4.3
Lungo	Yes	0:55:40	6	Yes	1330	Yes	300	10	3.3
Lungo	Yes	1:02:27	1	Yes	1545	Yes	300	15	5
Lungo	Yes	0:34:31	1	Yes	903	Yes	300	3	1
Pankreas	Yes	1:13:05	7	Yes	1330	Yes	300	9	3
Pankreas	Yes	1:08:25	2	Yes	1843	Yes	300	7	2.3

yes/no buttons for functionality extremes. If you have been pane extremes in the column value is yes, otherwise not. The last three columns refer to testing extremes and their errors in the program. The first of these three columns is the number of extremes that have been evaluated. The second column is the number of errors found in these extremes. Whether it was the extreme, the program skipped or wrongly diagnosed. Number of false extremes falls on previous data tables. The last column is then calculation error in %.

6 Conclusion

In this work, we were created by an algorithm that has been designed for the analysis of breathing curve motion bearings with correction therapeutic dose. When testing the algorithm, it was found that the error of irradiation ranging up to 5%, which is due to the large number of motion data can be accepted. Of analyzed the respiratory movements can also determine how many times the treatment was interrupted, and how long it took each section as many times as there was a deep inhale or exhale, and how long it lasted the entire treatment cycle without interruption.

Acknowledgements The work and the contributions were supported by the project SV4506631/2101 'Biomedical Engineering Systems XII'. This work was supported by The Ministry of Education, Youth and Sports from the National Programme of Sustainability (NPU II) project "IT4Innovations excellence in science - LQ1602". This article has been supported by financial support of TA ČR, PRE SEED Fund of VSB-Technical university of Ostrava/TG01010137.

References

1. Yamada, S.M., Ishii, Y., Yamada, S., Kuribayashi, S., Kumita, S., Matsuno, A.: Advanced therapeutic strategy for radiation-induced osteosarcoma in the skull base: a case report and review. Radiat. Oncol. **7** (2012)
2. Xie, Y., Xing, L.: Intrafractional motion of the pancreas during cyber knife radiation therapy. Int. J. Radiat. Oncol. Biol. Phys. **81**, S789–S789 (2011)
3. Kim, J.B., Hwang, Y., Bang, W.C., Kim, J.D.K., Kim, C.: IEEE: real-time moving organ tracking in ultrasound video based on a 3D organ model (2012)
4. Sherwood, J.T., Brock, M.V.: Lung cancer: new surgical approaches. Respirology **12**, 326–332 (2007)
5. Bogart, J.A.: Stereotactic body radiotherapy for poor-risk lung cancer: "More cyber, less knife?". Cancer J. **13**, 75–77 (2007)
6. Karaman, K., Dokdok, A.M., Karadeniz, O., Ceylan, C., Engin, K.: Intravascular placement of metallic coils as lung tumor markers for cyberknife stereotactic radiation therapy. Korean J. Radiol. **16**, 626–631 (2015)

7. Rashid, A., Mone, P., Sarfaraz, M.: Predicting late lung complications following lung tumor radiosurgery with cyber knife using biologically effective doses and normalized dose-surface histograms. Med. Phys. **34**, 2482 (2007)
8. Tezcanli, E.K., Goksel, E.O., Yildiz, E., Garipagaoglu, M., Senkesen, O., Kucucuk, H., Sengoz, K.M., Aslay, I.: Does radiotherapy planning without breath control compensate intra-fraction heart and its compartments' movement? Breast Cancer Res. Treat. **126**, 85–92 (2011)
9. Wang, J., Zhong, R.M., Bai, S., Lu, Y., Xu, Q.F., Zhou, X.J., Xu, F.: Evaluation of positioning accuracy of four different immobilizations using cone-beam ct in radiotherapy of non-small-cell lung cancer. Int. J. Radiat. Oncol. Biol. Phys. **77**, 1274–1281 (2010)
10. Wong, V.Y.W., Tung, S.Y., Ng, A.W.Y., Li, F.A.S., Leung, J.O.Y.: Real-time monitoring and control on deep inspiration breath-hold for lung cancer radiotherapy-combination of ABC and external marker tracking. Med. Phys. **37**, 4673–4683 (2010)
11. Bettinardi, V., Picchio, M., Di Muzio, N., Gianolli, L., Messa, C., Gilardi, M.C.: PET/CT for radiotherapy: image acquisition and data processing. Q. J. Nucl. Med. Mol. Imaging **54**, 455–475 (2010)
12. Brock, J., McNair, H.A., Panakis, N., Symonds-Tayler, R., Evans, P.M., Brada, M.: The use of the active breathing coordinator throughout radical non small-cell lung cancer (NSCLC) radiotherapy. Int. J. Radiat. Oncol. Biol. Phys. **81**, 369–375 (2011)
13. Cole, A.J., Hanna, G.G., Jain, S., O'Sullivan, J.M.: Motion management for radical radiotherapy in non-small cell lung cancer. Clin. Oncol. **26**, 67–80 (2014)
14. Hu, W.G., Ye, J.S., Wang, J.Z., Xu, Q., Zhang, Z.: Incorporating breath holding and image guidance in the adjuvant gastric cancer radiotherapy: a dosimetric study. Radiat. Oncol. **7** (2012)
15. Kovacs, A., Hadjiev, J., Lakosi, F., Antal, G., Vandulek, C., Ezer, E.S., Bogner, P., Horvath, A., Repa, I.: Dynamic MR based analysis of tumor movement in upper and mid lobe localized lung cancer. Pathol. Oncol. Res. **15**, 269–277 (2009)
16. Macrie, B.D., Donnelly, E.D., Hayes, J.P., Gopalakrishnan, M., Philip, R.T., Reczek, J., Prescott, A., Strauss, J.B.: A cost-effective technique for cardiac sparing with deep inspiration-breath hold (DIBH). Physica Medica-Eur. J. Med. Phys. **31**, 733–737 (2015)
17. Nemoto, K., Oguchi, M., Nakajima, M., Kozuka, T., Nose, T., Yamashita, T.: Cardiac-sparing radiotherapy for the left breast cancer with deep breath-holding. Jpn. J. Radiol. **27**, 259–263 (2009)
18. Roth, J., Maleika, A., Engenhart-Cabillic, R., Strassmann, G.: Radiotherapy of the breast under breathhold. Geburtshilfe Und Frauenheilkd **70**, 812–816 (2010)

Measuring Improvement in Access to Complete Data in Healthcare Collaborative Database Systems

Nurul A. Emran, Fathin N.M. Leza and Noraswaliza Abdullah

Abstract Accessing complete data is crucial especially in healthcare domain. Nevertheless, within multi data providers context, accessing complete data is a challenge because not only data must be collected and integrated, we must also seek collaborative effort among the data providers. In this paper, we present the result of implementing a framework called as Collaborative Integrated Database System (COLLIDS) in terms of the degree of improvement in accessing complete data it offers for data providers. Our experiment is based on real healthcare data sets taken from a collaborative system in Malaysia where the population-based completeness (PBC) is adopted as a measure. The results that are evaluated using Wilcoxon Sign Rank Test show that COLLIDS is of benefit for most data providers as increment of more than 50% data completeness can be observed in the results set. We conclude with the cases where COLLIDS will be worth (and not worth) to be implemented based on the characteristic of data providers that participate in the collaborative system.

Keywords Data completeness · Population-based completeness · Collaborative systems · Wilcoxon sign rank test · Healthcare

N.A. Emran (✉) · F.N.M. Leza · N. Abdullah
Computing Intelligence Technologies (CIT), Centre of Advanced Computing
Technology (C-ACT), Universiti Teknikal Malaysia Melaka (UTeM),
Hang Tuah Jaya, 76100 Melaka, Malaysia
e-mail: nurulakmar@utem.edu.my

F.N.M. Leza
e-mail: fathinnabilla19@gmail.com

N. Abdullah
e-mail: noraswaliza@utem.edu.my

© Springer International Publishing AG 2017
D. Król et al. (eds.), *Advanced Topics in Intelligent Information
and Database Systems*, Studies in Computational Intelligence 710,
DOI 10.1007/978-3-319-56660-3_11

117

1 Introduction

Access to complete data is crucial in many domains, especially in healthcare. Unnecessary, repetitive and painful medical procedures are usually experienced by many patients due to the inability for healthcare operators (clinics, hospitals and etc.) to access complete health records that reside in healthcare providers databases where patients sought their past treatments. In addition, without a systematic model, knowledge on coverage of specific data sets (i.e. medical supplies, medicines, patients) is beyond reach of healthcare providers. This information is usually crucial for healthcare providers to perform operational collaboration required to run their business smoothly. For example, pharmaceutical departments might need to acquire medicines or blood with shortage of supplies from other hospitals whose coverage on the supplies is more complete. However, accessing complete data is a challenge, not only due to the effort to prepare the infrastructure for data integration, but also to seek consent of collaboration among data providers. Therefore, data accessibility model that can support seamless access to complete data where information on data coverage can also be provided is necessary.

Data accessibility problem has received considerable attention from many related researchers. For example, [1, 2] propose ways to securely access data where multilevels security, security tags and devices to detect security breaches are proposed [1, 2]. Biometric and face recognition methods are also proposed to secure data access [3]. Hara and Madria [4] and Lu et al. work on data replication to ease access to restricted mobile host network [4], while cloud computing technology is proposed to provide seamless access to data [5]. Deethshith et al. [6] work on duplication allocation to improve speed of data access within ad hoc mobile networks [6]. Within multi data providers environment, sharing method is initially proposed by [7] to control illegal access to data where the participants must agree to a unified view of data structure that is transparent to all. Collaborative data sharing is an idea proposed by Ives et al. to extend the sharing idea where a model called as ORCHESTRA is propose to describe the architecture [8]. In fact, data sharing has been shown useful to increase the chance in getting complete data sets through data integration [9]. This idea has been improved where data federation with virtual database concept has been introduced [10].

While considerable amount of work on data accessibility work are available, practically, details on how complete data can be accessed and measured among multi data providers is limited. We therefore proposed COLLIDS to address the gap (please refer [11]). In this paper, we present the result of implementing COLLIDS where an experiment is conducted to determine the usefulness of it within the context of healthcare data sets under study. Several cases of data providers completeness are presented in order to understand the characteristics of data providers that will gain benefit from implementing COLLIDS. Statistical evaluation using Wilcoxon Sign Rank test is used to verify the significant improvement on the access of complete data sets.

This paper is organized as follows. Section 2 presents the related work on data completeness; Sect. 3 covers the design of the experiment. The experimental results and statistical evaluation are presented in Sect. 4. Finally, Sect. 5 concludes the paper.

2 Related Work

Similar work proposed by Felix et al. measures information sources completeness for mediator-based information system using coverage and density measures [12]. Their work focus on supporting the mediators in query planning based on sources (or combination of sources) completeness. We extend their work by introducing access tool layer to support seamless access to the integrated database, where data sets from multiple providers are combined through union operator to form complete data sets. Furthermore, we adopt population-based completeness (PBC) measure that supports multi-dimensional coverage of data sets (please refer [13]).

Razniewski and Nutt proposed Table Completeness (TC) and Query Completeness (QC) [14]. They developed a framework to describe completeness of databases and query answers. In their framework, completeness of database is described in two ways: Table Completeness (TC): which classified certain parts of a relation as complete, and Query Completeness (QC): which classified a set of query answers as complete. Alkharboush and Li [15] proposed a decision rule method to measure completeness and consistency of data [15]. The decision rule method is proposed incorporated with missing (error) ratio method to find defective value in databases. Through this method, the users are able to precisely identify defective and normal data and compare the quality change across different databases.

Early works on data completeness focused on the presence of nulls that were denoted as non-existed case or unknown case [16]. Several categories of data completeness can be observed, according to the items that are considered as missing18. These categories are Null-based Completeness (NBC), Tuple-based Completeness (TBC), Schema-based Completeness (SBC) and Population-based Completeness (PBC) [13]. NBC focuses on measuring the values that are missing, normally presented as nulls; TBC measures the tuples or records that are missing; SBC measures the missing schema elements like attribute and entities and PBC measures the missing individuals from datasets under measure [13].

PBC is useful to measure data coverage, where the concern is not to quantify null values, the number of records or schema elements. For example, in a genetic study, given a dataset a bioinformatician might be interested to know the coverage of genes that have been studied. In this case, neither they care about the records of genes that are missing nor the values that are missing. Instead, they care about 'gene individuals' that are missing. Because PBC measures focus on individuals, multi-dimensional completeness is supported. For example, bioinformatician inquiry regarding dataset completeness can be extended to species coverage and genus coverage. In addition, PBC is useful to be applied in multi-data providers environment.

A simple ratio formula is commonly applied to measure completeness where completeness of data set under measure can be defined as:

$$Completeness(D, R) = \frac{|(D \cap R)|}{|R|} \in [0, 1], \tag{1}$$

where D is the data set under measure and R is the reference data set [13]. Depending on the types of data completeness of interest, this simple ratio formula can be adapted. To use PBC measure, we must first determine the dimensions of interest. According to PBC measure, the population must be defined according to the dimension of interest [13]. For example, if the dimension of interest is the coverage of genes, then the population is individual genes present in a data provider. This population needs to be compared to a reference population to compute the data providers completeness. Reference population in this example is the union of genes collected from all data providers. Close world assumption is adopted where reference population is considered as complete data set.

3   Experiment Design

In the experiment, COLLIDS is implemented as a system prototype where healthcare domain is the chosen as the case study to measure completeness of data providers. In this research, we are interested in measuring the coverage of 'individuals' that are present in multi-data providers' datasets from specific dimension. Therefore, PBC measure chosen as the measure to support our objective.

In this research we argue that COLLIDS will improve access to complete data. In other words, we hypothesized that there will be an increment in the ratio of completeness of data accessible by data providers (who participate in the collaborative system). Therefore, we evaluate COLLIDS in terms of increment of access to complete data for all data providers. Even though data integration solution will usually improve data completeness, studies on whether this will always be the case is limited. In addition, we would like to validate how significant is the ratio (especially in the case where improvement on data completeness is present). Thus, a statistical method is used to evaluate COLLIDS.

In order to determine the appropriate statistical test to use, normality test must be conducted. Normality assumption is the core of most standard statistical procedures. The normality test defines a criterion and gives its sampling [17]. Furthermore, normality test is more precise because actual probabilities are calculated [18]. SPSS Statistical tool supports Kolmogorov-Smirnov (K-S) and the Shapiro-Wilk (S-W) test. Both Kolmogorov-Smirnov (K-S) and the Shapiro-Wilk (S-W) test are the prerequisite for other statistical tests, commonly used to check assumptions of normality [19]. Both of the test methods require testing the null hypothesis if data (for each group of data) is normally distributed. In addition, these tests are sensitive to the size of the sample. For larger datasets greater than 50 Kolmogorov-Smirnov (K-S) is used and Shapiro-Wilk (S-W) is used for smaller datasets less than 50. Therefore, small deviations from normality are reported as significant [19]. If the result of the test is significant (which is p <0.05) the null hypothesis is rejected. This means the assumption of normality distributions is also rejected. If data is normally distributed, parametric-test such as T-Test can be used for further analysis of data. For data that are not normally distributed, non-parametric test [18] such as Wilcoxon test is used for further analysis.

To evaluate COLLIDS, we determine sample data that are relevant before the null hypothesis is defined. By using SPSS tool, we must check the significance level before we run the normality test. The results of normality test is used to decide the type of further test (parametric test or non-parametric test) to use. The purpose of the further test is to quantify and to determine whether the increment of data completeness is significant for all data providers.

Data is collected from 106 clinics (ranged from the year 2012 to 2014) which is visited by a universitys staff and their dependents (staffs family members). The sample data collected consists of patient treatment information (i.e. staff ID, staff name, patient name, diagnosis received, diseases, and drugs). In the experiment, we chose to measure the coverage of patients (staffs) treated by the clinics. Using PBC measure, the dimension of interest is patient, which is represented by their staff IDs. Thus, we extract patients data set (which consists of 45, 634 tuples). We adapt PBC by introducing the following definitions to suit our needs. Given n clinics (as data providers), p_i is the population of patients under measure for clinic i and reference population (RP) is:

$$RP = p_1 \cup p_2 \cup p_3 \cup \cdots p_n$$

Completeness of Clinic i, (in percentage) is measured as:

$$Completeness(p_i, RP) = \frac{|(p_i \cap RP)|}{|RP|} \times 100$$

3.1 Data Completeness Cases

We analyze data completeness cases in order to determine the type of data providers under measure. Three types of relationship for the population of interest are classified: (1) superset-subset relationship (2) subset-overlap relationship (3) disjoint-overlap relationship. The formula in (1) is revisited. We use examples taken from healthcare domain to illustrate the case. Suppose that we would like to measure the coverage of diseases treated by healthcare providers, using PBC, disease is translated into the population of interest. We present the completeness cases by using two clinics under measures, Clinic 1 and Clinic 2 as data providers as shown in Fig. 1a–d.

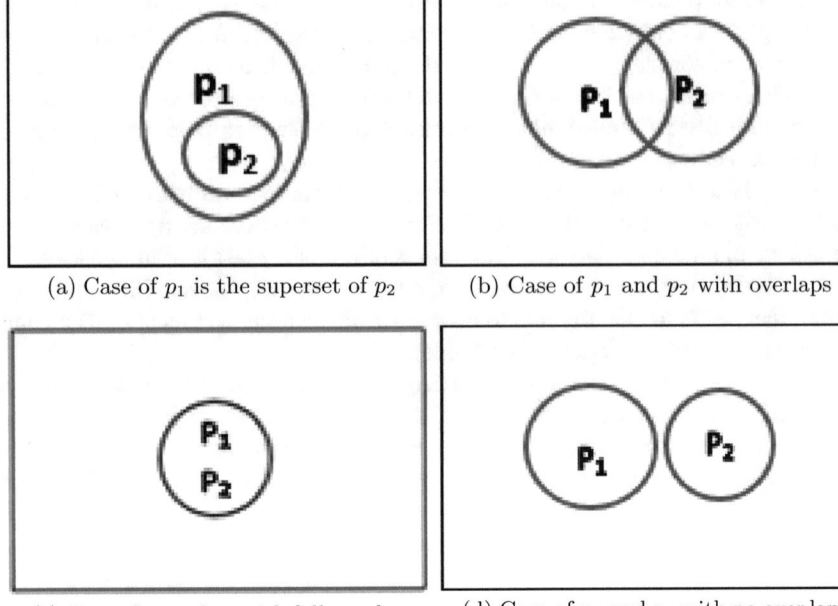

(a) Case of p_1 is the superset of p_2 (b) Case of p_1 and p_2 with overlaps

(c) Case of p_1 and p_2 with full over laps   (d) Case of p_1 and p_2 with no over laps

Fig. 1 Data completeness cases

We adapt the definition of reference population (*RP*) in this example as a set of diseases covered by both clinics. The population of interest for Clinic 1, p_1 is defined as a set of diseases recorded in Clinic 1; the population of interest for Clinic 2, p_2 is defined as a set of diseases recorded in Clinic 2. In case 1 where p_1 is the superset of the p_2, notated as $p_1 \supset p_2$ (as shown in Fig. 1a), the completeness of p_1 is greater than p_2 as p_1 has all diseases covered by p_2, where $RP = p_1$. COLLIDS is useful for Clinic 2 but not for Clinic 1. In case 2 where $p_1 \subseteq p_2$, two possibilities are considered: (1) there is an overlap between p_2 and p_1 (as shown in Fig. 1b), and (2) p_1 is equal to p_2 (as shown in Fig. 1c). For the case depicted in Fig. 1b, COLLIDS will benefit both clinics as neither p_1 nor p_2 is equal to *RP*. For the case depicted in Fig. 1c, COLLIDS does not benefit both clinics as both p_1 and p_2 are equal to *RP*. In case 3, where p_1 and p_2 are disjointed, notated as $p_1 \neq p_2$ (as shown in Fig. 1d). Similar to the case depicted in Fig. 1b, COLLIDS will be useful for both clinics as neither p_1 nor p_2 is equal to *RP*.

From the analysis of these cases, we can see that the usefulness of COLLIDS depends on the characteristics of the data sets of contributed by data providers relative to the reference population. We will use the cases identified in this section to conclude the usefulnesss of COLLIDS based on the experiment results.

4 Results and Discussion

4.1 Normality Test Results

The normality test is conducted to test the normal distributions of the data. The normality test does not test the difference between groups of data but instead, to verify whether the data is no different than a normal distribution. Hence, it will accept (or reject) the null hypotheses. The sample data are divided into two groups. The first group of data is the set of all healthcare providers PBC result; denoted as As-Is completeness. This dataset represents individual completeness of clinics before COLLIDS is implemented, where access to patient records in other participating clinics are restricted. 106 clinics are measured to get the datasets. The second group of data is the set of increment values of completeness for all 106 clinics; denoted as Completeness Increment. This data represent the amount of improvement (in percentage) in accessing complete patient records after COLLIDS is implemented. The hypothesis (H_1) and the null hypothesis (H_0), which must be formulated prior to the normality test, are as follows:

- Null Hypothesis (H_0): *The distributions of data for both groups under measure are normal.*
- Alternative Hypothesis (H_1): *The distributions of data for both groups under measure are not normal.*

High significance value in normality test (which is more than 0.05) indicates that the data are normally distributed. In this normality test, we focus on the significant result in Kolmogorov-Smirnov as the sample dataset observed (N) has more than 50 samples. The normality histogram of normality test conducted on the first group of data is shown in Fig. 2, where the significance values (Sig.) (which is p-value) in

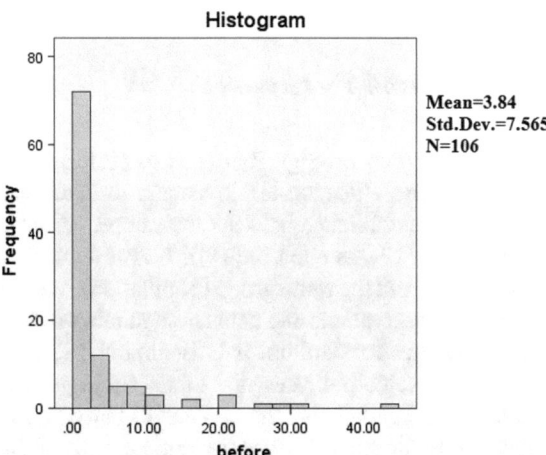

Fig. 2 Normality test result for As-Is completeness

Mean=3.84
Std.Dev.=7.565
N=106

Fig. 3 Normality test result
for completeness increment

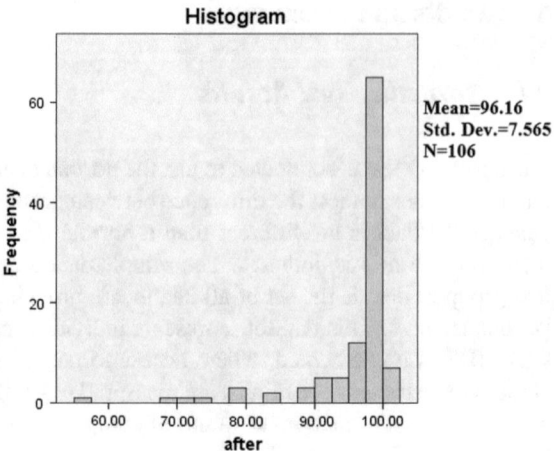

Kolmogorov-Smirnov is 0.000. This indicates the result is significantly not normal (for $p < 0.05$).

The result of normality test conducted for the second group also yielded the same significance values (Sig.) (which is p-value) in Kolmogorov-Smirnov of 0.000 (the normality histogram is shown in Fig. 3).

This indicates that the result is also significant and the distribution of the second groups dataset is also not normal. As both normality tests for both groups are significant, thus the null hypothesis (H_0) is rejected. The results support H_1 indicates that the distributions of data in both groups are not normal. The results suggest that non-parametric test method (such as Wilcoxon-test, Friedman test, McNemar test and etc.) must be used for further test to quantify the difference between the two groups of data (whether increment in access to complete data is significance after COLLIDS is implemented).

4.2 Wilcoxon Test Results

We choose Wilcoxon-Sign Rank test to perform non-parametric statistical significance test. The Wilcoxon-test is usually used to determine and evaluate if there is a significance difference between the samples or groups in the context of treatment comparison ([17], as cited in [18]). It also tests the null hypothesis where the test samples are from the same group (population). Wilcoxon sign test compares the two dependent observations and counts the number of negative and positive differences. It also uses the standard normal distributed z-value to test the significance. A low significance value indicates that there is a significant difference between the two results. This signifies that the two results came from the two samples or groups are significantly different. We use the same groups of datasets in the normality test for

Fig. 4 Rank results

		N	Mean Rank	Sum of Ranks
Completeness Increment – As-Is Completeness	Negative Ranks	0[a]	.00	.00
	Positive Ranks	106[b]	53.50	5671.00
	Ties	0[c]		
	Total	106		

a. Completeness Increment < As-Is Completeness

b. Completeness Increment > As-Is Completeness

c. Completeness Increment = As-Is Completeness

Wilcoxon-Sign Rank test which are: As-Is Completeness and Completeness Increment. These are the hypotheses that are formulated in this test:

- Null Hypothesis (H_0): *COLLIDS gives no effect on improvement of access to complete data.*
- Alternative Hypothesis (H_1): *COLLIDS improves access to complete data.*

Figure 4 shows the ranks results from the Wilcoxon-Signed Rank test which consists of the results of an average number of negative and positive ranks. The figure shows the sample size and the sum of ranks, including the mean rank (which is not necessary to calculate the W-value but helps in the interpretation of data).

From the figure, the total number of observation (N) is 106.

In positive ranks row, the number of observations (N) is 106, which represents the Completeness Increment after applying COLLIDS is greater than As-Is Completeness. The result in positive ranks row indicates that all clinics have a positive increment in access to complete data. In negative ranks row, the number of observations (N) is zero, which referred as the Completeness Increment after applying COLLIDS is less than As-Is Completeness. This means that there is no clinic that has increment of completeness less than the As-Is completeness. In ties row, the result shows there are no tied ranks between both groups. This means that there is no clinic that has the same score (in Completeness Increment and As-Is Completeness). Both results in negative and tied rows indicate that none of the clinics has no (or less) increment of access to complete datasets after COLLIDS is applied. Therefore, these ranks results indicate that all clinics enjoy improvements in terms of access to complete datasets with COLLIDS.

The mean rank column shows an increase for improvement of access to complete data (average rank of 0.0 vs. average rank of 53.50). The mean is ranked in increasing order based on the mean of the differences between two groups, where the mean rank for the positive row is 53.50. This represents the order of the ranks of difference in completeness among clinics, where there is positive increment of access to complete data. In Wilcoxon Signed-Rank test, the lowest value of both ranks (positive and negative ranks) is converted to z-value in SPSS (which is the lowest is fell into negative rank as shown Fig. 4). This allows exact significance values to be calculated

Fig. 5 Test statistics$_a$

	Increment – As Is
Z	-8.966[b]
Asymp. Sig. (2-tailed)	.000

a. Wilcoxon Signed Ranks Test
b. Based on negative ranks.

based on the normal distribution. Figure 4 shows that the statistic is based on the negative rank that calculates the result of z-value and significant value in Fig. 5. The result in Fig. 5 shows the test statistics result for Wilcoxon Signed Rank Test. The p-value associated with the Wilcoxon test given at the intersection of the row labeled as Asymp. Sig. (2-tailed) (asymptotic significance, 2-tailed) is referred as the difference between both groups that correspond to the means in the hypothesis. In Fig. 5, the z-value test shows that there is approximately normal distribution which is -8.966 (based on the negative ranks). The result of p-value Significance (2-tailed) shows the comparison between two groups is $p = 0.000$. Based on the result, Wilcoxon Signed Rank test achieved to reveal there is a significant difference between both groups of data where ($p = 0.000$) for significance level ($p < 0.01$). Thus, we accept hypothesis (H_1) and reject the null hypothesis (H_0). We further analyzed the case of completeness where our data provider is fallen to. We also found that all healthcare providers have the increment completeness greater than 50% after applying COLLIDS (in terms of patient records coverage). This validates that access to complete data has significantly improved for all data providers, especially for patients coverage in COLLIDS. In addition, those clinics that have increment almost to 100% have the lowest coverage of patients. Hence, COLLIDS is useful for those clinics. We also compute the distribution of the clinics based on the increment of access to complete data (two group of datasets used in normality test and Wilcoxon-Sign Rank test). We found that most healthcare providers in COLLIDS have more than 90% increment of access to complete data. This indicates that most healthcare providers in this case study have low coverage of patients datasets which makes COLLIDS useful.

5 Conclusion

In conclusion, within the context of healthcare data sets under study, COLLIDS has been shown as useful as improvement on access to complete data is significant. The result indicate that none of the data providers (clinic) falls under Case 1 (refer Sect. 3.1 for cases analysis). The clinics under measure are more likely to be data providers with characteristics defined in Case 2 and or Case 3. This indicates that none of the clinics has complete coverage on their patients data set and thus, collaborative integration advocated by COLLIDS will benefit all participants (data providers).

Acknowledgements The authors would like to thank the support provided by the Universiti Teknikal Malaysia, Melaka (UTeM) during the course of this research. This research is sponsored by the Ministry of Education Malaysia (under Fundamental Reserach Grant FRGS/1/2015/ICT04/FTMK/02/F00289).

References

1. Sun, Y.G.: Access control method based on multi-level security tag for distributed database system. In: Proceedings of 2011 International Conference on Electronic & Mechanical Engineering and Information Technology, vol. 5, pp. 2509–2512 (2011)
2. Ashtaputre, N., Bhutkar, S., Patil, P., Sathe, H.: Data access and retrieval for portable devices. In: Proceedings of the 8th International Conference on Computer Science and Education, ICCSE 2013, pp. 357–360 (2013)
3. Dahiya, N., Kant, C.: Biometrics security concerns. In: Proceedings—2012 2nd International Conference on Advanced Computing and Communication Technologies, ACCT 2012, pp. 297–302 (2011)
4. Hara, T., Madria, S.: Data replication for improving data accessibility in ad hoc networks. IEEE Trans. Mob. Comput. **5**, 1515–1532 (2006)
5. Kumar, A., Bhattacharya, I., Bhattacharya, J., Maskara, S., Kung, W.M., Wang, Y.C., Chiang, I.J.: Deploying cloud computing to implement electronic health record in Indian healthcare settings. Open J. Mobi. Comput. Cloud Comput. **1**, 35–47 (2014)
6. Deethshith, N., Prakash, N.J., Gunaseelan, A., Santhoshkumar, S.P.: Peak monitoring the egotistic nodes in MANET during duplication allocation. Int. J. Eng. Trends Technol. (IJETT) **9**, 454–459 (2014)
7. Smith, G.D.: Increasing the accessibility of data. Br. Med. J. **308**, 1519–1520 (1994)
8. Green, T.J., Karvounarakis, G., Taylor, N.E., Biton, O., Ives, Z.G., Tannen, V.: ORCHESTRA: facilitating collaborative data sharing. In: SIGMOD 07 Proceedings of the 2007 ACM SIGMOD International Conference on Management of Data, pp. 1131–1133 (2007)
9. Lans, R.V.D.: Clearly Defining Data Virtualization, Data Federation, and Data Integration (2010)
10. Danyaro, K.U., Jaafar, J., Liew, M.S.: MetOcean data to linked data. In: 2014 International Conference on Computer and Information Sciences (ICCOINS), pp. 1–5 (2014)
11. Leza, F.N.M., Emran, N.A.: Data accessibility model using QR code for lifetime healthcare records. World Appl. Sci. J. **30**, 395–402 (2014)
12. Naumann, F., Freytag, J.C., Leser, U.: Completeness of integrated information sources. Inf. Syst. J. **29**, 583–615 (2004)
13. Emran, N.A.: Data completeness measures. Adv. Intell. Syst. Comput. **355**, 117–130 (2015)
14. Nutt, W., Razniewski, S.: Completeness of queries over SQL databases. In: Proceedings of the 21st ACM International Conference on Information and Knowledge Management—CIKM'12, p. 902 (2012)
15. Alkharboush, N., Li, Y.: A decision rule method for assessing the completeness and consistency of a data warehouse. In: Proceedings—2010 IEEE/WIC/ACM International Conference on Web Intelligence and Intelligent Agent Technology—Workshops, WI-IAT 2010, pp. 265–268 (2010)
16. Zaniolo, C.: Database relations with null values. J. Comput. Syst. Sci. **28**, 142–166 (1984)
17. Wilcoxon, F.: Individual comparisons of grouped data by ranking methods. J. Econ. Entomol. **39**, 269 (1946)
18. Bellera, C.A., Marilyse, J., Hanley, J.A.: Normal approximations to the distributions of the Wilcoxon statistics: accurate to what N ? Graph. Insights. J. Stat. Educ. **18**, 1–17 (2010)
19. Rose, S., Spinks, N., Canhoto, A.I.: Management Research: Applying the Principles (2015)

Rough Set Theory for Supporting Decision Making on Relevance in Browsing Multilingual Digital Resources

Jolanta Mizera-Pietraszko and Jolanta Tancula

Abstract Browsing digital library (DL) collections seems to pose a challenge for a user owning to the number of factors like for instance, operability of the system, interface readability or clarity, and retrieval efficiency directly related to it, or the number of digital items within the user's domain. However, when it comes to searching for an item in a foreign language to the user, the number of the factors arises even more which translates proportionally to the growing number of clicks aimed to retrieve the target item. Such a procedure usually leads to disheartening the user from browsing the digital collections. Our study into the user's behavior interacting with multilingual DL system is set out to propose a rough set theory based model which automatically generates a decision rule based on the minimum number of the decision factors. Analyzed is a set of the predefined factors specifically influencing the user's decision on clicking an item of his special interest. We aim to limit the number of the factors however, without losing the precision of the final user's decision. To our best knowledge, rough set theory has not been implemented for multilingual decision making purposes.

Keywords Rough set theory · Decision support systems · Multilingual digital libraries · Information retrieval

J. Mizera-Pietraszko (✉) · J. Tancula
Institute of Mathematics and Computer Science, Opole University,
Oleska 48, 45-052 Opole, Poland
e-mail: jmizera@math.uni.opole.pl

J. Tancula
e-mail: jtancula@math.uni.opole.pl

© Springer International Publishing AG 2017
D. Król et al. (eds.), *Advanced Topics in Intelligent Information and Database Systems*, Studies in Computational Intelligence 710,
DOI 10.1007/978-3-319-56660-3_12

129

1 Introduction

Decision making support is becoming the more challenging task in case of uncertainty and incompleteness of information needed to take an action whether in industry, office or academia. We apply rough set theory to support the user's decision making.

Rough set theory created by Pawlak [1] captures a concept of a finite set with a lower and upper approximation defined by a human. The lower approximation of the concept is a set of the attributes sufficient to make the reliable decision about the object while the upper approximation is a finite set of other attributes which can be classified with some probability to the set (fuzzy set) allowing to support the same decision making on the comparison basis [2]. Both the upper and lower approximation, which are non-empty sets, can be called rough sets with imprecise boundary region, on the contrary to crisp sets, when the boundary region is precise.

We define the set approximations to determine the minimum number of factors allowing the user to make the decision on the selection of the full-text items from multilingual digital resources. As browsing digital collections profiles the information need, making decision on which of the numerous digital items is relevant or not, without taking into consideration the selection criteria, is a challenge. Such a motivation justifies our approach to applying rough set theory to support the user in making the crucial decision on clicking the full-text items available for download.

The remainder of this paper is structured as follows: Sect. 2 introduces the literature overview in the field of the rough set theory, section three presents our decision table of supporting decision on relevance while browsing multilingual digital resources, other sections discuss methodology of creating fuzzy sets and reduct sets from the multilingual project real data to support decision making on relevance. As the last part of this paper, the conclusion remarks are made. We plan to extend this project.

1.1 Fuzzy Idea About the User's Need

Relevance does not refer to the information retrieval system, which makes a kind of guesses while building the ranking list of the responses to the particular query, but it refers just to the user who expresses his or her information need by clicking on the item which hopefully will be relevant. Boolean model serves as an example. Thus, fuzzy idea about the user's real information need refers not only to the system, but also to the user, however from a different perspective, since while browsing a digital collection, the user models his knowledge depending on so called random walk. Fulfilling the information need by making the shortest path of the random walk is a key problem widely studied in the area of information retrieval. Hoenkamp [3] considers intuitive nature of information need by proposing lattice theory to formalize the notion with the aim at enabling the user to express the need as independent of a query and of a language model, even when the relevant results to such a query do not exist. Analysis of information need based on geo-localization positioning systems integrated with the user's profile while browsing cultural heritage digital resources is discussed from the context-aware perspective of the benefits [4]. Chronological recommendation of highly-cited papers assumes that the initial information need evolves with the time progressing while browsing research papers. The process can be modeled twofold: dynamic ranking feature construction and dynamic evolving feature weight. Some experimental studies with the ACM corpus reveal the recommendation of the highly-cited papers can be enhanced by time-series ranking [5]. Another approach relies on bound with query versus without it to measure the user's need domain. Making decision on relevance can improve precision of the information [6].

The information need evolves from generic, meaning the user has only heard about something without any knowledge about the subject matter, to more and more specific, by expanding the knowledge during the searching process which then transfers to the higher precision of the information retrieval system.

1.2 Conditional Attributes as Criteria in Decision Making Under Uncertainty

We define the following attributes contributing to support the user's decision making:

- Digital Library Project
- Languages other than English that is usually treated as lingua franca
- Target language of the digital item
- Number of the collection items for each language
- Average number of the query matches
- Number of the collection items within the user's domain
- Average target language competence is accessed by the user within the scale 0–5 according to the Common European Framework of References for Languages proposed by the Council of Europe

All the attributes are computed in the target language collection as we assume that the user who does not feel that the foreign language competence is sufficient to browse the collection, withdraws from the task. The attribute *Average number of the query matches* indicates the extent to which it is worth to undertake the task including the system bandwidth. Also, the attribute *Number of the collection items within the user's domain* has been found essential from the perspective of the language semantics, in particular the user's familiarity of technical terminology.

2   Related Work

Application of rough set theory to analysis of relationship between the language competence and accessibility to multilingual digital collections with the support on relevance is the first approach to the problem presented.

In addition to the substantial collection of the original works on rough set theory authored by its creator Pawlak [2, 7, 8], around three hundred thousand works follow his research. According to him, a system is defined as a quadruple $S = (X, A, V, \delta)$, where X is a set of objects with their upper and lower bounds, A—their attributes, V is a set of the values of these attributes and δ is a decisive function, whose values are Yes, or No. Pawlak proposed to express the values of the decision function δ: $X \times A \rightarrow V$ by the means of Boolean model in which discernible binary relations form reduct sets and decision rules [9]. Going further, he introduced topological operations like approximation space $(U, R^{(U)})$ of the universe U of objects and $R^{(U)}$: $U \times U$ being an indiscernibility relation between the attributes [10].

As the objects can be attributed to some data which share the same information, they are indiscernible—the reasoning allows to apply the theory to pattern recognition, or natural language processing [11]. In our work, we approximate the query vagueness language.

Other works authored by the Pawlak's followers discuss granular computing applications like information processing, or decision analysis. The application areas include association rules, concept representation, approximate knowledge, scoring and ranking information retrieval results [12]. Multigranulation of rough sets is widely studied under incomplete environment in decision making, incomplete information or neighborhood systems [13]. Thangavel et al. [14] discuss meta-heuristic algorithms of the rough set theory.

Customer satisfaction analysis, probabilistic decision making or quality of rough approximation in classifying problems are some examples of their application areas [15]. A work by Polkowski [16] dedicated to the rough set theory creator, highlights the partition concept $\{A, X \backslash A\}$, where A is a set of objects and $X \backslash A$ is a problem decidable based on the knowledge. Such a paradigm transfers the incremental knowledge to the partitions belonging to the universe U and consequently making the crisp notion A a non-crip one. Both lower and upper approximations solve this paradigm analyzed by our model in which the user's knowledge is becoming incremental while browsing digital collections.

3   Preliminaries

Following a founder of set theory, Georg Cantor, "A set is the result of collecting together certain well-determined objects of our perception or our thinking into a single whole objects called the elements of the set" [1]. As opposed to set theory, where the core concept of set is a collection of entities being either a real objects or the conceptual entities as its elements, in rough set theory, we assume that having some data we create a set of elements. However, the elements of the set which contain the same information and therefore are similar, create elementary sets $E^{(S)}$. The sum of any elementary sets creates a definable set $S^{(D)} = \sum_{S \in N} E^{(S)}$. On the contrary to it, an undefined set is called a rough set, which can be described by two sets definable by its upper $U^{(A)}$ and lower approximations $L^{(A)}$. The difference between the upper $U^{(A)}$ and lower $L^{(A)}$ approximation is called the set boundary $S^{(B)} = U^{(A)} - L^{(A)}$. The set is called a rough set R if and only if the edge of the set $E^{(S)}$ is a non-empty set $E^{(S)} \neq \{\varnothing\}$ such that $R = \{r_n : r_i \in E^{(S)}, E^{(S)} \neq \varnothing, L^{(A)} \subseteq S^{(B)} \subseteq U^{(A)}, 1 \leq i \leq n \in N\}$. Rough set theory has been applied in many fields. In this paper, we apply this theory to approximate an access to digital collections of multilingual library systems depending upon the user's competence in the target language different from English.

3.1   Decision System Grounded upon Rough Sets

Information system is a multilevel structure for recording and storing data that enables to construct the models which describe some processes being analyzed. In this section we are going to focus on decision support systems grounded upon the rough sets. Let us first introduce a formal definition of information system.

Definition 1 Let S be an information set such that $S = (U, A)$, where U is a non-empty finite set of objects, A is a non-empty finite set of attributes, $a \in A$ an $a: U \rightarrow V_a$ and V_a is a domain of attribute a. In addition, let's denote set B: $B \subseteq A$ as a vector of information for object $x \in U$.

3.2   Decision Table on Supporting Relevance

Data is represented in many ways. We create decision table which is called information table built from any arbitrary data. Then, the data is divided into attributes called conditional attributes and decision attributes called decisions.

Definition 2 Decision table is a set denoted by $S = (U, A \cup \{dec\})$ where U is a set of decision objects $U = \{u_1, \ldots, u_n\}$ called the universe and A is a set of attributes denoted by $a_i: U \rightarrow V_i$, where d is a decision $dec = \{d_1, \ldots, d_n\}$.

Table 1 serves as an example of our decision table created based on multilingual digital Project Gutenberg.

In Table 1 we define a set of decision objects $U = \{u_1, u_2, \ldots, u_{15}\}$, where $u_1 = $ "Chinese", $u_2 = $ "Danish",..., $u_{15} = $ "Tagalog", a set of conditional attributes $A = \{A^{(1)}, A^{(3)}, A^{(4)}\}$, where $A^{(1)} = $ "Number of the collection items", $A^{(3)} = $ "Number of the collection items within the user's domain" and

Table 1 Decision table of relevance in browsing multilingual Project Gutenberg

Language of the collection	Number of the collection items $A^{(1)}$	Average number of matches $A^{(2)}$	Number of the items within the user's domain $A^{(3)}$	Target language competence $A^{(4)}$	Relevance $A^{(5)}$
Chinese	958	61	71	0	Y
Danish	98	22	12	1	N
Dutch	1260	45	214	2	Y
Esperanto	286	59	54	0	Y
Finish	1487	79	211	1	N
French	2079	62	268	4	Y
German	1785	89	311	5	Y
Greek	653	27	48	2	Y
Hungarian	206	11	17	1	N
Italian	1168	72	63	3	Y
Latin	342	46	32	1	N
Portuguese	786	89	18	1	Y
Spanish	705	127	26	3	Y
Swedish	265	35	15	3	Y
Tagalog	71	12	3	0	N

$A^{(4)}$ = "Target language competence" and one-element set of decision attribute is $A^{(5)}$ = "Relevance".

The objects in Table 1 are divided into classes related to the user's final decision on whether or not to click on the digital item based upon the attributes considered $Rel^{(Y)}$ = {1, 3, 4, 6, 7, 8, 10, 12, 13, 14} (positive decision) and $Rel^{(N)}$ = {2, 5, 9, 11, 15} (negative decision).

Figure 1 shows relationships between some attributes for each of the Project Gutenberg collection languages. The collections with the greatest number of the full-text digital items, specifically those within the user's domain which determine the users' positive decision, are in German perhaps because the project is German,. Still, the most dominant attribute is the user's competence in the target language, since otherwise browsing the collection seems counterproductive. The attribute marked in red indicates the system's precision.

4   Indiscernible Sets

In this section defined are indiscernible sets of objects in relation [17] to our example in Fig. 1.

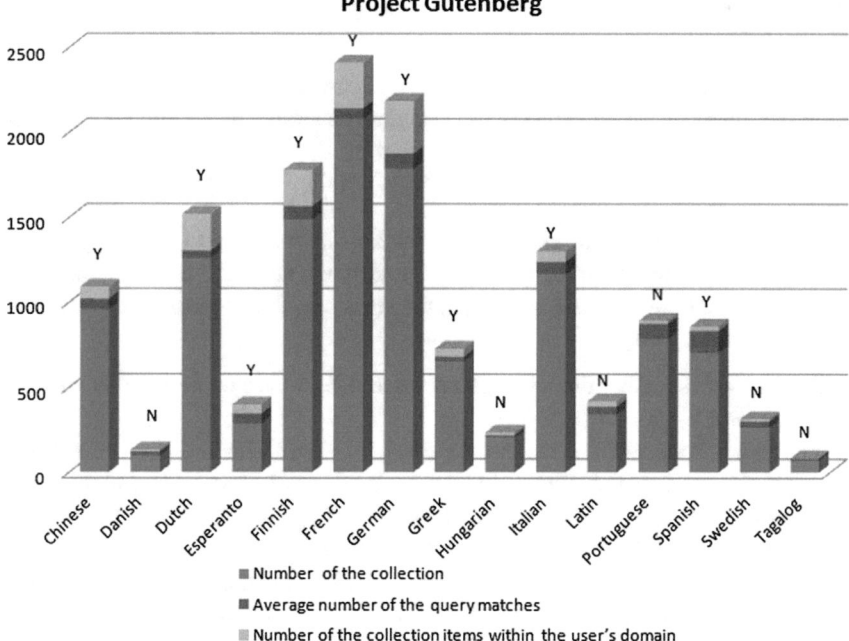

Fig. 1 Object-related comparative stacked column chart of some attributes

Definition 3 Objects x and y belonging to U and a set of attributes $B \subseteq A$ are called discernible if and only if there exists such an $a \in B$ that $a(x) \neq a(y)$, otherwise x and y are indiscernible such that $a(x) = a(y)$.

Thus, the decision support system can be denoted

$$IND_S(B) = \{x, y \in U \times U \mid \bigvee_{a \in B} a(x) = a(y)\} \tag{1}$$

Indiscernibility relation $IND(B)$ is said to be an equivalence relation or alternatively reflexive relation, when an object is in relation with itself $xIND(B)x$, symmetric (if $xIND(B)y$ then $yIND(B)x$) and transitive (if $xIND(B)y$ and $yIND(B)z$ then $xIND(B)z$).

With reference to our example, it is for attribute A1, then A1 and A2 and finally A1, A2 and A3

$IND(\{A1\}) = \{1, 4, 8, 11, 12, 13\}$
$IND(\{A1, A2\}) = \{\{1, 4, 12, 13\}_{SD}, \{2, 14\}_{MS}, \{5, 6, 7, 10\}_{DD}, \{9, 15\}_{MM}, \{8, 11\}_{SS}\}$
$IND(\{A1, A2, A3\})$
$= \{\{1, 4\}_{SDS}, \{2, 14\}_{MSM}, \{5, 6, 7\}_{DDD}, \{8, 11\}_{SSS}, \{9, 15\}_{MMM}, \{12, 13\}_{SDM}\}$

The types of the relations in our example refer to their attributes described in Table 1.

4.1 Approximation of the Set Boundaries

A set X of objects described by the attributes A can be defined by a lower (positive region) or upper approximation (negative region) according to the following formula

$$\underline{B}X = \{m | [m]_B \subseteq X\} \text{ or } \overline{B}X = \{m | [m]_B \cap X \neq \varnothing\} \tag{2}$$

where $[m]_B$ is a set of indiscernible objects. On the contrary to boundary region, which does not allow to associate unambiguously the attributes to set X, the inside region does. In case of non-empty edge, the set is called an approximate, otherwise it is called a crisp edge.

Definition 4 A set X is called a rough set when $BN_B(X)$ is a non-empty set meaning $BN_B(X) = \overline{B}(X) - \underline{B}(X) \neq 0$.

The upper bound of our set is then $B = \{A1, A2, A3\}$ for the class of negative decision on clicking the item by the user $Rel^{(N)} = \{2, 5, 9, 11, 15\}$ such that $\underline{B}(X) = \{2, 14, 9, 15\}$, and consequently $\overline{B}(X) = \{2, \underline{8}, 9, 11, 12, \underline{13}, 14, 15\}$.

Since $BN_B(X) = \overline{B}(X) - \underline{B}(X) = \{8, 13\}$ is not empty, we have here a rough set.

In Fig. 2, the area in dark orange denoted with U is the universe, the area in light orange is the upper bound $\overline{B}(X)$, the area in yellow is the lower bound $\underline{B}(X)$, whereas the area in blue is our set X.

4.2 Precision of Approximation

Both crisp and approximate sets are comparable in terms of their power based on coefficient of approximation precision.

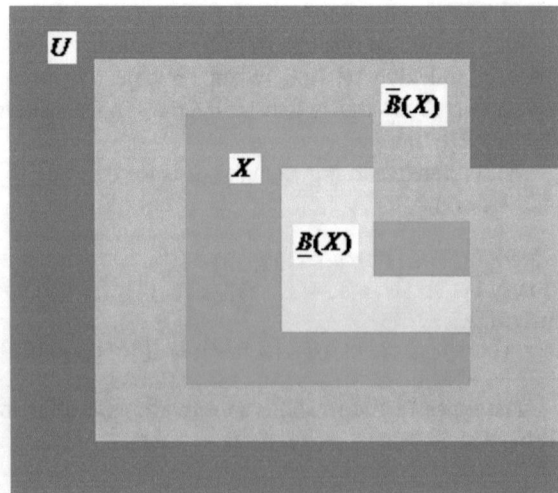

Fig. 2 Graphical representation of approximation of our set X

$$\alpha_B(X) = \frac{|\underline{B}X|}{|\overline{B}X|} \text{ where } 0 \le \alpha_B \le 1 \tag{3}$$

For a crisp set it is $\alpha_B(X) = 1$, otherwise X is a rough set $0 \le \alpha_B(X) < 1$.

For the attributes in Table 1, coefficient of approximation precision is $\alpha_B(X) = \frac{4}{8} = \frac{1}{2}$. This coefficient $\alpha_B(X) = \frac{|\underline{B}X|}{|\overline{B}X|}$ denotes a power for the set class of $Rel^{(N)} = \{2, 5, 9, 11, 15\}$.

5   Reduct Sets in Decision Making

In a decision support process not all the attributes are necessary to make a decision. Reduct sets or simply reducts RED(S) are the granular subsets that enable to define the characteristics of the whole set of its attributes. One of such reducts in decision table is called a decision reduct.

Definition 5 A set of attributes is called a reduct when for any of two objects x and y satisfied is a condition $a(x) \ne a(y)$ and both x and y are discernible by A and B simultaneously on condition that B is a granular discernible set.

Definition 6 Core is a set of attributes belonging to each of the reducts and denoted by

$$COR(B) = \bigcap_{B \in RED(S)} B \tag{4}$$

For computing the reducts, we are going to apply Boolean algebra and Boolean functions. Following the data in Table 1, we create a discernible matrix.

Using the data from Table 2, we create a Boolean function $f(B)$

$$f(B) = (A1)(A2)(A3)(A4)(A1 \cup A2)(A1 \cup A3)(A1 \cup A2 \cup A3)(A1 \cup A3 \cup A4)$$
$$(A1 \cup A2 \cup A4)(A2 \cup A3 \cup A4)(A1 \cup A2 \cup A3 \cup A4)$$

then we apply an absorption rule such that $p \wedge (p \vee q) \equiv p$, so we get

$$f(B) = (A1)(A2)(A3)(A4)$$

following, we use conjunction rule with respect to alternative $p \wedge (q \vee r) = p \wedge q \vee p \wedge r$ like here $f(B) = (A1A2A3A4)$, so we get one reduct only $R_1 = \{A1, A2, A3, A4\}$ On applying the reduct C = X − R1 = {4, 8, 13} we get our reduct set presented in Table 3.

Reduction decision table allows to narrow down the set of searching to a few sets of objects only and associated to them attributes, as it is shown in Table 3. Our approach proves to produce quite promising results in the support of decision

Table 2 Table of discernible set X

F	2	9	11	12	14	15
1	A1, A2, A3	A1, A2, A3	A2	A3	A1, A2, A3, A4	A1, A2, A3
3	A1, A3, A4	A1, A2, A3, A4	A1, A2, A3, A4	A1, A2, A3, A4	A1, A3	A1, A2, A3, A4
4	A1, A2, A3	A1, A2, A3	A3	A1	A1	A1, A2, A3
5	A1, A2, A3	A1, A2, A3	A1, A2, A3	A1, A3	A1, A2, A3, A4	A1, A2, A3
6	A1, A2, A3, A4	A1, A2, A3, A4	A1, A2, A3, A4	A1, A3, A4	A1, A2, A3, A4	A1, A2, A3, A4
7	A1, A2, A3, A4	A1, A2, A3, A4	A1, A2, A3, A4	A1, A3, A4	A1, A2, A3, A4	A1, A2, A3, A4
8	A1, A3, A4	A1, A2, A3, A4	–	A2, A3, A4	A1, A3	A1, A2, A3, A4
10	A1, A2, A3, A4	A1, A2, A3, A4	A1, A2, A4	A1, A3, A4	A1, A2, A3	A1, A2, A3, A4
13	A1, A2, A4	A1, A2, A4	A2, A3, A4	A4	A1, A2	A1, A2, A4

Table 3 Reduction decision table for Project Gutenberg digital library

Language of the collection	Number of the collection items $A^{(1)}$	Average number of matches $A^{(2)}$	Number of the items within the user's domain $A^{(3)}$	Target language competence $A^{(4)}$	Relevance $A^{(5)}$
Esperanto = 4	286	59	54	0	y
Greek = 8	653	27	48	2	y
Spanish = 13	705	127	26	3	y

making on how far the user can limit the searching by reducing the number of sets with no impact on the user's decision accuracy of the relevance.

6 Conclusion

Our model shows that Boolean reasoning integrated with decision tables built on the basis of rough set theory allows to determine mutual relationships between the attributes such as language competence, the number of the items in a target language and the system efficiency, expressed as the average number of matches.

In our further work, we plan to extend the model to some more attributes related to searching for multilingual items, but especially we want to add different weights to the attributes with the aim to study their influence on the multilingual ranking list.

References

1. Inuiguchi, M., Hirano, S., Tsumoto, S.: Studies in Fuzziness and Soft Computing: Rough Set Theory and Granular Computing. Springer, Berlin (2003)
2. Pawlak, Z.: Rough sets. Int. J. Comput. Inf. Sci. **11**, 341—356 (1982)
3. Hoenkamp, E.: On the notion of an information need, In: Hoenkamp, E. (ed.) Advances in Information Retrieval Theory, In: 2nd International Conference on the Theory of Information Retrieval, ICTIR 2009, LNCS, vol. 5766, pp. 354–356, Springer (2009)
4. Jailani, A.K., Kusakabe, S., Araki, K.: Adaptive contex-awareness model for cultural haritage information based on user needs. In: 4th International Congress on Advances in Applied Informatics, Okayma, Japan, pp. 339–342. IEEE Computer Society (2015)
5. Jiang, Z., Xiaozhong, L., Liangcai, G.: Chronological citation recommendation with information-need shifting, In: 24th ACM International Conference on Information and Knowledge Management, pp. 1291–1300. ACM (2015)
6. Wang, B., Gao, G.: Bound on information need in information retrieval, In: Proceedings— 2010 International Conference on Intelligent Computing and Cognitive Informatics, ICICCI 2010, Kuala Lumpur, Malaysia, pp. 75–78. IEEE Computer Society (2010)
7. Pawlak, Z.: Rough Sets, Rough Sets & Data Mining, pp. 1–7. Kluwer Academic Publishers (1997)
8. Pawlak, Z.: Rough Sets; Theoretical Aspects of Reasoning About Data. Springer Science & Business, Media BV (1991)
9. Pawlak, Z., Skowron A.: Rough sets and boolean reasoning. Int. J. Inf. Sci. (Elsevier) **177**, 41–73 (2007)
10. Pawlak, Z., Skowron A.: Rough sets; some extensions. Int. J. Inf. Sci. (Elsevier) **177**, 41—73 (2007). Elsevier
11. Pawlak, Z., Skowron A.: Rudiments of rough sets. Int. J. Inf. Sci. (Elsevier) **177**, 28—40 (2007). Elsevier
12. Lin, T.Y., Yao Y.Y., Zadeh L.A.: Data mining, rough sets & Granular computing. In: Kacprzyk, J. (Ed.) Studies in Fuzziness & Soft Computing. Springer (2002)
13. Yang, X., Yang, J.: Incomplete Information Systems & Rough Set Theory Models and Attribute Reductions. Science Press, Springer, Beijing (2012)
14. Thangavel, K., Pethalaksmi, A.: Dimensionality reduction based on rough set theory: a review. Appl. Soft Comput. (Elsevier) **9**, 1–12 (2009)
15. Greco, S., Hata, Y., Hirano, S., Inuiguhi, M., Miyamato, S., Nguyen, H.S., Slowinski, R.: Rough sets & current trends in computing. In: 5th Conference on Rough Sets and Current Trends in Computing (RSTCT 2006), LNAI, vol. 4259. Springer, Heidelberg (2006)
16. Polskowski, L.: Rugh sets; mathematical foundations. In: Kacprzyk, J. (dd.) Advances in Soft Computing, p. 303, Springer (2002)
17. Nguyen, H.S.: Applied Mathematics: Decision Systems. Warsaw University (2011)

The Social Influence on the Behavioral Intention to Use Mobile Electronic Medical Records

Yi-Horng Lai

Abstract Past research on information systems primarily focused on system quality. However, with rapid development and popularity of information systems, discussion on the electronic medical record should not be centered merely on technological perspectives, since technologies continuously evolve and will reach maturity in the end. In contrast, enhancing the usage of mobile electronic medical records requires everlasting efforts. Moreover, prior researches ignored the influence of social interaction on behaviors because they contended that individual actions are isolated from one another. In fact, however, social influence plays an important role in determining users' intention to use mobile electronic medical records. This study is therefore based on the conformity of behavior perspective and explores factors that influence the intention to use the electronic medical record system. This study extends the technology acceptance model with contagion model. Results showed that (1) perceived usefulness and social influence affect behavioral intention to use the electronic medical record system, (2) perceived ease of use and social influence positively influence perceived usefulness as proposed in TAM. Overall, this model provides acceptable explanation on the use of the electronic medical record system.

Keywords Social influence · Influence of conformity · Mobile electronic medical records · Contagion model

1 Introduction

The medical record is a document that contains an individual patient's health-related information and histories. It is part of a patient's health record and is written or transcribed by physicians. Its content includes various illnesses, injuries

Y.-H. Lai (✉)
Department of Health Care Administration, Oriental Institute of Technology,
New Taipei City, Taiwan
e-mail: FL006@mail.oit.edu.tw

© Springer International Publishing AG 2017   141
D. Król et al. (eds.), *Advanced Topics in Intelligent Information
and Database Systems*, Studies in Computational Intelligence 710,
DOI 10.1007/978-3-319-56660-3_13

inoculations, allergies, treatments, and prognosis…etc. It also includes other information, such as health information about a patient's parents and siblings, occupation, and military service. The record may be reviewed by physicians when diagnosing a condition.

The earliest medical records are paper-based. The paper-based medical records can only be used by one physician at a time, and it is burdensome. Because paper-based medical records are not efficient for clinic service and it is not easy to process medical procedure, the electronic medical record is becoming widely used in recent years. Many clinical organization managers believe that the electronic medical records can achieve better efficiency and accuracy. The most important contribution of the electronic medical records is that it can be viewed by more than one physician at a time. Unlike the paper-based medical records, the electronic medical records can reserve the complete medical record, and it is helpful for clinic research.

Some researches point out that the adoption of electronic medical records can improve the quality of medical service and reduce the Medicare costs. However, some important issues still need to be discussed before adopting electronic medical records, such as the interoperability privacy, older record incorporation, social and organizational barriers, preservation technology limitations, and legal status.

Many studies indicate that social influence plays an important role in technology acceptance model (TAM). Venkatesh and Davis [1] integrated social influence processes (subjective norm, voluntariness, and image) and TAM in technology acceptance model 2 (TAM2). Venkatesh and Davis [1] stated in their study that there have been two distinctive approaches to the study of attitude towards a new technology and its acceptance. They are the technology acceptance model (TAM) and the social influence theory (or social information processing). In Venkatesh and Davis [1] and Venkatesh and Bala's studies [2], social influence (subjective norm) was added into TAM. And this concept was later applied to studies on innovative technology acceptance. Many studies have shown that social influence has a significant effect on the behavioral intention to use technology [1]. But some studies exhibit a non-significant effect, especially studies on professionals with high autonomy [3–8].

Individual cognition and behavior will not separate from social influence [9]. Social influence not only provide information exchange, but also the important factors for cognition shaping, and information technology adoption. Therefore, this study investigates the social influence on the technology acceptance based on social influence.

1.1 The Technology Acceptance Model

Based on the theory of reasoned actions (TRA) [10], TAM addresses factors influencing a user's attitude toward using and their intention to use technology [11]. TAM has been widely adopted in studies exploring technology acceptance due to its parsimonious nature and highly reliable constructs. Examples include studies testing user acceptance of word processors [11], spreadsheet applications [12], email [13], and websites [14].

TAM considers perceived usefulness (PU) and perceived ease of use (PE) as two major factors influencing a user's behavioral intention (BI) in using new technology [15]. The former refers to the perceived effectiveness of improving the user's performance, while the latter refers to how effortless a user perceives using the technology to be. Prior research has found that PE mostly influences attitude and intention indirectly through PU [16]. Perceived usefulness and user attitude in turn influence intention to use, which predicts actual usage of technology. This study proposes that the relationships of TAM will be the same when it is applied to the use of electronic medical record in Taiwan. Based on these studies, the hypotheses in this study are:

H1   Perceived usefulness is positively associated with behavioral intention in the use of mobile electronic medical records.
H2   Perceived ease of use is positively associated with behavioral intention in the use of mobile electronic medical records.
H3   Perceived ease of use is positively associated with perceived usefulness in the use of mobile electronic medical records.

1.2   Social Influence Theory

Lee et al. [17] stated in their study that there have been two distinctive approaches to the study of attitude towards a new technology and its acceptance. They are the technology acceptance model (TAM) and the social influence theory (or social information processing). In Venkatesh and Davis [1] and Venkatesh and Bala's studies [2], social influence (subjective norm) was added into TAM. And this concept was later applied to studies on innovative technology acceptance. Prior researches on technology use have found that user's attitude toward new technology is influenced by the objective characteristics of the system, the extent of use, and individual user differences [18].

Studies continuously report that people are not always rational in selecting and using technologies, and attitudes toward using technology are influenced by culture, norms, social contexts, or salient others [17, 19, 20]. To explain such confounding results, Salancik and Pfeffer [21] proposed the social information processing. According to the social information processing [21], individuals' perceptions of technologies are also influenced by the opinions, information, and behaviors of people they communicate with. In a study using the social information processing, Fulk and Stienfiedf [19] reported that technology-related attitudes are often influenced by social interactions and psychological processes rather than directly by objective and independent assessments of technical characteristics [17].

Kelman's research [22] on social influence attempted to understand the changes brought about in individual's attitude or behavior by external stimuli, such as information communicated to them. He investigated whether these changes were temporary superficial changes or more lasting changes that became integrated in the person's value system. Kelman proposed that individuals can accept influence through three processes of social influence: compliance, identification, and internalization [22].

Compliance means that an individual adopts an induced behavior not because they believe in its content but rather due to a desire to gain rewards or avoid punishments. Identification means that an individual accepts influence because they want to establish or maintain satisfying relationships with another person or group. Internalization means that an individual accepts influence because there is a perceived congruence between their values and the values exemplified by another person or group.

By distinguishing between these processes, one can ascertain whether usage behavior is caused by the influence of referents on one's intent or by one's own attitude. Kelman [22] observed that each of the above three processes is characterized by a distinctive set of antecedent conditions corresponding to a characteristic pattern of internal responses (thoughts and feelings) in which the individual engages while adopting the induced behavior. Similarly, each of the three processes is characterized by a distinctive set of consequent conditions, involving a particular qualitative variation in the subsequent history of the induced response. For instance, behavior induced through compliance tends to be performed under surveillance by the influencing agent. In contrast, behavior induced through identification tends to be performed under salience of one's relationship with the agent; and behavior induced through internalization tends to be performed under conditions of the relevance of the issue, regardless of surveillance or salience.

Being applied to the use of a new information system, these social influence processes determine an individual user's commitment, or more specifically, psychological attachment [23], to the use of any new information technology. Users who perceive the use of the information system to be congruent with their values are likely to be internalized (committed and enthusiastic) in their system use [24]. Users who accept the use of the information system due to respect its values and accomplishments without adopting them as his or her own are likely to be identified (affiliated) in their system use [24]. Individuals who perceive such use merely as a means to obtain rewards and avoid punishments are likely to be compliant in the system use [25].

According to the research of Malhotra and Galletta [24], the use of an information system needs to be viewed as a continuum instead of the dichotomy of use and non-use. This continuum defines the range from avoidance of use to meager and unenthusiastic use (compliant use) to skilled, enthusiastic and consistent use (committed use) [24].

Based on the result of these studies, the hypotheses of this study are:

H4   Social influence is positively associated with behavioral intention in the use of mobile electronic medical records.

H5   Social influence is positively associated with perceived usefulness in the use of mobile electronic medical records.

Many studies have shown a significant effect of social influence on behavior intention to use technology in the acceptance studies [1], but some studies exhibit a non-significant effect, especially studies on professionals with high autonomy [3–8]. The respondents experimented in the study are physicians in a medical center in Taiwan. While they have the autonomy to decide whether to use the electronic medical record, social influence tend to impact their behavioral intention to use it.

Yueh et al. [8] investigated what factors affected students' adaptation and continued use of a Wiki system for collaborative writing tasks through an extension of the Unified Theory of Acceptance and Use of Technology. They found that social influence did not affect students' actual use of Wiki system for individual and collaborative learning. Alkhunaizan and Love's study [6] aimed to empirically examine some of the factors affecting the acceptance of m-commerce within the context of Saudi Arabia, and they discovered that social influence did not have a significant influence on behavior intention to use m-commerce. Lee's study [5] investigates how stock investors perceive and adopt online trading in Taiwan, and the result showed that subjective norm had a positive effect on intention to trade online. Nawaz et al. study [7] aimed to address the factors influencing teachers' intention to use e-learning systems in a Sri Lankan school, and the result of this study was that social influence did not have significant influence on attitude towards using e-learning systems.

2   Materials and Research Method

With perceived ease of use (PEOU), perceived usefulness (PU) and social influence integrated, Fig. 1 is the research framework of this study in a TAM model in which perceived ease of use, perceived usefulness, and social influence effect users' behavioral intentions.

2.1   Methodology

The validity of the proposed relationships in the research model presented above was investigated through this study. This study is to explore the use of electronic medical record system in a Taiwan's medical center and it was carried out from September 1, 2015 to October 31, 2015. Feedbacks from 140 users were collected.

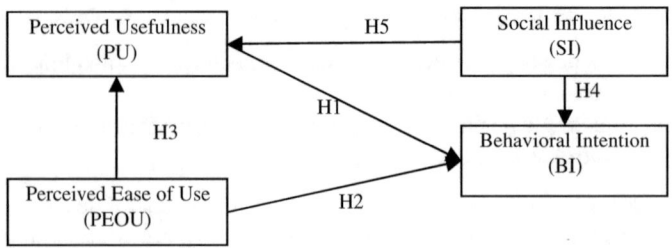

Fig. 1 Research framework

Table 1 Scale reliability

Scale	N of items	Mean	S.D.	Cornbach's α
Perceived usefulness (PU)	4	2.705	0.596	0.637
Perceived ease of use (PE)	4	2.784	0.628	0.788
Behavioral intention (BI)	3	2.721	0.659	0.624

This present study was developed in a way that the model constructs in TAM were adapted to the context of using mobile electronic medical records. Scale items on the survey were designed to measure perceived usefulness (PU), perceived ease of use (PE), and a user's behavioral intention (BI). The questionnaire contains no identifying information about individual participants.

The questionnaires of PU, PE, and BI in this study are modified from David's study [11]. All variables exhibit a high level of reliability with the Cronbach's alpha values (Table 1) exceeding the recommended 0.6 [26]. There are 3 parts in the research questionnaire, and each item has 5 scales.

The regression model is investigated using R 3.3.2. Path analysis was performed on the model using standardized maximum likelihood estimation. The advantage of the path analysis method is that it can test the overall model fit by combining multiple endogenous variables based on the priori hypothesis.

2.2 Contagion Model

Contagion model was built by Marsden and Friedkin for measuring social influence with social network [23]. This approach based on the traditional practice of social metrology, and it emphasizes that social influence is from structural cohesion. In other words, social influence constrains and influences these entities in the social network. This model can verify that the influence from others is significant.

3   Results

Feedbacks from a total of 120 mobile electronic medical records system users in a medical center in Taiwan were collected. Among the 120 users, 81 were males (67.50%) and 39 were females (32.50%).

Each physician offers five best friends' name within the medical center. The data can be integrated into social network structure with Ucinet 6 computer software (Fig. 2).

3.1   Social Influence in Behavioral Intention

The result of the effect of social influence, PU, and PEOU on behavioral intention to use mobile electronic medical records with contagion model is presented in Table 2. It can be found that the effects of social influence and PU on behavioral intention are significantly correlated (t-value = 2.288; t-value = 2.445), but the effect of PEOU on behavioral intention is not significant (t-value = − 0.590).

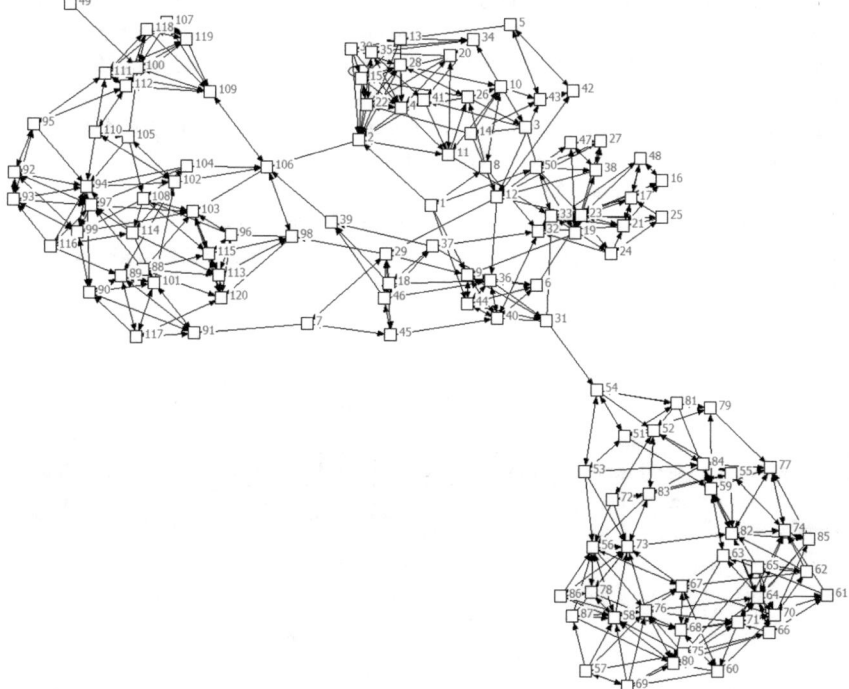

Fig. 2 Social network structure

Table 2 The result of regression analysis in behavioral intention in the use of mobile electronic medical records with contagion model

	Estimate	Std. error	t value
Intercept	1.109	0.275	4.036*
Social influence	0.154	0.067	2.288*
PU	0.596	0.244	2.445*
PEOU	−0.130	0.220	−0.590

*:P-Value <0.05

The result of this study confirms the hypothesis that perceived usefulness is positively associated with behavioral intention (H1), and this is in accordance with Davis et al. [11], and Chen and Hsiao [15]. However, the hypothesis that perceived ease of use is positively associated with behavioral intention (H2) is not supported, and this is in agreement with Davis et al. [11], and Chen and Hsiao [15].

The hypothesis that social influence is positively associated with behavioral intention (H4) is supported, and this is consistent with Venkatesh and Davis [1], and Venkatesh and Bala [2].

3.2 Social Influence in Perceived Usefulness

The result of the effect of social influence and PEOU on perceived usefulness in the use of mobile electronic medical records with contagion model is shown in Table 3. It can be found that the effects of social influence and PU on behavioral intention are significantly associated (t-value = 2.123; t-value = 34.333).

Based on the result, the hypothesis of perceived ease of use is positively associated with perceived usefulness (H3), and this is in accordance with Davis et al. [11], and Chen and Hsiao [15].

The hypothesis that social influence is positively associated with perceived usefulness (H5) is supported, and this is the same with Venkatesh and Davis [1], and Venkatesh and Bala [2].

Table 3 The result of regression analysis in perceived usefulness in the use of mobile electronic medical records with contagion model

	Estimate	Std. error	t value
Intercept	0.225	0.101	2.235*
Social influence	0.052	0.024	2.123*
PEOU	0.860	0.025	34.333*

*:P-Value <0.05

4 Conclusion

Findings of this study have proved that TAM with social influence is an applicable model in examining factors influencing users' behavioral intention to use the electronic medical record system.

This study has shown that the perceived usefulness is positively associated with users' behavioral intention to use the electronic medical record system. In addition, a useful and easy usage of the system can improve users' behavioral intention to use it. Furthermore, the perceived ease of use is positively associated with the perceived usefulness. It means that easy operation of the electronic medical record system is very important for the users, and the easier it is, the more people think it is useful. However, the effect of perceived ease is different from that of Davis' result. In Davis' study, perceived ease of use had a significant effect on intention, while perceived ease of use had no effect on behavioral intention in this study.

Due to the widespread use of information technology in hospitals, mobile electronic medical records have become necessary tools in hospitals, and it is time to make use of them to provide patients with more services. Based on the effect of social influence, this study demonstrates that it plays an important role in the electronic medical record adoption. In order to increase healthcare service quality, more and more hospitals in Taiwan have implemented the electronic medical record system. The process of adopting the system is conducted by information technology, and its primary function is to help physicians and hospital administrators manage individual patients' information in digital storage devices. Since the development of information technology in Taiwan has become ripe and stable, it is time to launch healthcare services with digital medical records for patients, physicians and hospital administrators. Better service quality in medical care can be fulfilled with mobile electronic medical records.

References

1. Venkatesh, V., Davis, F.D.: A theoretical extension of the technology acceptance model: four longitudinal field studies. Manage. Sci. **46**(2), 186–204 (2000)
2. Venkatesh, V., Bala, H.: Technology acceptance model 3 and a research agenda on interventions. Decis. Sci. **39**(2), 273–315 (2008)
3. Chau, P.Y.K., Hu, J.H.: Investigating healthcare professionals' decisions to accept telemedicine technology: an empirical test of competing theories. Inf. Manage. **39**, 297–311 (2002)
4. Schaper, L.K., Pervan, G.P.: ICTs & OTs: a model of information and communications technology acceptance and utilization by occupational therapists (part 2). Stud. Health Technol. Inf. **130**, 91–101 (2007)
5. Lee, M.C.: Predicting and explaining the adoption of online trading: an empirical study in Taiwan. Decis. Support Syst. **47**, 133–142 (2009)
6. Alkhunaizan, A., Love, S.: What drives mobile commerce? An empirical evaluation of the revised UTAUT model. Int. J. Manage. Mark. Acad. **2**(1), 82–99 (2002)

7. Nawaz, S.S., Thowfeek, M.H., Rashida, M.F.: School Teachers' intention to use E-Learning systems in Sri Lanka: a modified TAM approach. Inf. Knowl. Manage. **5**(4), 54–59 (2015)
8. Yueh, H.P., Huang, J.Y., Chang, C.: Exploring factors affecting students' continued Wiki use for individual and collaborative learning: An extended UTAUT perspective. Australas. J. Educ. Technol. **31**(1), 16–31 (2015)
9. Kilduff, M., Tsai, W.: Social networks and organizations, Thousand Oaks. Sage, CA (2003)
10. Ajzen, I., Fishbein, M.: Understanding Attitudes and Predicting Social Behavior. Prentice-Hall, Englewood Cliffs, NJ (1980)
11. Davis, F.D., Bagozzi, R.P., Warshaw, P.R.: User acceptance of computer technology: comparison of two theoretical models. Manage. Sci. **35**(8), 82–1003 (1989)
12. Mathieson, K.: Predicting user intentions: comparing the technology acceptance model with the theory of planned behavior. Inf. Syst. Res. **2**(3), 173–191 (1991)
13. Szajna, B.: Empirical evaluation of the revised technology acceptance model. Manage. Sci. **42**(1), 85–92 (1996)
14. Gefen, D., Karahanna, E., Straub, D.W.: Trust and TAM in online shopping: an integrated model. MIS Q. **27**(1), 51–90 (2003)
15. Chen, R.F., Hsiao, J.L.: An empirical study of physicians' acceptance of hospital information systems in Taiwan. Telemed. E-Health **18**(2), 120–125 (2012)
16. Hu, P.J., Chau, P.Y.K., Liu, S.O.R., Tam, K.Y.: Examining the technology acceptance model using physician acceptance of telemedicine technology. J. Manage. Inf. **16**(2), 91–112 (1999)
17. Lee, J., Cho, H., Gay, G., Davison, B., Ingraffea, T.: Technology acceptance and social networking in distance learning. Educ. Technol. Soc. **6**(2), 50–61 (2003)
18. Davis, F.D., Bagozzi, R.P., Warshaw, P.R.: User acceptance of computer technology: comparison of two theoretical models. Manage. Sci. **35**(8), 982–1003 (1989)
19. Fulk, J.C., Stienfiedf, J.: Schmitz, & Power, J.G.: A social information processing model of media use in organizations. Commun. Res. **14**(5), 529–552 (1987)
20. Rice, R.E., Love, G.: Electronic emotion: Socio-emotional content in a computer-mediated communication network. Commun. Res. **14**(1), 85–105 (1987)
21. Salancik, G.R., Pfeffer, J.: A social information processing approach to job attitudes and task design. Adm. Sci. Q. **23**, 244–252 (1978)
22. Kelman, H.C.: Compliance, identification, and internalization: three processes of attitude change?". J. Confl. Resolut. **2**, 51–60 (1958)
23. O'Reilly III, C., Chatman, J.: Organizational commitment and psychological attachment: the effects of compliance, identification, and internalization on prosocial behavior. J. Appl. Psychol. **71**, 492–499 (1986)
24. Malhotra, Y., Galletta, D.F.: Extending the technology acceptance model to account for social influence: theoretical bases and empirical validation. In: Proceedings of the 32nd Hawaii International Conference on System Sciences, IEEE, Hawaii, pp. 1–14 (1999)
25. Klein, K.J., Sorra, J.S.: The challenge of innovation implementation. Acad. Manage. J. **21**, 1055–1080 (1996)
26. Nunnally, J.C.: Psychometric Theory. McGraw Hill, New York (1978)

Stock Prices Growth Pattern by the Emergency Demand After the Great East-Japan Earthquake

Kenji Yamaguchi, Yuriko Yano and Yukari Shirota

Abstract In many Japanese companies, stock prices were declining just after the Great East Japan Earthquake in Fukushima prefecture on March 11th, 2011. Especially, the North-East Japan companies were a lot damaged. However, on the other hand, some companies' stock prices rose in the prefectures, such as YAMAYA. YAMAYA is an alcohol wholesaler and retailer company, and the headquarters are located in Sendai-city, the capital and biggest city of Miyagi Prefecture next to Fukushima. From our previous analysis, we concluded that the reason for the rapid growth of YAMAYA would be an increase of alcohol consumption by building construction workers, and others who gathered in Sendai-city. Therefore, their alcohol consumption was considered to be one of the triggers for the economic reconstruction in Sendai-city. In the paper, we conducted an analysis by the Random Matrix Theory to find growth companies in Japan similar to YAMAYA's pattern. As a result, we found out that the emergency demand for drinking water had promoted the growth of drinking water makers like Coca-Cola CENTRAL JAPAN. The paper describes our Random Matrix Theory based approach to find out the similar growth pattern to YAMAYA.

Keywords Singular value decomposition · Random matrix theory · Great East-Japan earthquake · Emergency demand · Stock data analysis · Water server companies

K. Yamaguchi (✉)
Infrastructure Department IT Center, Ochanomizu University, Tokyo, Japan
e-mail: yamaguchi.kenji@ocha.ac.jp

Y. Yano · Y. Shirota
Faculty of Economics, Department of Management, Gakushuin University, Tokyo, Japan
e-mail: yuriko.yano@gakushuin.ac.jp

Y. Shirota
e-mail: yukari.shirota@gakushuin.ac.jp

© Springer International Publishing AG 2017

151

D. Król et al. (eds.), *Advanced Topics in Intelligent Information and Database Systems*, Studies in Computational Intelligence 710,
DOI 10.1007/978-3-319-56660-3_14

1 Introduction

In this paper, we will analyze effects of the 2011 Great East-Japan Earthquake on Japanese economics. Many Japanese companies were devastated by the earthquake. To analyze the damages, we shall use the Random Matrix Theory [1–3] on stock data. There, we find the correlations among companies' stock data to obtain the similar fluctuation pattern companies.

Just after the East-Japan earthquake, lots of Japanese companies, especially the North-East Japan (the Tohoku Province) companies, were damaged so much. However, on the other hand, some companies' stock prices rose in the Province such as YAMAYA. Then, the remarkable growth of YAMAYA, an alcohol wholesaler and retailer company, was impressive. The YAMAYA headquarters are located in Sendai-city, the capital and biggest city of Miyagi Prefecture. From our previous analysis, we concluded that the reason for the rapid growth of YAMAYA would be an increase of alcohol consumption of building construction workers and others who gathered in Sendai-city. Hence, their alcohol consumption was considered to be one of triggers for the economic reconstruction in Sendai-city [4].

In the paper, we conducted a stock price analysis by the Random Matrix Theory to find growth companies in Japan similar to YAMAYA' pattern. As the stock price data, we used data of the *1620 companies in the first section of Tokyo Stock Exchange (TSE)*. The 1620 companies are Japanese representative major companies. The data period is from *11th March to 13th September in 2011*. The number of days is 127. As an analysis result, we found out that the emergency demand for drinking water had promoted the growth of drinking water makers like Coca-Cola CENTRAL JAPAN.

The paper describes our Random Matrix Theory based approach to find out the similar growth pattern to YAMAYA. In the next section, we shall explain the Random Matrix Theory and its math process named SVD. In Sect. 3, *the YAMAYA's rapid growth will be described*. Then we describe the analysis results to find the similar pattern companies to YAMAYA in Sect. 4. And then, in Sect. 5, we will find the eigenvector which expresses a company group risen by the emergency demand of drinking water. Finally, we conclude the paper.

2 Random Matrix Theory and SVD

In the section, we will explain our methods. Our research objective is measurement of natural disaster effects on Japan's economy conditions. The disasters include Japan's earthquakes and other countries' earthquakes and floods. In the methods, first we analyze Japan's stock price data. The analysis method we adopted is Random Matrix Theory and the Random Matrix Theory math process is SVD (Singular Value Decomposition) [5, 6]. The SVD is used in various kinds of applications; For example, in text mining, LSA (Latent Semantic Analysis) uses the SVD. Concerning the SVD math process, Shirota et al. visually explain the intrinsic meanings and interpretation of the eigenvectors/pricipal components, so that readers can easily and deeply understand SVD [7, 8]. In this analysis, we conduct the SVD on the standardized return values of stock price data. The return value is the ratio between today's price and the previous day's one and defined as follows: $G_{i,j} = \ln(S_{i,j}/S_{i,j-1})$ where $S_{i,j}$ is the ith company's stock price on j-th day and $G_{i,j}$ is the return value on j-th day. Because different stock values have varying levels of volatility, the return value should be standardized. Then, in the stock data analysis, each company's data during the period is standardized, so the mean value becomes 0 and the standard deviation becomes 1.

The proposed method is conducted as follows: First, we conduct SVD on the time series data matrix X of stock price return values, so that we can obtain the matrices U, W, and V^T. Then, we have gotten two kinds of eigenvectors obtained by U W and by S V^T. We call the eigenvectors (1) *Brand-Eigenvector*, obtained by U W, and (2) *Dailymotion-Eigenvector*, obtained by S V^T. In the paper, we use only Brand-Eigenvectors which are represented just by *eigenvectors* or *principle components*.

In general, the SVD method is, in a financial analysis, called the random matrix theory, and it is utilized to find the stable company classes [1, 3, 9]. Using the extracted stable classes, they make an excellent performance portfolio [2, 10]. Our usage of eigenvalues/eigenvectors in the SVD is identical to one by Plerou's proposed cross correlation analysis [2, 10]. The conversion method between both was described in our paper [11]. Our research goal is, however, completely different from theirs, and we would like to find and investigate the time series effects of the disaster on stock prices. Plerou et al. and other financial researchers are interested in stable company class [2, 10]. On the other hand, we are investigating the dynamic changes by the same the random matrix theory. A disaster triggers a stock downfall, and one industry's breakdown inflicts harm and transmits on others like a supply chain breakdown. The effects are dynamic and not stable; Some industries will recover soon, and the damage will gradually diminish, and others are not. We would like to investigate the time seris changes from a viewpoint of time series data analysis because we are interested in the effect duration time and its magnitude.

3 YAMAYA's Rapid Growth

In the section, we explain our approach. We would like to find out unexpected growing companies after the earthquake. In many Japanese companies, stock prices were declining just after the Great East Japan Earthquake. Especially, the North-East Japan companies were a lot damaged. However, some companies' stock prices drastically rose in the region. This was unexpected things for us. The statistics report in March 2012 says that personal consumption in Tohoku increased. This kind of special demand for personal consumption is an expected one. Figure 1

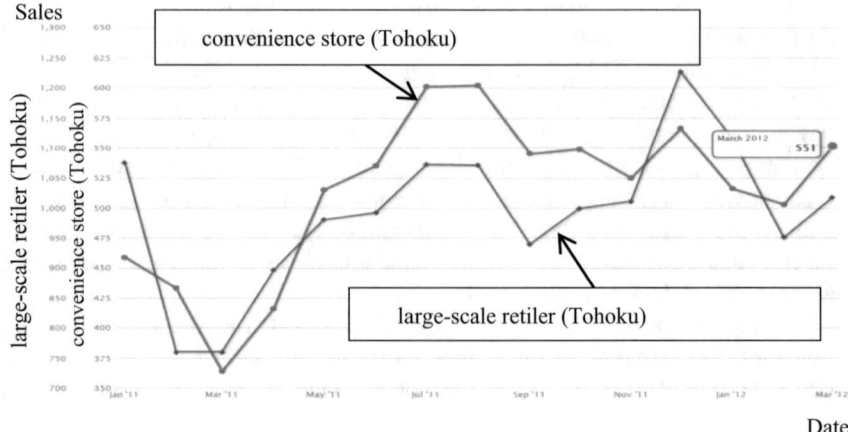

Fig. 1 Sales of large-scale retailers and convenience stores in Tohoku (monthly data from Jan 2011 to Mar 2012)

illustrates movements of sales amounts of large-scale retailers and convenience stores in Tohoku 6 prefectures [12]. The sales amount of convenience stores in March 2012 was 551 hundred millions JPY. This increase was due to construction engaged workers' consumption and selling of bus tickets [6]. Such a bounce back was an expected one.

We were interested in very rapid growth companies in Tohoku as unexpected growth companies. Then we surveyed stock prices of large companies which were headquartered in Tohoku. The number of large companies headquartered in Tohoku was 17. Among the 17 companies, we focused attention on the rapid growth of YAMAYA because the rapid growth was extraordinary [4]. YAMAYA is a liquor wholesaler and retailer company. Figure 2 illustrates stock price movements of YAMAYA. YAMAYA is headquartered in Sendai-city. The interesting point is that

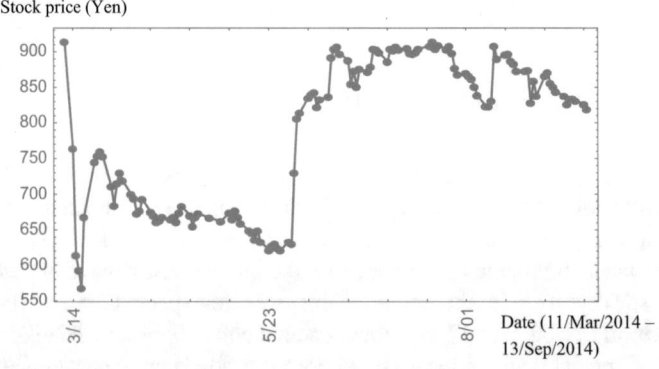

Fig. 2 YAMAYA stock price fluctuation

first the stock prices were decreased due to the disaster and then drastically increased the stock prices in June. The rapid growth was unexpected one. First we investigated various reasons such as the marketing strategy. Finally we found out that the SAKE (Japanese alcohol) consumption by reconstruction engaged workers transferred from other areas must have affected the YAMAYA's rapid growth. For example, that is a steeplejack who is a construction engaged person with high skills. The emergency demand for them had gotten their salaries higher. As a result, they could afford to drink expensive SAKE. In addition, the construction engaged persons had to stay in Sendai for a long time apart from their home. Sendai was the only big city which could become a base camp for remediation activities at that time. We heard that from the staff member of Tohoku Economic Federation when we conducted the hearing at the Tohoku Economic Federation headquarter. From the analysis, we conclude that SAKE consumption increased in Sendai by gathered people, especially construction engaged workers, was one of triggers which stimulated the Sendai mini-bubble [4]. In the next section, we describe the SVD analysis to find another unexpected growth company similar to YAMAYA.

4   Similar Pattern Companies to YAMAYA

In the section, we describe the result of the SVD. As the stock price data, we used data of the *1620 companies in the first section of Tokyo Stock Exchange (TSE)*. The 1620 companies are Japanese representative major companies. The data period is from *11th March to 13th September in 2011*. The number of days is 127. Then we can get 127 eigenvector from the return value matrix because Min(1620, 127) = 127. Each Brand-Eigenvector consist of 1620 company elements. For example, the 1619th company element is the YAMAYA's element. Figure 3 shows

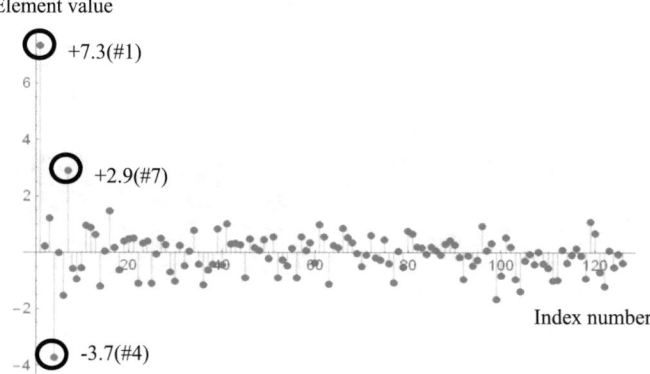

Fig. 3 Each element value of YAMAYA in 126 principal components. The absolute values of #4 and #7 are larger than others

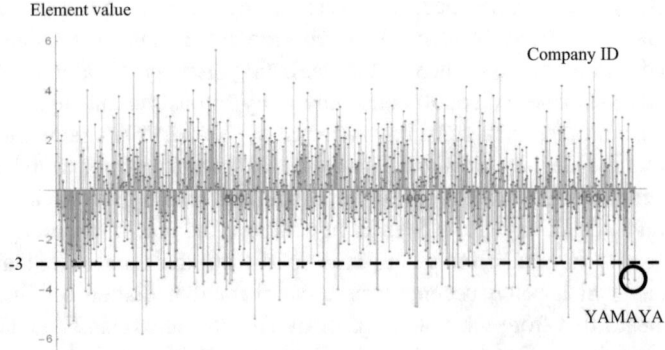

Fig. 4 The 1620 company elements in the #4 BrandEigenvector. The YAMAYA value is −3. The number of companies of which values are less than −3 is over 91

the YAMAYA element values among 126 eigenvectors. The absolute values of YAMAYA elements in #1, #4 and #7 are larger than others. The values are +7.3 (#1), −3.7 (#4), and +2.9 (#7). The #1 eigenvector (principal component) shows the average trend/movement, in general. Therefore, we ignore the #1 eigenvector and we take the #4 and #7 as candidate group which express a similar tendency to YAMAYA's.

First of all, we shall examine the #4 eigenvector. Figure 4 shows the 1620 company elements in the #4 eigenvector. The YAMAYA element value is −3 and the 91 companies have element values less than −3. In other words, so many companies have the same feature as YAMAYA in the #4 principal component that we cannot extract the YAMAYA feature. Then we shall examine the #7 eigenvector. Figure 5 shows the 1620 company elements in the #7 eigenvector.

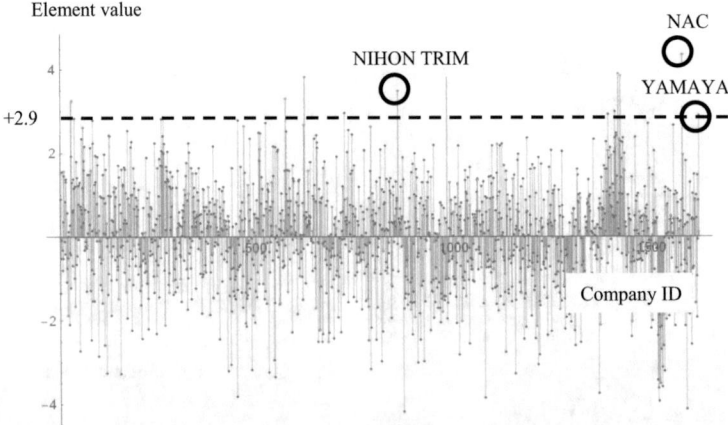

Fig. 5 The 1620 company elements in the #7 BrandEigenvector. The YAMAYA value is +2.9. The number of companies of which values are greater than +2.9 is 8

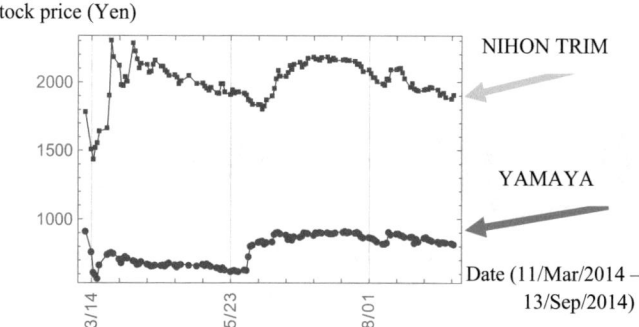

Fig. 6 Stock price fluctuations of NIHON TRIM and YAMAYA

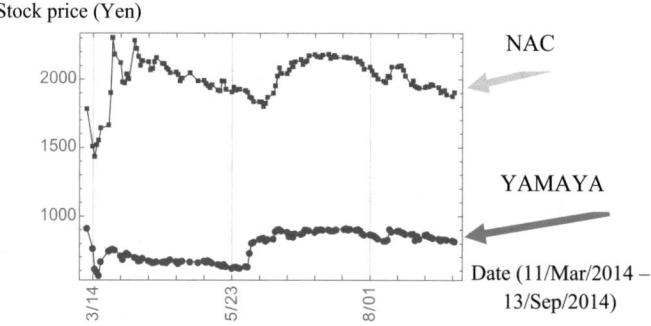

Fig. 7 Stock price fluctuations of NAC and YAMAYA

The YAMAYA element value is +2.9 and the 12 companies have element values greater than +2.9. In other words, we can guess that the 12 companies have the same feature as YAMAYA in the #7 principal component. The resultant 12 companies data include the following two companies:

- NIHON TRIM (ID 857), +3.5
- NAC (ID 1578), +4.3

The both companies are companies selling water servers. The stock prices are shown in Figs. 6 and 7. As #7 eigenvector figures show, the stock price fluctuations are similar to one of YAMAYA. We evaluated the similarity levels of the return values (See Fig. 8). The similarity level can be calculated as the Pearson's correlation coefficient because the return values were in advance standardized as follows:

$$\frac{1}{n}\sum x_i y_i = 1 - \frac{1}{2n}\sum (x_i - y_i)^2$$

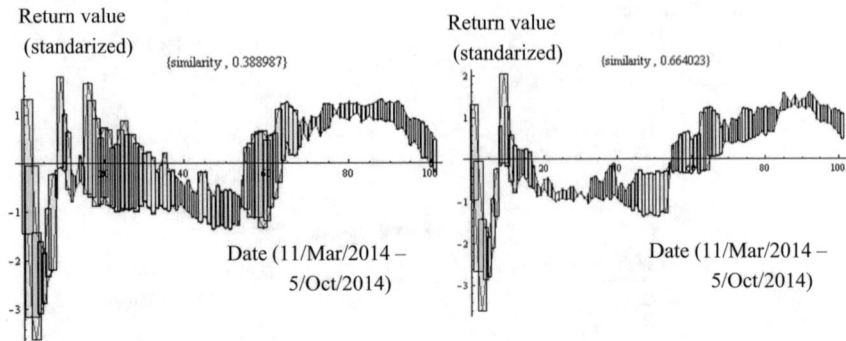

Fig. 8 The measurement of standardized return value similarity between YAMAYA and NIHON TRIM (*the left figure*) and one between YAMAYA and NAC. We found that NAC and YAMAYA shows a great similarity as the similarity level 0.66

The transformation has been shown and its meaning has explained by Kuboyama in [13].[1]

The similarity levels are 0.39 between YAMAYA and JAPAN TRIM and 0.66 between YAMAYA and NAC. Therefore, the NAC movement is more similar to the movement of YAMAYA. This can be seen visually from Fig. 8 where the pink rectangle length expresses the distance from the YAMAYA's return value.

5   Emergency Demand of Drinking Water

In the section, we will investigate effects by the emergency demand of drinking water. In the precious section, we found two water server service companies as the similar pattern company to YAMAYA. They are NIHON TRIM and NAC. We can guess that just after the earthquake, the demand of drinking water was soaring, and that the stock prices of NIHON TRIM and NAC were rising. It is impressive that the SAKE demand in Sendai-city rose the YAMAYA stock price, and on the other hand, the drinking water demand rose the stock prices of NIHON TRIM and NAC.

Then, we shall investigate whether other drinking water companies exist or not. Therefore, we see the NIHON TRIM elements in 127 eigenvectors first (See Fig. 9). The NIHON TRIM element value in the #6 eigenvector is +5.7 which is much higher than others. Then we shall see the Brand-Eigenvector #6 (See Fig. 10). Among the #6 eigenvector positive parts, we found the following three drinking water companies.

---

[1]The idea of the visualization was given by Prof. Kuboyama through our private communication.

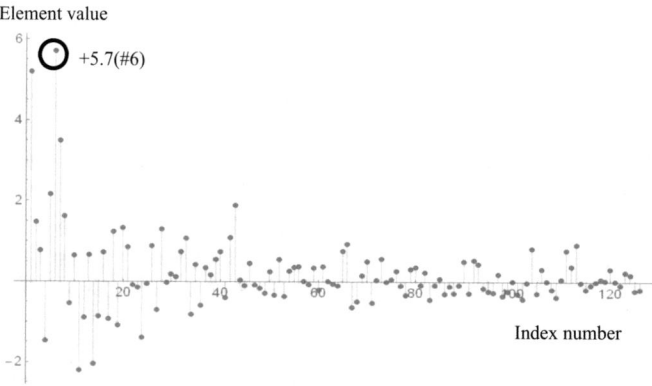

Fig. 9 Each element value of NIHON TRIM in 124 principal components. The absolute values of #6 is much larger than others

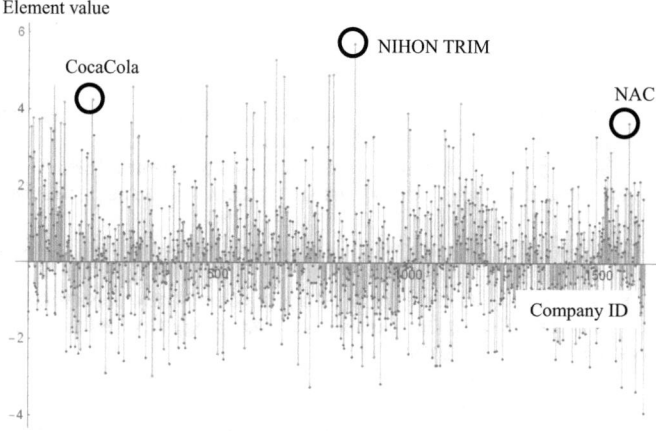

Fig. 10 The 1620 company elements in the #6 BrandEigenvector

- CocaCola CENTRAL JAPAN (ID 170), +4.2
- NIHON TRIM (ID 857) +5.7
- NAC (ID 1578) +3.7

It is a plausible guess that CocaCola CENTRAL JAPAN stock prices rose owing to the emergency demand of drinking water. Therefore we think that the #6 eigenvector expresses the rapid growth company group by the water demand.

6 Conclusion

In the paper, we conducted the SVD analysis to find the similar pattern to YAMAYA's rapid growth after the Great East-Japan Earthquake. As the similar feature companies, we found out the two water server companies NIHON TRIM and NAC. We conclude that the two companies' stock prices were risen owing to the emergency demand of drinking water. In addition, from the NIHON TRIM's SVD analysis, we finally found the drinking water company group belongs to the #6 eigenvector which included CocaCola CENTRAL JAPAN, NIHON TRIM, and NAC.

Acknowledgements This research was partly supported by funds from the Telecommunications Advancement Foundation research project in 2015–2016. In addition, this study was partly supported by a grant from the Japanese Society for the Promotion of Science from 2015–2017 (15K03619). We sincerely express our gratitude to the Society for its support.

References

1. Potters, M., Bouchaud, J.-P., Laloux, L.: Financial Applications of Random Matrix Theory: Old Laces and New Pieces, arXiv preprint physics/0507111 (2005)
2. Plerou, V., Gopikrishnan, P., Rosenow, B., Amaral, L.A.N., Stanley, H.E.: A random matrix theory approach to financial cross-correlations. Physica A: Statistical Mechanics and its Applications, vol. 287, pp. 374–382, 12 Jan 2000
3. Bouchaud, J.-P., Potters, M.: Financial Applications of Random Matrix Theory. Oxford University Press (2011)
4. Shirota, Y., Hashimoto, T., Sakura, T.: Trigger of economic reconstruction from the East-Japan earthquake disaster. In: Humanitarian Technology Conference (R10-HTC), 2013 IEEE Region 10, pp. 67–72 (2013)
5. Bishop, C.M.: Pattern Recognition and Machine Learning. Springer (2006)
6. Efron, B.: Large-Scale Inference. Cambridge University Press (2010)
7. Shirota, Y., Chakraborty, B.: Visual Explanation of eigenvalues and math process in latent semantic analysis. Inf. Eng. Express **2**, 87–96 (2016)
8. Shirota, Y., Chakraborty, B.: Visual explanation of mathematics in Latent semantic analysis. In: Proceedings of IIAI International Congress on Advanced Applied Informatics 2015, 12–16 July, 2015, Okayama, Japan, pp. 423–428 (2015)
9. Anderson, G.W., Guionnet, A., Zeitouni, O.: An Introduction to Random Matrices (Cambridge Studies in Advanced Mathematics). Cambridge University Press (2009)
10. Plerou, V., Gopikrishnan, P., Rosenow, B., Amaral, L.A.N., Guhr, T., Stanley, H.E.: Random matrix approach to cross correlations in financial data. Phys. Rev. E **65**, 066126 (2002)
11. Lubis, M.F., Shirota, Y., Sari, R.F.: Thailand's 2011 flooding: its impacts on Japan companies in stock price data. Gakushuin Econ. Pap. **52**, 101–121 (2015)
12. Tohoku Economic Federation, Economic Climate in Tohoku. TOKEIREN (Tohoke Economic Federation), pp. 18–20, June 2012 (written in Japanese)
13. Okuse, Y., Kuboyama, T.: Practical Data Analysis for Economics and Management (written in Japanese). Kodansha (2012)

Structure-Based Virtual Screening for Novel Modulators of Human Orexin 2 Receptor with Cloud Systems and Supercomputers

Rafael Dolezal, Eugenie Nepovimova, Michaela Melikova
and Kamil Kuca

Abstract Narcolepsy is a chronic neurologic disease characterized by excessive and inappropriate daytime sleepiness. Its pathophysiology is closely associated with the neuropeptide orexin, the absence of which is now believed to be responsible for most of the symptoms of the disease. Currently, no therapeutics against narcolepsy are available and afflicted patients are treated only symptomatically. Therefore, development of small-molecules able to penetrate into the brain and activate the orexin receptors represents a hopeful way. In this work we describe a computational approach which applies structure-based virtual screening of 1 million ligands to find novel potential modulators of orexin 2 receptor (OX2R). So-called rational computer-aided search for OX2R modulators was performed on a software-as-a-service (SaaS) cloud platform using iDock molecular docking program as a virtual screening engine. The results of the cloud-based calculations with iDock are analyzed and compared with the results of high-throughput flexible molecular docking in AutoDock Vina employing a pleasingly parallelized computation scheme on a peta-flops-scale supercomputer.

Keywords Virtual screening · Orexin · Narcolepsy · Clouds · Supercomputers

R. Dolezal (✉) · E. Nepovimova · M. Melikova · K. Kuca
Faculty of Informatics and Management, Center for Basic and Applied Research,
University of Hradec Kralove, Rokitanskeho 62, 50003 Hradec Králové,
Czech Republic
e-mail: rafael.dolezal@uhk.cz

E. Nepovimova
e-mail: eugenie.nepovimova@uhk.cz

M. Melikova
e-mail: michaela.melikova@uhk.cz

K. Kuca
e-mail: kamil.kuca@uhk.cz

R. Dolezal · E. Nepovimova · M. Melikova · K. Kuca
Biomedical Research Center, University Hospital Hradec Kralove, Sokolska 581,
50005 Hradec Králové, Czech Republic

© Springer International Publishing AG 2017 161
D. Król et al. (eds.), *Advanced Topics in Intelligent Information
and Database Systems*, Studies in Computational Intelligence 710,
DOI 10.1007/978-3-319-56660-3_15

1 Introduction

Computational approaches introduced into the drug design and discovery arena have often been praised for their capability to gently disclose not only how strongly a drug molecule can bind to a biological target, but also for elucidating how they bind together. Due understanding of the approaches' ability in research has been proved eloquently by several hundred novel drugs approved for clinical practice which have been discovered thanks to significant assistance of in silico simulations (e.g. HIV protease inhibitors nelfinavir and amprenavir) [1]. Methodologically, modern computer-aided drug design (CADD) methods have evolved from classical Hansch correlation analysis of the relationships between structure and biological activities (QSAR) by extending to more complex mathematical modeling and computational chemistry simulations [2]. Increasing the success ratio in hit discovery and subsequent lead structure optimization, actual advances in computer technologies have enabled carrying out sophisticated data mining of many thousands intercorrelated molecular descriptors (e.g. by partial least square regression, principal component analysis, pattern recognition methods, artificial neural networks, nearest neighbor, cluster analysis, SIMCA, etc.), to deeply investigate electronic structures and interactions of large molecular systems such as proteins and nucleic acids or to perform structure-based virtual screening (SBVS) of large virtual ligand libraries for a selected biological target using supercomputers [3].

Philosophy of the current rational drug discovery is to accelerate the whole process and to reduce its costs by proper utilization of various in silico methods. However, accurate computational methods are also very complex and time consuming, and the necessary and underlying computer power has to be paid for. In comparison with more or less simple experimental methods like combinatorial synthesis, bioisosteres synthesis, scaffold variations, peripheral lead modifications, multi-target directed lead combinations, Topliss's tree procedures for lead optimization, or in vitro high-throughput screening, even advanced in silico methods can hardly promise a quicker and neater way to novel drugs [4]. That is the reason why CADD methods are often simplified to a critical theoretical level to achieve as easily as possible some satisfying calculation outputs. It is truly a matter of hot scientific discussions how to judge the benefits of, in majority over-simplified, molecular mechanics-based methods accompanying drug research.

In the present study, we will focus on the issue of searching for novel/unknown ligands of human orexin 2 receptor (OX2R) by SBVS. The selected receptor OX2R is substantially related with a rare disease called narcolepsy, which is characterized by excessive daytime sleepiness accompanied with one or more of three additional symptoms (cataplexy or sudden loss of muscle tone, vivid hallucinations, brief periods of total paralysis) linked to the occurrence of rapid eye movement (REM) sleep at inappropriate times [5]. On biochemical level, narcolepsy manifests itself by decreased production of orexin peptides by orexin neurons in the lateral hypothalamus, which is the cause of relaxation of orexinergic neurons distributed throughout the central nervous system, consequently leading to misbalance of sleep-wake cycle regulation, impairment of food intake and pleasure-seeking behavior. Hopefully, narcolepsy could be treated pharmacologically through activating OX2R by suitable synthetic agonists, but no such agonist is known at present. Only several antagonists (e.g. suvorexant) are used in practice as hypnotics.

Since an X-ray model of OX2R has already been determined, SBVS can now initiate the discovering process of small molecule agonists by evaluating firstly the in silico binding energy of a set of chemical structures towards the receptor [6]. To accomplish SBVS, we utilized an open-source software-as-a-service (SaaS) platform iStar with iDock as a molecular docking engine. In total, 1 million ligand molecules were docked in OX2R model on a small-size cloud system employing ultra-fast calculation capabilities of iDock program. The 1000 top-scoring compounds resulting from iDock were flexibly re-docked in OX2R model by AutoDock Vina program employing a pleasingly parallelized operation scheme deploying separate jobs over a peta-flops-scale supercomputer. The results of both approaches are compared, generalized and interpreted with respect to their usefulness for drug discovery of novel OX2R modulators.

2 Human Orexinergic Neuronal System

Human orexinergic neuronal system is composed of two homologous receptors—orexin 1 receptor (OX1R) and orexin 2 receptor (OX2R)—and two activating neuropeptides—orexin A (OXA) and orexin B (OXB). The orexin peptides are formed by hydrolytic cleavage of prepro-orexin, containing 131 amino acids, which is expressed in orexin neurons of lateral hypothalamus. From thence they are distributed to the central nervous system (CNS). The scaffold of OXA is built of 33 amino acids stabilized with two disulfide bridges (Cys6–Cys12, Cys7–Cys14), having a pyroglutamoyl function attached to the N-terminus and an amidic group at the C-terminus. Its structure is folded into three helical sections. OXB is a shorter chain containing 28 amino acids with an aminated C-terminus. Unlike OXA, OXB lacks the strengthening disulfide bridges and comprises only two helices. Nonetheless, both orexin peptides are relatively flexible structures with many alternating conformations, as was determined experimentally by nuclear magnetic resonance (NMR) [7].

On the other hand, orexin receptors are localized throughout the whole CNS, even though they can marginally be found also in pancreas, gastrointestinal system, kidney and adipose tissue. From the structural point of view, orexin receptors belong to G-protein-coupled receptor (GPCR) family, consisting of a seven-fold helical transmembrane domain interconnected with a C-terminal globular functional unit localized in the cytosol. OX1R is built of 425 amino acids, while OX2R is composed of 444 amino acids. Both orexin receptors contain a disulfide bridge binding the third transmembrane segment (TM3) with the second extracellular loop (ECL2) and share a high degree of structural similarity (i.e. 63.23% of pairwise

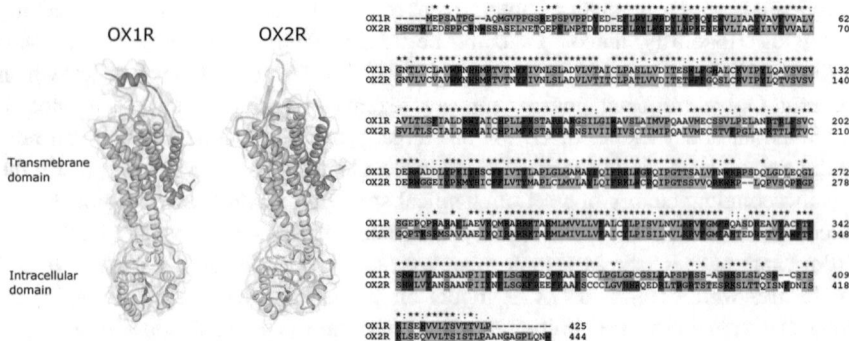

Fig. 1 X-ray models of OX1R (PDB ID: 4ZJ8) and OX2R (PDB ID: 4S0 V). Both receptors share 63.23% of pairwise sequence identity (marked by asterisks over the residues)

identity, 282 identical positions, 81 similar positions determined by alignment in Clustal Omega program) (Fig. 1).

OXA exhibits strong activation potency towards both OX1R and OX2R, while OXB activates preferentially OX2R but with 10 times high an efficiency as OX1R. OX2R pathways are predominantly associated with wakefulness and arousal regulation, whereas OX1R subsystem is involved in feeding control, coordination of reward, coping nociception and stress. Generally, activation of orexin receptors in the CNS brings about neuroexcitation by closing K^+ channels and activation of Na^+/Ca^{2+} exchange [8].

At present, an intensive research of orexin receptors has revealed many details about the orexinergic signaling cascades. By X-ray and NMR, 3D structures of OXA, OXB, OX1R, OX2R have been determined and profoundly analyzed. It was proved that modulation of orexin receptors can be useful in the treatment of sleep disorders, narcolepsy, cataplexy, obesity, hypophagia, attention deficit, depression, bipolar disorders, and, moreover, in colon cancer and Parkinson's disease. These studies are very important for development of new ligands capable to modulate the orexin receptors activity. Since discovery of orexin receptor antagonists has been successfully started by suvorexant, the main attention in this research area is moved especially to development of small molecule agonists able to cross blood-brain barrier which might be deployed to tackle narcolepsy.

3 Problem Definition

Theoretically, it is quite easy to propose a method for discovery of OX2R antagonists since the only issue herein is to block interactions between OXA/OXB and the receptor. Once the natural agonist is prevented from interaction with the receptor, the signaling process cannot develop and the orexinergic system stays relaxed. Of course, finding efficient antagonists of OX2R is not straightforward, but in comparison with searching for agonists it seems to be considerably easier. In case of OX2R agonists, one needs to mimic peptide-protein interactions within OXA/OXB-OX2R complex, which is generally understood as a challenging task. Particularly with GPCR, the activation process depends on a variety of binding interactions and subsequent conformational changes, which makes the agonists' development extremely difficult [9].

Fortunately, a common property of both agonists and antagonists is a significant affinity for the target receptor. Because design of novel OX2R agonists from scratch would require very complex and time consuming studies of the activated receptor, we reduced the objective of the present work to searching for promising OX2R modulators. This task was accomplished by SBVS via molecular docking. The principle of such SBVS can be concisely defined as computational evaluation of binding energies of a huge set of compounds towards OX2R model. After estimating the binding energies, the top-scoring candidates can be regarded as potential modulators of OX2R.

SBVS coupled with flexible molecular docking is nowadays a common computational chemistry approach which has been profoundly described in the literature [10]. However, some innovations and improvements of the methods still emerge, especially with regard to calculation speed, accuracy, user-friendliness and availability. In order to investigate the benefits of different computation technologies in SBVS, we utilized an open source SaaS platform iStar which employs iDock program as an engine for molecular docking, and free AutoDock Vina program running in "high performance computing" (HPC) mode on a peta-flops-scale supercomputer. At first, we compared the performance of both platform methods in SBVS of 1000 ligands representing a random selection of FDA-approved drugs. Finally, we launched a process of 1 million docking jobs, corresponding to 1 million drug-like ligands, on an iStar-iDock cloud to obtain insight into binding energy population in the chemical space close to OX2R. Because the calculations in AutoDock Vina were deliberately set to higher precision, we performed re-docking only for 1000 top-scoring ligands resulting from the first stage docking of 1 million ligands on the iStar-iDock cloud. In all studies, we used an X-ray model of OX2R (PDB ID: 4S0 V) as the target receptor. The following section provides a brief description of the undertaken calculations and achieved results.

4   SBVS Using iStar and iDock Cloud System

SBVS is substantially associated with a more or less demanding computational method known as molecular docking which aims at finding such geometrical position of a ligand and a receptor/enzyme molecule that exhibits the lowest potential energy, thus the strongest mutual attraction. The task of molecular docking is solved as a minimization problem using various potential energy gradient-driven methods (e.g. steepest descent, conjugate Polak-Ribiere algorithm, Fletcher-Reeves algorithm, eigenvector following, etc.). For expressing the potential energy of a molecular system, a mathematical definition of each energy contributors: (1) bond stretching (E_s); (2) bond bending (E_b); (3) dihedral torsion (E_t); (4) van der Waals interactions (E_{vdW}); (5) hydrogen bonding (E_{hb}); (6) electrostatic interactions (E_e) and a set of atom-and-bond-specific constants (i.e. force field) are necessary Eq. (1) [3].

$$E_s = \sum_{i=1}^{n} K_s \left(R_i^a - R_i^0\right)^2 ; E_b = \sum_{i=1}^{n} K_b \left(\theta_i^a - \theta_i^0\right)^2 ; E_t = \sum_{i=1}^{n} \frac{V_n}{2} \left(1 + \cos\left(n\varphi_i^a - \varphi_i^0\right)\right);$$

$$E_{vdW} = \sum_{i,j \in vdW; i \neq j}^{n} \left[\frac{A_{ij}}{R_{ij}^{12}} - \frac{B_{ij}}{R_{ij}^{6}}\right] ; E_{hb} = \sum_{i,j \in hb; i \neq j}^{n} \left[\frac{C_{ij}}{R_{ij}^{12}} - \frac{D_{ij}}{R_{ij}^{10}}\right] ; E_e = \sum_{i,j \in e; i \neq j}^{n} \left[\frac{q_i q_j}{eR_{ij}}\right].$$

$$(1)$$

SBVS is simply a high-throughput extension of molecular docking which performs quickly, often through parallelized and distributed calculations, an evaluation of the binding energy of many ligands to a selected receptor.

For the present study, we utilized an open source SaaS platform iStar employing iDock as a molecular docking engine. This web platform is constructed of several elements which together bring highly effective SBVS capabilities to the user (Fig. 2.). Briefly, the web client is implemented with Twitter Bootstrap, jQuery, jQuery UI, three.js, zlib.js, jquery-dateFormat and jquery_lazyload functions. It is compatible with Google Chrome 30, Mozilla Firefox 25, MSI Explorer 11, Apple Safari 6.1 and Opera 17. The web server uses node.js, mongodb, express and spdy modules. The files used for SBVS are managed on MongoDB platform [11].

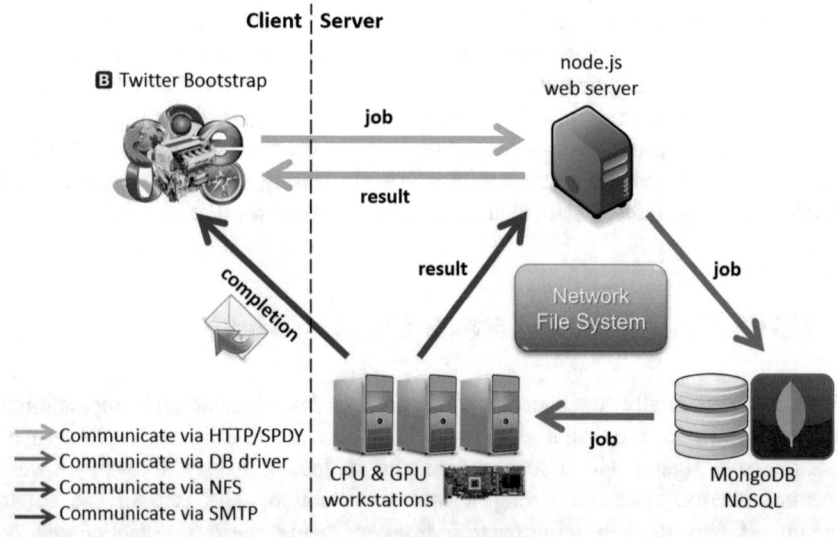

Fig. 2 Architecture of the open source platform iStar. It is a software-as-a-service system designed for bioinformatics and chemometrics [11]

Functionally, iStar enables scheduling, deployment and monitoring of SBVS jobs, messaging and storing the resulting data. Beside iGrep, iCUDA, iView applications, it is associated with iDock program developed for parallelized molecular docking by Hongjian Li. The ligands for virtual screening, originally obtained from ZINC database, are converted to PDBQT format and stored at the web server database (http://istar.cse.cuhk.edu.hk/idock/). The ligands used for SBVS can be selected from a huge pool of 17 million compounds by setting limits on molecular weight, calculated logP, apolar desolvation, polar desolvation, number of hydrogen bond donors and acceptors, topological surface area tPSA, net charge, and number of rotatable bonds. At present, the iStar & iDock platform provides the ligands which are stored at the ZINC database as All Clean subset. Nonetheless, to accomplish SBVS, this platform needs an externally imported PDB file with the receptor 3D model. Mostly, the free database of proteins rcsb.org can be utilized for these purposes.

Beside the iStar platform which secures user-friendliness and computational power management, the most crucial element for SBVS is the iDock program. iDock was developed from open source code of AutoDock Vina, the benchmark program in flexible molecular docking, and, thus, it borrows many substantial features from its predecessor. Although iDock tries to improve AutoDock Vina, it implements only calculations of flexible ligands and rigid receptors. By default, iDock utilizes grid maps of granularity 0.15625 Å to screen the global ligand-receptor interactions by distributing independent Monte Carlo tasks to separate threads [12].

The scoring function of iDock is exactly the same as that used in Vina, giving the binding energy estimate of the ligand-receptor system by summation of five terms (i.e. Gaussian energy 1, Gaussian energy 2, repulsion, hydrophobic interaction, and hydrogen bonding). The optimization algorithm is also the same in iDock and Vina. It is divided into two parts: global and local. The global search utilizes Monte Carlo principle with random mutation of the current solution, while the local search is based on Broyden-Fletcher-Goldfarb-Shanno (BFGS) algorithm, approximating the inverse Hessian of the scoring function. Compared with Vina, iDock increases the default number of parallel Monte Carlo runs from 8 to 64 and stops the BFGS local optimization only if no other allowed step can be performed. Due to these changes, iDock can assure finding the energy minimum with somewhat higher probability comparing to Vina. The improvement of iDock also consists in revision of CPU and memory utilization of Vina. iDock evaluates the capacity of every vector structure and employs R-value reference of the C++ 11 standards.

The iStar web platform along with iDock engine allows users to easily start extensive SBVS without the necessity to solve computational issues. The system automatically proposes the grid box center and size in the active site of the receptor and offers simple filters for designing a custom virtual ligand library. Therefore, one only needs to upload the biological target model and to set parameters for selecting the screened ligands. Once the SBVS is completed, the user is notified by e-mail about the results, which can be downloaded from the web server. The results involve a list of docked compounds, their binding energy estimates and binding modes in PDBQT format.

5   SBVS with AutoDock Vina and Salomon Supercomputer

The underlying principles implemented in AutoDock Vina program have already been mentioned in the previous chapter. Nowadays, Vina still represents a classical tool in computational chemistry and biology for evaluating interactions between ligands and enzymes or receptors. Fortunately, Vina is natively implemented as a multithreading application since it deals with ideally computationally separable tasks. Up to this date, many Vina variations have been published providing some better features than the original C++ code. However, Vina remains a principal robust standard which easily helps medicinal chemists in understanding molecular level of drug action. With some effort, Vina can be extended to SBVS regimen by incorporation into distributed calculation schemes using computer grid systems or supercomputers equipped with a scheduler for managing the submitted jobs.

For the present study, we developed a code implementing job arrays which sends individual flexible molecular docking tasks to PBS scheduler and collects the results from computer nodes. The ligands are stored in LIGANDS directory, grid-box parameters, CPUs, exhaustiveness, rigid and flexible receptor parts are defined in conf.txt:

```
#!/bin/bash
i=1;
for a in ~/LIGANDS
do
    echo "./vina -config ./conf.txt --ligand $a -out
         ./results_$i.pdbqt" >> ./job_$i.sh
    i++;
done
echo "content=(\$(ls -1 ./job_*.sh))" > array.sh
echo "\${content[\$PBS_ARRAY_INDEX - 1]} " >> array.sh
qsub -q mygrid -J 1-1000 -l
select=1:ncpus=24,walltime=00:24:00 ./array.sh
```

For this SBVS, we utilized Salomon (Czech Republic) supercomputer consisting of 1008 compute nodes with 24,192 CPUs and 129 TB RAM in total. The peak performance of the system is over 2 peta-flops. Each node has two Intel Xeon E5-2680v3 CPUs with 24 cores. The system, running CentOS Linux, is interconnected by 7D enhanced hypercube InfiniBand network and affords interesting power for medium-sized virtual screenings.

Although supercomputers offer great computational power, they are considered too complicated for laymen in information technologies. Comparing with clouds systems, supercomputers are generally more suitable for high performance computing (HPC), but traditional medicinal chemists will prefer cloud solutions which can carry out many elementary steps automatically without disturbing the user.

6 Results and Discussion

Utilizing the iStar & iDock platform operating on four powerful computers, each with four Intel Xeon E7-4830 v2 processors and 512 GB DDR3 RAM (i.e. 320 cores in total), we performed SBVS of 1000448 ligands in OX2R (PDB ID: 4S0 V). The calculations implemented only multithreading. The center of the gridbox of size of $17 \times 14 \times 16$ Å was automatically placed to x = 52 Å, y = 8 Å, z = 53 Å. The calculations were completed after 125 h. One docking job is done approximately in 30 s. Some elementary statistics of the estimated binding energies within SBVS are given in Table 1.

The top-scoring candidate for OX2R exhibited the binding energy of -13.08 kcal/mol. Its binding mode in the central tunnel of the transmembrane domain of OX2R is displayed in the left part of Fig. 3.

The top scoring compound outlined in Fig. 3. represents a ligand with a relatively strong affinity for OX2R receptor. It occupies the place in the opening of the transmembrane helical domain, which can hinder the activation by OXA/OXB. However, as has been already mentioned, we cannot deduce from this binding mode that the compound might be a potential OX2R agonist. Further investigation is necessary to elucidate subsequent steric interactions in the ligand-receptor complex.

Table 1 Basic statistics of SBVS on iStar & iDock platform for OX2R (PDB ID: 4S0 V)

N	Mean (kcal/mol)	Min (kcal/mol)	Max (kcal/mol)	STD (kcal/mol)
1000448	−8.3671	−13.080	10.200	1.0083

 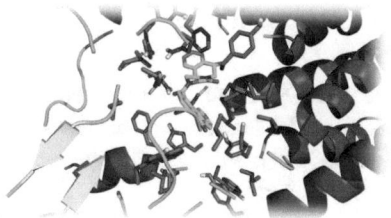

Fig. 3 The top-scoring candidates in OX2R (PDB ID: 4S0 V) provided by iDock (*left*) and by AutoDock Vina (*right*)

Table 2 Basic statistics of SBVS by AutoDock Vina and iStar & iDock for OX2R (PDB ID: 4S0 V). 1000 top-scoring candidates from complete SBVS by iStar & iDock

Program	N	Mean (kcal/mol)	Min (kcal/mol)	Max (kcal/mol)	STD (kcal/mol)
AutoDock Vina	1000	−12.4172	−15.4000	−9.9000	0.7943
iStar & iDock	1000	−11.7469	−13.080	−11.4520	0.2810

1000 top-scoring candidates resulting from SBVS in iStar & iDock calculations were also submitted to flexible molecular docking in the same OX2R model using AutoDock Vina program and Salomon supercomputer. In the configuration file, the gridbox of 30 × 30 × 30 Å was centered at the same point as in the case of iStar & iDock calculations. Unlike iStar & iDock calculations, 38 amino acid residues encompassed by the gridbox were set as flexible structures for docking in Vina program. Further, the calculations in Vina were set to utilize 24 CPUs in multi-threading mode with the exhaustiveness parameter equal to 24. Completing one docking task in Vina took 4.5 h on average, although all 1000 jobs were finished on Salomon after 6 h. The 1.5 h delay was caused by the scheduler which starts different tasks without any preference for jobs of a single supercomputer user. The results are summarized in Table 2.

Interestingly, the resulting binding energy estimates from AutoDock Vina and iStar & iDock correlate rather weakly, although significantly (**Pearson's R = 0.3677, p = 2.2437e-33; Spearman's R = 0.3558, p = 3.2961e-31**). The best candidate scored with −13.08 kcal/mol by iDock (Fig. 3.) provided in Vina a binding energy of −10.7 kcal/mol. Conversely, the top-scoring candidate marked with binding energy of −15.4 kcal/mol by AutoDock Vina was characterized only with energy of −12.393 kcal/mol by iDock. From this it is evident that iDock is much faster than Vina, but because of omitting the receptor flexibility it probably does not assign best scoring to the docked ligands. This discrepancy can be properly arbitrated only by calculations on a higher level of theory or experimentally. However, iStar & iDock cloud platform remains attractive even while the accuracy of both iDock and AutoDock Vina has to be properly investigated.

7   Conclusions

We have reported on very demanding simulations of non-covalent interactions between 1 M chemical compounds and OX2R receptor employing molecular docking and a SaaS cloud system. 1 k top-scoring ligands resulting from this phase of SBVS were re-docked in OX2R applying flexible molecular docking to obtain more accurate estimates of Gibbs free energies. For these purposes, we developed a SBVS protocol to distribute the computational jobs in a supercomputer. We have proved in the present article that cloud systems may be equipped with fast docking algorithms but the achieved results might be contradictory with the outputs of practically identic calculations. Since iDock and AutoDock Vina differ especially in handling flexible residues, we may suppose that low correlation between the binding energies is caused by improper optimization in iDock program. Nonetheless, the top-scoring candidate revealed in AutoDock Vina (−15.4 kcal/mol) seems to be worthy of further research because scoring lower than −12.5 kcal/mol is empirically taken as a significant in silico level.

Acknowledgements The support of the Specific research project at FIM UHK is gratefully acknowledged. This work was also supported by long-term development plan of UHHK, by the IT4Innovations Centre of Excellence project (CZ.1.05/1.1.00/02.0070), and Czech Ministry of Education, Youth and Sports project (LM2011033).

References

1. Kubinyi, H.: Success stories of computer-aided design. Computer Applications in Pharmaceutical Research and Development, pp. 377–424. Wiley (2006)
2. Waisser, K., Dolezal, R., Palat, K., Cizmarik, J., Kaustova, J.: QSAR study of antimycobacterial activity of quaternary ammonium salts of piperidinylethyl esters of rafael.dolezal phenylcarbamic acids. Folia Microbiol. **51**, 21–24 (2006)
3. Dolezal, R., Ramalho, T., França, T.C., Kuca, K.: Parallel flexible molecular docking in computational chemistry on high performance computing clusters. In: Núñez, M., Nguyen, N. T., Camacho, D., Trawiński, B. (eds.) Computational Collective Intelligence, vol. 9330, pp. 418–427. Springer International Publishing (2015)
4. Topliss, J.G.: A manual method for applying the Hansch approach to drug design. J. Med. Chem. **20**, 463–469 (1977)
5. Akintomide, G.S., Rickards, H.: Narcolepsy: a review. Neuropsychiatr. Dis. Treat. **7**, 507–518 (2011)
6. Yin, J., Mobarec, J.C., Kolb, P., Rosenbaum, D.M.: Crystal structure of the human OX2 orexin receptor bound to the insomnia drug suvorexant. Nature **519**, 247–250 (2015)
7. Takai, T., Takaya, T., Nakano, M., Akutsu, H., Nakagawa, A., Aimoto, S., Nagai, K., Ikegami, T.: Orexin-A is composed of a highly conserved C-terminal and a specific, hydrophilic N-terminal region, revealing the structural basis of specific recognition by the orexin-1 receptor. J. Pept. Sci. **12**, 443–454 (2006)
8. Kukkonen, J.P., Leonard, C.S.: Orexin/hypocretin receptor signalling cascades. Br. J. Pharmacol. **171**, 314–331 (2014)
9. Heifetz, A., Bodkin, M.J., Biggin, P.C.: Discovery of the first selective, Nonpeptidic Orexin 2 receptor agonists. J. Med. Chem. **58**, 7928–7930 (2015)
10. Sliwoski, G., Kothiwale, S., Meiler, J., Lowe Jr., E.W.: Computational methods in drug discovery. Pharmacol. Rev. **66**, 334–395 (2014)
11. Li, H., Leung, K.S., Ballester, P.J., Wong, M.H.: istar: a web platform for large-scale protein-ligand docking. PLoS One B, e85678 (2014)
12. Li, H., Leung, K.S., Wong, M.H.: idock: A Multithreaded virtual screening tool for Flexible Ligand Docking, In: (N.E.) Computational Intelligence in Bioinformatics and Computational Biology (CIBCB), pp 77–84. IEEE (2012). doi:10.1109/CIBCB.2012.6217214

Supply Chains of Cross-Border e-Commerce

Arkadiusz Kawa

Abstract A feature of e-commerce is worldwide coverage. Almost any person or company can be a customer of an online shop. However, this common availability is in practice quite apparent. Despite the dynamic development of e-commerce, communication in other languages, the form of payment, currency, legal and tax conditions, as well as the delivery of products remain barriers to the free cross-border flow. The article focuses on the last factor mentioned above. The lack of delivery of goods to a distant place or a relatively long time and high cost of providing the purchased product hinders further development of e-commerce. This problem can be solved by introducing an intermediary that consolidates shipments from many retailers and delivers them to many clients scattered in different corners of the world. The main contribution of this article is to develop a model facilitating cooperation between online shops dealing with cross-border trade. The purpose of the idea is to reduce costs and accelerate the delivery of goods ordered abroad via the Internet.

Keywords e-Commerce · Cross-border · Supply chain · CEP (courier express and postal) industry

1 Introduction

The rapid development of the Internet, and thus also e-commerce, has created new distribution channels for many trading, service and manufacturing companies. According to the European Commission, e-commerce is one of the main factors leading to better prosperity and competitiveness of Europe. It has significant potential that may contribute to economic growth and employment [2]. It is expected that its further development will have far-reaching effects, perhaps even

A. Kawa (✉)
Poznan University of Economics and Business, al. Niepodległości 10,
61-875 Poznan, Poland
e-mail: arkadiusz.kawa@ue.poznan.pl

© Springer International Publishing AG 2017
D. Król et al. (eds.), *Advanced Topics in Intelligent Information
and Database Systems*, Studies in Computational Intelligence 710,
DOI 10.1007/978-3-319-56660-3_16

exceeding the changes that concerned trade over the past several decades. Physical presence while shopping is becoming less and less important. Customers buy products, placing orders electronically, and the purchased goods are delivered to their workplaces, homes, click and collect points and parcel lockers. Placing ordering in this way replaces the trip to a store, and the delivery of the consignment eliminates the way back with the purchased goods.

In contrast with traditional trade, online shopping is inseparably associated with the delivery to the final customer (so-called last mile), i.e. the most complicated and costly process in the whole supply chain. Internet retail businesses carry out a very large number of small orders. Unfortunately, there are delays in deliveries about which buyers are not informed at all. Customers often do not have too much influence on the choice of the company that will deliver the goods, either. The delivery of the goods is most frequently performed by CEP (courier, express and postal) companies.

Additionally, more and more attention has recently been paid to expanding business activities beyond the borders of a single country. Sellers look for new buyers abroad, while customers want to have a greater choice of suppliers. A trend in e-commerce arises, then, which is defined as cross-border trade. It is particularly evident in the countries of the European Union. It is based on selling products to customers who are located in another country. However, it is related to several problems, such as a high cost and long delivery time, language barriers, different legal regulations and taxes, etc. [8]. The high cost and long delivery time are, in turn, associated with the aforementioned problem of the last mile, but also with the problem of the relatively small flow of goods between countries which is realized by a single CEP operator. So the economies of scale do not take place yet.

Therefore, there is a real need to offer e-commerce to retailers and to, indirectly, recommend comprehensive services to their customers, which would include, on the one hand, logistics services in Europe, and, on the other hand, full information on the quality of the service.

The aim of this article is to develop a model of an intermediary facilitating cooperation between online shops dealing with cross-border trade. This model is expected to contribute to cost reduction and acceleration of the delivery of goods ordered abroad via the Internet.

The structure of the article is as follows. Section 2 describes the electronic cross-border trade in Europe. Section 3 presents logistical problems in e-commerce. Section 4 proposes the above-mentioned model. Section 5 summarizes the article and points to future directions of the research.

2 Cross-Border e-Commerce in Europe

Currently, e-commerce can be divided into several trends in the field of logistics, which will determine further development of the CEP industry. These are: reverse logistics, same-day delivery, development of new models of cooperation in logistics (dropshipping, fulfillment, one-stop e-commerce), broker services and cross-border transport [7]. This article focuses on the latter trend.

Cross-border e-commerce still has a relatively small share in the whole market of e-commerce. In 2014, approx. 15% of the EU inhabitants made a purchase from sellers from a different country. This represents an increase in the share of this type of trade by 25% compared to the previous year. Not everywhere, however, is cross-border e-commerce equally developed. For example, in 2014 only 4% of Poles made a purchase on the Internet from a seller located in another country, which placed Poland on the penultimate place in the European Union. Most foreign shopping is done by Luxembourgers (65%) and Austrians (40%), and the least by Romanians (1%). The EU average is 15% [1].

The total value of the commodity circulation in e-commerce within individual countries and among the EU Member States is estimated at about €241 bn. Of this amount, €197 bn (80%) are traded on domestic markets. Only about €44 bn (18%) cross the borders between the EU Member States, and another €6 bn (2%) come from import from countries outside the EU [9].

It can be seen from these data that the potential of electronic cross-border trade within the EU still remains unexploited. Only 8% of companies are involved in cross-border selling. Managers of these enterprises argue that it is too complicated and too expensive. As part of the efforts to unleash the potential of e-commerce, the European Commission has adopted a package of proposals to stop the unjustified geo-blocking, increase the transparency of package delivery prices, and improve the enforcement of consumer rights [4].

3 Logistics Problems of Cross-Border e-Commerce

The logistics of products offered by online stores is one of the basic factors influencing the consumer's decision about making purchases in them. Deliveries and product returns are one of the most important issues for both online shoppers and online stores in the EU. The European Commission indicates that the problem lies in particular in cross-border deliveries of packages realized for the needs of small and medium enterprises and those sent to the less developed and less accessible regions. Therefore, it puts a lot of effort into increasing the availability of e-commerce for all EU citizens and businesses, regardless of their size and location [2].

Another problem is the relatively little access to information about the CEP market, in particular about the available services, operators and prices. Many customers know only certain operators whose services they could use. In the case of cross-border transport, they can choose between an international courier service or a common service provider, so the postal operator. This makes it difficult for new entrants to gain market share and reduces the competitive pressure on the existing operators, which in turn limits the incentives to improve the service quality and leads to higher prices [6].

Currently, online stores selling their products abroad incur a very high cost of shipping—depending on the country it is up to 5 times higher than the cost of a consignment realized within the country. The lower price of the product sold does not often compensate for the cost of delivery, which discourages buyers from abroad. It is one of the greatest barriers to the development of cross-border trade conducted via the Internet [5]. Consumers and small enterprises claim that the problems with the delivery, in particular the high prices, prevent them from increasing the sales or purchases in other Member States. Foreign exchange in e-commerce could be completely different if these costs were significantly reduced.

Apart from the cost of delivery, another barrier to the development of cross-border e-commerce is the delivery time. It results mainly from the distance between the vendor and the customer. In most cases (mainly outside the border regions) it will be much greater than in the case of domestic shipments. In international trade, shipments often have to undergo additional operations, go through a greater number of hubs and branches, which further prolongs the time of delivery [8].

Operating activities of CEP companies are based on the hub and spoke concept. It is a system used for the distribution of small size or weight loads. In contrast to direct deliveries, hubs are used that connect the individual places where shipments are posted and received. The hub and spoke (H&S) concept minimizes storage costs and reduces the individual costs of transportation. Although a single consignment is transported over a long distance, the total distance for all shipments counted separately is shorter than in the case of direct deliveries. This solution works very well for a large number of items that are posted and received in multiple locations. An example is distribution within a country where most large cities are connected with one another by means of one or more hubs. Figure 1 illustrates the delivery distribution system within a country X using the H&S system. In this case, customer A places an order for selected products at store S. In the next step, S performs pick and pack operations, and orders a courier service from company C. The courier collects the shipment and delivers it to the local cargo terminal C1X. Then, the consignment, together with items from nearby cities, is transported to hub C. Shipments from all branches across the country are delivered to hub C. They are then sorted and transported by linehaul (usually at night) to local branches. In this case, the merchandise goes to local cargo terminal C2X. In the morning the shipment is picked up by a courier from the local branch and delivered to customer A.

As is shown in Fig. 1, the distance travelled by a shipment is much longer than in a direct connection from point S to point A. This extends the delivery time, but significantly reduces the unit cost thanks to the consolidation with other consignments. Customers must wait for the ordered goods until the next working day, but, in return, the cost of delivery is a dozen to several hundred times lower than in the case of direct delivery.

A problem with the H&S system occurs in the case of routes along which few consignments are transported. Underutilization of the vehicle cargo space causes

Fig. 1 Hub and spoke
system in distribution within a
single country

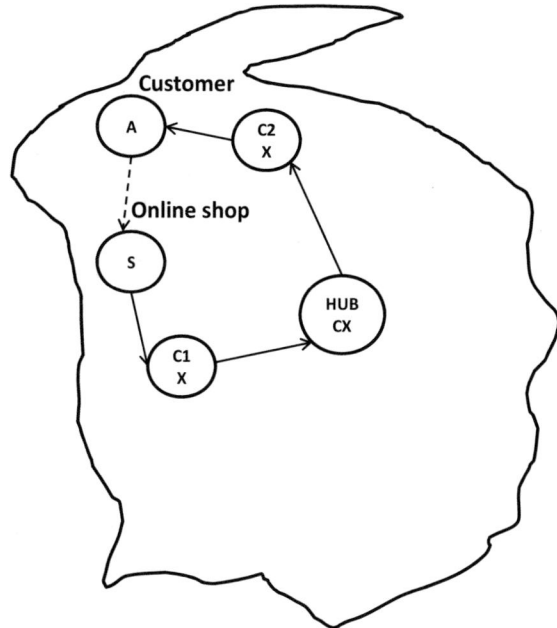

the unit cost of transportation to increase significantly. Moreover, in the case of small packages (which prevail in e-commerce) the total cost of delivery rises considerably when the consignment passes through many local terminals and hubs. It is associated with additional costs of sorting and handling. Such a complex and costly system occurs in the case of cross-border transportation.

Figure 2 presents the route of delivery of the goods ordered by customer A in store S. In relation to Fig. 1, here hub CY has been added. Although points A and S are close to each other, the product passes through the individual points in the H&S system, which increases the total cost of the delivery. Due to the fact that there is a very little flow of goods between the CX hub and the CY hub, the cargo space in the means of transport is not fully utilized. In addition, the freight rates in international transport are higher than in domestic transport. Furthermore, relatively little competition (there are only a few enterprises) in express cross-border deliveries causes the CEP operators to use their bargaining power. It all makes the cost of cross-border delivery several times higher than that of distribution within a single country. This discourages customers from ordering goods from foreign online stores, which deepens the problem of under-used cargo space. Therefore, a solution is needed to overcome this problem, reduce the number of the sorting and handling operations, and thus reduce the costs of cross-border deliveries.

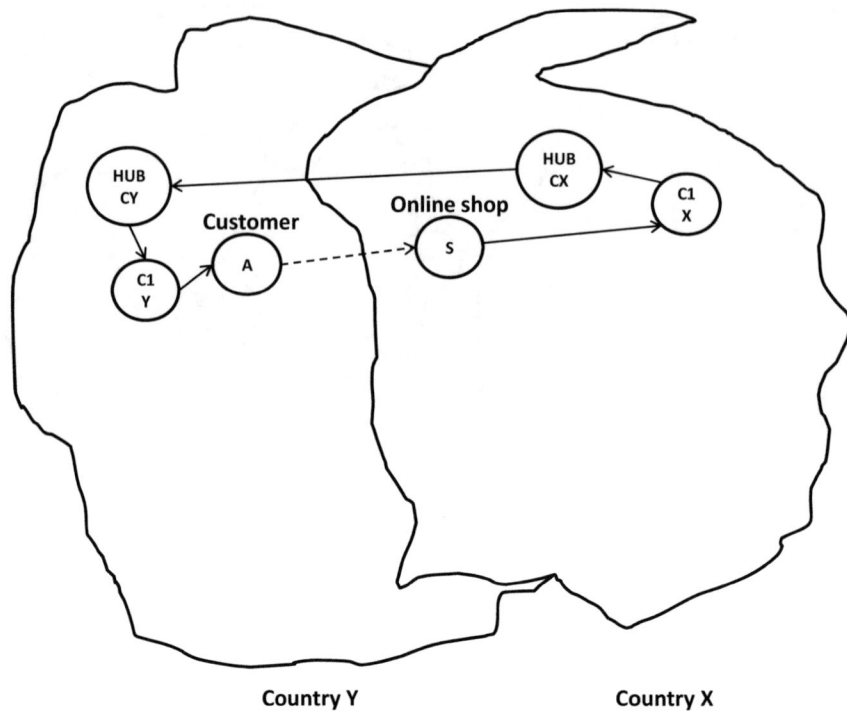

Fig. 2 Hub and spoke system in distribution between two countries

4 Exemplification of the Model Facilitating Cooperation Between Online Shops Dealing with Cross-Border Trade

Figure 3 shows a simplification of the cross border e-commerce market. There are two online stores located in country X (S1 and S2) and two customers in country Y (A1 and A2). A1 orders a product in shops S1 and S2, and A2 orders in S2. The stores are separately served by two independent CEP operators (C1 and C2). C1 delivers the shipments to A1 through its H&S system, while C2—to customer A2. C2 benefits from the economies of scale [10] and delivers the goods together to A1 and A2 from point S2 to hub C2Y. Then the shipments are separated and delivered to points C2Y1 and C2Y2.

In the case of a small flow of shipments between hubs C1X and C1Y, C2X and C2Y such a system of distribution of goods in cross-border e-commerce is ineffective. Relatively high costs of delivery of products to customers appear due to the underutilization of the cargo space and a large number of the sorting and handling operations.

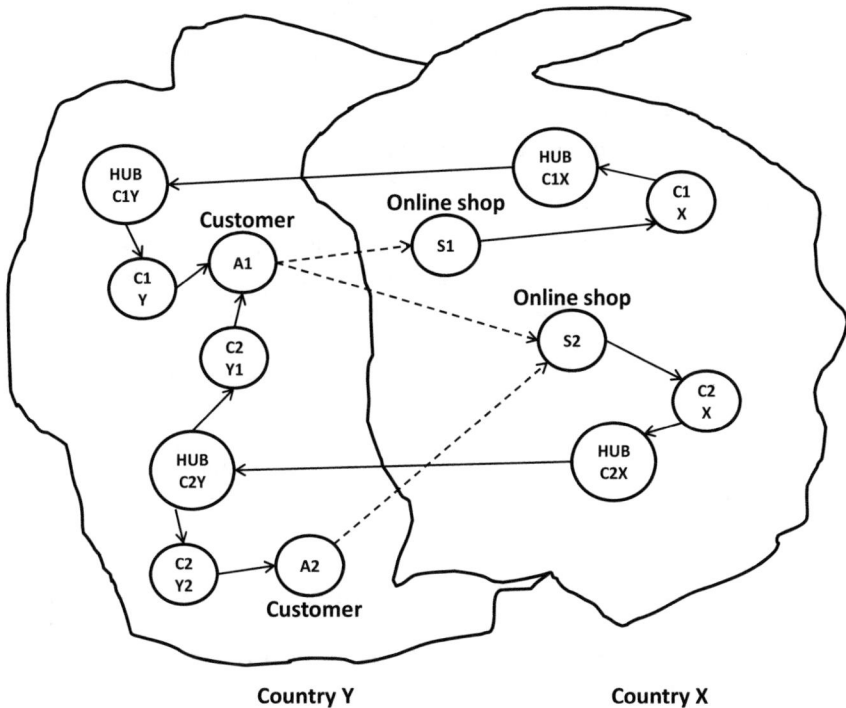

Fig. 3 Hub and spoke system in distribution between two countries with two customers and two online shops

This problem can be solved by introduction of an additional entity to the cross-border e-commerce in the form of a consolidator. In the literature, such an entity is defined as the fourth party logistics (4PL). It manages the flow of information between the supplier, customer and logistics service provider. The consolidator proposed in this study acts like the CEP brokers, already present on a number of national markets for several years. The difference between them is that the broker only wins transport orders and passes them on to the CEP operator which decides how to transport the consignments; the consolidator, in turn, additionally selects the carriers for the service. The consolidator does not possess any means of transport. It can be said that it configures a temporary supply chain for the needs of a single transaction.

The consolidator has a website which enables to find offers, compare them, monitor shipments and make payments. However, the consolidator automates their business with continuously cooperating customers by providing the API (application programming interface) and integrating with sales platforms. Such platforms group and systematize up-to-date information about CEP services and prices, which helps make the decision about the company that delivers shipments. On the basis of specific criteria such as the place of origin and delivery, dimensions and weight of the parcel, the user is given appropriate cross-border transport offers by the system. The consolidator's system automatically recommends the shipping options that are adjusted to the ordered products to the customer of the online store. For example, for a larger package courier or mail services are suggested rather than delivery to a parcel locker. Depending on the planned date of delivery, the system may offer different prices. Express deliveries by air freight will be more expensive than the economical road transport.

Moreover, the system automatically generates the shipping documents (picking list to the warehouse, labels to be stuck on packages), monitors the realization process and informs the e-seller and the customer about the current status of the delivery.

For customers, besides time, certainty of delivery of the product is important. Ordering in foreign stores, customers express concerns not only about when, but also whether at all and in what condition they will receive the shipment. They must therefore have constant access to the information about where the consignment is located and what the expected date of delivery is. This will be possible thanks to the track and trace system.

The consolidator does not need to invest in infrastructure, because it uses the resources of other organizations. Its key task is the right choice of carriers assigned to the individual routes and time synchronization of the operation of individual vehicles in the region and between regions and of the work in the terminals and hubs. The consolidator, collecting orders from a number of senders, becomes a "big" customer of courier and postal companies. This increases the bargaining power and allows to get much better cooperation conditions than individual customers are offered, sending small numbers of shipments.

Managing the consolidator's activities organized in this way requires application of complex IT systems. Such a system should integrate all the terminals and hubs of many different logistics service providers. This requires interoperability between the systems, and so mutual access to necessary data. In addition, standardization of the processes and the used infrastructure is needed. For example, shipments are transported in certain loading units, and the barcode labels describing the shipment (details of the sender and recipient, terms of delivery, etc.) must be processed by the various entities dealing with the shipments.

All the data concerning the shipments and carriers are placed in a data cloud by the consolidator. This ensures access to the system for all stakeholders anywhere in the world. Moreover, each driver is equipped with an electronic device which is used to scan the code from the shipment, receive information about the shipment and send the data.

Customers of consolidators may mainly be micro, small and, partially, medium-sized companies that run their business on the Internet, i.e. online shops and sellers at online auctions.

Figure 4 shows the pattern of a consolidator's operation in cross-border e-commerce. In every country it has access to the hubs (IX and IY) which are connected to the local terminals. In practice, this may be more than one hub, and they may belong to more than one CEP operator. Hubs between individual countries are connected by linehauls. The process in Fig. 4 differs from the one shown in Fig. 3 in that stores S1 and S2 are operated by one consolidator I which selects appropriate carriers to collect shipments from shops S1 and S2 and deliver them to hub IX through the terminals of these carriers—I1X and I2X. Then, the consolidated shipments are transported from the IX hub to the IY hub. Loads from shops S1 and S2 are transported to customers A1 and A2 by a single means of transport. Organization of the transport between the hubs is done by the consolidator, but it can also be done by the CEP operator itself if it is a better solution. In the next step, unloading, sorting and shipment of goods from the IY hub by local courier companies (IY1 and IY2) to customers A1 and A2 takes place. As a result of this process, hubs C1X and C1Y and the carrier terminal C1Y have been eliminated (compare Figs. 3 and 4). This makes it possible to achieve the benefits in the form of fewer handling and sorting operations. Thanks to the selection of offers competitive to those of the CEP operators by the consolidator the costs of transportation between the terminals and hubs can, in turn, be reduced.

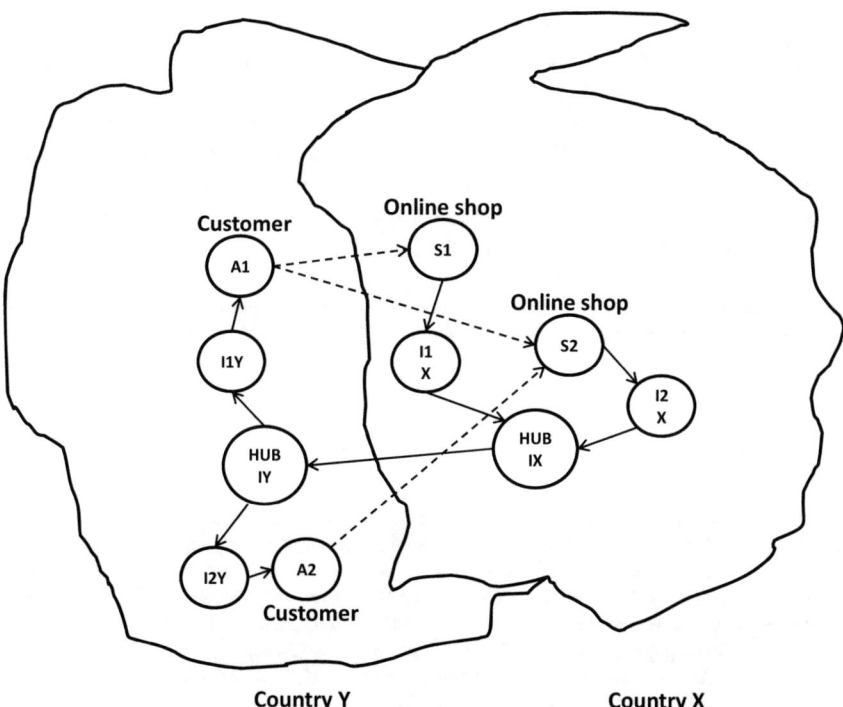

Fig. 4 Cross-border e-commerce with a consolidator

5   Conclusion

The model proposed in the study significantly reduces the number of the sorting and handling operations. It solves the problems of the organization of international logistics, and in particular the one with the high cost of deliveries, by consolidating shipments from various senders depending on the country of delivery. This will help to achieve the economies of scale—the CEP operator can offer better price conditions for a larger number of shipments. Additionally, thanks to the support of the supply chain by a single system it will be possible to track the shipments. Apart from the possibility to lower the costs, the limited number of operations reduces the risk of the goods being damaged during loading, unloading and so on.

Moreover, the proposed solution is consistent with the assumptions of the Green Paper of the European Commission [3], according to which the attractiveness of purchases made over the Internet is determined by three main factors: the price of the product together with the cost of delivery, ensured quality of the delivery of the product, and access to information on the order status. In addition, the European Commission places great emphasis on integration of the systems of companies throughout the whole e-commerce supply chain, particularly among smaller CEP operators in the field of cross-border transport. Increased interoperability can accelerate the exchange of information, facilitate the consolidation of the needs for transportation, parcel delivery and invoicing, develop multimodal transport and reduce administrative costs [3].

The study hereby proposed a general concept of cross-border e-commerce using an integrator. The further direction of the research will be the development and verification of the model in practice. For example, a larger number of countries can be included in the model, the reverse logistics process may be added, or the crowdsourcing solutions can be used for local courier services etc.

Acknowledgements This paper has been written with financial support of the National Center of Science [Narodowe Centrum Nauki]—grant number DEC-2015/19/B/HS4/02287.

References

1. Ecommerce Europe: European B2C e-commerce report 2015. Brussels (2015)
2. European Commission: A roadmap for completing the single market for parcel delivery Build trust in delivery services and encourage online sales, http://www.eur-lex.europa.eu/legal-content/EN/TXT/PDF/?uri=CELEX:52013DC0886&from=EN. Brussels (2013)
3. European Commission: An integrated parcel delivery market for the growth of e-commerce in the EU. Green paper., European Commissions. Brussels (2012)
4. European Commission: Boosting e-commerce in the EU, http://www.europa.eu/rapid/press-release_MEMO-16-1896_en.htm. Brussels (2016)
5. European Commission: Commission presents roadmap for completing the Single Market for parcel delivery, Press release, http://www.europa.eu/rapid/press-release_IP-13-1254_en.htm. Brussels (2016)

6. European Commission: Proposal for a regulation of the European parliament and of the council on cross-border parcel delivery services, http://www.eur-lex.europa.eu/resource.html? uri=cellar:8eec1e90-2330-11e6-86d0-01aa75ed71a1.0002.02/DOC_1&format=PDF. Brussels (2016)
7. Kawa, A.: Logistyka e-handlu w Polsce (Logistics of e-commerce in Poland), http://www. media.poczta-polska.pl/file/attachment/612453/bb/logistyka-e-handlu-w-polsce.pdf. Poznań (2014)
8. Kawa, A., Zdrenka, W.: Conception of integrator in cross-border e-commerce. LogForum **12** (1), 63–73 (2016)
9. Martens, B., Turlea, G.: The drivers and impediments for online cross-border trade in goods in the EU, Digital Economy Working Paper, http://www.ftp.jrc.es/EURdoc/JRC78588.pdf (2012)
10. Okholm, H.B. et al.: Principles of e-commerce delivery prices, Copenhagen Economics, http://www.posteurop.org/NeoDownload?docId=456601. Copenhagen (2016)

Towards a Personalized Virtual Customer Experience

Bartłomiej Pierański and Sergiusz Strykowski

Abstract In an experience economy, customers are no longer satisfied with products and services themselves. Products and services are required to create memorable events, perfectly suited to individual needs and expectations of each customer. Such personalization that can potentially lead to an increased customer experience is currently recognized as the most important strategic goal for retailers. To reach this goal, the substantial use and contribution of new technologies seems to be of paramount importance. In this paper, a conceptual model of a virtual retail store for a personalized customer shopping experience is proposed. The model employs process mining, recommender systems, and big data analysis to create both personalized offers and personalized virtual shopping spaces which customers can immerse themselves in when making purchase decisions.

Keywords Personalization · Virtual retailing · Virtual reality · Customer experience

1 Introduction

At the end of 2014, Forbes predicted [1] that personalization would not be a trend; it would be a marketing tsunami that will entirely transform all thinking about marketing, shifting from globalization and generalization to localization and individualization. This tsunami has been very evident in retailing activities in 2015 and 2016.

An experience economy, a concept introduced in 1999 by Joseph Pine and James Gilmore, is becoming a must in today's business. The concept assumes that

B. Pierański (✉) · S. Strykowski
Poznan University of Economics and Business, Al. Niepodleglosci 10,
61-875 Poznan, Poland
e-mail: b.pieranski@ue.poznan.pl; bartlomiej.pieranski@ue.poznan.pl

S. Strykowski
e-mail: s.strykowski@ue.poznan.pl

© Springer International Publishing AG 2017   185
D. Król et al. (eds.), *Advanced Topics in Intelligent Information
and Database Systems*, Studies in Computational Intelligence 710,
DOI 10.1007/978-3-319-56660-3_17

products and service themselves do not satisfy customers any longer. They need to create memorable events—"the experience"—and that is what customers are willing to pay for. According to one survey [2], 86% of customers are ready to pay more for a better customer experience.

Unfortunately the "ordinary" experience is not enough to attract customers. It has to be a personalized experience, perfectly suited to each individual customer that they can fully immerse in and recognize as created especially for them.

In this paper, a conceptual model for personalized virtual customer experience is proposed. The model employs process mining and recommender systems to create both personalized offers and personalized virtual shopping spaces which customers can immerse themselves in while making purchase decisions.

The reminder of this paper is organized as follows. Section 2 provides an overview of related work in personalization and shopping customer experience. In Sect. 3 the proposed conceptual model of personalized virtual customer experience is presented. Finally, Sect. 4 concludes the paper and outlines the future work.

2 Related Work

It is widely accepted in the literature that virtual technology can enhance the customer experience [3–5]. The "customer experience" could be defined as a set of interactions between a customer and a product, brand, service, company or part of its organization which provoke a reaction as a result of an ordinary or extraordinary experience [6]. This experience implies the customer's involvement at different levels, such as the emotional, affective, spiritual, physical, sensorial, behavioral (e.g. lifestyle), intellectual, cognitive, rational (functional and utilitarian), as well as relational or social. For many researchers this enhancement of customer experience in virtual reality can be done by so called telepresence, a high level of interactivity and multisensory feedback [5, 7–9]. Great emphasis is put on telepresence that is conceptualized as the sensation experienced in virtual technology [10], or in other words this is the ability of virtual technology to induce a sense "of being there" [11, 12]. It is suggested that telepresence is partly related to the consumer's state of mind and partly to the virtual technology itself [5]. To make consumers even more involved in virtual reality or to increase their sense "of being there" the virtual technology should be immersive, which means that it should be capable of creating an immersive virtual environment [5]. With the development of technology, including faster processor speed and higher-resolution graphics, virtuality has becoming more and more deeply engaging.

On the other hand, it is not only technology that can induce a sense "of being there". Very promising possibilities lay within the concept of personalization. Based on the definition of personalization, it can be concluded that this term means the adjustment of a company's offering to the individual needs of each customer [13] or is a form of product differentiation which enables meeting the individual needs of customers [14]. Personalization can also be understood as a process of delivering to each customer the right product, in the right place and at the right time [15]. Information technology has fundamentally changed the possibilities in the area of personalization. The breakthrough concerns the way enormous volumes of data concerning customers can be gathered and analyzed. Information technology also allows for the personalization of offers without customer involvement. In such cases, customers do not have to define their needs or communicate them to the offer provider. The only activity that is required, for example, is to do shopping at e-stores. Each such activity is a source of data for creating customer profiles [16].

In retailing, personalization is regarded as the highest priority and at the same time the biggest challenge for management [17]. The major prerequisites for personalization are the following [13]:

- Customer expectations;
- Direct access to data relating to the purchasing behavior of customers;
- Advanced technological possibilities for collecting and processing information, leading to the creation of customer profiles.

The essence of a personalized offer is matching the characteristics of a product to the individual needs of customers. However, it must be pointed out that personalization is implemented in a different way within the retail sector than in the manufacturing sector. Retailers individualize their offer not by changing the physical form of individual goods, but by providing consumers with ready-made purchasing combinations. This means that the customer does not need to browse all the products from the offered range, but can be presented only with a selection that corresponds to their current and/or potential needs [13]. Although it has not been widely investigated, it can be hypothesized that a personalized virtual environment can enhance telepresence and, as a consequence, customer experience. Even though personalization is a broad concept, recent studies have focused only on avatar-like sales assistants as way to personalize the virtual environment. In fact, these assistants provide tailored recommendations based on data mining techniques [18]. It is not surprising that such assistants are not the only possibility for virtual reality personalization. Within a virtual reality environment the possibilities and areas of personalization seem to be much higher, not only when compared with bricks-and-mortar retailers but also with e-commerce or m-commerce.

3 Personalized Virtual Customer Experience

In the conceptual model for a Personalized Virtual Customer Experience, personalization is defined as a process of tailoring customers' experiences in three dimensions:

- Assortment;
- Space;
- Ambient conditions.

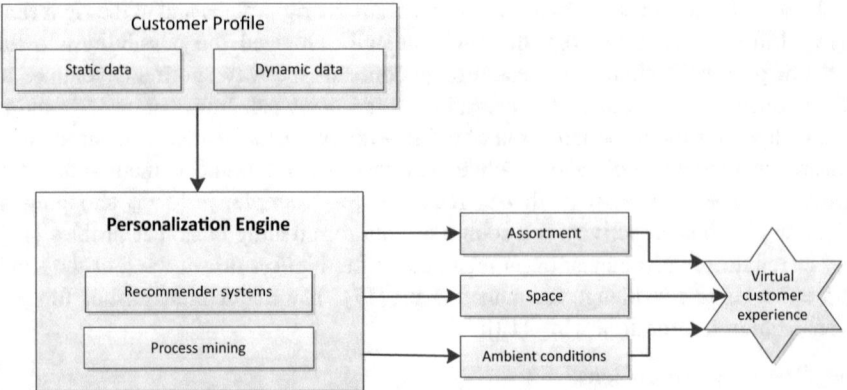

Fig. 1 Conceptual model for the personalized virtual customer experience

The model is presented in Fig. 1. It consists of the following main components:

- Customer Profile;
- Personalization Engine;
- Three dimensions of personalization: Assortment, Space, and Ambient conditions;
- Virtual customer experience.

The Customer Profile includes two main types of data:

- Static data: this describes the characteristics of a customer that are constant during a specific period of time; e.g. marital status;
- Dynamic data: this relates to activities performed by a customer; e.g. purchase history.

The data included in the Customer Profile is as follows:

- Behavioral data: purchase history (products or services purchased by the customer); products viewed but not purchased; products added to the cart but eventually abandoned; products being searched for;
- Demographic data: age, gender, residential area (address), education, occupation;
- Social profile: interests (movies, music, books, hobbies), friends;
- Social media profile: activities on social media (e.g. Facebook likes and dislikes, Twitter followings, etc.);
- Lifestyle data: type of property owned, pets;
- Family details: marital status, children;
- Device-related data: e.g. smartphone brand;
- Psychographics: religious and political views;
- Personal wishes: expectations and interests expressed directly by the customer;
- Contextual data: e.g. customer's location-related data such as current weather or social events being held.

The data stored in the Customer Profile is used by the Personalization Engine to generate the various aspects within the three dimensions of the Virtual Customer Experience.

The Personalization Engine employs two main technologies:

- Recommender systems;
- Process mining.

Recommender systems use data filtering algorithms to adapt the content—assortment and virtual space—to the needs and requirements of each individual customer. Because these systems are entirely autonomous, they therefore require no human supervision; which makes it possible to create personalized offers to all customers regardless of their number and the number of products and services. There are two fundamental algorithms used in creating recommendations [19]:

- Collaborative filtering—filtering based on the similarity of behavior;
- Content-based filtering—filtering based on the similarity of content.

Modern recommender systems typically take a hybrid approach; i.e. they use a combination of both types of filtering. They are often also supported by additional technologies like domain ontologies and inference engines.

Collaborative filtering assumes that customers with similar behavioral profiles have similar tastes. This is based on the following rule: "we enjoy the same experiences that are enjoyed by people like us". The approach is based on building customer profiles; including data on activities, preferences, opinions, and beliefs. This data is then used to predict that assortment and its arrangement in virtual space which will bring a positive experience to the customer based on an analysis of positive ratings on similar assortments and their arrangements by customers with similar profiles. Ratings are collected through both explicit and implicit means:

- Explicit means: a customer is directly asked to provide a rating on a particular item (the product, its arrangement in a virtual space, etc.);
- Implicit means: a rating is inferred from customer's activities; e.g. the length and frequency of viewing a specific product.

The main advantage of collaborative filtering is that it is not necessary to analyze the structure of customer experiences—therefore it is possible to express fine recommendations regarding multi-faceted items without the need to "understand" their internal structure and content.

Content-based filtering is based upon an assumption that customers generate positive responses to experiences that are similar to those appraised positively in the past. This is based on the rule: "we have a definite taste therefore we like experiences of a similar kind". The approach compares prospective experiences to those that the customer liked in the past (purchased, positively rated, they belong to a liked category, etc.) and recommends similar ones. The key issue in this approach is the necessity to recognize customer preferences regarding activities within one type of content and apply these preferences to another type. The task is straightforward but the results are less valuable if the recommendations apply to experiences in the same category; e.g. recommending yet another detective movie based on information that in the last week the customer watched five movies of this genre. Much more valuable is to recommend a movie or music based upon previously read and positively rated press releases.

Process mining is an approach to analyzing logs, including data describing time-based events, and turning them into process models. An event is defined as anything that happened that was of some importance, and thus data on it was stored in an information system. Examples of such events are as follows: a customer placing an order, a user liking a picture on a social media website, a user writing a comment on a blog post, a customer looking at a product description on an e-commerce site, and a customer withdrawing a specific amount of money from an ATM.

There are three classes of process mining techniques [20]:

- Discovery
- Conformance checking
- Enhancement

Discovery techniques refer to creating process models based on event log data only; i.e. there is no a priori information about what the process model should look like.

Conformance checking techniques refer to comparing event log data to existing process models and analyzing the discrepancies.

Enhancement techniques refer to enriching an existing process model by including additional information in the event log and describing the contextual aspects of performing the process in a specific environment. An example is enriching a process model with performance data.

3.1 Ambient Conditions

Ambient elements create so called atmospherics [21]. Kotler [22] defined this term as "the effort to design buying environments to produce specific emotional effects in the buyer that enhance purchase probability". Atmospherics or ambient elements have an impact on consumers in a way that they are not fully aware of. The reason for this being it is at a more subconscious level of influence. That is why atmospherics is recognized as background stimuli [21]. Various ambient elements (atmospherics dimension) can be distinguished; they are presented in Table 1 [21].

Technological limitations pose certain obstacles to using all the ambient elements within virtual reality (olfactory and taste, for instance). On the other hand the same reality permits the designing of some ambient elements more easily. One of them is the visual dimension. VR allows one to choose from an almost unlimited range of colors and lighting levels; as well as something that especially creates new

Table 1 Five atmospheric dimensions

Dimension	Description
Visual	Color, lighting level, appearance of objects (size and shape)
Aural	Volume, pitch, tempo and style of sounds
Olfactory	Nature and intensity of sound
Tactile	Temperature, texture and contact
Taste	Nature and intensity of taste sensations

possibilities, the size and shape of store fixtures. Another dimension is the aural. The appropriate music can be played not through speakers as it is in bricks-and-mortar stores but through earphones. The latest technology developments seem to be very promising in the area of including the tactile dimension in VR. More and more sophisticated virtual gloves are available on the market that make it possible for instance to feel the texture or weight of a given product.

All the ambient elements can influence shoppers within a range that is limited by two elements: stimulus awareness and stimulus overload. If the intensity of an ambient element (lighting level, music volume, etc.) is too low (lower than the level of stimulus awareness) shoppers will not be affected by them. On the other hand, however, if the intensity is higher than the stimulus overload point shoppers will experience perceptual overloading. Ambient elements have a positive impact on consumers between stimulus awareness and stimulus overload. It is not surprising that for each customer the level of acceptable intensity, as well as the level of stimulus awareness and overload, can be different. For this reason in an impersonalized store environment, it is recommended to consider a zone of maximum effectiveness that will be large enough to meet the requirements of the highest possible number of customers (Fig. 2).

VR makes it possible to personalize each of the above mentioned ambient elements. The possible ways to personalize the ambient elements are presented in Table 2.

One has to stress that VR, apart from allowing the creation of personalized ambient elements, can overcame obstacles that retailers have to face when creating perfect ambient condition inside bricks-and-mortar stores.

To summarize, one can conclude that within VR each consumer can enjoy the purchasing process in a different environment, tailored to his/her needs. Therefore the zone of maximum effectiveness presented in Fig. 2 can be considered redundant; being be replaced by a point of maximum effectiveness. Each point can represent the personalized atmospheric that meets the needs of each customer.

3.2 Assortment

VR seems to be the perfect environment for assortment personalization for two reasons: unlimited selling space and unlimited possibilities in the assortment

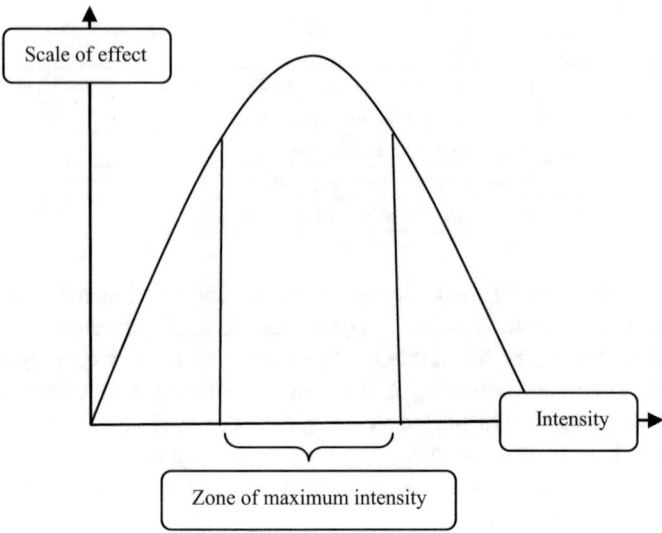

Fig. 2 Zone of maximum effectiveness [21]

Table 2 Examples of personalization for selected ambient elements

Ambient elements	Examples of personalization
Visual	Color of the floor, walls, ceiling and fixtures tailored to each consumer's aesthetic needs
	Color and intensity of the light tailored to each consumer's needs
Aural	Type and volume of the music tailored to each consumer's needs
Tactile	Texture of the floor, walls, ceiling and fixtures tailored to each consumer's needs

combinations offered to each consumer. Bricks-and-mortar retailers have to struggle with a finite amount of selling space. Obviously this limits the number of products retailers can offer to their customers. VR overcomes these obstacles as the selling area is not limited and can be extended and arranged according to the needs of consumers. As a consequence the number of products carried by a virtual store can be very high. This gives retailers the possibility to have more products with a greater diversity of products to choose from when providing the consumer with ready-made purchasing combinations.

This issue is closely related to the possibility of unlimited assortment combinations. Combinations can be made by adjusting the depth and width of the assortment in order to meet each consumer's needs. Assortment width is defined as the number of different product types offered by a retailer, while assortment depth is considered to be the number of product varieties offered [21]. According to each shopper's preferences a specific number of different products, as well as a specific number of product varieties, can be provided.

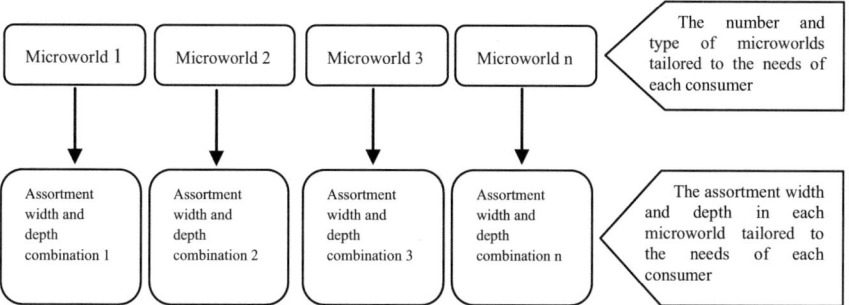

Fig. 3 Assortment personalization in VR stores

3.3 Space

In the endeavor to personalize space the concept of microworlds can be employed. Microworlds can serve as a first step in the process of personalization. In this scenario the type and number of microworlds would first be determined. Then in each microworld the width and the depth of the assortment can be tailored to each consumer (Fig. 3).

VR also allows for so called dynamic personalization. This means that the assortment offered can be different (in terms of its width and depth) every single time a given consumer 'enters' the VR store. In this scenario not only would the assortment provided to each shopper be unique, but also every visit by each consumer in the VR store would result in unrepeated assortment combinations. In other words, this way of assortment personalization can lead to unrepeatable customer experiences.

4 Conclusions and Future Work

In this paper, a conceptual model for Personalized Virtual Customer Experiences has been proposed. The model defines customer experience as a personalized combination of three aspects: assortment, space, and ambient conditions. The personalization is performed by the Personalization Engine based on data collected in the Customer Profile. The data is of two kinds: static and dynamic. Static data refers to customer characteristics that are constant during a specific period of time; dynamic data refers to activities performed by customers; e.g. products purchased, products added to the cart but eventually abandoned. To handle these two types of data, the Personalization Engine uses two main technologies: recommender systems and process mining.

Future work includes the operationalization of the model. The internal structure of the Personalization Engine needs to be elaborated, both at the technological and business level. Specific algorithms within recommender systems and process mining should be developed and implemented. Also, the concept of the customer experience as a combination of assortment, space, and ambient conditions should be investigated and elaborated in detail.

Acknowledgements The work presented in this paper has been partially supported by the National Science Centre, Poland [Narodowe Centrum Nauki]—grant No. DEC-2015/19/B/ HS4/02287.

References

1. Dan, A.: 11 Marketing Trends To Watch For in 2015. http://www.forbes.com/sites/avidan/2014/11/09/11-marketing-trends-to-watch-for-in-2015 (2015)
2. The Customer Experience Impact Report, 2011. http://www.oracle.com/us/products/applications/cust-exp-impact-report-epss-1560493.pdf (2011)
3. Ha, S., Stoel, L.: Online apparel retailing: roles of e-shopping quality and experiential e-shopping motives. J. Serv. Manag. **23**(2), 197–215 (2012)
4. Grewal, D., Levy, M., Kumar, V.: Customer experience management in retailing: an organizing framework. J. Retail. **85**(1), 1–14 (2009)
5. Lau, H., Kan, Ch., Lan, K.: How consumers shop in virtual reality? How it works? Adv. Econ. Bus. **1**(1), 28–38 (2013)
6. Walls, R.A., Okumus, F., Wang, Y., Joon, D., Kwun, W.: An epistemology view of consumer experiences. Int. J. Hospitality Manag. **30**, 10–21 (2011)
7. Park, J., Stoel, L., Lennon, S.J.: Cognitive, affective and conative responses to visual simulation: the effects of rotation in online product presentation. J. Consum. Behav. **7**(1), 72–87 (2008)
8. Haans, A., IJsselsteijn, W.A.: Embodiment and telepresence: toward a comprehensive theoretical framework. Interact. Comput. **24**, 211–218 (2012)
9. Jin, S.-A.A., Yongjun, S.: The roles of spokes-avatars' personalities in brand communication in 3D virtual environments. J. Brand Manag. **17**(5), 317–327 (2010)
10. Riva, G., Waterworth, J.A., Waterworth, E.L., Mantovani, F.: From intention to action: the role of presence. New Ideas Psychol. **29**(1), 24–37 (2011)
11. McCreery, M.P., Schrader, P., Krach, S.K., Boone, R.: A sense of self: the role of presence in virtual environments. Comput. Hum. Behav. **29**(4), 1635–1640 (2013)
12. Kober, S.E., Neuper, C.: Personality and presence in virtual reality: does their relationship depend on the used presence measure? Int. J. Hum.-Comput. Interact. **29**(1), 13–25 (2012)
13. Borusiak, B., Pierański, B.: Forms of food distribution. In: Food Retailing. Wageningen Academic Publishers (2017)
14. Hanson, W.: Principals of Internet Marketing. South West, Cincinnati, OH (2000)
15. Hart, C.W.L.: Mass-customization: conceptual underpinnings, opportunities and limits. Int. J. Serv. Ind. Manag. **6**(2) (1995)
16. Romanowski, R., Borusiak, B., Strykowski, S., Pierański, B.: Personalization of advertisement as a marketing innovation. Manag. Sci. Educ. **1**, 18–20 (2014)
17. Grant, G.: Personalization: Retail Marketing's Priority for 2015. http://www.blog.demandware.com/intelligence/personalization-retail-marketings-priority-for-2015 (2015)
18. Corvello, V., Pantano, E., Tavernise, A.: The design of an advanced virtual shopping assistant for improving consumer experience. In: Pantano, E., Timmermans, H. (eds.) Advanced Technologies Management for Retailing: Frameworks and Cases, pp. 70–86. IGI Global (2011)

19. Brusilovsky, P., Kobsa, A., Nejdl, W. (eds.): The adaptive web. Methods and strategies of web personalization. In: Lecture Notes in Computer Science, vol. 4321. Springer, Berlin (2007)
20. Aalst van der, W.: Process Mining. Data Science in Action, 2nd edn. Springer, Berlin (2016)
21. Sullivan, M., Adcock, D.: Retail Marketing, p. 150. Thomson (2002)
22. Kotler, P.: Atmospheric as marketing tool. J. Retail. **49**, 48–61 (1973)

Towards Big Management

Marcin Hernes and Andrzej Bytniewski

Abstract Actual paradigms of management and the emergence of the concept of Big Data creates the need to introduce changes in the classical approach to management. The aim of the paper is to specify the actual approach to management, called Big Management. The first part of paper characterizes an actual paradigms for management and Big Data phenomenon. Next, the Big Management concept have been defined and main technologies for its support have been described. Last part of paper presents challenges and case studies of Big Management.

Keywords Paradigms of managements · Big data · Big management

1 Introduction

In the management process, organizations need to take into consideration actual paradigms noticeable especially in the globalization phenomenon, following customers' needs, making permanent changes or treating managing people as a form of leadership [1]. However, the most important paradigm refers to the need of taking decisions in near to real time based on the most up-to-date and the most valuable information. Nowadays, there is abundance of information, the information is changeable and unstructured, and the situation is referred to as Big Data phenomenon [2].

Defining in reference works actual paradigms of management and the emergence of the concept of Big Data creates the need to introduce changes in the classical approach to management.

M. Hernes (✉) · A. Bytniewski
Wrocław University of Economics, ul. Komandorska 118/120,
53-345 Wrocław, Poland
e-mail: marcin.hernes@ue.wroc.pl

A. Bytniewski
e-mail: andrzej.bytniewski@ue.wroc.pl

© Springer International Publishing AG 2017   197
D. Król et al. (eds.), *Advanced Topics in Intelligent Information
and Database Systems*, Studies in Computational Intelligence 710,
DOI 10.1007/978-3-319-56660-3_18

The aim of the paper is to specify the actual approach to management, called by the authors Big Management. The first part of paper characterizes an actual paradigms for management and Big Data phenomenon. Next the Big Management concept has been defined and main technologies for its support have been described. Last part of paper presents challenges and case studies on Big Management.

2 Actual Paradigms of Management

Analyzing reference works one may notice that many authors [e.g. 1, 3, 4] state that it is necessary to reanalyze and create actual, fundamental principles thanks to which the development of theory and management practices will be continued. Drucker also points out that the so far existing paradigms of management expressed in the following theses are no longer valid:

- the concept of management refers to the essence of an enterprise and rules of its operation/functioning,
- there is or should be one ideal organizational structure,
- there is or should be one proper method of managing people.

That is why, he formulated actual management paradigms which have been presented in Table 1 (in relation to old, classical paradigms).

In paper [3] it has also been observed that there appeared a discordance between the current theory of management and the nature of things. The authors claim that people who are successful at managing are those who create new rules thanks to which they achieve good results, as well as those who take advantage of the informatization process to improve the functioning of their companies in the information economy, or in other words the information-technological paradigm, i.e. global cross-linked informational capitalism [5].

It is also worth stressing that a dominating employee in terms of produced value is a knowledgeable employee, or in other words, an employee working in the information sector [3]. The category has been defined in various ways in reference books.

Porat [6] connected it with activities performed on symbols (knowledge producer, knowledge distributor, market coordination and research specialist, information processing entity). Other authors [e.g. 7, 8] distinguish four types of information-related professions connected with the following areas:

- information generating (e.g. science and technology, market research, collecting information),
- information processing (e.g. business operations),
- information distribution (e.g. education),
- information infrastructure service and maintenance.

Table 1 Actual paradigms of management in relation to old paradigms

No.	Old paradigm	Actual paradigm
1	Management is business management	Management principles apply to all organizations
2	There is, or there must be one right organizational structure	Look for, develop and test the organization structure that fits the task
3	There is, or there must be one right way to manage people	Don't manage people. Lead them
4	Employees are just what they are, employees. What they need is a pay check and some motivation. You get the work done by commanding and controlling	You need to treat employees as volunteers, not just employees. They want more than a paycheck; they seek interesting and rewarding work. You inspire them by leading, not commanding
5	Innovations in your industry come from your own industry	Innovations in your industry don't necessarily come from within your own industry. They are likely to come from outside as well
6	The economy is defined by national boundaries	National boundaries restrain but don't define
7	Technologies, markets and end-users are predefined for each industry	Customers, with their increasing disposable incomes dictate policy and strategy. Technologies crisscross
8	National boundaries define economies	National boundaries restrain but don't define
9	Cheap labor is a major competitive advantage	Cheap labor won't give a company a substantial advantage as manual labor is becoming smaller and smaller part of total costs. Labor productivity and not cheap labor per say will be a competitive advantage
10	Change has to be managed	You cannot manage change. You can only stay ahead of change to win at the marketplace

Motivating information sector employees is particularly difficult as it is difficult to directly supervise the results of their work. Most often you can only rely on social control and trust [3] or the system of ideology, and on the identity of individual actors, which specify standards with which employees fully identify [9].

The paper [3] has also defined the sustainability paradigm as the ability of an enterprise to continue learning, adapting and developing, revitalizing, reconstructing, reorienting to maintain lasting and outstanding position on a market by offering above average values to buyers today and in the future (in compliance with the paradigm of innovative growth), thanks to the limited changeability constituting business models, resulting from creating new possibilities and targets, and responding to them while balancing out interests of various groups.

Many authors [1, 10] claim that management in the 21st century has taken on a new orientation. It is to a greater extent based on the ability to cope with constantly changing environment, not on stability; it is organized around networks instead of hierarchies; it is built on shifting cooperants, partnerships and alliances (not on self-reliance), on creating thanks to technological superiority [11]. The authors emphasize that the only stable and lasting thing in the modern economic world is change.

Organizations introducing new changes in management restructure their processes and apply a new approach to management around rapidly changing knowledge (information, technology). It needs to be clearly stressed that actual management paradigms relate not only to international corporations but also to enterprises operating on local or regional markets. In terms of actual management paradigms, the paper [3] has formulated fundamental principles of enterprise operation:

1. The principle of creative destruction—company successes must be ready to undergo cannibalism so as to catch up with changes. Companies must be ready to destroy old approach and solutions, which are still successful, if they want to build new ones, before the new ones start being successful. If a company does not destroy itself, it will be destroyed by others (by competitors).
2. Companies which want to guarantee themselves quick growth and high profits must take advantage of technological imbalance (technological imbalance is the result of radical changes in production technology), search for developmental imbalance (developmental imbalance is the result of diversified level of income in individual countries/regions) or create sociological imbalance (sociological imbalance results from changes in human habits, it results from creating new needs, it reflects movement of existing assets, and not creating new ones) [9, 11]. Other forms of activity have a slow pace of growth, low return rates, and they require the lowest possible costs to survive. Imbalance is driven by changes.
3. "Understanding, recognizing and accepting limitations imposed onto organizations by "inherent" weaknesses is the beginning of wisdom for all organizations".
4. "There are no permanently perfect companies. Companies, just like branches of industry go through their better or worse times". Their ability to recreate enables them to survive and "be reborn anew".

3   Big Data Phenomenon

The term Big Data appeared in the first decade of the 21st century. The idea encompasses a series of data attributes which exist in too large quantities, are too highly unsystematised, and are subject to changes which are too rapid to apply to them traditional data management methods [12].

Big Data, however, is not only the volume, but also diversity of data and speed of information flow (stream of information). More and more often, economic activity issues are discussed in internet forums or discussion blogs, and they are positively or negatively assessed on social networking sites. Here data comes in more quickly than in case of traditional, systematized information, for example in data warehouses. In order to understand customer's needs and their experience in using new products and services, companies must make use of new forms of data (unstructured, disorganized) and methods of their processing and analyzing.

Nowadays, the environment of the enterprise in the knowledge economy is characterized by the following phenomena, which are determinant of Big Data [2, 13]:

1. A high volume of data—means the mass nature of the data in terms of terabytes or even petabytes, this may have an impact on the incidence of problems related to the creation of economic models including those stemming from the large number of factors or criteria influencing the decision-making process.
2. Speed (variability in the data)—is the phenomenon of stream of data within an organization in real-time or near real time. If the structure of IT is not prepared for such rapid flow, accumulation and often extraction of large amounts of data, it may occur (as a consequence of this phenomenon) delays in the decision-making process (a problem with the storage of these data).
3. Diversity—addresses the problem of the diversity of structures or features of the data, which storing in traditional relational databases can cause the loss of some data or become at all impossible (e.g. in the case of unstructured data). Therefore, these data should be stored in the structures of NoSQL, understood as "not only SQL". An example of this kind of data can be business documents written in natural language, the content of web pages, e-mail or Internet forums.
4. Value—is the need for processing such data before using them, to make them valuable for analytical purposes.

The computer system implementing analytical functions should, among other things [14]:

- be efficient and flexible in terms of loading large volumes of data and to exercise their joints and enrichment (Data Explorer),
- contain a set of analytical functions as tools in the work of business analysts,
- respond quickly to questions and to pursue the analysis, improving the work of analysts and designers,
- allow the creation of interactive data analysis, the results of which are also supplied by mobile devices.

4   Definition and Determinants of Big Management

Taking into account actual paradigms of management, and the Big Data phenomenon (characterized in the previous parts of the paper) Big Management can be defined as execution of management processes including actual paradigms, combined with the Big Data concept.

Big Management determinants are specified in the following way:

1. Traditional (old) management paradigms are insufficient.
2. Globalization—the possibility of functioning on international markets, companies do not depend on time or place of carrying out their activities. Greater competition—customers' orders need to be executed fast and satisfactorily otherwise customers will choose other competitive offers.
3. Turbulent nature of the environment—which may change in various ways. There are always signs of coming changes, however not everyone is capable of acknowledging and interpreting them correctly. Most of essential changes take place very quickly and come as a surprise to companies which underestimate the importance of monitoring the environment, which results in inadequate perception of the surrounding.
4. New organizational forms—new organizations are networks of tightly "weaved spider's webs" which are based on virtual rather than on vertical integration, on interdependence rather than on independence, on mass customization rather than on mass production [3].
5. Rapid pace of developmental changes—the pace is so high that a man's experience very often becomes insufficient as it is characterized by a high level of inertia—it becomes indispensable to apply technology which enables facilitating decision making processes based not only on knowledge but also on experience.
6. New technologies, innovations—it is assumed that lasting, permanent competitiveness may be achieved only by constantly developing and implementing innovations, creating new customers and new markets, findings new opportunities, entering new markets and areas.
7. Large number of information, diversity, multitude of streams of information (changeability), lack of structure of the information.
8. The need to manage knowledge—it may be said that effective functioning of an organization in the current socio-economic conditions is based on proper management of knowledge. The process of knowledge management involves first of all application of new information technologies. There is a constant increase of investments into the technologies. They allow for rapid processing of information, which results in the ability to respond to changes taking place in a given environment, and improved quality of taken decisions. Thanks to the technologies companies become more competitive as they are capable of reaching a wider number of customers and expanding their area of activity.
9. Common wisdom (collective intelligence)—making use of an organization's external source of knowledge, namely the society, to solve problems. The main assumption of the whole strategy is that a group has greater knowledge than individuals. The so-called synergic effect is used here. Cooperation of individuals brings better results than the sum of their individual actions [15]. Customers' opinions about a product on the basis of which it is possible to forecast sales volumes and improve a product may serve as an example here.

10. Multi-agent systems—consisting of several or several dozen agent software (agents) whose role is to support making decisions concerning a given problem.
11. Cooperation between a man and an agent. The process of supporting making financial decisions with the use of a multi-agent system requires permanent cooperation between humans and a software agents. The cooperation may take various forms. One of them may involve a situation when software agents generate various variants of a decision while a man takes the final decision. The cooperation may also involve software agents automatically taking final decisions on the basis of criteria defined by a man which specify the level of his or her satisfaction from taken decisions. The form of cooperation may also be connected with deciding on a final decision on the basis of variants created by a man (an expert) and variants generated by a software agent.
12. New information and commutation technologies (cloud computing, in-memory databases, NoSQL—described in greater detail in the following item of the paper).
13. Mobility—the use of mobile technologies, however, enables quick transfer of up-to-date data indispensable to take effective and fast decisions, and to organize and control companies more efficiently irrespective of the time or place a person managing a company is at. Efficiency of managing knowledge also increases as access to it is also easier. Management becomes more flexible, and an organization keeps up with changes occurring in the turbulent economic environment.
14. Internet of Things—an idea which focuses on connecting all possible devices to the Net wirelessly and with the use of wires. Objects connected with one another within a given infrastructure are capable of identifying each other, communicating with one another, and cooperating. Internet of Things involves most of all communication between devices and their autonomous functioning on the basis of information exchanged between devices. An important role in the kind of communication is played by software layer of high accessibility, which enables Internet of Things to be used practically in all branches of industry, creating information foundations in management processes.

5 Technologies for Big Management

The execution of Big Management is possible only thanks to the use of many types of information and communication technologies. The most important technologies are considered to be the following ones:

1. Cloud computing which can be defined as computing services offered by external entities, available upon request at any given time, dynamically adjusting themselves in response to changing demands of their users [16]. The functioning of IT systems supporting management within cloud computing

services will enable effective use of resources, which will consequently enable to lower costs connected with IT infrastructure covered by companies.

2. Grid computing, also referred to as parallel computing executed by a network of computers, in which computing applications are divided and then each section is performed simultaneously by a set/pool of machines defined in a grid environment.

3. Data processing by database engine (in-Database)—computing applications are done simultaneously within a database so as to use the mechanisms and architecture of massive parallel processing offered by some databases.

4. In-memory databases processing—relational system of data management suited to process data and events in RAM, without the use of HDD. Application of modern structural solutions and extensive simplification of the databases enables to reach effectiveness and capacity inaccessible for traditional databases.

5. Cognitive agent software—it performs cognitive and decisive functions which take place in the human brain, thanks to which it is capable of understanding the real meaning of observed business phenomena and processes. The software enables not only quick access to information, fast selection of information one is looking for, its analysis and drawing conclusions, but also, apart from reacting to stimuli from the environment, it has cognitive capacity which enables learning through empirical experience gained through direct interaction with the environment, which allows taking automatic decisions and executing them [17, 18].

Also following technologies (described in [16]) can be used for Big Management: MapReduce paradigm (e.g. framework Apache Hadoop, Apache Spark), distributed data file systems, data analytics, and machine learning techniques.

6 Big Management—Challenges

Fundamental challenges Big Management is facing include:

- lowering costs,
- increasing quality,
- beating competitors,
- gathering and processing information in real time,
- taking decisions in next to real time,
- taking decisions automatically,
- improving the quality of economic forecasts,
- increasing qualifications of employees.

Big Management must enable lowering costs of functioning of organization with a simultaneous increase of the quality of offered products and services. It is possible by automatically taking quick and accurate decisions on the basis of information gathered and processed in a near real time.

There appear new areas of company's activity, the possibility to reach to a wider circle of customers. The range of possibilities offered by the Big Management makes it easier to find suppliers and recipients, starting business relations with them, negotiating and coordinating cooperation irrespective of their location. New ways of analyzing company's activities emerge because the access to information is virtually unlimited and quite fast. Big Management also facilitates contacts or relations with customers and suppliers, and enables analysis of the efficiency of such contacts. An increase in the quality of economic forecasts also constitutes a challenge. As an example we can look at the analysis of customers' opinions about products, which is done automatically by agent programs. As a result of the analysis, an organization receives information concerning opinions (whether they are positive, neutral, or negative) and features which are preferred by customers. Thus, it is possible to forecast the rise or fall of sales of a given product, and what kind of features should be improved so as to meet demands of customers (in case of classical management facilitating systems, information about the rise/fall of sales was available with a certain delay—after registering sales figures or volumes in a transaction system, which was usually too late to take effective decisions concerning further action with respect to a given product in question). An important challenge for the Big Management is to permanently improve employees' qualifications. The process is supported thanks to the use of knowledge management systems which, among other things, facilitate transfer of knowledge between organizations and among employees.

7   Case Studies

In order to determine the impact of Big Management on the effectiveness of organizations, the activities of three companies was analyzed: **Google**, **eBay** and **Nokia**.

The subject of analysis was the use of actual management paradigms and technologies to support Big Data. The analysis was performed using the following research methods:

1. Cases study—analysis of the functioning of selected enterprises in the years 2002–2007.
2. Studies of related works—used in articles containing content related to the subject of analysis [19–27].
3. The observation of phenomena—concerns study the effects of actual management paradigms to net income (in 2006–2015) of the surveyed companies.

The results of analysis are presented in Table 2, while Fig. 1 presents net income of analyzed companies.

Taking into consideration actual paradigms of management it needs to be said that Google and eBay have used them in the full scope, which may have contributed to an increase or stabilization of their net income in the years 2006–2014. Low net

Table 2 Results of big management utilization in Google, Amazon and Nokia

Paradigm/technology	Google	eBay	Nokia
Management principles apply to all organizations	+	+	−
Look for, develop and test the organization structure that fits the task	+	+	+
Don't manage people. Lead them	+	+	+
You need to treat employees as volunteers, not just employees. They want more than a paycheck; they seek interesting and rewarding work. You inspire them by leading, not commanding	+	+	+
Innovations in your industry don't necessarily come from within your own industry. They are likely to come from outside as well	+	+	−
National boundaries restrain but don't define	+	+	+
Customers, with their increasing disposable incomes dictate policy and strategy. Technologies crisscross	+	+	−
Cheap labor won't give a company a substantial advantage as manual labor is becoming smaller and smaller part of total costs. Labor productivity and not cheap labor per say will be a competitive advantage	+	+	+
You cannot manage change. You can only stay ahead of change to win at the marketplace	+	+	+
Sustainability	+	+	±
Cloud computing and Grid computing	+	+	+
In-Database processing and In memory Databases	+	+	+
Software agents (intelligent technologies) and others	+	+	+

income of eBay in 2014 resulted from the fact that in May 2014 eBay experienced a major cyberattack. Because of the attack, eBay had to force its users to change their passwords, which cost the company traffic. eBay also experienced lower traffic on its site because of a change in SEO (search engine optimization) by Google [27]. Nokia, on the other hand, used most of actual paradigms of management. However, none of the following paradigms had been used by this company:

- management principles apply to all organizations
- innovations in your industry don't necessarily come from within your own industry; they are likely to come from outside as well,
- customers, with their increasing disposable incomes dictate policy and strategy.

Technologies crisscross, was caused by the following reasons [20]:

- Nokia moved too slowly (Nokia failed to respond to the iPhone and the shifting consumer demand that came with it),
- Android Paid Off for Samsung and Windows Phone Hasn't … Yet for Nokia (Nokia, on the other hand, spent its time focusing on Symbian until the company's recent partnership with Microsoft. But Nokia's flagship Lumia Windows Phones haven't paid off yet, as evidenced by Nokia's Q1 earnings),
- Hurting on Both Ends (Not only did Nokia move too slowly in the smartphone market, it didn't anticipate competition in the lower end of the market),

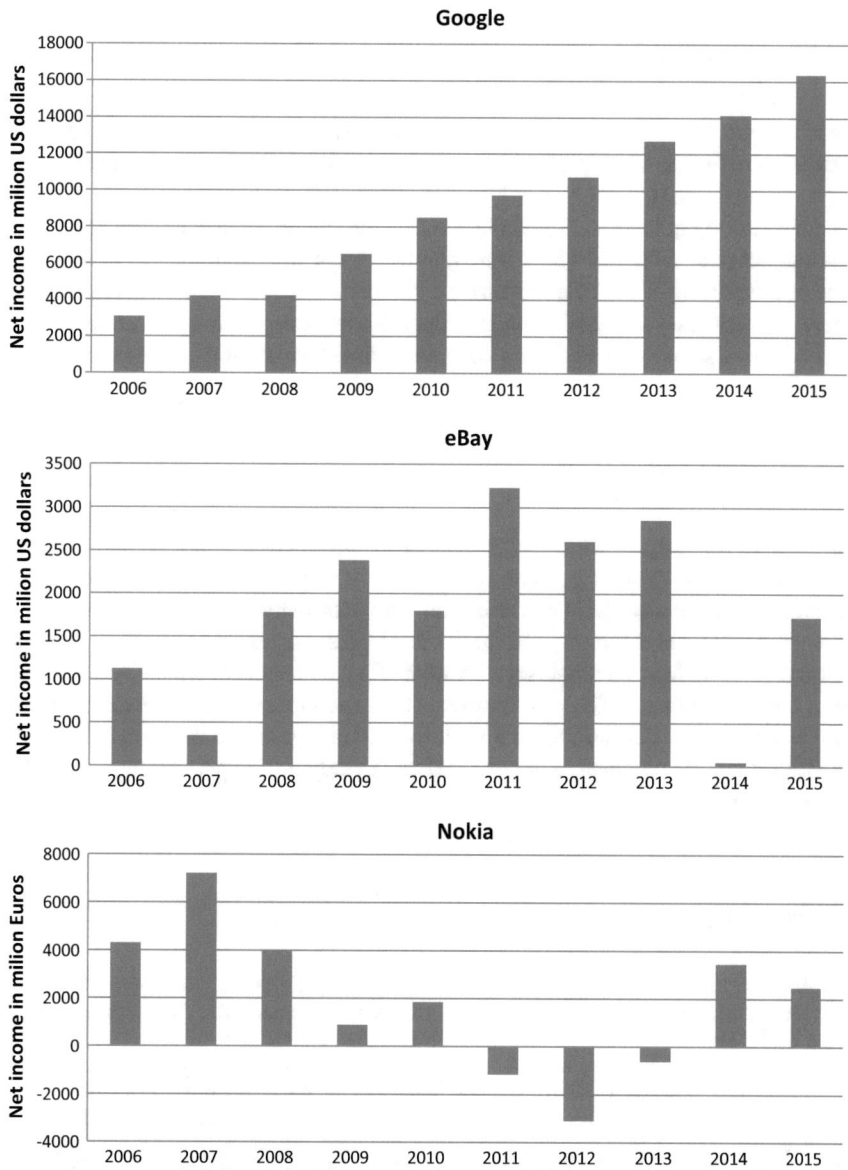

Fig. 1 Net income of Google, eBay and Nokia. *Source* Statistica 2016

- Nokia Didn't Have the Panache (Nokia was sort of an older brand, there wasn't a new panache to it. Samsung, as a marketed brand, was perceived as an innovator. Nokia has a legacy baggage—they are the traditional brick, candybar phone maker),

- Execution Is Key (Nokia's execution has been shoddy in recent years, it doesn't mean it can't make a comeback with Windows Phone).

The fact that Nokia had not used these paradigms might have resulted in the drop of Nokia's net income in the years 2009–2013. Taking into account the use of Big Data supporting technologies, it needs to be said that all the examined companies have applied them, which is why it is difficult to determine the effect of the application on net income. Since there are many references in literature concerning the subject [e.g. 1, 2, 17] which state that the use of the mentioned technologies is one of the preconditions of maintaining a company's competitive position, one may draw the conclusion that in the mentioned companies, the application of new technologies supporting Big Data constitutes the foundation of their functioning.

8   Conclusions

In order to properly execute the process of management it is not enough to apply actual paradigms of management, or to use only new technologies supporting Big Data. It is indispensable to combine the activities, which in this paper has been defined as Big Management. On the basis of case studies one may draw the conclusion that Big Management has a positive effect on results obtained by companies (e.g. their net income). The paper presents only preliminary results of researches on Big Management. Further research shall focus, among other things, on the analysis of Big Data determinants/parameters and on developing IT systems supporting Big Management.

References

1. Drucker, P.F.: Management Challenges for 21st Century. Harper Business, New York (2000)
2. Davenport, T.H., Paul, B., Bean, R.: How 'Big Data' is different. MIT Sloan Manag. Rev. **54** (1) (2012)
3. Grudzewski, W.M., Hejduk I.K.: Enterprise of the future. Changes of management paradigms. Master of Business Administration, vol. 19, no. nr 1, pp. 95–111 (2011)
4. Hamel, G.: The way, What, and How of management innovation. Harvard Bus. Rev. **84**(2) (2006)
5. Castells, M., Cardoso, G. (eds.): The Network Society: from Knowledge to Policy. Johns Hopkins Center for Transatlantic Relations, Washington, DC (2005)
6. Porat, M.U.: The information economy: definition and measurement. In: Cortada, J.W. (ed.) Rise of the Knowledge Worker, Resources for the Knowledge-Based Economy. Butterworth-Heinemann, Boston (1998)
7. Reinhardt, W., Schmidt, B., Sloep, P., Drachsler, H.: Knowledge worker roles and actions— results of two empirical studies. Knowl. Process Manag. **18**(3), 150–174 (2011)
8. Palmer, N.: Empowering Knowledge Workers. Future Strategies Inc. (2014)
9. Jemielniak, D.: Jobs in companies knowledge on the example of the organization of high-tech, WAiP (2014) (in Polish)

10. Kaplan, R.S., Norton, D.P.: Mastering the management system. Harvard Bus. Rev. **86**(1) (2008)
11. Carnall, C.: Managing Change in Organizations. Pearson Education Limited Harlow (2003)
12. Robak, S., Franczyk, B., Robak, M.: Applying big data and linked data concepts in supply chains management. Annals of computer science and information systems. In: Proceedings of Federated Conference Computer Science and Information Systems (FedCSIS), Kraków, pp. 1203–1209 (2013)
13. Dumbill, E.: What is big data? An introduction to the big datalandscape, Strata O'Reilly, 11 Jan 2012
14. Big Data What it is and why it matters. http://www.sas.com/en_us/insights/big-data/what-is-big-data.html. Access: 15.06.2016
15. Howe, J.: Crowdsourcing. Why the Power of the Crowd is Driving the Future of Business. Three Rivers Press, New York (2008)
16. Bello-Orgaz, G., Jung, J.J., Camacho, D.: Social big data: recent achievements and new challenges. Information Fusion, vol. 28, pp. 45–59. Elsevier (2016)
17. Bytniewski, A., Chojnacka-Komorowska, A., Hernes, M., Matouk, K.: The implementation of the perceptual memory of cognitive agents in integrated management information system. In: Barbucha, D., Nguyen N.T., Batubara, J. (eds.) New Trends in Intelligent Information and Database Systems. Studies in Computational Intelligence, vol. 598. Springer International Publishing Switzerland (2015)
18. Molnár, B., Benczúr, A.: Modeling information systems from the viewpoint of active documents. Vietnam J. Comput. Sci. **2**(4), 229–241 (2015)
19. Bouwman, H., Carlsson, C., Carlsson, J., Nikou, S., Sell, A., Walden, P.: How Nokia failed to nail the Smartphone market. In: 25th European Regional ITS Conference, International Telecommunications Society (ITS), http://www.EconPapers.repec.org/RePEc:zbw:itse14: 101414. Brussels (2014)
20. Chang, A.: 5 Reasons Why Nokia Lost Its Handset Sales Lead and Got Downgraded to 'Junk. http://www.wired.com/2012/04/5-reasons-why-nokia-lost-its-handset-sales-lead-and-got-downgraded-to-junk/
21. Hoff, T.: eBay Architecture. http://www.highscalability.com/ebay-architecture
22. Korkosz-Gębska, J.: Creativity and innovation in the management of a modern enterprise. In: Innovation in Management and Production Engineering, Oficyna Wydawnicza Polskiego Towarzystwa Zarządzania Produkcją (2014) (in Polish)
23. Landeweerd, M., Spil, T., Klein, R.: The success of Google search, the failure of Google health and the future of Google plus. In: Grand Successes and Failures in IT Public and Private Sectors, pp. 221–239. Springer (2013)
24. Mishra, M.K.: Why is eBay the most successful online auction? Global J. Manag. Bus. Res. **10**(9), 62–65 (2010)
25. Schory, G.: An Inside Look at eBay's Global Innovation Strategy. http://deloitte.wsj.com/cio/2014/08/25/an-inside-look-at-ebays-global-innovation-strategy/
26. Spalding, G., Razon, R.: 18 In Memory Database. http://www.butleranalytics.com/18-in-memory-database/
27. Garner, P.: Why did eBay's operating margins continue to decline in 4Q14. http://marketrealist.com/2015/02/ebays-operating-margins-continue-decline-4q14/

Part III
Computer Vision Analysis, Detection, Tracking and Recognition

Combination of Collision Detection and Visibility Algorithms in Simulation of the Effective Placement of Anti-air Elements

Dalibor Cimr, Richard Cimler and Hana Tomášková

Abstract The computer simulation of the effective placement of air defense elements is presented in this paper. The proposed algorithm used for plane detection is based on a combination of collision detection algorithms and visibility algorithms. The method is designed to be computationally effective in order to enable the smooth running of the simulation. The optimal placement of selected air defense elements is computed for any given place on Earth with marked defended structure and selected air-raid routes. The map height profile is loaded from Google Maps API during simulation initialization. Range and visible area for given air-defense elements are computed based on the height profile. Different placements of anti-air elements are tested and the success rate of air raid scenarios gives the optimal locations for the placement of such elements.

Keywords ABM · Collision detection · Visibility algorithm · Air defense · Simulation · Agent-based modeling · Military simulation

1 Introduction

In the real world, it is not possible to carry out tests and experiments of various military procedures. One of these is testing radar system placement in a vast area, because there are huge financial demands on radar operations. Additionally, there are limited options for testing, because it is impossible to place air defense elements just anywhere in the world. So it is important to find another way to test various procedures or strategies. One of the possible choices is to use simulations and Agent-Based Modeling (ABM), which is a computational method that enables the user to create

D. Cimr · R. Cimler (✉) · H. Tomášková
Faculty of Informatics and Management, University of Hradec Kralove,
Hradec Kralove, Czech Republic
e-mail: richard.cimler@uhk.cz

D. Cimr
e-mail: dalibor.cimr@uhk.cz

© Springer International Publishing AG 2017
D. Król et al. (eds.), *Advanced Topics in Intelligent Information
and Database Systems*, Studies in Computational Intelligence 710,
DOI 10.1007/978-3-319-56660-3_19

and experiment with models consisting of agents that interact within a stochastic or deterministic environment [1]. It gives the possibility to carry out tests and simulations of different situations.

In this paper, the simulation of air-raids on selected defended objects is presented. The simulation has been created in AnyLogic 7.2. The aim of the simulation is to find the optimal placement of anti-air elements, viz., radar. Radar systems represent the group of radar installations or antennas supporting better target estimation [2]. Information from a radar system is an important part of the decision-making process of an air defense commander.

The research in this paper is focused on finding the right setup of a collision detection algorithm (CDA) and a visibility algorithm (VA) in a given environment. The collision detection algorithm is used to determine whether an air attack element is in a space where it can be detected by radar. The model needs a detection method with low computational complexity to ensure the smooth running of the simulation. A visibility algorithm is necessary to determine whether an air attack element can be detected by radar or is hidden in the environment.

2   Problem Definition

There are a few existing publications solving some parts of the given problem. It is possible to implement collision detection using different methods. It can be represented by axis-aligned bounding boxes (AABB) or oriented bounding boxes (OBB) [3]. There is a similar scheme for exact collision detection between complex models. The comparison of these methods [4] shows that AABB is slower than OBB, but OBB needs more storage space for representation by scalars (AABB does not need to represent the orientation). Another way is to do continuous collision detection (CCD) [5], which is the method used in physics-based simulations or computer-aided design. Each object is represented as a triangle mesh which is tested during deformation, e.g., a car crash with about 1.1–1.2 million triangles [6] in Fig. 1. Collision detection is very often used in interactive rigid body simulation [7]. It is an important part of modern computer programs, such as computer games or animation software. It is not only used for collision detection, but also for some reaction to it, like collision resolution.

Fig. 1   Triangle mesh and testing during deformation [6]

In computer graphics, there is also used an analytic visibility method for triangular meshes [8]. It is represented as an edge–triangle intersection algorithm to define all visible triangles. It is possible to fully analyze a scene against an older method like depth-buffered rasterization. These old methods are still used because of their computational simplicity. The hierarchical z-buffer visibility algorithm [9] quickly rejects hidden geometry and generates a coherence of image. This method is based on saving the distance of the objects in the scene from the observer to a two-dimensional array. When some object or part is closer to the observer's view, the information about the depth is rewritten and the closest pixels in the positions are used for the rendering in a rasterization pipeline. This is used to image mappings, such as texture mapping, bump mapping, and environment mapping [10].

3 Combination of CDA and VA

A classical visibility algorithm determines the visibility of all objects, which means many edges and pixels from the observer's view. In a given simulation, it is important to find the visibility of airplanes from the radar's position. The first part of the algorithm is to define a 3-dimensional linear function between the radar and the airplane. Because the visibility algorithm is used in uneven terrain, there is the need for the declaration of an elevation grid. This grid contains numeric information about the model environment. These values are compared with the elevation values of a function at the given coordinates. If the result is lower than the elevation grid value, this means that the radar agent can see the airplane agent in the model.

In simulation, it is necessary to find an effective algorithm to detect collisions between an airplane, representing an attack element, and the detection space of the radar. A simplified radar space to detect airplanes is shown in Fig. 2, where R represents the radius and H represents the height. The airplane is represented only as a point, for an effective simulation run. The real size of an airplane is a few meters, and the radar detection space is in kilometers.

Fig. 2 Radar range in model

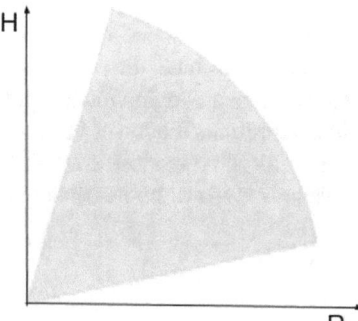

The detection method consists of three stages. The accomplishment of the previous stage is the prerequisite for continuing with the next stage. If any stage is impossible to accomplish, that means that the airplane cannot be detected by the radar. These stages are:

1. Altitude of airplane A higher than altitude of radar R;
2. Airplane is in the detection space;
3. Radar is turned in the direction of the airplane.

3.1 First Stage

The first stage is only a comparison of two values to leave out undetectable airplanes which are flying under the range of the radar. If $z_A > z_R$, where z_A is the altitude of the airplane and z_R is the altitude of the radar, then the airplane is not under elevation the limit and can be detected by the radar.

3.2 Second Stage

The second stage starts with the calculation of the radar's distance and the distance of the airplane. The following formulas are used:

$$\tan \alpha * \sqrt{(x_R - x_A)^2 + (y_R - y_A)^2} > z_A$$

$$z_A > \tan \beta * \sqrt{(x_R - x_A)^2 + (y_R - y_A)^2}$$

$$(1)$$

$$\sqrt{(x_R - x_A)^2 + (y_R - y_A)^2 + (z_R - z_A)^2} < \text{radius}, \qquad (2)$$

where x_R, y_R and z_R are the coordinates of the radar, x_A, y_A and z_A are the coordinates of the airplane, the radius is the distance within which the radar can detect an airplane, β is an angle which represents the upper vertical limit, and α is an angle which represents the lower vertical limit.

If an airplane is close enough to the radar, there is still the possibility that the airplane is lower or higher than the limit of the detection space.

3.3 Third Stage

In the third stage, the vector of the radar's rotation u and the vector in the direction from the radar to the airplane v are calculated. In the next step, the angle γ between these vectors is set to be

$$\gamma = \frac{u_1 \cdot v_1 + u_2 \cdot v_2}{\sqrt{u_1^2 + u_2^2} + \sqrt{v_1^2 + v_2^2}}, \tag{3}$$

where u_1, u_2, v_1 and v_2 represent the coordinates of u and v. The numerator represents multiplication of vectors, and the denominator, multiplication of the vectors' distance. If this angle is lower than the permissible value of the model, all stages of detection are a success and the airplane is detected by the radar.

4 Model Description

Agent-based simulation using the formulas described above has been created with the Anylogic multimethod tool. Anylogic is a simulation tool supporting Agent-based modeling, as well as discrete event and system dynamics. It includes object-oriented programming for the construction of large models. The software was created by the Anylogic company in 2000. The last version of the program is 7.3.4, which was published in 2016. Simulation can be used as a decision support system which can be used during the deployment of radar systems in the terrain.

4.1 Model Parameters

The airplane flight is defined using discrete event simulation modeling [11]. Each route is a list of points representing the coordinates of the next move of the airplane. The height of the airplane has been calculated from an environment elevation for the prevention of collision with the environment.

The radar agent is described by the radius R representing a range of detection, the angle α representing a lower vertical limit, and the angle β representing an upper vertical limit.

The overall success of airplane detection is evaluated by three basic criteria. The first criterion is the percentage of detection successes. The second is the critical distance, which is represented by the minimum distance from the protected object at the moment of the first detection. Thus, if the percentage of successes is less than 100%, then the critical distance is zero because at least one airplane has not been detected. The model includes also a third criterion, which is the average of the detection distances for all the planes.

4.2 Environment

The environment of model is represented by a simplified real map. It is implemented as a x3d file. The size of the map is 50×50 km. The data are loaded from the Google Maps API during the initialization of the simulation. The information about elevation is loaded, processed, and saved to a file. The scale of the map is 1 km.

5 Case Study

The algorithm has been tested in a situation where the location, air strike routes, and protected object have been selected by an expert, who formulated five prescribed airstrike scenarios on the protected object. As a result of the simulation, the optimal placement of air defense elements has been found.

The first scenario pretends to fly in a different direction than the position of the object and later, using one of the valleys, an airstrike is executed. The second scenario tries to use a different valley to hide from the radar. The third and fourth scenario represent a predictable attack using the valley from different positions. The fifth scenario is a straight flight to the object, which is the simplest variant.

The testing area has been selected in a hilly environment in the mountains of Switzerland, with GPS coordinates (44°36'18.1"N 6°30'23.9"E, 44°09'35.5"N 6°57'6.5"E). A fictitious protected object has been placed by the expert at a strategic position. The airstrikes' routes have been designed to correspond to a common military strategy, in order to keep the planes well hidden and safe. The environment with the scenarios is shown in Fig. 3. In the experiment, the radar characteristics α and β were set to 2°, and R was set to 20 km.

5.1 The Experiment

The impact of the number of radar installations on the detection success has been tested in an experiment. The reason for testing is the high price of radar. The goal of the experiment is to find the optimal number of radar installations to protect the area. All possible options of radar placement were tested.

5.2 Results

Small differences in the results of three and four radar installations means that the fourth installation's added value is negligible. Using three radar installations is the

Fig. 3 Environment with airstrike scenarios

Table 1 The average experimental results

Number of radar installations	1	2	3	4
Probability of detection (in %)	27.77	47.82	62.17	65.12
Probability of 100% detection (in %)	0.07	0.17	0.27	0.28

Table 2 The best experimental results

Number of radar installations	1	2	3	4
Probability of detection (in %)	100	100	100	100
Minimal critical distance	3.99	18.19	19.56	19.23
Average of detection distances	6.84	22.80	29.10	29.59

best choice in the given environment. Table 1 describes the average results of an experiment. The results are confirmed in Table 2, which represents the best results of an experiment.

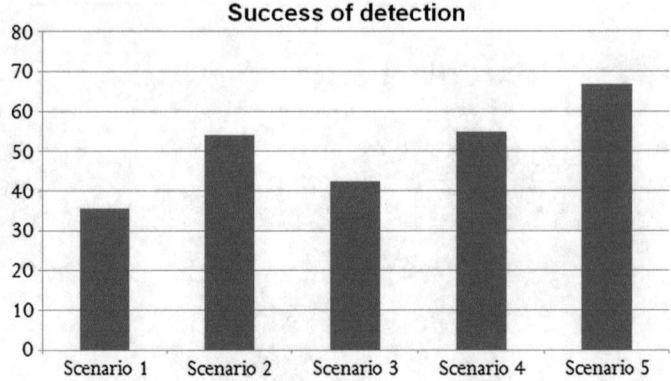

Fig. 4 Percentage success of detection. Scenario number is on *x*-axis. Success rate in on *y*-axis

Figure 4 describes the success rate of the attack scenarios. Straight flight has been shown to be the worst option for an air raid. The first scenario, which pretends to fly in a different direction than the position of the object and then uses the cover of the mountains, is the best strategy.

6 Conclusion

In this paper, a novel approach to combining a collision detection algorithm and a visibility algorithm in the simulation of effective radar placement was presented. The optimal placement of radar installations in a given place on the Earth with a selected object to be defended and with designated air-raid routes was found. The created application can be used as a decision support system in situations where what is needed is to evaluate the optimal placement of radar systems in an environment. The parameters of the radar can be adjusted in order to simulate specific types of radar, such as mobile stations.

Future work is focused on the creation of an extension of the simulation in order to be used as a decision support system for leaders of air-strikes. The optimal air-raid route will be computed for an environment with a given radar placement. A further extended version is also planned, with the creation of a heat map of positions which are difficult to cover by radar. Such a map can be used during the planning of air-raids in an environment with an unknown placement of radar systems.

Acknowledgements The support of the Specific research project at FIM UHK is gratefully acknowledged.

References

1. Gilbert, G.N.: Agent-Based Models. SAGE (2008)
2. Godrich, H., Petropulu, A., Poort, H.V.: Estimation performance and resource savings: trade-offs in multiple radars systems. In: 2012 IEEE International Conference on Acoustics, Speech and Signal Processing (ICASSP), pp. 5205–5208. IEEE (2012)
3. van den Bergen, G.: Efficient collision detection of complex deformable models using aabb trees. J. Graph. Tools 2(4), 1–13 (1998). Jan
4. Gottschalk, S., Lin, M.C., Manocha, D.: Obbtree: A hierarchical structure for rapid interference detection. In: Proceedings of the 23rd Annual Conference on Computer Graphics and Interactive Techniques, pp. 171–180. ACM (1996)
5. He, L., Ortiz, R., Enquobahrie, A., Manocha, D.: Interactive Continuous Collision Detection for Topology Changing Models Using Dynamic Clustering, pp. 47–54. i3D '15, ACM (2015). http://doi.acm.org/10.1145/2699276.2699286
6. Tang, M., Manocha, D., Yoon, S.E., Du, P., Heo, J.P., Tong, R.F.: Volccd: fast continuous collision culling between deforming volume meshes. ACM Trans. Graph. 30(5), 1–15 (2011). Oct
7. Bender, J., Erleben, K., Trinkle, J.: Interactive simulation of rigid body dynamics in computer graphics. Comput. Graph. Forum 33(1), 246–270 (2014). Feb
8. Auzinger, T., Wimmer, M., Jescke, S.: Analytic visibility on the gpu. Comput. Graph. Forum 32(2pt4), 409–418 (2013)
9. Onishi, K., Noborio, H., Koeda, M., Watanabe, K., Mizushino, K., Kunii, T., Kaibori, M., Matsui, K., Kon, M.: Virtual liver surgical simulator by using z-buffer for object deformation. In: International Conference on Universal Access in Human-Computer Interaction, pp. 345–351. Springer (2015)
10. Park, W.C., Lee, K.W., Kim, I.S., Han, T.D., Yang, S.B.: An effective pixel rasterization pipeline architecture for 3d rendering processors. IEEE Trans. Comput. 52(11), 1501–1508 (2003). Nov
11. Myskova, H.: Dynamic systems with inexact data: algorithm and theory. EQUILIBRIA (2016)

A Combination of Deep Learning and Hand-Designed Feature for Plant Identification Based on Leaf and Flower Images

Thi Thanh-Nhan Nguyen, Thi-Lan Le, Hai Vu, Huy-Hoang Nguyen
and Van-Sam Hoang

Abstract This paper proposes a combination of deep learning and hand-designed feature for plant identification based on leaf and flower images. The contributions of this paper are two-fold. First, for each organ image, we have performed a comparative evaluation of deep learning and hand-designed feature for plant identification. Two approaches for deep learning and hand-designed feature that are convolutional neuron network (CNN) and kernel descriptor (KDES) are chosen in our experiments. Second, based on the results of the first contribution, we propose a method for plant identification by late fusing the identification results of leaf and flower. Experimental results on ImageClef 2015 dataset show that hand designed feature outperforms deep learning for well-constrained cases (leaf captured on simple background). However, deep learning shows its robustness in natural situations. Moreover, the combination of leaf and flower images improves significantly the identification when comparing leaf-based plant identification.

Keywords CNN · KDES · Plant identification

1 Introduction

Plant identification aims at determining the name of species based on observations. This task is time consuming and difficult even for the botanist experts. Recently, the advanced research in computer vision community allows building automatic plant

T. Thanh-Nhan Nguyen (✉) · T.-L. Le · H. Vu · H.-H. Nguyen
International Research Institute MICA, HUST-CNRS/UMI-2954-GRENOBLE INP,
Hanoi University of Science and Technology, Hanoi, Vietnam
e-mail: nttnhan@ictu.edu.vn

T. Thanh-Nhan Nguyen
Thai Nguyen University of Information and Communication Technology,
Thai Nguyen, Vietnam

V.-S. Hoang
Vietnam Forestry University, Hanoi, Vietnam

© Springer International Publishing AG 2017
D. Król et al. (eds.), *Advanced Topics in Intelligent Information
and Database Systems*, Studies in Computational Intelligence 710,
DOI 10.1007/978-3-319-56660-3_20

identification based on images analysis. However, a large number of automatic plant identification methods still work with only one sole organ of the plants. Among different organs of the plant, leaf and flower are the most widely used. Leaf is usually flat and easy to collect almost the whole year while flower has a high distinguishing capacity. However, using one sole organ for plant identification is not always relevant because one organ cannot fully reflect all information of a plant due to the large inter-class similarity and large intra-class variation.

Recently, there are two new trends in plant identification. First, for each organ-based plant identification, instead of using hand-designed feature, learning feature has been investigated more and more [1, 2]. Second, the plant identification moves from one sole organ to multi-organ [3]. This is motivated by a real scenario where one user tries to identify a plant by observing its different organs. This also can reflect the process of plant identification of botanists. Observing simultaneously several organs allows the botanists to disambiguate species that could be confused when using only one organ.

This paper has two main contributions. First, for each organ image, we have performed a comparative evaluation of deep learning and hand-designed feature for plant identification. Two approaches for deep learning and hand-designed feature that are convolutional neuron network (CNN) and kernel descriptor (KDES) are chosen in our experiments. Second, based on the results of the first contribution, we propose a method for plant identification by late fusing the identification result of leaf and flower.

2   Related Work

There are two main approaches for plant identification. The first one uses hand-designed feature while the second one employs the deep learning strategy. A number of hand-crafted (or hand-designed) features have been used for plant identification, such as edge features [4], KDES [5] and shape properties [6]. The hand-designed feature approach requires knowledge about the application. Moreover, approaches using hand-crafted features also require pre-processing techniques. For the leaf-based images, LeafSnap is a noticeable application that achieved more than 80% accuracy rate. For the flower-based images, by using a dataset of 120 species, [7] obtains the recognition rate of 76.3%. Recently, convolutional neural network obtains state-of-the-art results in many computer vision applications [8, 9]. Typically CNNs are AlexNet, VGG, GoogLeNet and ResNet. Some studies try to apply CNN such as [1] to learn unsupervised feature representations for 44 different plant species collected at the Royal Botanic Gardens, Kew, England. However, it still lacks a detail analysis and comparative evaluation of hand-designed feature and deep learning for plant identification.

Recently, the plant identification has been expanded with multi-organ/multi-images such as leaf-scan, leaf, flower, fruit so that the obtained results are better. The common way to do it is to combine recognition results by late-fusion techniques.

Such kind of approaches are listed in technical report of the LifeCLEF competitions 2015 [3]. However, to the best of our knowledge, there is no plant identification technique combining two most important organs which are leaves and flowers, to achieve better performance for the plant identification.

3   Proposed Method

3.1   Overview

The overview of plant identification based on of multi-organ images is illustrated in Fig. 1. The system consists of three main components: Preprocessing, Single organ-based plant identification and Multiple organ-based plant identification. In our work, we focus on two most important organs of the plant that are leaf and flower.

- Preprocessing: depending on the characteristic of the image, we propose corresponding preprocessing techniques.
- Single organ-based plant identification: In this component, we propose to use two approaches: hand-designed feature and learning feature. For hand-designed feature, we employ KDES with multi-class SVM (Support Vector Machine) while for learning feature approach, we propose to use GoogLeNet because of its remarkable result for object recognition.
- Multiple organ-based plant identification: we perform late fusion with Sum rule technique to combine results of single organ-based plant identification.

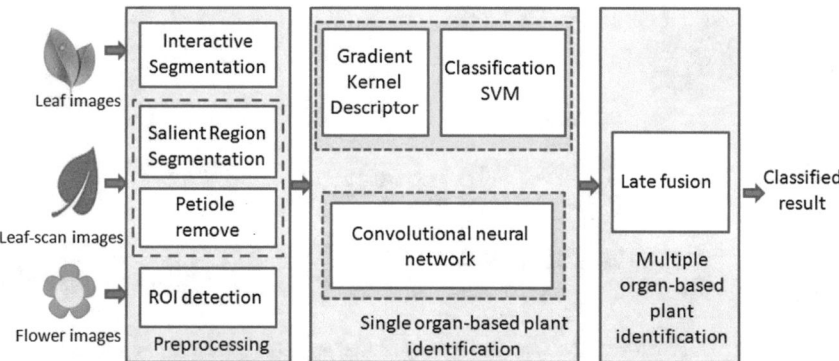

Fig. 1   Overview of multiple organ-based plant identification

3.2 Preprocessing

The images of object of interest (e.g. leaf and flower) can be captured in different conditions. In our work, we use ImageClef2015 dataset [3]. In this dataset, leaf images are divided into two categories: leaf (leaf image captured on natural background) and leaf-scan (leaf image captured on simple background) while flower images are normally captured in natural background.

For leaf images, an interactive segmentation method [5] based on Watershed algorithm is first applied to segment leaf from background regions. Then, the main direction of the leaves are normalized based on moment calculations.

With leaf-scan images, a saliency extraction method and a common segmentation technique (e.g., mean-shift algorithm) [10] are employed. A segmented region is selected based on a condition that its saliency value is large enough. The connected-region techniques then are applied to connect the selected regions. Since leaf-scan images usually contain simple background, we can obtain relatively good segmentation results for leaf-scan images.

Concerning flower images, as flower images are normally taken in complex background, it is not easy to have stable and good segmentation results. Therefore, saliency-segmentation procedure that is similar for leaf-scan image is chosen for determining ROI (Region Of Interest) from flower images. Flower ROI is determined by top-left and bottom-right points on boundary of the connected-regions. Detail information on these techniques can be found in [11]. Some examples of leaf, leaf-scan and flower images after applying the preprocessing techniques are illustrated in Fig. 2.

(a)   **(b)**   **(c)**

Fig. 2 Examples of leaf-scan, leaf and flower images after applying corresponding pre-processing techniques

3.3 Hand-Designed Feature for Single Organ Identification

Kernel descriptor (KDES) [12] has been applied in our previous work for leaf-based plant identification [5]. In this paper, we compare the performance of this feature with an effective deep neural network that is convolutional neural network. KDES is extracted through 3 steps: pixel-level feature, patch-level feature and image-level feature extraction.

Pixel-level features extraction

For each pixel z, its gradient vector consists of two components: the magnitude m(z) and the orientation $\theta(z)$ where the orientation vector is defined as $\tilde{\theta}(z) = [\sin(\theta(z)) \cos(\theta(z))]$.

Patch-level features extraction

The patch-level feature is extracted through two steps. The first step aims at generating a set of patches from image while the second one allows computing patch feature. Derived from match kernel representing the similarity of two patches, we can extract feature vector for the patch using approximate patch-level feature map, given a designed patch level match kernel function. The gradient match kernel, defined in Eq. 1, is formed from three kernels that are gradient magnitude kernel $k_{\tilde{m}}$, orientation kernel k_o and position kernel k_p. These kernels are defined in [12].

$$K_{gradient}(P, Q) = \sum_{z \in P} \sum_{z' \in Q} k_{\tilde{m}}(z, z') k_o(\tilde{\theta}(z), \tilde{\theta}(z')) k_p(z, z') \tag{1}$$

where P and Q are patches of two different images that we need to measure the similarity. z and z' denote the position of a pixel in the image patch P and Q respectively. $\tilde{\theta}(z)$ and $\tilde{\theta}(z')$ are gradient orientations at pixel z and z' in the patch P and Q respectively. Then, the approximative feature over image patch P is constructed as:

$$\bar{F}_{gradient}(P) = \sum_{z \in P} \tilde{m}(z) \phi_o(\tilde{\omega}(z)) \otimes \phi_p(z) \tag{2}$$

where $\otimes$ the Kronecker product, $\tilde{m}(z)$ is normalized $m(z)$, $\phi_o(\tilde{\omega}(z))$ and $\phi_p(z)$ are approximate feature maps for the kernel k_o and k_p, respectively. The basic idea of representation based on kernel methods is to compute approximate explicit feature map for kernel function.

Image-level features extraction

After extracting patch-level feature, the feature of whole image will be computed as follows. First, a spatial pyramid is built by dividing the image into cells at several layers. Then, the feature map is defined as:

$$\bar{\phi}_S(X) = \frac{1}{|X|} \sum_{x \in X} \phi(x) \tag{3}$$

where C is a cell that has a set of patch-level features $X = \{x_1 \ldots x_p\}$ and $\phi(x)$ is approximative feature maps for the kernel $k(x, y)$.

After computing feature at image level, we apply multi-class SVM as classifier.

3.4 Deep Learning for Single Organ Identification

GoogLeNet of Szegedy et al. that won the classification and object recognition challenges in the ILSVRC 2014 [13] used a new variant of convolutional neural network called "Inception" for classification with the intention of increasing network depth with computational efficiency. Figure 3 shows a schematic view of GoogLeNet network. GoogLeNet is a very deep neural network model with 22 layers when counting only layers with parameters (or 27 layers if we also count pooling). In this work, we fine-tune GoogLeNet with plant flower and leaf images with the following parameters: initial learning rate: 0.001; batch size: 5; number of iteration: 50,000. GoogLeNet output score is employed to produce the ranked list of relevant plant species.

3.5 Result Fusion

We combine identification results from leaf and flower image as follows:

$$score(q, species) = \sum_{i=1}^{N} score(I_i, species) \tag{4}$$

where q is query observation, N is number of images of this observation, *score* $(I_i, species)$ is obtained similarity score when using image I_i of the observation. Note that the number of images of each observation can be different.

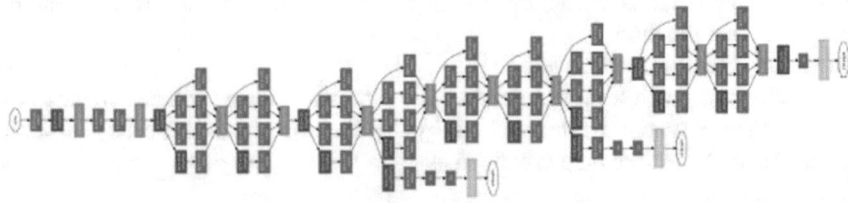

Fig. 3   A schematic view of GoogLeNet network (adapted from [13])

4 Experimental Results

4.1 Dataset

In our experiment, we use a large dataset of plant named PlantCLEF 2015 [3]. This dataset has more than one hundred thousand images belonging to 41,794 observations of 1000 plants species living in West Europe. One observation is a set of images depicting the same individual plant, observed by the same person in the same day with the same device. For each observation, images of seven different organs including leaf, leaf-scan (leaf image captured on simple background), flower, fruit, stem, entire plant and branch are captured. In this paper, we use leaf including leaf and leaf-scan and flower images. Table 1 gives detail information of training and testing set. Some examples of the flower and leaf images are illustrated in Fig. 4.

Table 1 Training and testing sets provides by PlantCLEF2015

	Leaf	Leafscan	Flower	Total
Train	8,885	13,198	27,975	50,058
Test	2,690	221	8,327	11,238
All	11,575	13,419	36,302	61,296

Fig. 4 Examples of flower and leaf with simple and complicated background images on PlantCLEF 2015

4.2   Evaluation Measures

We evaluate our proposed approach at two different levels: single-organ based plant identification named image level and multi-organs plant identification named observation level. For this, we compute the accuracy at rank k at image and at observation level as follows: $Accuracy = \frac{T}{N}$ where T and N are the number of correct recognition and the number of queries respectively. One image or one observation is considered as correct recognition if the relevant plant belongs to the k first plants of the retrieved list. In our experiments, accuracy at rank 1 (k = 1) and rank 10 (k = 10) are used.

4.3   Results and Discussions

Our system is implemented in C++, Matlab with the use of three libraries that are OpenCV, KDES and Caffe.

Comparative evaluation for single organ identification In order to compare the performance of KDES and CNN, we have performed two experiments. The aim of the first experiment is to evaluate KDES and GoogLeNet on the preprocessed images. Accuracies obtained at Rank 1 and Rank 10 are shown in Table 2. With leaf and flower images, result of GoogLeNet is much better than KDES. However, with leaf-scan, KDES outperforms GoogLeNet. This result allows recommending using hand-designed feature if the working images are images of leaf captured on a well-constraint condition. Otherwise, deep learning is a good choice.

The effect of preprocessing step for deep learning is analyzed in the second experiment. For this, we compare accuracy of GoogleNet of raw and preprocessed database. Table 3 shows results of GoogLeNet with and without applying preprocessing technique. For leaf and leafscan, the result on raw images is lower than that on preprocessed images. This means that preprocessing is important step to improve identification accuracy. However, with flower images, the obtained results on raw images and preprocessed images are relatively similar. This shows the capacity of deep learning in recognizing objects of nature scene.

Table 2 Comparison of hand-designed feature (KDES) and deep learning (GoogleNet) for plant identification at image level on preprocessed images

	Leaf		LeafScan		Flower	
	Rank 1	Rank 10	Rank 1	Rank 10	Rank 1	Rank 10
KDES	24.26	46.36	**78.28**	**92.76**	10.95	24.62
GoogLeNet	**47.30**	**72.70**	69.78	86.22	**66.60**	**90.23**

Table 3 Accuracy of plant identification obtained with GoogleNet at image level on images with and without preprocessing

	Leaf		Leafscan		Flower	
	Rank 1	Rank 10	Rank 1	Rank 10	Rank 1	Rank 10
Without preprocessing	35.39	62.79	67.56	83.11	67.45	90.82
With preprocessing	**47.30**	**72.70**	**69.78**	**86.22**	66.60	90.23

Table 4 Accuracy for plant identification based on only leaf, and combining leaf and flower

	Accuracy (%)	
	Rank 1	Rank 10
Only leaf	33.28	68.81
Only flower	61.67	86.76
Fusion of leaf and flower	**64.46**	**90.77**

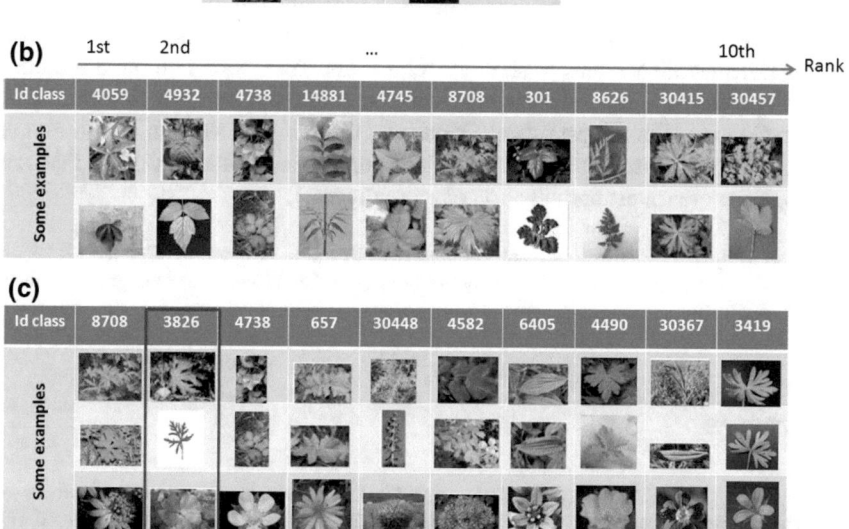

Fig. 5 Example of multiple organ-based plant identification **a** Input: observation 11012 of class 3826 having two images: one image of leaf and one of flower, **b** 10 first retrieved results when using only leaf, **c** 10 first retrieved results when combining leaf and flower images. The correct species is bounded by a *red* rectangle

Plant identification result by combining leaf and flower From testing dataset, we extracted 574 observations that contain images of both flower and leaf in order to evaluate multiple organs plant identification. The obtained results are shown in Table 4. We can observe that when combining flower and leaf images, the accuracy increases significantly for both Rank 1 and Rank 10. The result also shows that flower is a distinguishing organ for plant identification. However, this organ is not always available. An example of retrieval results for observation 11012 is illustrated in Fig. 5. This observation has two images: one of leaf and one of flower (see Fig. 5a). Figure 5b, c show the 10 first retrieved results sorted in descending order of confidence score when using only leaf and fusing leaf and flower respectively. We may observe that due to the large inter-class similarity of leaf images, this observation is not correct identified. However, when adding information of flower, the correct plant is returned at second rank.

5 Conclusions

In this paper, we have performed a comparative evaluation of two approaches for plant identification. The obtained results show that for the well-constrained case, hand-designed feature outperforms deep learning. However, in weak-constrained situation, deep learning seems to show its robustness. Then, based on the evaluation, we have proposed a plant identification method combining leaf and flower images. The obtained results are promising. However, the fusion technique is still simple and does not take into account the role of each organ. In the future, we will focus on developing more robust combination of these organs.

Acknowledgements The authors thank Collaborative Research Program for Common Regional Issue (CRC) funded by ASEAN University Network (Aun-Seed/Net), under the grant reference HUST/CRC/1501.

References

1. Lee, S.H., Chan, C.S., Wilkin, P., Remagnino, P.: Deep-plant: plant identification with convolutional neural networks. In: 2015 IEEE International Conference on Image Processing (ICIP), pp. 452–456, IEEE (2015)
2. Choi, S.: Plant identification with deep convolutional neural network: Snumedinfo at lifeclef plant identification task 2015. In: Working Notes of CLEF 2015 Conference (2015)
3. Goëau, H., Bonnet, P., Joly, A.: lifeclef plant identification task 2015. In: CEUR-WS, ed. CLEF: Conference and Labs of the Evaluation forum. Volume 1391 of CLEF2015 Working Notes, Toulouse, France, Sept (2015)
4. Kumar, N., Belhumeur, P.N., Biswas, A., Jacobs, D.W., Kress, W.J., Lopez, I.C., Soares, J.V.: Leafsnap: a computer vision system for automatic plant species identification. In: Computer Vision–ECCV 2012, pp. 502–516. Springer (2012)
5. Le, T.L., Duong, N.D., Nguyen, V.T., Vu, H., Hoang, V.N., Nguyen, T.T.N.: Complex background leaf-based plant identification method based on interactive segmentation and kernel

descriptor. In: Proceedings of the 2nd International Workshop on Environmental Multimedia Retrieval, pp. 3–8. ACM (2015)

6. Cope, J.S., Corney, D., Clark, J.Y., Remagnino, P., Wilkin, P.: Plant species identification using digital morphometrics: a review. Expert Syst. Appl. **39**(8), 7562–7573 (2012)

7. Nilsback, M.E., Zisserman, A.: A visual vocabulary for flower classification. In: 2006 IEEE Computer Society Conference on Computer Vision and Pattern Recognition (CVPR'06), vol. 2, pp. 1447–1454. IEEE (2006)

8. Yoo, H.J.: Deep convolution neural networks in computer vision. IEIE Trans. Smart Process. Comput. **4**(1), 35–43 (2015)

9. Krizhevsky, A., Sutskever, I., Hinton, G.E.: Imagenet classification with deep convolutional neural networks. Adv. Neural Inf. Process. Syst. 1097–1105 (2012)

10. Achanta, R., Hemami, S., Estrada, F., Susstrunk, S.: Frequency-tuned salient region detection. In: IEEE Conference on Computer Vision and Pattern Recognition, CVPR 2009, pp. 1597–1604. IEEE (2009)

11. Le, T.L., Dng, D., Vu, H., Nguyen, T.N.: Mica at lifeclef 2015: multi-organ plant identification. In: Working Notes of CLEF 2015 Conference (2015)

12. Bo, L., Ren, X., Fox, D.: Kernel descriptors for visual recognition. In: Adv. Neural Inf. Process. Syst. 244–252 (2010)

13. Szegedy, C., Liu, W., Jia, Y., Sermanet, P., Reed, S., Anguelov, D., Erhan, D., Vanhoucke, V., Rabinovich, A.: Going deeper with convolutions. In: Proceedings of the IEEE Conference on Computer Vision and Pattern Recognition, pp. 1–9 (2015)

An Efficient Defect Classification Algorithm for Ceramic Tiles

Khaled Ragab and Nahed Alsharay

Abstract The main aim of this paper is to reduce the required computing time to detect and classify defects in ceramic tiles. Consequently, this paper proposed an algorithm that divides the ceramic image into partitions and identify the defected partitions. Furthermore, the classification algorithm is applied only to the defected partition. As a result, the required time to classify defects is reduced. Simulation results indicate a significant improvement in terms of classification time in comparison to the current technique.

Keywords Image partition · Spot and crack defect detection and classification

1 Introduction

In the last decades, an exponential growth of the daily-generated images which required being processed and enhanced has been witnessed. This growth encouraged the development of new techniques in the image processing and enhancement field. One of the most interesting issues in image processing is the detection of the defects on an image. Automated defects detections play a crucial rule in many industries such as biscuit bake color [23], ceramic tile manufacturing [21], the color of potato chips [19], food products [14], textile fabrics [11], electronic chip manufacturing [12], and wood [2]. This paper focuses on detecting defects in ceramic tile industry. In ceramic tiles manufacturing, a mix of clays that have been shaped and fired at high temperature is the cornerstone of producing a good ceramic tile. Consequently, the quality of the ceramic tile can be easily changed depend on the

K. Ragab (✉) · N. Alsharay
College of Computer Sciences and Information Technology,
King Faisal University, Hofuf, Saudi Arabia
e-mail: kragab@sci.asu.edu.eg; 215141950@student.kfu.edu.sa; kabdultawab@kfu.edu.sa

K. Ragab
Faculty of Science, Computer Science Division Mathematics Department,
Ain Shams University, Cairo, Egypt

© Springer International Publishing AG 2017
D. Król et al. (eds.), *Advanced Topics in Intelligent Information and Database Systems*, Studies in Computational Intelligence 710,
DOI 10.1007/978-3-319-56660-3_21

235

environment. This makes ceramic tile subject to defects [20]. Image defects can be detected manually by human labor, however, it will be slow and subject to human made errors. Digital image processing is a research area in which computer programs are designed to automatically extract image features and defects [15]. With the increasing demand consumers, from a quality and quantity perspective, the use of automated inspection is one of the key technology in the manufacturing industry. There are many techniques of automated inspection techniques to detect different defects in textured and plain tiles [18]. In this paper, we address the following question: *Can we design and develop an accurate defects classification algorithm while reducing the required computing time?*. This paper focuses on two defect classification algorithms that were introduced in [18, 20] to detect and classify spot and crack defects. These algorithms applying two steps on the whole image. The first step is to detect either there is a defect or not. The second step classifies the type of defect. In this research work, the authors identified that some parts in the ceramic image has no defect however, these algorithms apply the second step (classification algorithm) needlessly. Certainly this increases the computing time. Therefore, the main contribution of this paper is designing and developing an algorithm that detects and classifies spot and crack defects. It divides the ceramic image into non-overlapped blocks (partitions). Similarly to [18, 20] it has two steps. The first step identifies the defective partitions. Moreover, the second step applies the classification algorithm only to the defective partitions. The rest of this paper is organized as follows. Section 2 introduces a review of related works. Defect detection and classification algorithms for detecting crack and spot defects are introduced in Sect. 3. Section 4 illustrates the proposed algorithm. Section 5 discuses the experiment results of the spot and crack defect classification. Section 6 concludes this paper.

2   Related Work

Image processing is a major used in the production industry to enhance and protect products against any failure during production process. Ceramic tile industry carried out many steps for creating ceramic tiles passing through heating and pressure. On the final product, there are different types of surface defects appear as a result of the variation of chemical and mechanical processes in the ceramic tile production. Therefore, many processing algorithms have been proposed to detect various types of defects. These algorithms can be categorized into four different categories based on defect detection mechanisms [10, 21]. These approaches are filtering methods, structural algorithms, model-based techniques, and statistical methods. Mathematical translation and filters are the basic operations used in filtering approaches for defect detection [5, 6, 22, 24]. In [1], C. Boukouvalas et al., the authors proposed new techniques to detect defects based on the type of defects. They applied optimal line filter to detect short and long cracks, optimal spot filer to detect spot defects, and chromatic structural defect detection to detect water drops, texture, and color defects. Computational time is a major issue in defects detection process. Consequently, using techniques based on separate filtering for different kinds of defects is not a preferable idea, and such techniques required execution of a large number of operations. Image processing and edge detection algorithms are the basic operations used in structural approaches. The primary operations of the structural approaches are extracted texture primitives and model generalizes special placement rules. Texture primitives can be a line segment, individual pixel, or region with uniform grayscale [4, 8, 18, 20, 21]. There are some trails to develop structural algorithm in parallel in [7, 13]. The Model-based approaches include a common image processing used to analyze the image [10]. It includes hidden Markov model (HMM) and autoregressive model (AM). The complexity of statistical computation is the main problem with this method. Statistical approaches usually used histogram curve properties to detect the defects [10, 21]. It is often used when we have problems that required classification. An image histogram is a graphical representation of the distribution of the pixels in an image. The defect can simply extract when the threshold point in the histogram is optimal as in [16]. The authors used *Otsu* method to find the threshold point that leads to reducing the intra-class variance. However, it produces a result with a large error. The similarity among image pixels is represented in term autocorrelation of the image. This method checks the correlation between image pixel and patterns in image processing [3].

There are many algorithms that are used to detect the defects, these algorithms can't handle all types of defects. Nevertheless, a given type of defects can be detected by more than one algorithms. Figure 1 presents defect detection algorithms. The key parameters of the defect detection algorithms are the accuracy and the efficiency. However, the more accurate algorithm the more computing time required which reduces the algorithm efficiency. This paradigm motivates us to introduce the following question *Can we design a defects classification algorithm with high accuracy using partition techniques?*.

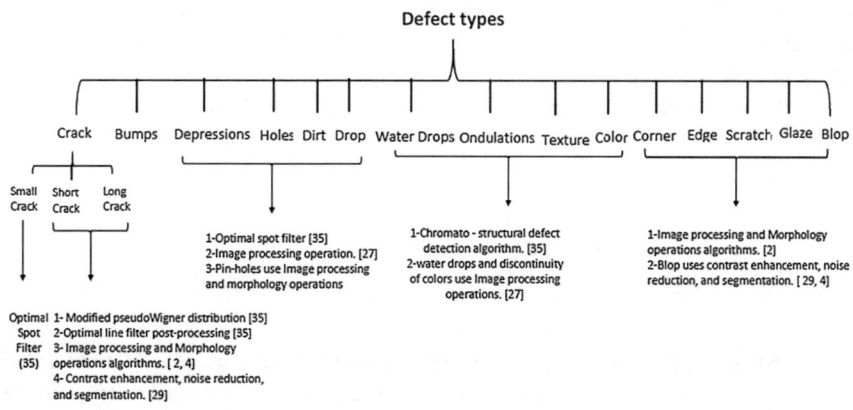

Fig. 1 Defects classification

3 Defect Detection and Classification Algorithm

The inspection process of a product is a major step in many maintenance tasks, especially in ceramic tile industrial. The stages of producing ceramic tile lead to various kinds of defects and faults on the final product. Therefore, there are different types of defect inspection used to detect and classify these defects. Defect inspection algorithms comprised of three steps, the preprocessing operation, defect detection, and defects classification. In this section, we introduce two defect classification algorithms for spot and crack defects based on algorithms proposed in two research papers [18, 20]. The work in this paper aims to reduce the computing time of these algorithms with maintaining the same accuracy of these algorithms. The following subsections are organized as follow. Section 3.1 introduces Preprocessing Operations. Defect Detection is discussed in Sect. 3.2. Defect classification is presented in Sect. 3.3.

3.1 Preprocessing Operations

The preprocessing operation is an important step on defect detection algorithms to eliminate light reflection and noises. Preprocessing operation carried out image enhancement, noise reduction, and edge detection. Image enhancement composes two steps to make the ceramic image clearer. First, convert the RGB ceramic tile images into grayscale images. After that apply the contrast stretching operation to enhance image brightness by stretching the intensity values from 0 to 255. Ceramic tile images are subject to various types of noise. Removing noise from the image by using filtering techniques is called noise reduction. Low pass filter or averaging filter is a linear filter used in removing the noise by replacing each pixel by the average value of the surrounding pixels. The median filter is an order-statistic non-linear filter, it orders the intensities in the neighborhood and updates the center pixel with the value determined by ranking. Edges in the image are curves where rapid changes occur in intensity or in the spatial derivatives of intensity. Applying filtering methods such as a median filter or averaging filters remove noises and make images more blurred [9].

3.2 Defect Detection

Whenever the preprocessing operations were completed, test image I_Test and reference image I_Ref were produced. Both I_Ref and I_Test will be compared to detect defects as follows. These images are stored as matrices of zeros and ones. We will compute the number of white pixels (one value) I_Test_No and I_Ref_No in the test image and reference image respectively. After that, we compare the overall number

Fig. 2 An example of the
I_Ref and I_Test matrices

$$I\_Ref = \begin{bmatrix} 0 & 1 & 1 & 1 & 1 \\ 1 & 0 & 1 & 0 & 1 \\ 0 & 1 & 1 & 0 & 1 \\ 1 & 0 & 1 & 1 & 1 \\ 1 & 1 & 1 & 1 & 1 \end{bmatrix} \quad I\_Test = \begin{bmatrix} 1 & 1 & 1 & 1 & 1 \\ 1 & 0 & 1 & 0 & 1 \\ 0 & 1 & 1 & 1 & 1 \\ 1 & 0 & 1 & 1 & 1 \\ 1 & 1 & 1 & 1 & 1 \end{bmatrix}$$

of white pixels of the test image I_Test_No with the reference image I_Ref_No. If the overall number of white pixels for test images I_Test_No is greater than the white pixels for reference image I_Ref_No then the defect is found. Otherwise, the test image doesn't have the defect [18, 20]. The following example shows two matrices I_Ref and I_Test as the reference and Test image respectively. Simply the overall number of white pixels in test image and reference image are 21 and 19 respectively. Therefore, we can decide the test image has defect because 21 is greater than 19 (Fig. 2).

3.3 Defect Classification

Defect classification plays the main part in defect detection algorithm that classifies the ceramic tiles automatically. After preprocessing operations and defect detection, defect classification specifies the type of defect if the defect was found. However, the defect classification step will not be carried out when the defect detection algorithm decided that there is no defect. In fact, there are various defect classification algorithms that are varied according to the defect type. In this section the author implemented two defect detection algorithms to detect spot and crack defects based on two research articles [18, 20]. Authors in [18, 20] had use the same preprocessing operations and defect detection algorithms for the ceramic tiles. However, the main difference between them in the pre-processing operations is the edge detection method. In [20] second derivative edge such as Laplacian of Gaussian (LOG) was implemented. First derivative edge detection such as Sobel operator implemented in [18]. The median filter are used to remove ceramic tile noise. The details of these algorithms are presented in the next subsections. Spot defect classification algorithm and crack defect classification algorithm are presented next.

3.3.1 Spot Defect Classification Algorithm

Defect classification algorithm for spot defect in [18, 20] are totally different. In [18] authors proposed algorithms to classify spot defect based on the size of the defect. They set the matrix as the size of the spot defect. After that, for each pixel in the ceramic tile, they computed the number of weighted pixels around the central pixel. Window size will be *matx* × *matx* used to specify the number of white pixels. Spot defect found when the number of white pixels around the current pixel is equal to *matx* × *matx*. In [20] the authors proposed algorithms to classify spot defects. They didn't mention the size of spot defect because they applied some methods such as LOG, a blurred image, and median filter. After that, they compared the total number of white pixels with 0.2 of image size. If the total number of white pixels greater than 0.2 of image size, then the spot defect is detected otherwise, spot defect is not detected.

3.3.2   Crack Defect Classification Algorithm

Defect classification algorithm for crack defect in [18, 20] are proposed the same algorithm based on the size of the crack defect. The algorithm was started by setting the length of the crack Crack_Length. After that, it moved from left to right and from top to bottom to comprehensive coverage all pixels in the ceramic tile. Connected pixels for each pixel are calculated and find the maximum of Connected Pixels. If the maximum Connected Pixels is greater than the Crack_Length, then the crack defect is detected. Otherwise, Crack defect is not detected.

4   Proposed Algorithm

In this section, we develop crack and spot defect algorithms similar to algorithms in [18]. However, the proposed algorithms are implemented by dividing reference image R_I and test image T_I into some blocks (e.g. 8 blocks in this paper) as shown in Fig. 4. After that, we deal with these blocks to check if defect found or not. Consequently, the classification algorithm will be applied only to the defective blocks in the Test image T_I. The pseudocode of the crack defect detection and classification with partition algorithm and spot defect detection and classification with partition algorithm are shown in algorithm 1 and algorithm 2 respectively.

Algorithm 1 illustrates crack defect detection and classification with partition. This algorithm starts with identifying the image size, number of blocks, and crack length. It includes three functions: the Image_Partition, the Crack_Defect_Detection and the Crack_Defect_Classification as shown in Fig. 3a. The Image Partition function divides the image into a number of blocks. The Crack_Defect_Detection function returns blocks of the ceramic images that have defects. In addition, the Crack_ Defect_Classification function identifies the blocks that have cracks. Algorithm 2 exhibits spot defect detection and classification with partition. It starts with determining the image size, a number of blocks, and spot size. In Addition, it includes three functions: Image_Partition, Spot_Defect_Detection and Spot_Defect_Classification as shown in Fig. 3b. The Image_Partition function divides the image into blocks. The Spot_Defect_Detection function identifies the blocks that have defects. In addition, the Spot_Defect_Classification function returns the blocks that have spot defect.

(a)   **(b)**

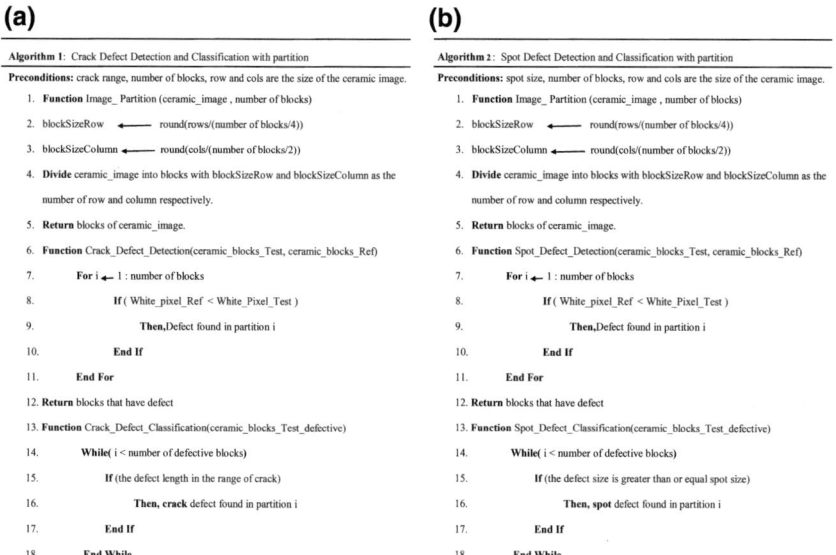

Fig. 3 **a** Crack defect detection and classification algorithm. **b** Spot defect detection and classification algorithm

5 Experiment Results

This paper developed experiments to evaluate the proposed partition algorithms for detecting both crack and spot defects. The setup of the developed experiments is introduced in the Sect. 5.1. Sections 5.2 and 5.4 introduce the performance criteria to check the effectiveness and accuracies of the defect detection and classification algorithms. Moreover, Sect. 5.3 discusses the results of the proposed algorithms and compared the proposed algorithms with the existing works in [18, 20].

5.1 Setup

The evaluation experiment was performed with Matlab 2015(a) installed on i7 machine, Windows 8.1 64-bit operating system,12 GB RAM and 2.4 GHz. This paper uses an offline test database of one hundred ceramic tile images. This paper used HP scanner to generate the database of different types of wall ceramic tiles with resolution varied from 100 to 200 DPI. Twenty percent of the database includes defect free ceramic images, while 40% of the database images have spot defects and the rest have crack defects. Moreover, to study the tolerance of the proposed algorithm against noises this paper added noises to about 10–20% of the ceramic tile images database.

5.2   Performance Criteria

Computing time is one the main criteria to study the performance of the proposed defect detection and classification algorithm with database of the ceramic tiles. The required time to process the detection and classification of defects for ceramic tiles in the industry is divided into detection time and classification time as shown in Eq. (1).

$$Total\_time = P_N + \sum_{i=1}^{N} Dt_i + Ct_i \tag{1}$$

where the P_N is required time to partition the test ceramic image into N blocks, Dt_i is the required time to detect either block i in the test ceramic tile has defect or not, Ct_i is the required time to classify the type of the defects in block i in the test ceramic tile. Ct_i equals zero if block i has no defects.

5.3   Discussion

The proposed system detects defects successfully. This section will show step-by-step the results of the proposed defect detection algorithm. Moreover, the proposed algorithm is compared with the existing techniques in [18, 20] in terms of detection rate and time. In addition, this section will discuss the tolerance of proposed algorithm against noises in ceramic tiles. To understand our proposed defect detection process, we apply the proposed method on a two RGB wall ceramic images. Then checks if any partition of the ceramic tile has defect or not. In advance, this paper applies the preprocessing operations on this image such as image enhancement, noise reduction and edge detection. After that, we also show the reference RGB image for that test image and the output after applying preprocessing operations on it. When defects exists in some partitions of the image, this paper applies a classification algorithm to classify the type of defects as shown in Figs. 4 to 5. Figure 4I(b-i) show eight partitions produced by applying the proposed algorithm on the test ceramic tile with crack defect that is shown in Fig. 4I(a). Partition one and three are the defective partitions that have a crack defect as shown in Fig. 5b. Figure 4II(b-i) shows eight partitions for the test image. These partitions have been produced by applying the proposed algorithm on the test ceramic tile Fig. 4II(a) with spots defect. Partition three and eight are the defective partitions that have spot defect as shown in Fig. 5a. We tested our proposed method for defect detection on about 100 ceramic tile images. In this case, the efficiency of the proposed defect detection algorithm is compared to the existing method [18, 20]. Table 1 shows the time comparison between the existing work and the proposed work. Table 1 comprises the reference and test ceramic images, the number of defects, the dimension of the image, time of existing work and our work, and speedup. The rightmost column in Table 1 shows that the proposed algorithm speedup the defect detection and classification processes. Figure 6

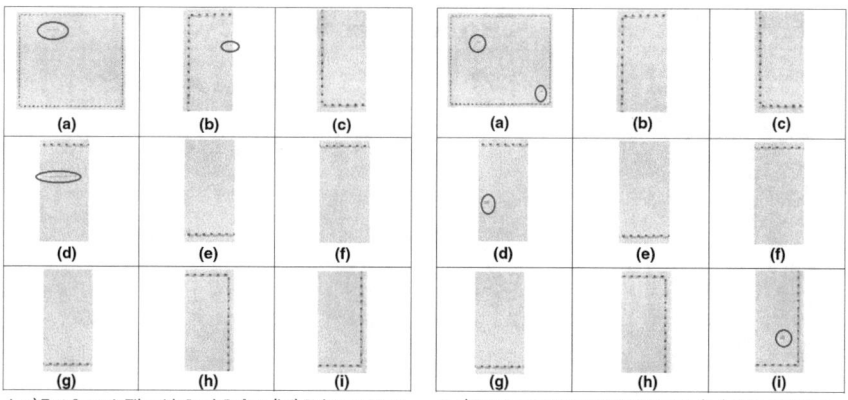

I - a) Test Ceramic Tile with Crack Defect, (b-i) Eight-partitions   II- a) Test Ceramic Tile with Spot Defect, (b-i) Eight-partitions

Fig. 4 Two samples of defective ceramic tiles

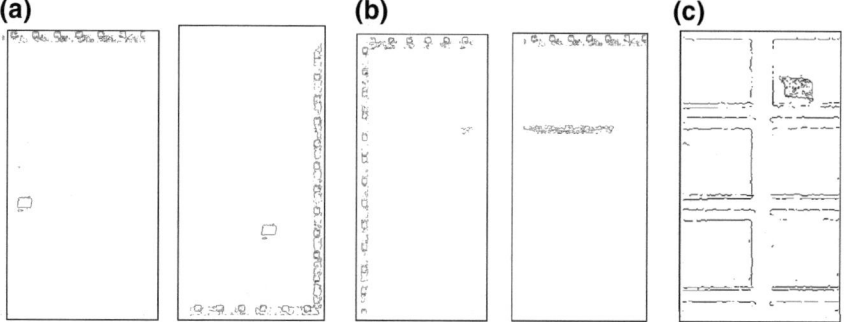

Fig. 5 **a** Spot defect exist in partition 3 and 8 **b** Crack defect exist in partition 1 and 3, **c** Spot defect in partition 1

shows the speedup of the proposed algorithm over the existing algorithms [18, 20] to detect and classify crack defect. This diagram plotted speedup with the dimension of ceramic tile image as shown in Table 1. In Fig. 6, crack and spot defect detection and classification speedup are green and red line respectively. Clearly, the size of each block is increasing due to the increasing of the size of ceramic tile image. Consequently, the speedup of crack defect classification goes down because the size of the block is 2374 × 1181. Moreover the time in such case will be increased because for each pixel in this block the connected pixels should be calculated. However, the speedup of spot defect goes up because spot defect classification depend on computed the number of weighted pixels around the central pixel for.

Table 1 Time comparison between existing work and our work

Ceramic Image (Reference - Test)	Number of defect	Algorithm	Dimensions	Time (Without Partition)	Proposed Algorithm	Speedup
	One		780 X 779	10.894433 s	1.617202 s	6.7
	Two	Spot Defect Detection	1560 X 1558	44.865346 s	4.894042 s	9.16
	One		4725 X4748	244.427150	11.673172	20.9
	One		780 X 779	103.27095 s	1.727216 s	59.79
	Two	Crack [27, 29]	1560 X 1558	595.760374s	4.816765 s	123.60
	One		4725 X4748	384.161847	5.301132	72.46

Fig. 6 Speedup for detecting spot and crack defects

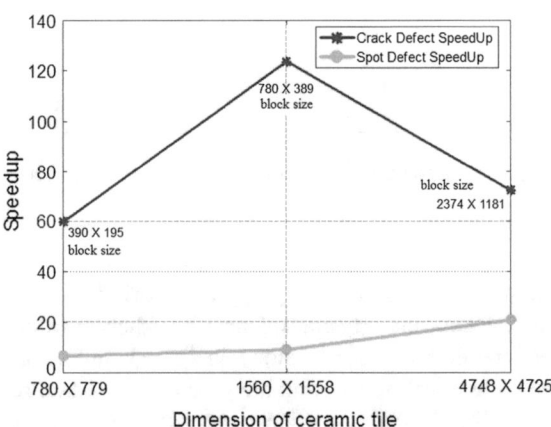

5.4 Evaluation Criteria of Defect Detection and Classification

The result of the defect detection algorithms may be fail in some cases. Therefore, various defect detection methods evaluated by different criteria to check the detection diagnosis and misdiagnosis. In this section, we introduce two evaluation criteria to check the effectiveness of the defect detection algorithms. The defect detection accu-

Table 2 Detection accuracy of esisting and proposed algorithms

Algorithm	True negative	True positive	False negative	False positive	Detection accuracy
Spot_defect_detection_Algorithm_[28]	20	27	34	19	0.47
Spot_defect_detection_Algorithm_[30]	20	53	14	13	0.73
Crack_defect_detection_Algorithm_[28, 30]	20	50	14	16	0.70
Proposed algorithm	20	52	14	14	0.72

racy as shown in Eq. 1 is used to determine the accuracy and the effectiveness of the defect detection algorithms [18, 20]. It is a classification rule which can be utilized by the following equation.

$$DA = \frac{TP + TN}{TP + TN + FP + FN} \tag{2}$$

where, TN True Negative, TP is True Positive, FN False Negative, and FP False Positive. True positive is referred to defective ceramic identified as defective. True negative is referred to defect-free ceramic identified as defect-free. False Positive is referred to defect-free ceramic identified as defective. False negative is referred to defective ceramic identifies as defect-free [17]. Table 2 shows the detection accuracy of the existing [18, 20] and proposed algorithms. The following results obtained by applying twenty defective free images, forty defective cracks, and forty defective spots. Detection accuracy of the proposed algorithm is the same as the existing work.

6 Conclusion

Ceramic manufacturing is a major industry. An automated ceramic defect detection system would naturally enhance the product quality. To reduce the computing time of the defect detection and classification this paper developed an efficient ceramic tile defect detection and classification algorithm. The experimental results verified that the developed solution succeeded to speed up the ceramic tile defects detection and classification in compared with the existing solutions. Moreover, the proposed solution retains the same accuracy as existed ones. The future work is to develop the proposed algorithm in parallel using the Graphical Processing Unit (GPU) and further work is to classify either the defect is spot or tail of crack in consecutive partitions.

References

1. Boukouvalas, C., Kittler, J., Marik, R., Mirmehdi, M., Petrou, M.: Ceramic tile inspection for colour and structural defects. In: Proceedings of Advanced Materials and Processing Technologies, pp. 390–399 (1995)
2. Brzakovic, D., Beck, H., Sufi, N.: An approach to defect detection in materials characterized by complex textures. Pattern Recognit. 23(1), 99–107 (1990)
3. Drobina, R., Machnio, M.S.: Application of the image analysis technique for textile identification. AUTEX Res. J. 6(1), 40–48 (2006)
4. Elbehiery, H., Hefnawy, A., Elewa, M.: Surface defects detection for ceramic tiles using image processing and morphological techniques. In: WEC (5), pp. 158–162 (2005)
5. Elbehiery, H.M., Hefnawy, A.A., Elewa, M.T.: Visual inspection for fired ceramic tile's surface defects using wavelet analysis. Int. J. Comput. Sci. Inf. Secur. 7(2), 67–74 (2005)
6. Fathi, A., Monadjemi, A.H., Mahmoudi, F.: Defect detection of tiles with combined undecimated wavelet transform and glcm features. Int. J. Soft Comput. Eng. (IJSCE) 2(2), 4–30 (2012)
7. Hocenski, Z., Aleksi, I., Mijakovic, R.: Ceramic tiles failure detection based on fpga image processing. In: IEEE International Symposium on Industrial Electronics, 2009. ISIE 2009, pp. 2169–2174. IEEE (2009)
8. Islam, M.M., Sahriar, M.R.: An enhanced automatic surface and structural flaw inspection and categorization using image processing both for flat and textured ceramic tiles. Int. J. Comput. Appl. 48(3) (2012)
9. Kang, G.: Digital Image Processing. Quest, vol. 1, pp. 2–20. Autumn 1977, 1, 2–20 (1977)
10. Karimi, M.H., Asemani, D.: Surface defect detection in tiling industries using digital image processing methods: analysis and evaluation. ISA Trans. 53(3), 834–844 (2014)
11. Kwak, C., Ventura, J.A., Tofang-Sazi, K.: Automated defect inspection and classification of leather fabric. Intell. Data Anal. 5(4), 355–370 (2001)
12. Lees, D.E., Henshaw, P.D.: Printed circuit board inspection-a novel approach. In: Cambridge Symposium_Intelligent Robotics Systems. International Society for Optics and Photonics, pp. 164–173 (1987)
13. Matić, T., Hocenski, Ž.: Parallel processing with cuda in ceramic tiles classification. In: Knowledge-Based and Intelligent Information and Engineering Systems, pp. 300–310. Springer (2010)
14. Mery, D., Pedreschi, F.: Segmentation of colour food images using a robust algorithm. J. Food Eng. 66(3), 353–360 (2005)
15. Mishra, R., Shukla, D.: A survey on various defect detection. Int. J. Eng. Trends Technol. 10(13), 642–648 (2014)
16. Ng, H.F.: Automatic thresholding for defect detection. Pattern Recognit. Lett. 27(14), 1644–1649 (2006)
17. Ragab, K.: Fast and parallel summed area table for fabric defect detection. Int. J. Pattern Recognit. Artif. Intell. (2016)
18. Rahaman, G., Hossain, M., et al.: Automatic Defect Detection and Classification Technique from Image: a Special Case Using Ceramic Tiles (2009). arXiv:0906.3770
19. Segnini, S., Dejmek, P., Öste, R.: A low cost video technique for colour measurement of potato chips. LWT-Food Sci. Technol. 32(4), 216–222 (1999)
20. Shire, A.N., Khanapurkar, M., Mundewadikar, R.S.: Plain ceramic tiles surface defect detection using image processing. In: 2011 4th International Conference on Emerging Trends in Engineering and Technology (ICETET), pp. 215–220. IEEE (2011)
21. Xie, X.: A review of recent advances in surface defect detection using texture analysis techniques. Electron. Lett. Comput. Vis. Image Anal 7(3) (2008)
22. Yang, X., Pang, G., Yung, N.: Robust fabric defect detection and classification using multiple adaptive wavelets. In: IEE Proceedings-Vision, Image and Signal Processing, vol. 152, pp. 715–723. IET (2005)

23. Yeh, J.C., Hamey, L.G., Westcott, T., Sung, S.K.: Colour bake inspection system using hybrid artificial neural networks. In: Proceedings of IEEE International Conference on Neural Networks, 1995, vol. 1, pp. 37–42. IEEE (1995)
24. Zhi, Y.X., Pang, G.K., Yung, N.H.: Fabric defect detection using adaptive wavelet. In: 2001 IEEE International Conference on Acoustics, Speech, and Signal Processing, 2001. Proceedings. (ICASSP'01), vol. 6, pp. 3697–3700. IEEE (2001)

Parking Assistant—Prediction of an Empty Parking Space in Time

Jan Tobola, Jan Dvorak and Ondrej Krejcar

Abstract This chapter concentrates on topic of intelligent helper for parking for mobile devices that are connected to internet and contain the GPS module. The helper focuses on the navigation of the user to the last known location of the vehicle, but also on the prediction and visualization of a possibly empty parking space at a given time. The assistant collects data about parking and processes them further. In this way, the application is able to recognize spaces where the user parks more commonly. After some time, it is also capable of predicting the area, where a free space could be found at a specific time. This area is then suitably visualized. Within this project, an application that contains parts of the described functionalities was created. It runs on the Google Android platform. Some suggestions mentioned in this chapter are currently only theoretical and can be further developed in the future research on this topic.

Keywords GPS navigation · Mobile applications · Parking assistant · Visualisation · Android

1 Introduction

Nobody likes driving around the city blocks and looking for a place to park the vehicle on the overfilled parking lots. In the same way, the search for the parked vehicle can be quite aggravating. Nowadays, people are surrounded with smart

J. Tobola · J. Dvorak · O. Krejcar (✉)
Center for Basic and Applied Research, Faculty of Informatics and Management,
University of Hradec Kralove, Rokitanskeho 62, 500 03 Hradec Kralove,
Czech Republic
e-mail: Ondrej@Krejcar.org

J. Tobola
e-mail: jan.tobola@uhk.cz

J. Dvorak
e-mail: jan.dvorak@uhk.cz

© Springer International Publishing AG 2017
D. Król et al. (eds.), *Advanced Topics in Intelligent Information
and Database Systems*, Studies in Computational Intelligence 710,
DOI 10.1007/978-3-319-56660-3_22

mobile devices, internet and GPS and therefore it should be possible to fully use the potential of those technologies in order to simplify the daily, but sometimes unpleasant tasks.

This chapter concentrates on the possibilities of using mobile devices [1] and their applications for simplifying vehicle parking and the vehicle's localisation using geographical latitude and longitude, obtained from the GPS satellites. However, the focus of this paper is not on the principle of GPS functioning (Global Positioning System), as it is discussed in the chapter [2].

Some of the applications that are available today use information about occupancy of the public parking lots from the paid services. However, these services do not offer data for all the cities and concentrate only on the paid parking lots, where the availability is observed by the parking lots' owners themselves. The proposal which is introduced in this chapter offers a different point of view. The application is used for remembering the current location of the parked vehicle by the user and for the navigation to return to it. There are many available applications which use such a function and they mostly function efficiently. However, the described parking assistant collects data and processes them further. In this way, the application recognizes places where the user often parks and after some time is able to estimate the area in which there is the highest probability to find a free parking space in the given time. Within these projects, an application that works on the Google Android platform and demonstrates a part of the described functionalities was created. Some suggestions mentioned in this chapter are currently only theoretical and can be further developed in the future research on this topic.

Another important part of this application, that should offer this sort of information, is clearly the suitable graphical representation. Therefore, the functionality for displaying the individual parking areas on the map, where the parking took place in the given time is included in the suggestion of the parking assistant application. In order to display this area (sector), it is necessary to firstly detect the cluster points in accordance to their density (Density-based clustering [3], DBSCAN algorithm [4]) and then find a convex hull of the given set of points. This can be achieved by various geometric algorithms, as the Graham Scan [5], Quick Hull [6], Jarvis March [7], and others which is showed on the figure (Fig. 1).

Fig. 1 Convex hull of points in detected clusters. *Source* [8]

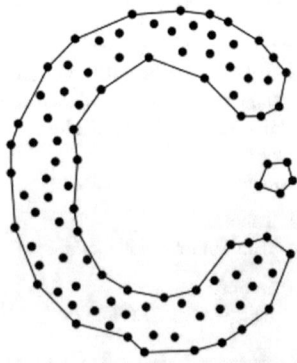

Fig. 2 Visualisation of the estimated area

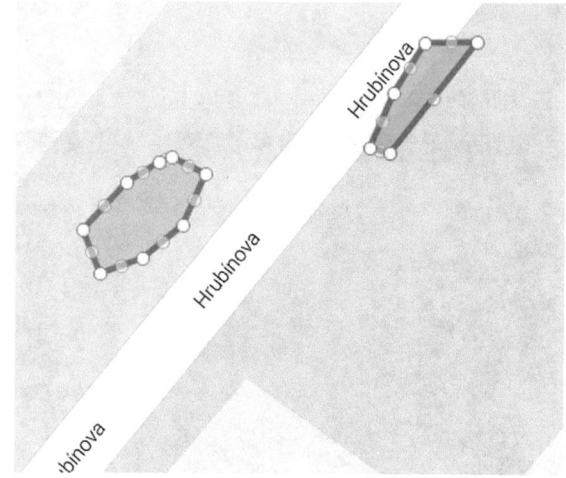

This method for searching for a parking space is only estimation and it can never provide exact results. The result is obtained from the data about previous parking. When a higher number of parking around the same place occurs and the algorithm parameters are suitably adjusted, it is possible to more precisely estimate the best parking space in the given moment. The visualisation of such space can be displayed in such a way, as it is shown on the figure (Fig. 2).

2 Problem Definition

As it was already mentioned in the previous chapter, multiple applications already exist which concentrate specifically on the parking and navigation to the free parking spaces. Some applications are available in the mobile store with Google Play applications. These function on the principle of remembering the current location and further launch of the navigation to the parked vehicle. Mostly, these application are quite simple and do not offer other improved functions or statistics.

The special type of applications is those that offer the possibility to search for the free parking spot and further reservation of this spot. These services are paid and the information about the status of the parking lots are provided by the partners. These include services such as Spot Hero, ParkWhiz and Parker. The preview of the user's interface [9] from the Spot Hero application is shown on the figure (Fig. 3). It is possible to see here the marking of the paid parking lots. In the second half of the image, the booking of the space is displayed.

The problem with such applications is the limited town database and the fact that they show only parking lots whose status is well known and regularly updated.

Another interesting and effective solution is the detection of the parking spaces using the algorithm from the area of computer vision. The detection of the parking

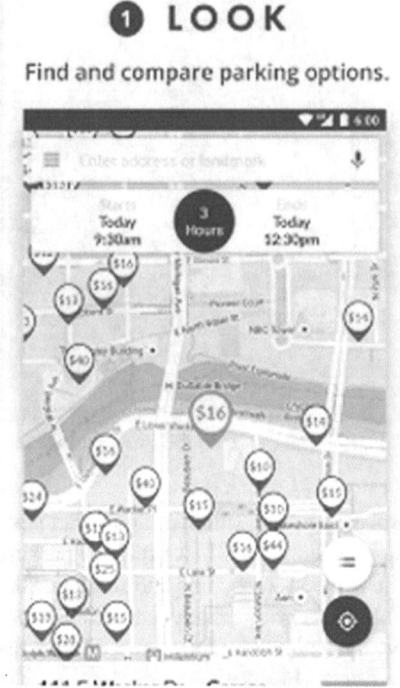

Fig. 3 Spot Hero. *Source* Google Play

space takes place from the static images made at a given time using the outside cameras [10], as shown on the figure (Fig. 4).

The project Street parking detection [11] concentrates on a similar approach. The disadvantages of this solution are the necessity to use other devices, more

Fig. 4 Detection of parking spaces from cameras placed outside [11]

complicated calculation for image processing and the inability to control all places, due to a limited image. Another potential disability could be the legal aspects which are connected to installing cameras.

In accordance to this problematic, there are a few theoretical suggestions in the form of patents that focus on the search of free parking spaces. These suggestions concentrate on the complex solution of data collection into geographical databases on a global level and offer sophisticated infrastructure and optimisation for providing better results. Patent [12] presents a suggestion for information offering about free parking spaces from the historical data that were collected by other users. For evaluation, it uses not only actual time, but also other known factors, which could influence the quantity of the vehicles for the given location, such as current weather, season, and information about traffic. The system compares the real-time data with historical data and executes the actualisations according to the suggested algorithm for specification of other enquiries about free parking spaces. This solution requires on global level significant amount of users, due to which the data collection can take place in order for these data to be of an asset to the other end users. Further, the access to above mentioned services for possible reservation of the parking spot is here included. Therefore, the infrastructure counts on the communication with other partners and their systems. However, the visualisation of concrete possible parking areas is not solved here.

The suggested solution in this chapter looks for the compromise between other solutions. It combines simple accesses for navigation to the parked vehicle, together with the prediction of free parking spaces from the historical data in repeatedly visited areas. However, the database is not on a global level as in the case of the previous solution. Therefore, each user builds his own database. The complex solutions require also a more sophisticated infrastructure proposal than in case of the suggested solution. In this chapter, the focus is on the simplicity of the suggested infrastructure and communication among other system components.

3 New Solution

The functioning principle of this solution was already briefly introduced in the previous chapters. In this chapter, the focus is laid on the possibilities, limitations, advantages and disadvantages of this solution.

From the basic point of view, it is a simple mobile application which works on the basis server-client (Fig. 5).

Fig. 5 Client-server communication

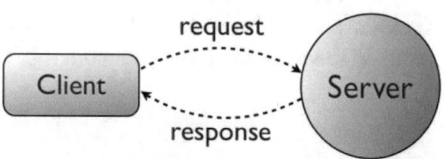

The communication between both ends runs through the REST web services [13]. The application could be implemented also without the server. The mobile device would save the needed data on its internal hard drive and would conduct also the data processing. The availability of the server has the advantage of saving the data outside of the mobile device [1]. This data can be then available through e.g. web interface with the possibility of extended statistics about parking [9].

The function of navigation to the parked vehicle will work in an easy way. After the launch of the application on the mobile device, the application forces the activation of GPS [14] and mobile data. It is important for the user to not have to manually turn these services on, because it would negatively influence the comfort of using this application. The user can make use of the button for saving the current location in the case when the vehicle is being parked. The device with GPS recognizes the current location of the device and receives two coordinates (latitude, longitude) which define the position on earth. Afterwards, the device sends a POST request on the server together with these coordinated, date and time of the parking. The next parameter would be the user's ID which would enable his clear identification on the server's site. This parameter is for now left out from the proposal for simplicity reasons. After receiving the message, the server parses the data and saves them into the database. If everything runs correctly, it sends a response with http status 200 and possibly with another own message. In this way the client receives a confirmation that his position was processed and saved by the server. Moreover, he should receive a notification from the application. The communication among components is described in the figure (Fig. 6).

The communication before the launch of navigation to the last saved place is run analogically. The device asks the server for return of the last saved coordinates, server reads the data from the database and returns them to the client. The device can then run the sub-programme which leads the user to the required location.

The parking assistant proposal also includes the function for displaying the sectors on a map, in which the parking took place in the given time. For displaying of such sector, it is necessary to firstly create sets of point clusters according to their density and further find a convex hull for each set. The algorithms which enable this processing were already introduced in the last chapter. These operations were conducted using suitable algorithms on the server. Only the found convex hulls would be sent back to the mobile device which would decrease the number of transferred data between the client and the server. The given sectors can be displayed from the obtained points, e.g. using Google Maps API [14], directly on the given device. Another advantage of delegating the task to the server is the moderate use of the mobile device [1], as it does not have to conduct the operations that require more challenging calculations.

The displayed sectors on the map provide the user with an overview of places where he already in the given time and season parked. The higher the density of the points in the area, the more probable it is that the user could again find in the given

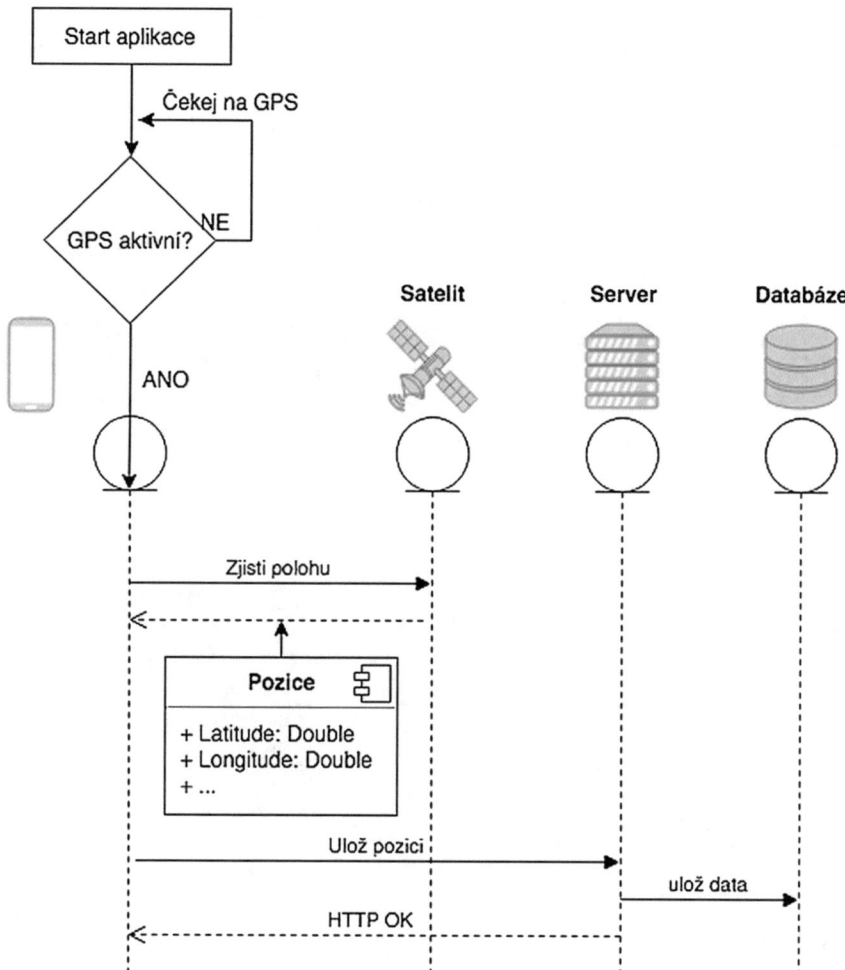

Fig. 6 Communication among components

time in this area a free parking space. The algorithms' parameters for data evaluation should be set empirically, probably in accordance to the chosen map scale.

Within this map visualisation, different views on the data could be seen. One of such displays would show only the points in the exact same way as they were saved. Another display would show individual sectors in different times (orders—hours, minutes) or for example the intersections of individual areas in the form of heat map during longer time sequences (months, seasons, years)—see figure (Figs. 7 and 8).

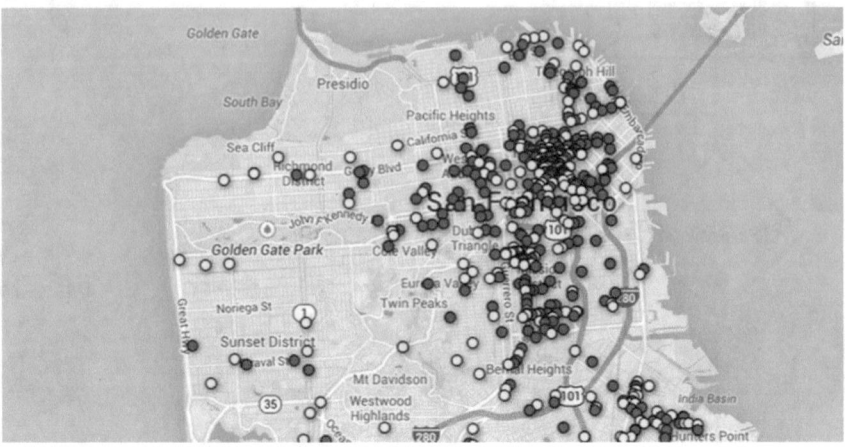

Fig. 7 Visualisation of parking history

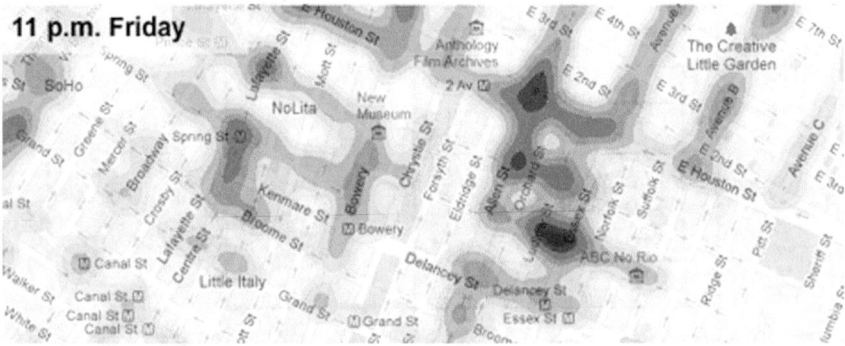

Fig. 8 Visualisation of parking blocks using heat map

4 Implementation

A trial mobile application was implemented on the Google Android platform and the server part was in Java language with the use of Spring Boot framework and MongoDB database. These modern server technologies were chosen due to the simple configuration that they offer. Apart from the product setting, it is also perfectly suitable for simple prototyping and testing of applications [15].

The introductory screen of the application contains three buttons in the middle and notification row under them. The first button saves the current position of the device by sending coordinates on the server. If the GPS is not active, the communication with the server is automatically disabled. The action of the first button is shown on the figure (Fig. 9).

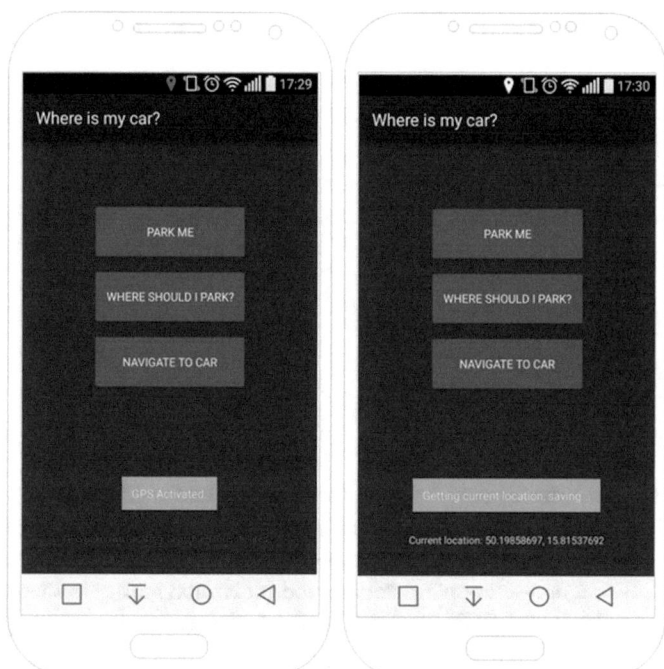

Fig. 9 Location saving

The generic class from the standard Android library—AsyncTask—was used for the sending of asynchronous request to the server. As the HTTP client, the Retrofit library was used.

A servlet has to be implemented on the server. It processes the clients' requirements and answers them in a suitable way (returns data). When using the Spring framework, this function is run by the DispatcherServlet. It is only necessary to create a Spring controller with the given mapping on URL and the HTTP method of the requirement.

For example the method for returning of the last saved location can be presented in the following way:

```
@RequestMapping("/last")
public LocationResponse getLastLoc() {
    ParkingEntity last =
parkingRepository.findFirstByOrderByParkingTimeDesc();
    LocationResponse response = new LocationResponse();
    response.setLat(String.valueOf(last.getLat()));
    response.setLon(String.valueOf(last.getLon()));
    return response;
}
```

LocationResponse is a subject, which is after the return from the controller method caught by the Jackson library and is serialised intro the JSON format. The resulting JSON that is received by the mobile device in the answer is displayed followingly:

$$\{"lat":"50.19859026","lon":"15.81536697"\}$$

The application uses there coordinates and other parameters to enter the foreign sub-programme Google Maps which launches the navigation to the final location. The Intent class is used for the launch of a new activity in the Android device.

```
Uri googleUri = Uri.parse("google.navigation:q="
        + locationResponse.getLat() + ","
        + locationResponse.getLon()
        + "&mode=w");
Intent mapIntent = new Intent(Intent.ACTION_VIEW, googleUri);
mapIntent.setPackage("com.google.android.apps.maps");
activity.startActivity(mapIntent);
```

The previous code launches the navigation in Google Maps in the w—walk mode, as it is expected that the user who is looking for his vehicle is walking. The figure (Fig. 10) demonstrates the navigation to the location that server returned.

MongoDB is a non-relational database and individual entries are saved in it as documents similar to the JSON format. This database enables the use of aggregation

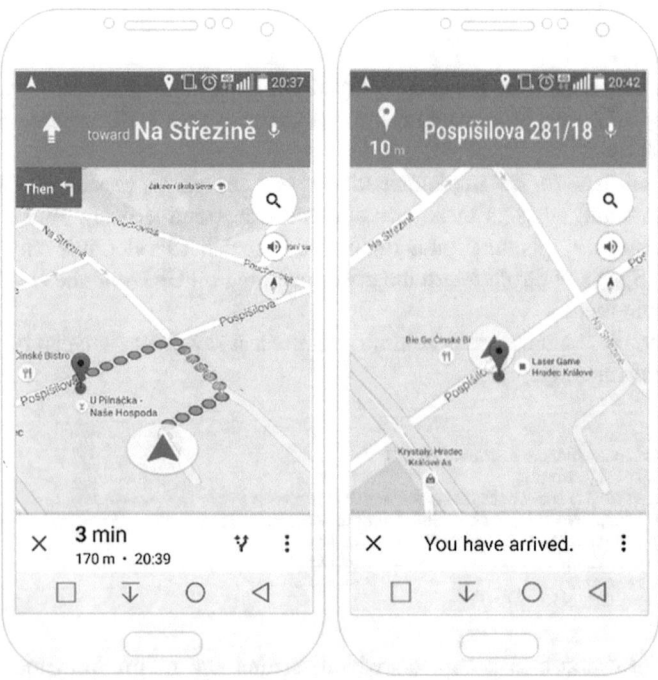

Fig. 10 Navigation to the vehicle

framework for aggregation and transformation of data. It functions on the principle of unix pipeline and simple modifiers which transfer data. In this trial application the data are aggregated according to hour-time intervals. This is enabled for example by the project operator for the structure change of the document together with other time aggregate operators. The aggregation provides simple filtering, data clustering, releasing, renaming or the changing of attribute embedding level. It is possible to obtain data from the database in a form in which they should arrive to the end device, when the enquiry is correctly written.

In order to create aggregating enquiries, the API from Spring Data was used. This API offers comfortable writing (generating) of enquiries using the Java method and objects. The next enquiry returns a list of coordinates about a defined maximal size in the given hour-time interval.

5 Conclusions

After the testing of the application, the navigation to the parked vehicle was found to be very practical. This finding is also supported by the number of downloads of similar application in Google Play store and App Store where it reaches few millions. The motivation lays in making a use of the nowadays available common technologies and in simplifying the overcoming of everyday tasks. The proposal with the use of the documented database offers a non-challenging data collection. Moreover, there is a great potential for evaluating of those data for their visualization and generation of various statistics in any geographical area. In the visualization area, it would be possible to cooperate with other current technologies, such as Google Earth or Street View, and process data for use in three-dimensional scope.

Acknowledgements This work and the contribution were supported by project "SP-2017—Smart Solutions for Ubiquitous Computing Environments" Faculty of Informatics and Management, University of Hradec Kralove, Czech Republic.

References

1. Behan, M., Krejcar, O.: Modern smart device-based concept of sensoric networks. EURASIP J. Wireless Commun. Netw. **155**, 2013 (2013)
2. Xu, G.: GPS: Theory, Algorithms and Applications. Springer Science & Business Media (2007)
3. Wicker, N., Dembele, D., Raffelsberger, W., Poch, O.: Density of points clustering, application to transcriptomic data analysis. Nucleic Acids Res. **30**(18), 3992–4000 (2002)
4. Ester, M., Kriegel, H.-P., Sander, J., Xu, X., Ester, M., et al.: A Density-Based Algorithm for Discovering Clusters in Large Spatial Databases with Noise. Institute for Computer Science, University of Munich, Munich (1996)

5. Kong, X., Everett, H., Toussaint, G.: The Graham scan triangulates simple polygons. Pattern Recognit. Lett. **11**(11), 713–716 (1990)
6. Barber, C.B., Dobkin, D.P., Huhdanpaa, H.: The Quickhull Algorithm for convex hulls. ACM Trans. Math. Softw. **22**(4), 469–483 (1996)
7. Atwah, M.M., Baker, J.W., Aki, S.: An Associative Implementation of Classical Convex Hull Algorithms. Mathematics and Computer Science, Kent State University, Kent (1996)
8. Duckham, M., Kulik, L., Worboys, M., Galton, A.: Efficient generation of simple polygons for characterizing the shape of a set of points in the plane. Elsevier (2008)
9. Behan, M., Krejcar, O.: Adaptive graphical user interface solution for modern user devices. In: Lecture Notes in Computer Science, LNCS vol. 6592, pp. 411–420 (2012)
10. True, N.: Vacant Parking Space Detection in Static Images. University of California, San Diego (2008)
11. Sergio Patricio Figueroa Sanz, J. Tran, Wang, C.Y.: Street Parking Detection. https://pdfs.semanticscholar.org/2783/e190b8077402e351a1aa1142442c1d06df48.pdf?_ga=1.81770263.1610325849.1481839284 (2015)
12. Kaplan, L.M., Hayes, H.R., Devries, S., Lindsay, M.G.: Method of operating a navigation system to provide parking availability information. Patent—US 7949464 B2 (2006)
13. Richardson, L., Ruby, S.: RESTful Web Services. O'Reilly Media, Inc. (2008)
14. Benikovsky, J., Brida, P., Machaj, J.: Proposal of user adaptive modular localization system for ubiquitous positioning. In: LNCS, vol. 7197, pp. 391–400 (2012)
15. Hustak, T., Krejcar, O., Selamat, A., Mashinchi, R.: Principles of usability in human-computer interaction driven by an evaluation framework of user actions. In: Mobile Web and Intelligent Information Systems, Lecture Notes in Computer Science, vol. 9228, pp. 51–62 (2015)

Recent Advances in the Field of Foreground Detection: An Overview

Ajmal Shahbaz, Laksono Kurnianggoro, Wahyono and Kang-Hyun Jo

Abstract Foreground detection is the classical computer vision task of segmenting out moving object in a particular scene. Many algorithms have been proposed in the past decade for foreground detection. It is often hard to keep track of recent advances in a particular research field with the passage of time. An overview paper is an effective way for the researchers to compare several algorithms according to their strengths and weaknesses. There are several overview papers in the literature; however, they are somewhat obsolete. This overview paper covers the recent algorithms proposed in past 3–5 years except Gaussian Mixture Models (GMM). The aim and contribution of this overview paper is as follows: First, algorithms are classified in three different categories on the basis of choice of picture's element, feature, and model. Then, each algorithm is summarized concisely. Furthermore, algorithms are compared quantitative and qualitatively using large realistic standard dataset. Paper is concluded with several promising directions for future research.

Keywords Foreground detection · Overview · Change detection dataset · Background maintenance · Background modeling

A. Shahbaz (✉) · L. Kurnianggoro · Wahyono
The Graduate School of Electrical Engineering, University of Ulsan,
Ulsan 44610, Korea
e-mail: ajmal@islab.ulsan.ac.kr

L. Kurnianggoro
e-mail: laksono@islab.ulsan.ac.kr

Wahyono
e-mail: wahyono@islab.ulsan.ac.kr

K.-H. Jo
The School of Electrical Engineering, University of Ulsan, Ulsan 44610, Korea
e-mail: acejo@ulsan.ac.kr

© Springer International Publishing AG 2017
D. Król et al. (eds.), *Advanced Topics in Intelligent Information
and Database Systems*, Studies in Computational Intelligence 710,
DOI 10.1007/978-3-319-56660-3_23

261

1 Introduction

Foreground detection aims at separating the moving object (called foreground object) from the static information called background. It is often considered backbone of multistage computer vision systems. It is primary step in many computer vision applications such as video surveillance systems [1, 2]. Therefore, the outcome of foreground detection directly affects the overall performance of the system.

The field of foreground detection has seen fast transition in the past few years. The reason might be many fold. First, development of the only large realistic dataset [3] consisting of almost 140,000 annotated frames with performance metrics. It helped researchers to test their algorithms with one concrete benchmark dataset. Secondly, Researchers investigated new ways to model background.

The overview papers are found to be very beneficial for researchers particularly novices. It makes easy to keep track of past and recent advances in a particular field. Several overview papers can be found in the literature [4, 5]. The first overview paper in the field of foreground detection can be tracked back to [4]. Mc Ivor [4] described 9 algorithms with the overview of steps of foreground detection. Piccardi [6] surveyed 7 algorithms with quantitative comparison of speed, computational cost, and accuracy. Cheung et al. [7] surveyed 11 algorithms and classified them into recursive and non-recursive methods. Elhabian et al. [8] provided a large survey in the background modeling. However, this classification in term of non recursive techniques is more adaptive for background maintenance than background modeling. Cristani et al. [9] classified algorithms in monocular and multiple sensor algorithms. Ajmal et al. [10] concise review of traditional algorithms with computational cost, accuracy, etc. for video surveillance applications. Bouwans et al. surveyed background subtraction algorithms according to different approaches [5, 11].

There were many overview papers published in the past. The latest survey paper was published in 2014 [12]. They do not covers most recent algorithms. Also, there is none overview papers that provides qualitative and quantitative comparison on recent algorithms on the basis of change detection dataset [3]. This overview paper is unique as it only compare most recent algorithms in field of foreground detection using change detection dataset [3]. Rest of the paper is organized as follows: Sect. 2 presents classification and concise summary of each algorithm. Experimental results and comparative evaluation are presented in Sect. 3. At the end, paper is concluded with future directions.

2 Summary of Recent Approaches

This section presents the classification of algorithms, and concise summary of working principle of algorithms. Each algorithm is summarized according to the type of model.

2.1 Classification of Algorithms

The algorithms are classified into three different categories:

1. Choice of picture's element (pixel or region): Usually pixel based methods gives good precision but not robust to noise than region based method.
2. Choice of feature type (color, edge, stereo, motion, or texture): Color features are most commonly used. But they have suffers from camouflage, shadows, illumination, etc.
3. Choice of model type (parametric, non-parametric, or hybrid): Parametric models are most commonly used. They rely on definition of several parameters for modeling background. Non-parametric methods only rely on observed data for modeling background.

2.2 Parametric Model Based Algorithms

Gaussian Mixture Models (GMM) [13, 14] is the most famous parametric based algorithm. It was proposed by Stauffer and Grimson [13] in 1999. Since then various improvements were published but the ones by Zivkovic [15] and Kawtrakulpong [14] are the most famous ones. In GMM [13, 14], each pixel is modeled as a mixture of Gaussian using mean and variance. The probability of observing a particular pixel value X_t is

$$P(X_t) = \sum_{i=1}^{K} \omega_{i,t} \eta(X_t; \mu_{i,t}, \sigma_{i,t}^2) \qquad (1)$$

where η, K, $\omega_{i,t}$, $\mu_{i,t}$, and $\sigma_{i,t}^2$ are probability density fucntion, number of Gaussian, estimate of weight, mean, and variance of the ith Gaussian in the mixture at time t. The decision criteria to mark particular pixel at time t as background or foreground is:

$$|X_t - \mu_{i,t}| > \lambda \sigma_{i,t} \qquad (2)$$

where λ is a constant threshold equal to 2.5. If a match is found with one of the K Gaussian components, the pixel is classified as background and its parameters are updated. If a pixel is labeled as foreground, only weight ω is updated.

Varadarajan et al. [16] proposed spatio-temporal framework named spatial GMM. It is specifically designed to handle dynamics backgrounds. Consider the data $X = x_1, x_2, \ldots, x_n$ where is number of data samples. Let i be number data samples in the region and k be number of classes a data sample could belong to. $z_{i,k} = [0, 1]$, be the membership of data sample to a cluster at position i. $\theta = [\omega_{i,k}, \mu_{i,k}, \sigma_{ik}^2]$, be the parameters of model.

$$E_p(z_{i,k}|x_q, \theta^{old})(z_{i,k}) = \sum_{q \in R_i} \gamma_q(z_{i,k}) \tag{3}$$

$$\gamma_q(z_{i,k}) = \frac{\omega_{i,k}\eta(x_q|\mu_{i,k}, \sigma_{i,k}^2)}{\sum_{q \in R_i} \sum_{j=1}^{K} \omega_j \eta(x_q|\mu_{i,j}, \sigma_{i,j}^2)} \tag{4}$$

Yingying et al. [17] uses the idea of traditional GMM with sharable mechanism to exploit the spatial temporal correlation between the pixels. Pixel values are modeled using Gaussian Mixture Models. The probability of observing certain pixel X at time t can be computed by Eq. (1). The sharable mechanism demonstrates many to one relationship between pixel and models. For each pixel, optimal model is searched from background and foreground models in $N \times N$ region. Pixel labeling decision follows Eq. (2).

Hasan et al. [18] proposes universal background subtraction system (UBSS). Their algorithm selects optimal color space (RGB or YCbCr) for the task of background subtraction. Initial frames without foreground object are used to model background named as Background Model Bank (BMB) which consists of single Gaussian. Initial frames are clustered into N groups based on correlation measure using K-means. Once background model is chosen, it is passed to background subtraction modules (also known as binary classifiers, BC) along with input image which produces binary mask D for each color channels of color space. All the background masks produced by BCs are aggregated into final foreground detection mask.

2.3 Non-parametric Model Based Algorithms

Hoffman et al. [19] developed non-parametric model called pixel based adaptive segmenter (PBAS) that models background by accumulating history of recently observed pixel values. PBAS [19] uses idea behind [20] and consists of many components. The main component, decision block, labels pixel as foreground based on per-pixel threshold on current image and background. The background model is updated over time to deal with gradual background changes. Sample consensus approach [20] determines whether a given observation can be considered as foreground or background based on its similarity with previously observed samples. First, background model B is formed through combination of pixels, which each contain a set of N recent background samples. A pixel is decided to be background if pixel value I is closer than a certain decision threshold R to at least $\sharp_{min}$ of the N background values. The decision labeling equation can be written as:

$$F(x_i) = \begin{cases} 1 & \text{if } (dist(I(x)_n, B(x)_n) < R, \forall n) < \sharp_{min} \\ 0 & \text{otherwise} \end{cases} \tag{5}$$

Thus, $F(x_i) = 1$ is foreground. The background model is updated for those pixels that are background i.e. $F(x_i) = 0$. This means that the pixel value in background model is updated by the current pixel value. This is only performed if the current pixel has high probability to be part of background. Furthermore, some random neighboring pixel can be updated as background even it is part of foreground. Further details can be found in [19].

SuBSENSE [21] is non-parametric model based algorithm that exploits feature space namely local binary similarity patterns (LBSP). It detect change by comparing a center pixel with neighboring pixels. LBSP can be considered counterparts of local binary pattern (LBP) and local ternary pattern (LTP). LBSP binary string can be computed using following equation:

$$LBSP(x) = \sum_{p=0}^{P-1} d(i_n, i_c).2^p \tag{6}$$

where i_n and i_c are neighboring pixels and center pixel respectively. The foreground and background is distinguished using sample consensus [20]. A pixel is decided to be background if pixel value I is closer than a certain decision threshold R to at least $\sharp_{min}$ of the N background values (Eq. 5).

PAWCS [22] is another non-parametric method based on [19–21]. This method's key advantages lies in its highly persistent and robust dictionary models based on color and local binary features as well as its ability to automatically adjust pixel level segmentation. A word based approach is implemented for monitoring of background representations at the pixel level without clustering. This appearances of pixels over time are termed as background words in local dictionaries using color and texture information. If the representation occurs persistently then it is termed as good representations of background. Infrequent representations are discarded and replaced by the better alternatives.

2.4 Hybrid Model Based Algorithms

Flux tensor with Split Gaussian models (FTSG) [23] is hybrid foreground detection method which uses motion, change, and appearance information. FTSG uses split Gaussian method to separately model foreground and background. It consists of three main modules: pixel level motion detection module, fusion module, and object level classification module. Multichannel flux tensor is used to detect motion. Trace of flux tensor matrix can be used to classify moving and non-moving regions. Split Gaussian are used to model background and foreground. Mixture of Gaussian is used to model background. Single Gaussian is used to model foreground. The foreground appearance model is used to distinguish static foreground from noise. Foreground mask obtained by flux tensor and split Gaussian are fused using rule based system to produce improved results. In object level classification module, removed and stopped objects are handled. Simone et al. [24] proposed a framework utilizing ability of genetic programming (GP) to combine several state of the art algorithms. GP algorithm automatically select the best algorithms, combine them in different ways, perform the most suitable post processing operations on the output of the algorithms. Unary, binary, and n-ary functions embedded into the GP framework for combining particular algorithms. In this way, the algorithm is ranked 1st in the change detection dataset [3]. The authors claims that they have combined 22 algorithms with the subsets of 3, 5, and 7 methods. As claimed by authors, the benefit of using genetic programming is threefold: (1) Automatic selection of algorithms that gives the best results. (2) Automatic deduction of ways to select the algorithms to generate intermediate masks. (3) Automatic selection of kind post processing by using unary, binary, and n-ary functions.

3   Experimental Results and Comparative Evaluation

Table 1 shows that quantitative comparison of seven performance metrics defined by the change detection dataset. Considering F measure IUTIS [24] is the most successful algorithm. GMM based algorithms such as Shared GMM [16] and UBSS [18] shows promising results. Overall, non-parametric model based algorithms performs very well. The criteria based on F-measure is used to declare if an algorithm performs well in particular category of change detection dataset. If F-measure of an algorithm in certain category is equal to or more than 0.60, than algorithms performs well in that category. All algorithms performed well in dynamic backgrounds category and all failed in night videos, low frame rate, and PTZ.

The foreground masks produced by the algorithms on the 11 categories of change detection dataset [25] were used for qualitative comparisons. The foreground masks were collected using change detection website [3] and background subtraction

Table 1 Quantitative analysis of all algorithms on 7 performance metrics defined by change detection dataset. Performance metrics are recall (Re), specificity (Sp), false positive rate (FPR), false negative rate (FNR), precision (P), F-measure (F), and percentage of wrong classification (PWC)

Methods	Re	Sp	FPR	FNR	PWC	F	P
GMM1 [18]	0.6846	0.9750	0.0250	0.3154	3.7667	0.5707	0.6025
GMM2 [19]	0.6604	0.9725	0.0275	0.3396	3.9953	0.5566	0.5973
GMM3 [20]	0.5072	0.9947	0.0053	0.4928	3.1051	0.5904	0.8228
SGMM [21]	0.5940	0.9865	0.0135	0.4060	2.9638	0.5735	0.6965
Shared GMM [22]	0.8098	0.9912	0.0088	0.1902	1.4996	0.7474	0.7503
UBSS [23]	0.7389	0.9927	0.0073	0.2611	1.2614	0.7288	0.7382
SUBSENSE [25]	0.8124	0.9904	0.0096	0.1876	1.6780	0.7408	0.7509
PAWCS [29]	0.7718	0.9949	0.0051	0.2282	1.1992	0.7403	0.7857
PBAS [28]	0.7840	0.9898	0.0102	0.2160	1.7693	0.7532	0.8160
FTSG [24]	0.7657	0.9922	0.0078	0.2343	1.3763	0.7283	0.7696
IUTIS [31]	0.7849	0.9948	0.0052	0.2151	1.1986	0.7717	0.8087

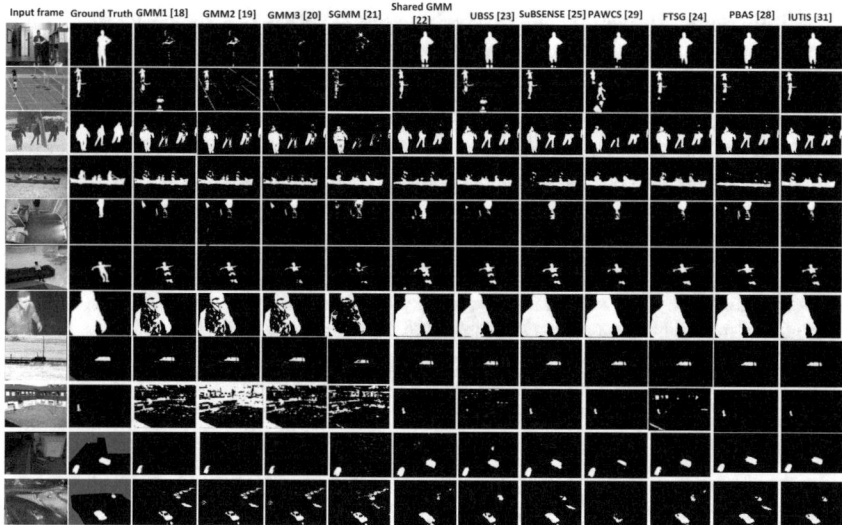

Fig. 1 Foreground mask obtained by all algorithms on the 11 categories of change detection dataset. eleven categories are listed *top* to *bottom* as *baseline*, camera jitter, bad weather, dynamic backgrounds, shadows, intermittent object motion, thermal, turbulence, PTZ, low frame rate, and night videos

libraries provided [26]. Figure 1 shows foreground masks obtained from eleven different categories of change detection dataset. IUTIS [24] performs well in all challenges. All algorithms performs well in categories such as bad weather, dynamic backgrounds, and shadow. Traditional GMM [13, 14] and SGMM [16] were unable to handle long static foreground objects. Such situation can be witnessed in office sequence (first row) where the foreground object stays static for longer time and it was dissolved in the background.

Categories such as PTZ, night videos, low frame rate, and turbulence seems to be most challenging. All algorithms seems to perform poorly in case of low frame rate and night videos. However, in case of PTZ, only UBSS performs well. Some common challenge that all algorithms suffers is camouflage. It can be seen in office (first row) and sofa sequence (sixth row) that legs of foreground object were labeled as background owing to camouflage effect.

4 Conclusion

This paper presents overview of recently published algorithms in the field of foreground detection. Most algorithms compared were published in past 3–5 years except GMM. This is first overview of its kind which covers latest algorithms and compared them quantitative and qualitatively using large realistic change detection dataset. The algorithms were compared quantitatively using seven performance metrics. Foreground mask were compared as part of qualitative analysis. The parametric model based algorithms such as shared GMM and UBSS shows promising results. They were only algorithms that were able to perform well in all categories as compared to other algorithms. In our opinion, non-parametric based methods such as PBAS, SuBSENSE, and PAWCS are new and offers more room for improvement. Therefore, the future advances in the coming time may foresee towards the non-parametric modeling methods.

Acknowledgements This work was supported by the National Research Foundation of Korea (NRF) Grant funded by the Korean Government (2016R1D1A1A02937579).

References

1. Wahyono, A.F., Shahbaz, A., Hariyono, J., Kang, H.D., Jo, K.H.: Integrating Multiple Tasks of Vision-based Surveillance System: Design and Implementation, FCV 2016, Takayama, Japan, pp. 91–94, 17 Feb 2016
2. Shahbaz, A., Jo, K.H.: Probabilistic Foreground Detector for Sterile Zone Monitoring, URAI 2015. Goyang City, Korea (2015)
3. Wang, Y., Jodoin, P.-M., Porikli, F., Konrad, J., Benezeth, Y., Ishwar, P.: CDnet 2014: An expanded change detection benchmark dataset, In: Proceedings of IEEE Conference on Computer Vision Pattern Recognition Workshops, June 2014
4. Mcivor, A.: Background subtraction techniques. In: International Conference on Image and Vision Computing, New Zealand, IVCNZ 2000, Nov 2010
5. Benezeth, Y., Jodoin, P.-M., Emile, B., Laurent, H., Rosenberger, C.: Comparative study of background subtraction algorithms. J. Elec. Imaging **19**(3), 112 (2010)
6. Piccardi, M.: Background subtraction techniques: a review. In: IEEE International Conference on Systems, Man and Cybernetics, Oct 2004
7. S. Cheung, C. Kamath, Robust background subtraction with foreground validation for urban traffic video, EURASIP J. Appl. Signal Process. (2005)
8. Elhabian, S., El-Sayed, K., Ahmed, S.: Moving object detection in spatial domain using background removal techniques-state-of-art. Recent Patents Comput. Sci. **1**(1), 3254 (2008)
9. Cristani, M., Farenzena, M., Bloisi, D., Murino, V.: Background subtraction for automated multisensor surveillance: a comprehensive review. EURASIP J. Adv. Signal Process. (2010) 24
10. Shahbaz, A., Hariyono, J., Jo, K.H.: Evaluation of Background Subtraction Algorithms for Video Surveillance. In: FCV 2015, vol. 28, Mokpo, Korea, Jan 2015
11. Bouwmans, T., El-Baf, F., Vachon, B.: Statistical background modeling for foreground detection: a survey. In: Handbook of Pattern Recognition and Computer Vision, vol. 4(2), World Scientific Publishing, pp. 181199, (2010)
12. Bouwans, T.: Traditional and recent approaches in background modeling for foreground detection: an overview. Comput. Sci. Rev. **11–12**, 31–66 (2014)
13. Stauffer, C., Grimson, W.E.L.: Adaptive background mixture models for real-time tracking. In: Proceedings of International Conference on Computer Vision and Pattern Recognition, vol. 2, IEEE, Piscataway, NJ (1999)
14. KaewTraKulPong, P., Bowden, R.: An improved adaptive background mixture model for real-time tracking with shadow detection. In: Proceedings of Workshop on Advanced Video Based Surveillance Systems (2001)

15. Zivkovic Z.: Improved adaptive gaussian mixture model for back-ground subtraction. In: Proceedings of International Conferene on Pattern Recognition, pp. 28–31, IEEE, Piscataway, NJ (2004)
16. Varadarajan, S., Miller, P., Zhou, H.: Spatial mixture of Gaussians for dynamic background modelling. In: 2013 10th IEEE International Conference on Advanced Video and Signal Based Surveillance (AVSS), pp. 63,68, 27–30 Aug 2013
17. Chen, Y., Wang, J., Lu, H.: Learning sharable models for robust background subtraction. In: 2015 IEEE International Conference on Multimedia and Expo (ICME), pp. 1–6 (2015)
18. Sajid, H., Cheung, S.C.S.: Background subtraction for static and moving camera. In: Accepted for IEEE International Conference on Image Processing (ICIP) (2015)
19. Hofmann, M., Tiefenbacher, P., Rigoll, G.: Background segmentation with feedback: the pixel-based adaptive segmenter. In: Proceedings of IEEE Conference on Computer Vision Pattern Recognition Workshops, pp. 3843 (2012)
20. Barnich, O., Van Droogenbroeck, M.: ViBe: a universal background subtraction algorithm for video sequences. IEEE Trans. Image Process. **20**(6), 17091724 (2011)
21. St-Charles, P.-L., Bilodeau, G.-A., Bergevin, R.: Flexible background subtraction with self-balanced local sensitivity. In: Proceedings of IEEE Conference on Computer Vision Pattern Recognition Workshops, pp. 408413, June 2014
22. St-Charles, P.-L., Bilodeau, G.-A., Bergevin, R.: A Self-adjusting approach to change detection based on background word consensus. In: IEEE Winter Conference on Applications of Computer Vision (WACV), Big Island, Hawaii, USA, 6–9 Jan 2015
23. Wang, R., Bunyak, F., Seetharaman, G., Palaniappan, K.: Static and moving object detection using flux tensor with split gaussian models. In: 2014 St-Charles, P.-L., Bilodeau, G.-A. (eds.) Proceedings of IEEE Workshop on Change Detection, Improving background subtraction using local binary similarity patterns, In Proceedings of IEEE Winter Conference on Application Computer Vision, pp. 509515, Mar 2014
24. Bianco, S., Ciocca, G., Schettini, R.: How far can you get by combining change detection algorithms? Submitted to IEEE Trans. Image Process. (2015). arXiv:1505.02921
25. Goyette, N., Jodoin, P.-M., Porikli, F., Konrad, J., Ishwar, P.: Changedetection. net: a new change detection benchmark dataset. In: Proceedings of IEEE Conference on Computer Vision Pattern Recognition Workshops, pp. 18 (2012)
26. Andrews, S.: BGSLibrary: an openCV C++ background subtraction library. In: IX Workshop de Viso Computacional (WVC'2013), Rio de Janeiro, Brazil, June 2013

Tracking of Bone Reparation Process with Using of Periosteal Callus Extraction Based on Fuzzy C-means Algorithm

Jan Kubicek, Marek Penhaker, Iveta Bryjova, Martin Augustynek, Tomas Zapletal and Vladimir Kasik

Abstract In the field of the clinical traumatology, bone reparation is one of the essential factors which are followed. Bone fractures are clinically evaluated on the base of the X-ray images providing relatively high contrast between bone and affected spot, on the other hand one important identifier of bone reparation (periosteal callus) is often badly observable from native data. Furthermore, periosteal callus is only clinically evaluated by naked eye without SW feedback providing supervised quantification. From the aforementioned reasons we established cooperation with Department of Traumatology on the automatic segmentation and modelling of a periosteal callus leading to tracking of bone reparation. We developed system serving for automatic extraction of a periosteal callus, consequently allowing for compute of callus area for time comparison in the form of the predictive model.

Keywords Periosteal callus · Fuzzy C-means · Traumatology · Image segmentation · Bone reparation · X-ray

J. Kubicek (✉) · M. Penhaker · I. Bryjova · M. Augustynek · V. Kasik
VSB–Technical University of Ostrava, FEECS, K450 17. Listopadu 15,
708 33 Ostrava–Poruba, Czech Republic
e-mail: jan.kubicek@vsb.cz

M. Penhaker
e-mail: marek.penhaker@vsb.cz

I. Bryjova
e-mail: iveta.bryjova@vsb.cz

M. Augustynek
e-mail: martin.augustynek@vsb.cz

V. Kasik
e-mail: vladimir.kasik@vsb.cz

T. Zapletal
Faculty Hospital, 17. Listopadu, 1790 Ostrava, Czech Republic
e-mail: tomas.zapletal@fno.cz

© Springer International Publishing AG 2017
D. Król et al. (eds.), *Advanced Topics in Intelligent Information
and Database Systems*, Studies in Computational Intelligence 710,
DOI 10.1007/978-3-319-56660-3_24

1 Introduction

Periosteal callus is key factor for starting of secondary reparation of the bone fracture. In the case of damaging periosteal vascular supplementing, it must not to be damaged endosteal supplementing during a treatment. Callus evolution goes through four phases:

- Proliferation phase (0–7 days)
- Differentiation phase (8–21 days)
- Ossification phase (since 4th week)
- Modeling and remodeling phase (8–12 weeks)

There is the typical example (Fig. 1) of the fractured bone with the manifestation of the periosteal callus indicating by arrows. As it is observable, the periosteal callus is badly distinguishable due to small size in comparison with the bone fracture. The formation of fracture callus occurs in the presence of secondary bone healing, and has relevance to the mechanical stability at the fracture. For instance, the amount of callus formation in long bone fractures is predictive of bending stiffness. Unfortunately, manipulation including measuring and modeling of periosteal callus is substantially subjective from native X-ray records, with inter-physician variability of 20–25%. Image processing algorithms have potential to render callus measurement objective and thereby reduce observer error. However, previous studies which measured callus with image processing protocols did not document the accuracy and objectivity of their methods. To establish credibility, it is critical that computational techniques be verified and validated. On the base of the native RTG records it is obvious that periosteal callus is badly observable. This object is not usually characterized by borders clearly differentiating callus from bone background. For this reason it is predictable that

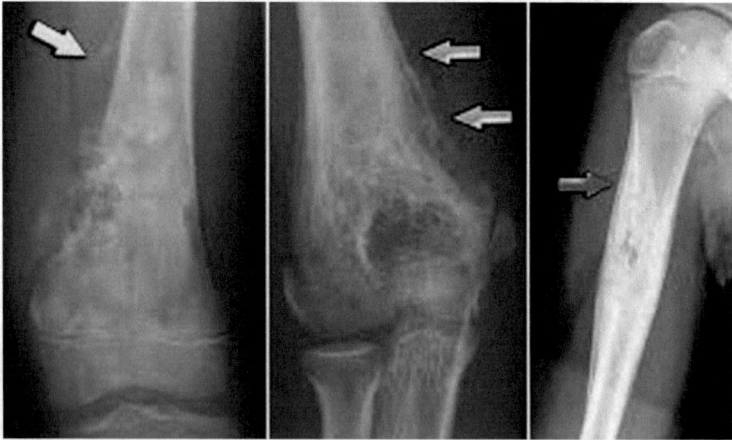

Fig. 1 The bone structure with periosteal callus

segmentation methods based on the edge detection would fail. An appropriate alternative is a multiregional segmentation model utilizing image separation into finite number of image regions. By this way, the periosteal callus can be modelled from each native RTG record over the time, consequently it would be possible to perform time evaluation of bone healing over the time throughout periosteal callus modelling [1–6].

2 Purpose and Data Analysis

The segmentation of bone structure is frequently discussed task in the field of traumatology and urgent medicine. The periosteal callus testifies about treatment efficiency and serves as reliable indicator of fracture prediction. Unfortunately, it is badly distinguishable as it apparent from Fig. 1. Physicians usually determine the diagnosis on the base of own experiences. This kind of diagnosis is strongly subjective, and it is influenced by a subjective error depending on practical experience of each physician. The proposed segmentation method is able to identify area of the periosteal callus from the bone background. On the base this method we obtain the mathematical model which reflects geometrical features of this object. By this

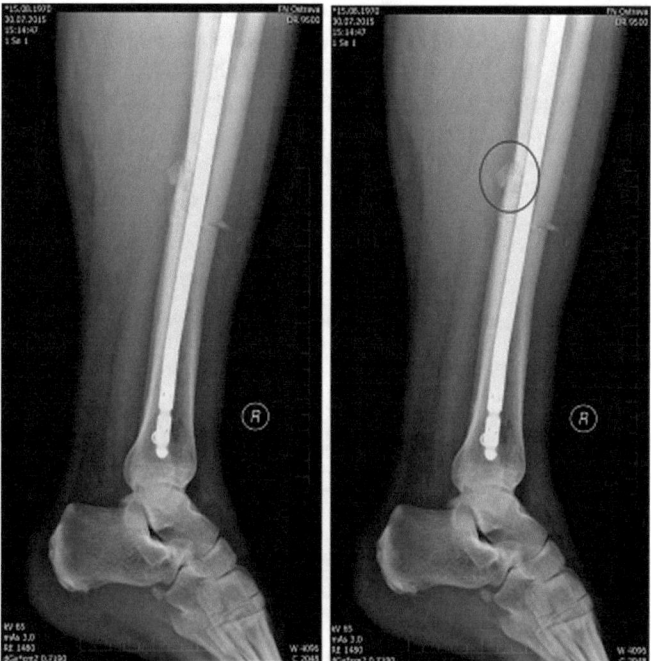

Fig. 2 Sample of analyzed data, periosteal callus is indicated by red circle

procedure we are able to track evolution of the periosteal callus, and obtain reliable prediction of the further fracture development. For practical reasons, data of 20 patients have been used. All selected patients have undergone X-ray imagining. Each examination was performed every 2 weeks [7–9].

Sample of analyzed data is depicted on the Fig. 2. Periosteal callus is indicated by red ellipse. This object is represented by slightly white color with weaker contrast in comparison with the bone structure [10–14].

3   Design of Segmentation Algorithm

The overall structure of segmentation process is illustrated on Fig. 3. Data of the bone fracture are acquired from X-ray. Firstly it is needed to perform selection of data where the periosteal callus is manifested. After taking region of interest (RoI), pixel's interpolation is applied. RoI extraction is needed for the increasing area of the periosteal callus. The important problem is that the periosteal callus takes relatively small part of the native image, and therefore individual structures are badly observable. For this particular task, the linear interpolation of tenth order is used. The interpolation technique significantly influences contrast of the extracted image, and mainly individual structures. If we used lower level of the interpolation, we would obtain badly recognizable image. On the other hand, greater level of the interpolation leads to higher computational time. The interpolation of tenth order is appeared as good compromise for achieving of sufficient image contrast. The low-pass filtering is used for the suppression of the higher frequencies which in some cases impair recognition of the periosteal callus. For this purpose the Gaussian filter has been used. After taking preprocessing steps, it is approached to the segmentation process. Firstly, number of centroids is selected. This step specifies number of output classes. We need to specify the first class for the periosteal callus,

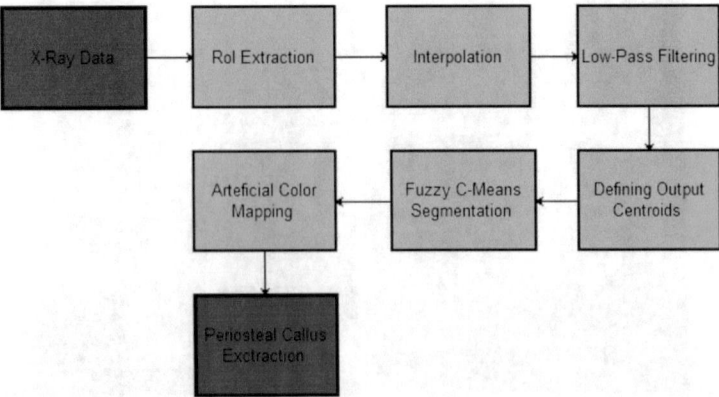

Fig. 3 Flow chart of proposed segmentation method for periosteal callus extraction

the second for the bone background, other classes represent adjacent structures. Namely, they are the soft tissues and the bones. In the concluding step, the filtration of adjacent tissues in the segmentation model is performed. The segmentation process separates individual tissues to the classes which are coded to color spectrum. By the filtration process we keep only the class representing of the periosteal callus.

The major benefit of the used segmentation approach is incorporating the neighborhood information into the C means centroid segmentation during the classification process. The segmentation approach uses the penalized objective function by regularization term in order to incorporate spatial context.

4 Penalized Version of C-means Segmentation

The major benefit of the used segmentation approach is incorporating of the neighborhood information into C means centroid segmentation during the classification process. The segmentation approach uses the penalized objective function by regularization term in order to incorporate spatial context. The objective function of the used algorithm is defined by the following equation:

$$J = \sum_{k=1}^{n} \sum_{i=1}^{c} (u_{ik})^q d^2(x_k, v_i) + \gamma \sum_{k=1}^{n} \sum_{j=1}^{n} \sum_{i=1}^{c} (u_{ik})^q (1 - u_{ij})^q w_{kj} \qquad (1)$$

The parameter $\gamma \geq 0$ controls effect of the penalty term. The major benefit of the fuzzy C means algorithm is that penalty term should be minimized in order to satisfy the principle of fuzzy C means (FCM) algorithm. The mentioned penalty term is minimized when the membership value for the particular class is large and the membership.

The objective function J can be minimized similarly to standard FCM algorithm. The derivation of the iterative algorithm for minimizing is performed by evaluating the centroids and the membership function that satisfy a zero gradient condition. The Lagrange multiplier is used for solving constrained optimization from the Eq. (1):

$$\varphi_q = \sum_{k=1}^{n} \sum_{i=1}^{c} (u_{ik})^q d^2(x_k, v_i) + \gamma \sum_{k=1}^{n} \sum_{j=1}^{n} \sum_{i=1}^{c} (u_{ik})^q (1 - u_{ij})^q w_{kj} + \lambda (1 - \sum_{i=1}^{c} u_{ik}) \qquad (2)$$

After taking first-order derivation according to u_{ik} we obtain the following expression:

$$\left[\frac{d\varphi_k}{du_{ik}} = q(u_{ik})^{q-1} d^2(x_k, v_i) + \gamma q(u_{ik})^{q-1} \sum_{j=1}^{n} (1 - u_{ij})^q w_{kj} - \lambda \right]_{u_{ik} = u_{ik}^*} = 0 \qquad (3)$$

The parameter u_{ik}^* is represented by following equation:

$$u_{ik}^* = \left(\frac{q(d^2(x_k, v_l) + \gamma \sum_{j=1}^{n} (1 - u_{ij})^q w_{kj})}{\lambda}\right)^{\frac{-1}{q-1}} \tag{4}$$

Since $\sum_{l=1}^{c} w_{lk} = 1$, this constraint equation is expressed by the form:

$$\sum_{l=1}^{c} \left(\frac{q(d^2(x_k, v_l) + \gamma \sum_{j=1}^{n} (1 - u_{ij})^q w_{kj}))}{\lambda}\right)^{\frac{-1}{q-1}} = 1 \tag{5}$$

The parameter λ from the Eq. (5) is given by following expression:

$$\lambda = \frac{q}{\left(\sum_{l=1}^{c} \left(\frac{1}{d^2(x_k, v_l) + \gamma \sum_{j=1}^{n} (1 - u_{ij})^q w_{kj}}\right)^{\frac{1}{q-1}}\right)^{q-1}} \tag{6}$$

By incorporating the zero gradient condition, the membership estimator is rewritten to the following term:

$$u_{ik}^* = \frac{1}{\sum_{l=1}^{c} \left(\frac{d^2(x_k, v_i) + \gamma \sum_{j=1}^{n} (1 - u_{ij})^q w_{kj}}{d^2(x_k, v_l) + \gamma \sum_{j=1}^{n} (1 - u_{ij})^q w_{kj}}\right)^{\frac{1}{q-1}}} \tag{7}$$

Similarly, by taking the Eq. (2) with respect to parameter v_i and setting the result to zero, we will obtain the expression:

$$v_i^* = \frac{\sum_{k=1}^{n} (u_{ik})^q x_k}{\sum_{k=1}^{n} (u_{ik})^q} \tag{8}$$

The expression is identical to that of the FCM because the penalty function from (1) is not depended on v_i. Generally, the penalized FCM algorithm is possible to summarize to the four essential stages:

Stage 1: Setting the centroids vi, parameter of fuzzification q and the value of c.
Stage 2: The calculation of the membership values from the Eq. (7).
Stage 3: The calculation of the cluster centroids.
Stage 4: Return back to the stage 2 and repeat it until the convergence is reached.

As soon as the algorithm converges, the defuzzification process is performed, in order to convert the fuzzy partition matrix to the crisp partition. The described method works on the principle of assigning of object k to the class C with the highest membership [15–18]:

$$C_k = arg_i\{\max(u_{ik})\}, \quad i = 1, 2, \ldots, c \qquad (9)$$

5 Periosteal Callus Modeling

The proposed algorithm works within the several consecutive steps. The algorithm structure can be summarized into following steps:

- Data acquisition
- Data preprocessing
- Segmentation procedure
- Extraction area of interest

The segmentation approach has been developed for the extraction of the periosteal callus. This structural object is important indicator of the bone reparation process. The evolution of this parameter is usually tracked during the healing period. The best observation of the periosteal callus is commonly achieved by X-ray imagining. For this reason the X-ray data has been processed for purposes of this analysis.

The main objective is segmentation of the periosteal callus with the target of the consequent periosteal callus extraction including of the geometrical parameters such as callus geometry and area. On the base mentioned parameters, it is possible tracking of the periosteal callus evaluation within the bone reparation. The overall segmentation procedure is depicted on the Fig. 4.

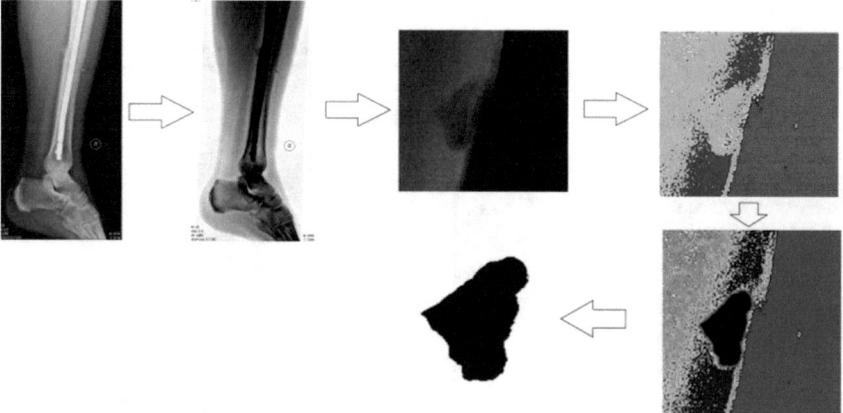

Fig. 4 Periosteal callus extraction process

The major disadvantage of the periosteal callus manifestation is the worse observability as it is visible from the native data. The process of the image inversion is applied on the native data. This procedure is especially important in the fact of the highlighting of the periosteal callus area. After that RoI with the interpolation is applied. This procedure is focused on the area where the object of interest is manifested in the same time this procedure significantly increases the image resolution which is important for the generating segmentation procedure. The segmentation procedure works on the principle region based methods. It means that the native data are separated into isolated areas. The significant benefit of the segmentation is fact that native data are represented by the artificial color map which is much more obvious than the native bone structures. In the final step the periosteal callus is extracted by the very simple thresholding procedure. In the output we generate mathematical model which fully reflects the geometrical parameters of the analyzed periosteal callus.

6 Algorithm Clinical Evaluation and Statistics

Within our analysis, twenty antero-posterior and lateral X-ray images of distal femur fractures were processed. Processing was performed immediately after surgery. We compared our results with expert opinions of three clinicians from Department of Traumatology from University hospital in Ostrava. Periosteal callus area was computed by digitally marked outlines (Fig. 5).

Five non-clinicians evaluated the same X-ray images using the proposed algorithm with the following standard operating procedure: (1) orient images to obtain vertical bone alignment with external cortex facing right; (2) ensure ROI encompasses the internal cortex and callus; and (3) reference the initial postoperative radiographs when tracing the external cortex.

Fig. 5 Periosteal callus outlined by manual way (five clinicians), and outlined with the proposed algorithm (five non-clinician). There is no significant statistical differences between individual groups

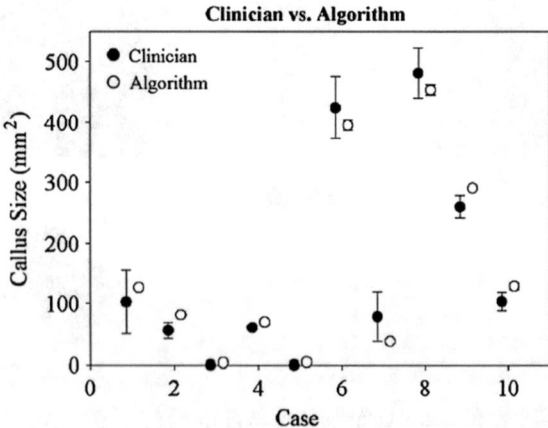

In the context of the statistical evaluation, we assessed effect of independent factors on preciseness. One-way and two-way ANOVA tests were used. In the case when test achieved significance level ($p < 0.05$), Tukey post hoc tests were determined significance between factor levels. Differences in callus area between the clinician and algorithm were tested with the Pearson r correlation, while case by case comparisons were examined with paired t-tests.

7 Conclusion

The Analysis of the periosteal callus is the commonly practiced procedure in the field of traumatology. Evaluation is usually undertaken subjectively on the base physician's opinion. This process is expectable influenced by subjective error. The major problem of the periosteal callus is a worse observation possibility. Due to this fact, methods based on the image edge detection fail. Therefore, we were focused on periosteal callus modeling on the base of the multiregional segmentation model differentiating image structures into isolated regions. This approach leads to the periosteal callus model, and other structures are suppressed. We propose the suitable method for the automatic extraction of the periosteal callus which has ambition to be applicable in the clinical practice. On the base this method, we obtain binary image reflecting of mathematical model of this object. Consequently, it is simple to identify of the geometrical parameters such area and perimeter corresponding with the fracture evolution. The mentioned geometrical features extraction is not particularly important in the case of the processing of the single images. This procedure is gaining importance itself, especially, in the case of the tracking the periosteal callus time evaluation. It is supposed that the bone reparation time process is accompanied by the evaluation of the periosteal callus. During this period especially shape and area manifestation is changing, therefore, this analysis is important.

Acknowledgments This article has been supported by financial support of TA ČR, PRE SEED Fund of VSB-Technical univerzity of Ostrava/TG01010137. The work and the contributions were supported by the project SV4506631/2101 'Biomedicínské inženýrské systémy XII'.

References

1. Raggatt, L.J., Wullschleger, M.E., Alexander, K.A., Wu, A.C.K., Millard, S.M., Kaur, S., Maugham, M.L., Gregory, L.S., Steck, R., Pettit, A.R.: Fracture healing via periosteal callus formation requires macrophages for both initiation and progression of early endochondral ossification. Am. J. Pathol. **184**(12), 3192–3204 (2014)
2. Epari, D.R., Lienau, J., Schell, H., Witt, F., Duda, G.N.: Pressure, oxygen tension and temperature in the periosteal callus during bone healing—an in vivo study in sheet. Bone, **43** (4), pp. 734–739 (2008)

3. Kubicek, J., Penhaker, M., Bryjova, I., Kodaj, M.: Articular cartilage defect detection based on image segmentation with colour mapping. Lecture Notes in Computer Science (including subseries Lecture Notes in Artificial Intelligence and Lecture Notes in Bioinformatics), vol. 8733, pp. 214–222 (2014)
4. Augat, P., Merk, J., Genant, H.K., Claes, L.: Quantitative assessment of experimental fracture repair by peripheral computed tomography. Calcif. Tissue Int. **60**, 194–199 (1997)
5. Gelaude, F., Vander Sloten, J., Lauwers, B.: Semi-automated segmentation and visualisation of outer bone cortex from medical images. Comput. Methods Biomech. Biomed. Eng. **9**, 65–77 (2006)
6. Kubicek, J., Bryjova, I., Penhaker, M.: Macular lesions extraction using active appearance method. Lecture Notes of the Institute for Computer Sciences, Social-Informatics and Telecommunications Engineering, LNICST, vol. 165, pp. 438–447 (2016)
7. Bryjova, I., Kubicek, J., Dembowski, M., Kodaj, M., Penhaker, M.: Reconstruction of 4D CTA brain perfusion images using transformation methods. Adv. Intell. Syst. Comput. **423**, 203–211 (2016)
8. Kubicek, J., Bryjova, I., Penhaker, M., Kodaj, M., Augustynek, M.: Extraction of myocardial fibrosis using iterative active shape method. Lecture Notes in Computer Science (including subseries Lecture Notes in Artificial Intelligence and Lecture Notes in Bioinformatics), vol. 9621, pp. 698–707 (2016)
9. Kukucka, M.: IEEE: Design of Experimental Fuzzy Diagnostic System (2007)
10. Augustynek, M., Pindor, J., Penhaker, M., Korpas, D.: Detection of ECG significant waves for biventricular pacing treatment. In: 2010 Second International Conference on Computer Engineering and Applications (ICCEA), pp. 164–167. IEEE (2010)
11. Penhaker, M., Stula, T., Augustynek, M.: Long-term heart rate variability assessment. In: 5th Kuala Lumpur International Conference on Biomedical Engineering (BIOMED 2011), pp. 532–535 (2011)
12. Peterek, T., Augustynek, M., Zurek, P., Penhaker, M.: Global courseware for visualization and processing biosignals. In: Dossel, O., Schlegel, W.C. (eds.) World Congress on Medical Physics and Biomedical Engineering, vol. 25, Pt. 12, pp. 404–407 (2009)
13. Tiedeman, J.J., Lippiello, L., Connolly, J.F., Strates, B.S.: Quantitative roentgenographic densitometry for assessing fracture healing. Clin. Orthop. Relat. Res. 279–286 (1990)
14. Whelan, D.B., Bhandari, M., McKee, M.D., Guyatt, G.H., Kreder, H.J., Stephen, D., et al.: Interobserver and intraobserver variation in the assessment of the healing of tibial fractures after intramedullary fixation. J. Bone Jt. Surg. Br. **84**, 15–18 (2002)
15. Kubicek, J., Penhaker, M.: Fuzzy algorithm for segmentation of images in extraction of objects from MRI. In: Proceedings of the 2014 International Conference on Advances in Computing, Communications and Informatics, ICACCI 2014, Art. no. 6968264, pp. 1422–1427 (2014)
16. Majernik, J., Jarcuska, P.: IEEE: Web-based delivery of medical education contents used to facilitate learning of infectology subjects. In: 2014 10th International Conference on Digital Technologies (Dt), pp. 225–229 (2014)
17. Penhaker, M., Kasik, V., Snasel, V.: Biomedical distributed signal processing and analysis. In: Saeed, K., Chaki, R., Cortesi, A., Wierzchon, S. (eds.) Computer Information Systems and Industrial Management, CISIM 2013, vol. 8104, pp. 88–95 (2013)
18. Penhaker, M., Klimes, P., Pindor, J., Korpas, D.: Advanced intracardial biosignal processing. In: Cortesi, A., Chaki, N., Saeed, K., Wierzchon, S. (eds.) Computer Information Systems and Industrial Management, vol. 7564, pp. 215–223 (2012)

Part IV
Data-Intensive Text Processing

Automatic Post-editing of Kazakh Sentences Machine Translated from English

Assem Abeustanova and Ualsher Tukeyev

Abstract Automatic post-editing of sentence from one language to another language is closely connected with a machine translation. Machine translation is a modern tool that helps to speed up the translation process and to reduce its cost. But still it is a problem to get a correct translation. Therefore now it is active investigates automatic post-editing improved quality of translation. In this paper is proposed method of automatic post-editing based on two stage procedure: on the first is defined incorrect words in a text by using translation memory (TM) technology, on the second stage is defined what alternative translation word is more better for defined incorrect words on first stage by using of maximum entropy model.

Keywords Automatic post-editing · Translation memory technology · Maximum entropy model

1 Introduction

As for machine translation, it should be noted that the computer does not have mind. It does not understand the nuances of language, hints in the text. Each new construction, phrase, idiomatic expression should be provided by the programmer and are included in the program. Depending on the style and purpose of the text, the same word can often have several meanings. In our time, machine translation software can recognize the style of text, and, depending on this, select the necessary dictionary. Also, the program itself can offer several translations to translator. Therefore, considering all these nuances the work was done.

The basic idea of this paper is to find the incorrect words in the Kazakh sentences that translated from English sentences and automatic correct the incorrect

A. Abeustanova (✉) · U. Tukeyev
Al-Farabi Kazakh National University, Almaty, Kazakhstan
e-mail: shormakovaassem@gmail.com

U. Tukeyev
e-mail: ualsher.tukeyev@gmail.com

© Springer International Publishing AG 2017
D. Król et al. (eds.), *Advanced Topics in Intelligent Information and Database Systems*, Studies in Computational Intelligence 710,
DOI 10.1007/978-3-319-56660-3_25

Kazakh words. As a result, the resulting work we get corrected right sentence in Kazakh sentence. For the finding in translated text a incorrect words is proposed to use translation memory technology [1] and for automated correct of incorrect words is proposed to use maximum entropy model [2]. The system was previously "trained" on the parallel corpus and then applied to "working" texts. Due to the preliminary "training" it is possible to achieve higher accuracy of the translation, more suitable terminology and to reduce the costs of post-editing.

2 Related Works

Among the many online translators PROMT [3] and Google [4] should be distinguished. These programs are used for comparative analysis. In order to understand the principles of actions and dictionary usage quality and grammar analysis, also the quality of the translation, the text was translated. Translation was carried out in the English-Kazakh direction (a text consisting of 51 words). The following two criteria were used in the analysis: (1) the proper selection of words's meaning by the system (lexical level), (2) the accuracy of matching words in a sentence (grammar level, the coordination of words in a sentence in gender, number, face, case, and punctuation). Take the following passage of a text as an example: "It would be hard to imagine a more evil piece of work than Robert Alton Harris. After a lifetime of vicious, random crime, in 1979 in California he murdered two teenage boys in cold blood for their car. As he drove away, he finished off the cheeseburgers they had been eating". Translation made by automatic translation system PROMT: "Edi elestetu kiyn astam zloe tuyndysy karaganda,Robert Al'ton Harris. Uzak jyldardan kein katygez kylmys, 1979 jyly Kaliforniada ol oltirgen eki jasospirim suyk kan oz avtokoligi ushin. Ol ketip kaldy, ol dobil chizburgery olar shyrsha.". The system has successfully coped with the search for the English equivalent of "in cold blood". Pay attention to the "workpiece". It is easy to understand that this error is caused by polysemy of words "piece" and "work". Furthermore, it was absolutely incorrect to translate the phrase "oz koligi ushin".

Moreover the author of the text within the meaning of the first sentence refers to the detail, the man would never have confused whereas it is animate or inanimate. The phrase "finished off the cheeseburgers" was translated by PROMT as "dobil chizburgery". But this phrase is unacceptable in the Kazakh language. A significant drawback is that in the output language word order is almost always the same as the input. Translation made by Google's online translator: Ol Robert Harris Alton Karaganda jumys negurlym jaman boligin elestetu kiyn bolar edi. Yzaly, kezdeisok kylmys omir kein, 1979 jyly Kaliforniada ol olardyn avtomobil'ge arnalgan suyk kan eki jasospirim ul oltirgen.Ol ketip, ol olar jep boldy chizburgery oshiru aiaktaldy." in contrast to the PROMT Google has translated idiom "in cold blood" and issued for translation "suyk kan", i.e. used the literal translation. Google also repeated the error in the PROMT translation "for" as "ushin". Instead of using the previous program's dobil chisburgery" Google used "jep boldy chizburgery", that is better. In the English-Kazakh translation online program PROMT made 6 errors (2 lexical and 4 grammatical). And Google also made 6 errors (3 lexical and 3 grammatical). From this study, it follows that the least number of errors in the translation from English into Kazakh was made by PROMT system. After completing the analysis of above mentioned modern systems of machine translation, we noticed that every machine translation system has its strengths and weaknesses. At this stage, the machine translation systems can not exist without a man, because we noticed that the translation is not perfect. Therefore, if we want to get high-quality translation, post-editing by a man is necessary [5].

The work of the Spanish group [6] focused on the sub-segments of the qualitative evaluation of machine translation (MTQE) at the word level. The main advantage of the verbal level MTQE is that it allows not only to estimate the effort needed for post-editing output from the MT system, and which words need to be edited after as the guidance for post-editors. In this article there is used the black box of bilingual resources from the Internet to the level of MTQE words. Namely, they combine two online MT systems, Apertium and Google Translate, and bilingual concordancer Reverso Context3 to detect a sub-segment of correspondence between sentences S in the original language (SL) and the translation of T hypothesis in the target language (TL). For this purpose, both S and T is segmented into overlapping sub-segments of variable length and they are translated to TL and SL, respectively, using a bilingual sources mentioned above. These matching sub-segments are used to retrieve a collection of functions that are then use a binary classifier to determine the word to be edited later. Their experiments confirmed that their method gives results comparable to level of technique, using significantly less capacity. Moreover, considering the fact that there are used (online) resources in their method, which are publicly available on the Internet and as soon as the binary classifier will be trained it can be used at the level of words on the MTQE doing new translations. Work inspired by the work Esplà-Gomis et al. [7], in which several MT-line systems are used to assess the quality of speech at a translation memory (TM) based on the automated translation system tasks. In the the paper of Esplà-Gomes and others (2011), taking into account the European Parliament (S, T) suggested the translator to the SL segment S, MT is used for the translation of the S sub-segments in the TL and TL sub-segments of the T sub-segment at the Sea Level is obtained through MT, which occur both in S and T are a evidence to the fact that they belong to. Alignment between S and S, together with a sub-segment transfers between S and T help to determine which words in T should be modified to obtain T, the desired translation S. There is considered the use of a general-purpose machine translation (MT) to assist users of computer translation (the CAT) based on translation memory systems (TM) to determine the target word in the target sentences to be changed (or changed or deleted) or kept unaltered the problem referred to as guidelines for the maintenance of the word. MT is used as a black box to align source and target sub-segments in the translation of units offered to the user. The source of language (SL) and the target language (TL) segments in matching TUs are segmented in overlapping sub-segments of variable length and translated into machine TL and SL, respectively. Bilingual subsegments are obtained and agreed between the SL segment in the segments TU and parts of translation are used to create functions, which are then used by binary Classifier to determine the target word to be changed, and those that will be saved unedited. Two approaches are presented in this paper: one using words to support recommendation system, those which can be trained on TM are used with CAT and more basic approach, which does not need any preparation. Experiments are hold by modeling text translation to several languages from corpuses, belonging to different aspects and with usage of 3 systems of MT. In this paper, by comparing previous works we suggest use and compose alignment technology with the method maximum entropy for automated post-editing of sentence from English to Kazakh to improve the quality of translation.

3   Method

3.1   Combining of Translation Memory Technology and Maximum Entropy Method

The approach we proposed consists of two modules. The function of defining incorrect words and post-editing of these incorrect words make analyses of words for post-editing and the module of text for the usage of carried out analysis. The method maximum entropy used for automatic post-editing of words and phrases of English-Kazakh sentences. The main objective of the training model Stage 1 is the definition of incorrect words and need to adjust these words. Initially we take the English sentences and translate them through a translator Apertium and in addition to manually write the expected correct translation of the English sentences. Eventually we have two files: the Kazakh sentences from translator and translated by person. These sentences of two files (translation of Apertium and the expected correct translation) are aligned by matrix phrase alignment method (Koehn, Statistical Machine Translation, p. 113) [8], and inappropriate segments and words of translated Kazakh text are determined what defined.

The main objective of Stage 2 is adjustment of sentence with predefined incorrect words. There are used a small number of parallel sentences on two languages (about 100,000) for initial data analysis. These sentences are used for building tables for training the system. There is we use method maximum entropy for generating cube tables. That is, the system is trained by cube tables created in advance. The table is constructed based on a bilingual parallel corpus and bilingual dictionary.

Indicators which give one unit as the remainder of the division by the number of classes, triggered (return true) only in the first grade, multiple to the two for the second, etc. This approach is not mandatory for the implementation of the classifier, but in order to understand the theory it is important to understand the difference between a sign and indicator, as well as differences in their numbering.

Classification takes place by the formula:

$$p(c\,|\,d,\lambda) = \frac{\exp \sum_i^{n\times k} \lambda_i f_i(c,d)}{\sum_{\tilde{c}\in C} \exp \sum_i^{n\times k} \lambda_i f_i(\tilde{c},d)} \tag{1}$$

In this formula [9]:

- f_i—i-th classification indicator (0 or 1);
- λ_i—the weight of the i-th classification indicator f_i;
- c—class hypothesis;
- C—the set of all possible classes;
- D—classified document.

Each indicator has a weight of f_i λ_i, which describes the relationship between the relevant classification criterion and class. The greater the weight, the stronger the connection. Thus, the numerator describes the exponential weights for class-hypothesis, and the denominator normalizes the value of the unit. The most difficult part of this formula—a set of weights λ.

3.2 Realization of Maximum Entropy Method

Maximum entropy method is very useful as well as the generation of tables and to determine the most probable word. That is, when using this method is given below the above description and in the end we get the equivalent word with the closest meaning to the context. The result of this attitude is not just a classification decision, it is the probability for a given class. One of the advantages of this classification is that it is much more accurately models the probable distribution of the classes. Using a machine translation from English into Kazakh translated text should be edited if there incorrect word in a sentence.

Then after finding the incorrect words in a sentence, alternative look at their translation so that we can insert and give post-edited correct translation and further is used on the basis of the method of maximum entropy. General description of the method is as follows:

$$f_i^j = \begin{cases} 1, & \text{if } d = w_i, c = AW_j \\ 0, & \text{in other cases} \end{cases} \tag{2}$$

where AW_j—alternative word (class), d—classified word.

In order to correct incorrect words multivalued dictionary databases and TM (Translation Memory) are used. The following items are shown as an example (Table 1).

Small case of sentence in 1235 was taken for accurate analysis. From this body it was defined that the word "ana" is incorrect and its equivalents were found:

Table 1 Two sense of the same words

Alternative words	Collocations
ana	anama kyzykty kitap al, ana jaksy koredi
mama	mamasynyn kuanyshyna ainalady
ene	enesi pisirgen
sheshe	sheshem balish pen bir kuty maimen
apa	apasy balish pisirip

1 *Anasy* ony ote jaksy koredi eken. (The *mother* was very fond of it)
2 Bir kuni *apasy* balish pisirip, kyzyna keshikpeuin aitady. (One day, her *mother* bake a cake, the daughter of late say)
3 Ogan *enesi* pisirgen balishinen jane bir kuty mai alyp bara jatyrmyn. (It'm going to my *mother's* bälişinen and a bottle of cooking oil)
4 Sen myna jolmen bar, men *ana* jolmen jurein. (You are this way, and the mother *in* a way)
5 Men sizge *sheshem* balish pen bir kuty maimen jiberdi. (I sent you my cake and a bottle of oil)
6 Uly akesinin maktanyshyna, *mamasynyn* kuanyshyna ainalady. (The pride of his father's, *mother's* joy becomes)
7 Sen *anama* kyzykty kitapty al. (You have an interesting book and a *mother*)

We used maximum entropy:

$$f^1 = \begin{cases} 1, & \text{if } d = \text{``}f_3 \wedge f_4 \wedge f_6\text{''}, c = AW_1 \\ 0, & \text{in other cases} \end{cases} \tag{3}$$

$$f^2 = \begin{cases} 1, & \text{if } d = \text{``}f_1 \wedge f_2 \wedge f_5\text{''}, c = AW_2 \\ 0, & \text{in other cases} \end{cases} \tag{4}$$

$$f^3 = \begin{cases} 1, & \text{if } d = \text{``}f_2 \wedge f_5\text{''}, c = AW_3 \\ 0, & \text{in other cases} \end{cases} \tag{5}$$

$$f^4 = \begin{cases} 1, & \text{if } d = \text{``}f_2 \wedge f_3 \wedge f_4\text{''}, c = AW_4 \\ 0, & \text{in other cases} \end{cases} \tag{6}$$

$$f^5 = \begin{cases} 1, & \text{if } d = \text{``}f_4 \wedge f_5\text{''}, c = AW_5 \\ 0, & \text{in other cases} \end{cases} \tag{7}$$

According to these rules, Table 2 was built.

Table 2 Calculations of probability for each case

		f_1	f_2	f_3	f_4	f_5	f_6
AW_1	f^1	0	0	1	1	0	1
AW_1	Weight	–	–	1/7 = 0.142	2/7 = 0.285	–	1/7 = 0.142
AW_2	f^2	1	1	0	0	1	0
AW_2	Weight	5/7 = 0.714	1/7 = 0.142		–	1/7 = 0.142	–
AW_3	f^3	0	1	0	0	1	0
AW_3	Weight	–	3/7 = 0.428	–	–	2/7 = 0.285	–
AW_4	f^4	0	1	1	1	0	0
AW_4	Weight	–	4/7 = 0.571	1/7 = 0.142	1/7 = 0.142	–	–
AW_5	f^5	0	0	0	1	1	0
AW_5	Weight	–	–	–	3/7 = 0.428	5/7 = 0.714	

As a result of calculating, the probability of incorrect words separated by parts of speech, were as follows:

$$P(AW_1) = 0.427$$
$$P(AW_2) = 0.998$$
$$P(AW_3) = 0.713$$
$$P(AW_4) = 0.855$$
$$P(AW_5) = 1.142$$

That is, the maximum entropy method selects value 1.142, and it gave better full decision using of the features and weights of different alternative words. That is only particular parts of sentence and the maximum meaning of probabilities, whereas after used the complemented generating cube tables we proposed considers the contexts of used incorrect words. That is, taking into account not only the parts of speech and probability of necessary words, but also the meanings of each needed word in the text. And this method is used in the Functions of defining incorrect words and post-editing of these incorrect words in Stage 2 in the construction of tables for each incorrect word.

4   Experimental Results and Discussion

4.1   Description of Function of Defining Incorrect Words and Post-editing of Them

The main work consists of two stage (modules). First stage is to find the right word in the Kazakh language translated from a person entered any English sentence. This part of the find and mark the incorrect words or segments of the sentence in the Kazakh language is based on the method of translation memory. The Function of defining some incorrect words and post-editing of these incorrect words Stage 1—is associated with a list of incorrect words, that is, if to make a detailed description, it is associated with the File consisting of three types of sentence (English sentences, Kazakh sentences, correct Kazakh sentences) obtained after algorithms of post-operators. In each line it looks as follows: *interesting subject, kyzykty sabak, kyzykty pan* etc. We find the wrong words, divide them, find the roots and compare them, that is, if to take from the example it will be *sabak*. Post-operator algorithm was edited and as a result defines only one wrong word and is written with a list of polysemantic words in the file. So, as a result, when there will be added new sentences in the initial three files, it automatically appears in this new file. And since we have to cover a list of incorrect words as more as possible we consider polysemantic words too. Morphological Analyzer Apertium is intended to Stemming algorithm for the Kazakh language to divide words from the roots and ends.

Function of defining incorrect words and post-editing of the incorrect words Stage 1—After determining the incorrect words we work with tables. That is, this table is ready for any new sentence. The wrong word found in defining incorrect words. We do following to reach readiness:

Stage 1:

1. search this wrong Kazakh word from the English-Kazakh dictionary, remember all the translations of this word and english version respectively.

2. then look for the english basis of the sentence (25,000 at the moment, it may be more) all sentences with this English word and consider Kazakh translation of these sentences.

3. make a table for each wrong word, and name each table by the English version of the word. That is, if the word *sabak* (subject) is incorrect and it occurs as a subject in the dictionary, then the table will be named *subject* for this word. And these all tables were saved as a file for each incorrect word. We remember all the synonyms of words found in English-Kazakh dictionary and write words nearby (only the roots of the word) from the found sentences which relate only to the wrong word, morphological analyzer of Apertium is used for these purposes. And calculate for each case how many times they appear in the 100,000 base of sentences. You can add even more sentences for corpus. And so it turns out a lot of tables with ready frequencies and incorrect words.

4.2 Description of Phase TEST

On the second stage, we fix already found in the first paragraph of incorrect words and for this purpose the generation of cube tables based on the maximum entropy method. That is found objectionable words are corrected based on the cube tables that have previously been generated. To there exists English-Kazakh dictionary which is required to determine the equivalents of the English translation of these words or segments that have been found incorrect. And they found the English and Kazakh equivalents of these words are searched for from a bilingual corpus. And this most communication takes place with the context and it can determine the meaning of a proposal based on context and word. From parallel English-Kazakh corpus considering Kazakh sentences computed by table of the cube is below described in more detail:

At this stage, we use the previous module to translate any incoming sentences in English.

$$\text{Sentence}_{english} \rightarrow \text{Sentence}_{Kazakh} \rightarrow \text{Sentence}_{correct}$$

1. Yandex [10] translator translates automatically.
2. The function phase of determining incorrect words and post-editing of these incorrect words, that is, the work of Stemming Algorithm is performed, and search for incorrect words from incorrect bad_slova.txt file from already finished table to calculate the probability for each found incorrect word.
3. probability algorithms are used.

When all the words are found from the table it is necessary to take into account the translated sentence that we correct. After that is we are not going to use an entire file, which means that we will take those words that are used only in this sentence. For example: the translated text will be: *Sabak kesh boldy.*

(The lesson was late.) That is we found incorrect word *sabak*. (Subject has some synonyms as lesson, object etc.) Now if you look at the table, let's say the file looks as follows: (The data are taken from one table)

3: men; sabak
2: men; takyryp
1: men; sub'ekt
1: keshe; sabak
4: keshe; takyryp
1: keshe; sub'ekt
1: ol; sabak
1: ol; takyryp
1: ol; sub'ekt
1: bol; sabak
1: bol; takyryp
1: bol; sub'ekt

Only those words which are found in translated sentence are used from this file. *Yesterday was a subject*, Apertium [11] translated as: ***Keshe sabak boldy*** and so we take only words *keshe* and *bol* with polysemantic words and calculate the probability for these words only:

1: keshe; sabak
4: keshe; takyryp
1: keshe; sub'ekt
1: bol; sabak
1: bol; takyryp
1: bol; sub'ekt

It is Calculated by the formula: $P(s) = P(s_1) + P(s_2) + \cdots + P(s_n)$ and use the probability to above words with their sentences, *Subject* has some synonyms as *lesson(саба), object(та ырып), subject (субьект)*:

$$P(\text{sabak}) = 1/25000 * 1/25000 = 0.00004 * 0.0004 = 0.0000000016$$
$$P(\text{takyryp}) = 4/25000 * 1/25000 = 0.00016 * 0.00004 = 0.0000000064$$
$$P(\text{sub'ekt}) = 1/25000 * 1/25000 = 0.0000000016$$

And select the maximum value and get a second probability which is *takyryp*, and paste this value into the sentence.

4. In order to paste the calculated right word it is necessary to find in which endpaste the word in sentence and the Morphological Analyzer Apertium is used. And as a result we get the full correct sentence with post-edited word or words.

The above mentioned and described an example of the result of the program is aimed at, that is, the phase of testing. That is, as a result we get the right corrected Kazakh sentence, and this sentence from English into Kazakh language translated through any interpreter translated sentence on working consistently with Stage 1 and Stage 2 to obtain high-quality translation.

In the development of the algorithm it is required to supplement the data, that is, the more sentences the more accurate information about post-editing of sentences. Base TM would be better updated and added.

Note, this may be:

1. if at least one root of the word is not there in bad_slova.txt we don't look to this sentence and explain that this word has no incorrect words.
2. it is possible to find this word in bad_slova.txt but not found in the list of tables from the stage of defining incorrect words and post-editing of them, we should take into account these words and algorithm of building the tables should add these new words.
3. it is possible that the sentence have several incorrect words, then we take all the incorrect words.

As a result of the made work using this technique it has been made the small analysis from small the sentence (the 100th offer). Different translators for check of the first stage have been used where is defined incorrect words or segments. By results of the carried-out small analysis it has been shown the following results.

As the result shows using the first stage of the offered method above described from these three translators the modified number of words corresponding to tables have been found. According to Table 3 it is visible that it is possible to find words from a context of which it is necessary to correct using offered by us by method translation memory. But as this inexact assessment, we can't be reliable that

Table 3 Percentage indicator finding of the translations of incorrect words from several translators

Google	Apertium	Promt
12%	10%	15%

Table 4 Percentage indicator coincidence of incorrect words between the translation of the expert and our system

Google	Apertium	Promt
6%	8%	11%

whether all found words everything can be changed. On it for the long analysis we use "the gold standard" from which could make a start and be sure more precisely above shown table. Rely on 3 experts who know source language well, and the source text is the English context. A task of experts to translate those texts. Sentences translated through experts are compared to offers which have been edited by our system. And as a result coincidence of words between experts and our method has the following values (Table 4).

Considering coincidence we can tell more precisely the system that we offered definitely corrected and improves translation quality.

5   Conclusion

The method consists of two parts: alignment method and method maximum entropy models. This paper focusing on the combination of those two methodology. To find the incorrect word use the method translation memory and on the next stage the maximum entropy method using for editing incorrect words.

As a result proceeding from the received results we can tell what offered post-editing methodic allows to find incorrect word in text and to edit these words. Proposed pos-editing methodic improves quality of machine translated Kazakh text. More exact assessment quality of the translated sentence is planned in the future. It is planned to assess editing the text on quality by using of standard evaluation methods.

References

1. Espla, M., Sanchez-Martinez, F., Forcada, M.L.: Using word alignments to assist computer-aided translation users by marking which target-side words to change or keep unedited. In: Proceedings of the 15th Annual Conference of the European Association for Machine Translation, pp. 81–89, Leuven, Belgium (2011)
2. https://en.wikipedia.org/wiki/Principle_of_maximum_entropy
3. Translator Promt. http://www.promt.ru
4. Translator Google. https://translate.google.kz/#kk/en
5. http://www.krugosvet.ru/enc/gumanitarnye_nauki/lingvistika/MASHINNI_PEREVOD.html
6. Espla-Gomis, M., Sanchez-Martinez, F., Forcada, M.L.: Using on-line available sources of bilingual information for word-level machine translation quality estimation. In: Proceedings of the 18th Annual Conference of the European Association for Machine Translation, pp. 19–26, Antalya, Turkey (2015)

7. Esplà-Gomis, M., Sánchez-Martínez, F., Forcada, M.L.: Using machine translation in computer-aided translation to suggest the target-side words to change. In: Proceedings of the 13th Machine Translation Summit, September 19–23, 2011, Xiamen, China, pp. 172–179
8. Koehn: Statistical Machine Translation, p. 113. http://www.statmt.org/book/
9. http://bazhenov.me/blog/2013/04/23/maximum-entropy-classifier.html
10. Translator Yandex. https://translate.yandex.kz/
11. http://wiki.apertium.org/wiki/Main_Page

Complex Technology of Machine Translation Resources Extension for the Kazakh Language

Diana Rakhimova and Zhandos Zhumanov

Abstract The paper is devoted to creating linguistic resources such as parallel corpora, dictionaries and transfer rules for machine translation for low resources languages. We describe the usage of Bitextor tool for mining parallel corpora from online texts, usage of dictionary enrichment methodology so that people without deep linguistic knowledge could improve word dictionaries, and we show how transfer rules for machine translation can be automatically learned from a parallel corpus. All describe methods were applied to Kazakh, Russian and English languages with a task of machine translation between these languages in mind.

Keywords Linguistic resources · Low resources languages · Parallel corpora · Dictionaries · Transfer rules

1 Introduction

Linguistic resources are an important part of any linguistic study. While languages that have been subjects of computational studies for a long time have a lot of resources ready to be used, other languages have an urgent need to develop such resources. Linguistic resources such as monolingual and parallel corpora, electronic dictionaries, and rule dictionaries are very important both for statistical and rule-based language processing. Development of the resources requires a lot of effort and time. It is only logical that low resourced languages need to take all possible opportunities to make that process easier and faster. In this paper, we describe how to use on-line texts and specialized tools to build and improve

D. Rakhimova · Z. Zhumanov (✉)
Laboratory of Intelligent Information Systems, Al Farabi Kazakh National University,
Almaty, Kazakhstan
e-mail: z.zhake@gmail.com

D. Rakhimova
e-mail: di.diva@mail.ru

© Springer International Publishing AG 2017
D. Król et al. (eds.), *Advanced Topics in Intelligent Information
and Database Systems*, Studies in Computational Intelligence 710,
DOI 10.1007/978-3-319-56660-3_26

linguistic resources. With tools and approaches described it is possible to create sufficient amount of different linguistic resources for low resources languages.

Contribution of this work consists of the following: it unites 3 technologies of linguistic resources extension: for parallel corpora, word dictionaries and transfer rules; the combined technology is applied to Kazakh-English and Kazakh-Russian language pairs. Combination of the three technologies allows using their results together for improvement of each other. Larger corpora help to increase coverage of dictionaries. Corpora and dictionaries together help to infer better transfer rules. In common, proposed complex (combined) technology of machine translation resources extension allow to improve of machine translation quality.

2   Related Works

Creation of linguistic resources that are being considered in this work has been an important task for all the languages. Techniques used in the work have been tried for different language pairs, but not for Kazakh-English or Kazakh-Russian.

Development of parallel corpora using Bitextor has been described in following works. Esplà-Gomis [1] describes Bitextor and apply it for collecting Catalan–Spanish–English parallel corpora. Esplà-Gomis and Forcada [2] describes creation of English-French parallel corpus. Rubino et al. [3] is devoted to Finnish-English parallel corpus. Esplà-Gomis et al. [4] deals with English-Croatian corpus. There are also similar works on Portuguese-English, Portuguese-Spanish, Slovene-English and Serbian-English language pairs.

Dictionary enrichment methodology for people without deep linguistic knowledge is described in [5] for Spanish and in [6] for Croatian. There are no similar works performed for Kazakh or Russian.

Structural transfer rules for Kazakh are described in [7, 8]. Sánchez-Cartagenaa et al. [9] shows an approach to automated generation of structural transfer rule for Spanish-Catalan, English-Spanish, Breton-French language pairs.

The complex technology that is described in the paper has not yet been used as such for one language pair. Only parts of it have been tested and applied to different languages.

3   Building Bilingual Parallel Corpora for Kazakh-English and Kazakh-Russian Language Pairs Using Bitextor

Creation of multilingual parallel corpora is one of the important tasks in the field of machine translation, especially for statistical machine translation. Today, the Internet can be considered a large multi-lingual corpus, because it contains a large number of websites with texts in different languages. Pages of the sites can be considered as parallel texts (bitexts). Bitextor tool is used to collect and align parallel texts from websites.

Bitextor is a free open source application for collection of translation memories from multilingual websites. The application downloads all HTML files from a website, then pre-processes them into a consistent format and applies a set of heuristics to select the file pairs that contain the same text in two different languages (bitexts). Using LibTagAligner library translation memories in TMX format are created from these parallel texts. The library uses HTML-tags and length of the text segments for alignment [1]. After cleaning the resulting translation memory from TMX format tags, we receive a parallel corpus with sentences in different languages aligned with each other.

A key element in Bitextor is the ability to compare file pairs and identify parallel texts in them. To do this, first of all, it uses file metrics (they can be called "fingerprints"), which are determined from numbered text segments. But before comparing file metrics, a set of heuristics is used. After applying heuristics, Bitextor does not need to process every pair of files to compare all of the metrics to each other. Metrics comparison is performed only if the file pair meets all the heuristics. List of heuristics:

1. Comparison of language of the text: if two files are written in the same language, one cannot be a translation of the other.
2. Comparison of file extensions: if within the same site one file is a translation of another file, they usually have the same extension.
3. File size coefficient: this parameter is relative and used to filter a pair of files whose size is different from each other.
4. Total difference between lengths of the texts: this option has the same function as the previous one, but it measures the size of plain text of every file in the symbols.

The process of creating corpora with Bitextor consists of several successive stages described below.

During download stage, website files are copied onto a computer using HTTrack application. This application downloads all HTML files from a multilingual website. Doing that it maintains directory tree structure.

During the next stage, all downloaded files are pre-processed in order to adapt them to the next stages. Bitextor uses LibTidy library to standardize possibly incorrect HTML files into valid XHTML files. It guarantees that tag structure within these files is proper. Original HTML file encoding is converted into UTF-8.

Once the files have been pre-processed, next step is to gather some information needed to compare files and generating the translation memory, such as name and file extension. The language of each text is determined using LibTextCat library. File metrics are also determined on this step.

Information obtained from files is stored in a list, organized in accordance with a position of analyzed file in the directory tree. It makes access to information easier, as file comparison is done level by level.

On the stage of comparing files and translation memory generation, comparison of files begins with a comparison of the levels. The user can limit the difference in depth of the directory tree during comparison. Parallel texts, as a rule, are at the same level in the tree or at very close levels, so there is no need to compare each file with all files at different levels.

Generation of translation memory in TMX format is done using LibTagAligner library. As with the metrics, Bitextor uses integer numbers for representing tags and text blocks in TagAligner.

Bitextor has beer run for following websites: http://www.kaznu.kz, http://www.bolashak.gov.kz, http://www.enu.kz, http://www.kazpost.kz, http://www.archeolog.kz, http://e-history.kz, http://inform.kz, http://egov.kz, http://primeminister.kz, http://tengrinews.kz and etc. (Fig. 1).

As a result of Bitextor's work from each site we obtained *.tmx file with the following format (Fig. 2).

In this format, tag <tu> includes a pair of aligned segments (in this case—sentences); tag <tuv>—separate sentences in two languages; tag <prop>—HTML file addresses from which these sentences have been extracted; tag <seg>—sentences themselves. In such *.tmx file sentence in one language corresponds to the sentence in another language. It should be noted that comparison quality depends on the website. Thus, we receive a file with parallel texts.

During cleaning of TMX files recurring segments, erroneous and meaningless sentence pairs were deleted. After removal of tags, we received Kazakh-English

```
  ▼              Terminal - apertium@apvb: ~/bitextor-code        − +

 File   Edit   View   Terminal   Tabs   Help
hashtable "cache_tests" summary: size=16 (lg2=4) used=4 stash-size=0 pool-size=
64 pool-capacity=256 pool-used=164 writes=4 (new=4) moved=0 stashed=0 max-stash
size=0 avg-moved=0 rehash=0 pool-compact=0 pool-realloc=1 memory=888
hashtable "hash->sav" summary: size=16384 (lg2=14) used=5738 stash-size=0 pool-
ize=0 pool-capacity=0 pool-used=0 writes=5835 (new=5738) moved=2361 stashed=1 m
x-stash-size=1 avg-moved=0.411467 rehash=10 pool-compact=0 pool-realloc=0 memor
=262520
hashtable "hash->adrfil" summary: size=16384 (lg2=14) used=5738 stash-size=0 pc
l-size=0 pool-capacity=0 pool-used=0 writes=5835 (new=5738) moved=3177 stashed=
 max-stash-size=2 avg-moved=0.553677 rehash=10 pool-compact=0 pool-realloc=0 me
ory=262520
hashtable "hash->former_adrfil" summary: size=16 (lg2=4) used=5 stash-size=0 pc
l-size=0 pool-capacity=0 pool-used=0 writes=5835 (new=5) moved=0 stashed=0 max-
tash-size=0 avg-moved=0 rehash=0 pool-compact=0 pool-realloc=0 memory=632
Done.
Thanks for using HTTrack!
apertium@apvb:~/bitextor-code$ svn update
Updating '.':
At revision 299.
apertium@apvb:~/bitextor-code$ bitextor -b 1 -v ru-kk.dic -q 0.2 -m 5 -d Bitext
r/ -u http://kaznu.kz/ -O kaznu.tmx -x ru kk
* kaznu.kz/kz/3312/page/About_Al-Farabi_Kazakh_National_University/Rector%e2%8€
* kaznu.kz/kz/14969/page/Science_and_innovations/Research_activity/ (33127 byte
)| - OK
```

Fig. 1 An example of running Bitextor for www.kaznu.kz

```
-<tmx version="1.4">
   <header adminlang="en" srclang="ru" o-tmf="PlainText" creationtool="bitextor" creationtoolversion="4.0"
   datatype="PlainText" segtype="sentence" creationdate="20151017T180048" o-encoding="utf-8"> </header>
   -<body>
     -<tu tuid="1" datatype="Text">
       -<tuv xml:lang="ru">
           <prop type="source-document">Bitextor/esep.kz/rus/showin/article/1964.html</prop>
           <seg>Счетный комитет - Структурные подразделения</seg>
         </tuv>
       -<tuv xml:lang="kk">
           <prop type="source-document">Bitextor/esep.kz/kaz/showin/article/1964.html</prop>
           <seg>Есеп комитеті - Құрылымдық бөлімшелер</seg>
         </tuv>
       </tu>
     -<tu tuid="2" datatype="Text">
       -<tuv xml:lang="ru">
           <prop type="source-document">Bitextor/esep.kz/rus/showin/article/1964.html</prop>
         -<seg>
           Трудовую деятельность начал в 1977 году экономистом-аналитиком в Опытном хозяйстве Казахской
           машиноиспытательной станции. С октября 1978 года по октябрь 1979 года - экономист совхоза
           «Алатау».
         </seg>
       </tuv>
```

Fig. 2 A format of obtained parallel corpus for Kazakh-Russian language pair

parallel corpus with 25000 sentences. This corpus is available at https://drive. google.com/drive/folders/0B3f-xwS1hRdDM2VpZXRVblRRUmM. For Kazakh-Russian language pair we received bilingual parallel corpus of 10,000 sentences.

For example, from the website kaznu.kz about 30 h of automatic work of a Bitextor more than 1500 parallel sentences have been collected. The volume of the results is directly dependent on the quantity and quality of the site's content. As it can be seen Bitextor allows saving human and time resources and obtaining parallel aligned corpora from multilingual websites. The received corpora can be used at creation of dictionaries, structural rules and solutions of problems of a polysemy of words in machine translation systems.

4 Automated Enrichment of Machine Translation System Dictionaries

Dictionaries are necessary for translation of texts from one language to another. There are thousands of translation dictionaries between hundreds of languages (English, Russian, Kazakh, German, and etc.) and each of them can contain many thousands of words. Usually, paper version of dictionary is a book of hundreds of pages for which a search for the right word is a fairly long and laborious process. Dictionaries used in machine translation may contain translations into different languages of hundreds of thousands of words and phrases, as well as provide users with additional features. Such as giving a user an ability to select the languages and translation direction, provide a quick search for words, ability to enter phrases, etc.

Today there are many methods of expanding dictionaries. We used method realized in Apertium by Miquel Esplà-Gomis. We used the tool to fill dictionaries for English-Kazakh, Kazakh-English language pairs in the free/open-source

```
▼                                                    Terminal - apertium@

File    Edit    View    Terminal    Tabs    Help

Reading file /home/apertium/apertium-testing/apertium-eng-kaz/apertium-eng-kaz.e
ng.dix
Is the word 'swimmers' possible? (y=yes, n=no, b=go back): y
Is the word 'swimmering' possible? (y=yes, n=no, b=go back): You have to type an
 answer y (yes), n (no) or b (go back):n
Is the word 'swimmers'' possible? (y=yes, n=no, b=go back): You have to type an
answer y (yes), n (no) or b (go back):y
Is the word 'swimmeri' possible? (y=yes, n=no, b=go back): You have to type an a
nswer y (yes), n (no) or b (go back):n
Is the word 'swimmera' possible? (y=yes, n=no, b=go back): You have to type an a
nswer y (yes), n (no) or b (go back):n
Is the word 'swimmer's' possible? (y=yes, n=no, b=go back): You have to type an
answer y (yes), n (no) or b (go back):y
Is the word 'quarts' possible? (y=yes, n=no, b=go back): You have to type an ans
wer y (yes), n (no) or b (go back):y
Is the word 'quarting' possible? (y=yes, n=no, b=go back): You have to type an a
nswer y (yes), n (no) or b (go back):n
```

Fig. 3 Example of using the method for adding words to English monolingual dictionary

Apertium machine translation system. English-Kazakh MT system has three types of dictionaries: English monolingual, Kazakh monolingual and English-Kazakh bilingual dictionary. All dictionaries, except Kazakh monolingual, have XML format, each word has tag showing which part-of-speech is it [10].

The method is used to assign stems and inflectional paradigms to unknown words if unknown word's paradigm (word pattern) does not appear in dictionaries. The tool needs a file with a list of unknown words that will be added into monolingual dictionary, monolingual dictionary, new dictionary that will be created with the new words added to special section marked "Guessed", information about a number of questions to be asked. For Kazakh language, it also needs the automation of second-level rules for MT system. The list of unknown words has to be pre-processed with the collection of scripts provided with the tool. After the tool is launched a user can choose among different combination of candidate stems and paradigms correct ones by answering questions asked by system. When a user confirms that the words have been detected correctly they get moved to appropriate dictionary section. In case when the system finds more than one solution for a word all possible options are written to the dictionary along with the number of found possible options (Figs. 3 and 4).

We have been using this methodology to extend a number of words in dictionaries, which mainly effects to the quality of the translation in machine

```
  </section>
  <section id="Guessed" type="Temporal">
<e r="" lm="swimmer" a="QueringUser" c="1 possible solutions; choose one and confirm it"><i>swimmer</i><par n="staff__n"/></e>
<e r="" lm="quart" a="QueringUser" c="2 possible solutions; choose one and confirm it"><i>quart</i><par n="house__n"/></e>
<e r="" lm="quart" a="QueringUser" c="2 possible solutions; choose one and confirm it"><i>quart</i><par n="Smith__np"/></e>
<e r="" lm="geographer" a="QueringUser" c="1 possible solutions; choose one and confirm it"><i>geographer</i><par n="staff__n"/></e>
<e r="" lm="purse" a="QueringUser" c="1 possible solutions; choose one and confirm it"><i>purse</i><par n="staff__n"/></e>
  </section>
```

Fig. 4 Generated dictionary entries

translation. The technology allows non-expert users who do not have a deep knowledge in computational representation of morphology but understand language being developed participate in building dictionaries. That means more people can add dictionary entries creating larger dictionaries in less time.

5 Automatic Generation of Structural Rules of Transformation of Sentences

Statistical methods and methods based on rules that are mutually reinforcing approaches to machine translation, which have different strengths and weaknesses. This complementarity appeared as a result growing interest in hybrid systems, combining statistical analysis and linguistic approaches. Therefore, the automatic generation of structural rules based on the transformation of small parallel corpora with further integration in MT system based on rules is a method of helping to solve the above problem in less time and more efficiently. This method avoids the need to manually write these rules of human. To use this method, the program will automatically generate the structural transformation rules for sentences has been adapted for English-Kazakh-Russian and Kazakh language.

Approach by means of which it is possible to receive automatically rules of superficial transfer from small parallel corpora uses the alignment templates (AT) which were originally used in statistical machine translation. The technology of the automated designing of structural rules of machine translation includes a number of stages:

1. Receiving lexical forms by transformation of two parties of the parallel corpus to intermediate representation using the machine translation system Apertium. The intermediate representation consists of lexical forms of words from the case.
2. The lexical form of source language is translated into target language, using the bilingual Apertium dictionary. One word can have several translations in this case rules of grammar of restrictions (Constraint Grammar—CG) [11] or the tagger for parts of speech based on the Hidden Markov Models (HMM), for the solution of a morphological polysemy, and the rule of the lexical choice, for the solution of a lexical polysemy are used.
3. To align, use IBM Model 1.3 and 4 and HMM alignment model for iteration 5 by Giza ++ for two directions of translation. Calculating the alignment Viterbi, according to the model for the two directions of translation. For aligned according to the sentence pairs, two sets of synchronized Viterbi alignments by finding intersections by Och and Ney (2003) [12].
4. The bilingual phrases corresponding to these alignments are taken.
5. For receiving more generalized AT, from bilingual phrases, lexical forms are removed.

6. For finding of optimum AT process of minimization of AT, by way of comparison of attributes and restrictions at the different levels is made, using algorithm of beam search.
7. Further, by use of the list of tags and groups of attributes, the corresponding pairs of languages generation of rules.

Some researchers devote more attention to the alignment step, considering that improving its quality will improve the quality of the entire system.

For the application of the method described above, the program has been adapted for the English-Kazakh-Russian and Kazakh language pair. Create a file with a complete list of classes and attributes of words that describe the morphological characteristics of the lexical forms.

The adapted program has been started on the test corpora of 300 parallel sentences for the Kazakh-Russian and 250 parallel sentences for the English-Kazakh language pair. As a result were received bilingual phrases and structural rules of transformation rules 13 for the Kazakh-Russian and 11 rules for the English-Kazakh language pair.

As seen in Table 1 of transfers hand-written rules and automatically generated rules are identical.

Automatic generation of structural rules of transformation of sentences is aimed to save time and development effort for creating structural rules. It relies strongly on parallel corpora that are result of using Bitextor as described in Sects. 3–4.

Table 1 Comparison of methods of transformation rules

Input text	Хорошая школа (Good school)	
	Level of superficial transfer (chunk)	
	Hand-written rules	The generated rules
	^adj-noun\<NP\>\<sg\>\<p3\>\<PXD\>\<CD\>{^жа сы\<adj\>$^мектеп\<n\>\<2\>\<4\>\<5\>$}$^sent\<SENT\>{^.\<sent\>$}$	^__adj___n_\<LRN\>{^жа сы\<adj\>$^мектеп\<n\>$}$^*executedtule16$^sent\<SENT\>{^.\<sent\>$}$
Translation	жа сы мектеп (Good school)	жа сы мектеп (Good school)
Input text	Football player	
	^noun\<NP\>\<sg\>\<p3\>\<PXD\>\<CD\>{^футбол\<n\>\<sg\>\<nom\>$^ойыншы\<n\>\<2\>\<px3sp\>\<5\>$}$^sent\<SENT\>{^.\<sent\>$}$	^__n___n_\<LRN\>{^футбол\<n\>\<nom\>$^ойыншы\<n\>\<px3sp\>\<nom\>$}$^*executedtule30$^sent\<SENT\>{^.\<sent\>$}$
Translation	футбол ойыншысы (Football player)	футбол ойыншысы (Football player)

6 Experiment Results

After implementing the technologies describe in the paper experiments on machine translation quality for the language pairs have been conducted. Collected resources were incorporated in Apertium machine translation platform. After that translation quality was compared with Sanasoft and Google Translate—both machine translation applications that support Kazakh-English and Kazakh-Russian language pairs. The results of experiments for Kazakh-English language pairs are shown in Tables 2 and 3.

By results of the translation it is possible to allocate with various machine translation systems the main mistakes:

1. Lexical errors in word formation in Kazakh and Russian. Not the right choice or the absence of the end (termination) of the word.
2. The wrong coordination and formation of the terminations in phrases (especially in case, personal and possessive inducement). Wrong systematization of the sequence of the terminations.
3. Violations in formation of syntactic sentence structure.
4. Wrong translation of words. This shortcoming can be explained with lack of this word or a basis of the word in the dictionary of the machine translation system. As a result of the translation quality evaluation showed different results. It may be noted that the translation from Russian into Kazakh most best result was shown by translation of Google Translate, and implemented of Apertium programs based on developed models and algorithms have good results. In the Kazakh language in Russian of Sanasoft were the highest.

For English-Kazakh pair MT systems have some mistakes in translations. The Google MT system gets a good BLEU score, but in translation of selected phrases,

Table 2 BLEU scores for English-Kazakh language pairs

MT application	Estimation %	
	For English-Kazakh translation	For Kazakh-English translation
Google Translate	20.57	33.08
Sanasoft	15.74	13.97
Apertium	58.97	34.5

Table 3 BLEU scores for Kazakh-Russian language pairs

MT application	Estimation %	
	For Russian-Kazakh translation	For Kazakh-Russian translation
Google Translate	9.16	6.01
Sanasoft	4.12	8.08
Apertium	8.49	5.66

it makes common mistakes such as not assigning the right possessive and case. The Sanasoft system has more errors as regards the translation of words, and many out-of-vocabulary words. Evaluation of English-Kazakh pair machine translation before of extension of linguistic resources was approximately 15%. More biggest evaluation of English-Kazakh after of extension of linguistic resources is explaned by more biggest volume increasing of linguistic resources.

7 Conclusion

In this paper, we describe complex (combined) technology of building linguistic resources for low-resourced languages. We show how to use Bitextor to create parallel corpora, a method of enriching dictionaries with new words without much of linguistic knowledge and how to collect transfer rules for machine translation from a parallel corpus. The results of applying described methods for Kazakh, Russian and English languages show that they allow to save human and time resources, to improve machine translation quality.

References

1. Esplà-Gomis, M.: Bitextor: a free/open-source software to harvest translation memories from multilingual websites. In: Proceedings of MT Summit XII, Ottawa, Canada, Association for Machine Translation in the Americas (2009)
2. Esplà-Gomis, M., Forcada, M.: Combining content-based and URL-based heuristics to harvest aligned bitexts from multilingual sites with Bitextor. Prague Bull. Math. Linguist. **93**, 77–86 (2010)
3. Rubino, R., Pirinen, T., Espla-Gomis, M., Ljubešic, N., Ortiz Rojas, S., Papavassiliou, V., Prokopidis, P., Toral, A.: Abu-MaTran at WMT 2015 translation task: morphological segmentation and web crawling. In: Proceedings of the Tenth Workshop on Statistical Machine Translation, pp. 184–191 (2015)
4. Esplà-Gomis, M., Klubicka, F., Ljubesic, N., Ortiz-Rojas, S., Papavassiliou, V., Prokopidis, P.: Comparing two acquisition systems for automatically building an English-Croatian parallel corpus from multilingual websites. In: LREC, pp. 1252–1258 (2014)
5. Espla-Gomis, M., Carrasco, R.C., Sánchez-Cartagena, V.M., Forcada, M.L., Sánchez-Martınez, F., Pérez-Ortiz, J.A.: An efficient method to assist non-expert users in extending dictionaries by assigning stems and inflectional paradigms to unknown words. In: Proceedings of the 17th Annual Conference of the European Association for Machine Translation, pp. 19–26
6. Ljubešic, N., Espla-Gomis, M., Klubicka, F., Preradovic, N.M.: Predicting inflectional paradigms and lemmata of unknown words for semi-automatic expansion of morphological lexicons. In: RANLP, p. 379 (2014)
7. Sundetova, A., Karibayeva, A., Tukeyev, U.: Structural transfer rules for Kazkah-to-English machine translation in the free/open-source platform Apertium. TÜRKİYE BİLİŞİM VAKFI BİLGİSAYAR BİLİMLERİ ve MÜHENDİSLİĞİ DERGİSİ, 7(1 (Basılı 8) (2014)

8. Sundetova, A., Forcada, M.L., Shormakova, A., Aitkulova, A.: Structural transfer rules for Kazakh-to-English machine translation in the free/opensource platform Apertium. Proceedings of the I International Conference on Computer processing of Turkic Languages (TurkLang'13), pp. 322–331, Astana, Kazakhstan (2013)
9. Sánchez-Cartagenaa, V.M., Pérez-Ortiza, J.A., Sánchez-Martínez, F.: A generalised alignment template formalism and its application to the inference of shallow-transfer machine translation rules from scarce bilingual corpora. Comput. Speech Lang. 32(1), 46–90 (2015)
10. Forcada, M.L., GinestíRosell, M., Nordfalk, J., O'Regan, J., OrtizRojas, S., PérezOrtiz, J.A., SánchezMartínez, F., RamírezSánchez, G., Tyers, F.M.: Apertium: a free/opensource platform for rulebased machine translation. Mach. Transl. 25(2), 127–144 (2011)
11. Karlsson, F., Voutilainen, A., Heikkilä, J., Anttila, A: Constraint Grammar: A Language Independent System for Parsing Unrestricted Text. Mouton de Gruyter (1995)
12. Och, F.J., Ney, H.: A systematic comparison of various statistical alignment models. Comput. Linguist. 29(1), 19–51 (2003)

Enhancing Latent Semantic Analysis by Embedding Tagging Algorithm in Retrieving Malay Text Documents

**Nurazzah Abd Rahman, Afiqah Bazlla Md Soom
and Normaly Kamal Ismail**

Abstract Latent Semantic Analysis (LSA) is a mathematical approach that uses Singular Value Decomposition to discover the important association of the relationship between terms and terms, terms and documents and also documents and documents. LSA adopted cosine similarity measure to calculate the similarity between the query and terms as well as the documents. This approach seem to be efficient if each of the term only have single meaning and a meaning only represent by single term. Unfortunately, there are terms that have multiple meanings and a single meaning that represent by multiple terms. If these terms are treated as a single word, it will lead the search engine to retrieve the irrelevant documents. The irrelevant documents that been retrieved will affect the effectiveness of the search engine. This paper propose to enhance LSA by embedding tagging algorithm. To investigate the effectiveness of LSA using tagging algorithm in retrieving documents from the Malay corpus, seven experiments are conducted. The first experiment conducted to compare the time taken for extracting the normal term list and the tagged term list, total number of both lists and also the time taken for the creation of term document matrix. All other experiments record all the results of the retrieval system by using different dimension and threshold value. The retrieval result averagely shows F-measure enhancement of approximate to 3.60% by using LSA with tagging algorithm (LSAT) compared to retrieval result of LSA.

Keywords Latent Semantic Analysis (LSA) · Tagging algorithm · Term document matrix · Threshold value

N.A. Rahman (✉) · A.B.M. Soom · N.K. Ismail
Faculty of Computer and Mathematical Science, Department of Computer Science,
Universiti Teknologi MARA, Shah Alam, Selangor, Malaysia
e-mail: nurazzah@tmsk.uitm.edu.my

A.B.M. Soom
e-mail: afiqahbazlla@tmsk.uitm.edu.my

N.K. Ismail
e-mail: normaly@tmsk.uitm.edu.my

© Springer International Publishing AG 2017
D. Król et al. (eds.), *Advanced Topics in Intelligent Information
and Database Systems*, Studies in Computational Intelligence 710,
DOI 10.1007/978-3-319-56660-3_27

309

1 Introduction

Information Retrieval (IR) is the process of retrieving information from a corpus to fulfill the user's interest, usually express in the form of query presented in natural language. Information retrieval actually manages with the organization, storage, representation, and access to information items [1–3]. While keyword searching can be a simple process to implement in any search engine, there exists an everlasting challenge of making a search engine that can accurately provide users with the exact information they are truly looking for rather than simply listing all of data containing matching keywords from their query.

Hence, throughout times since the beginning of the information age, computer scientists have been trying in many ways to improve capabilities of search engines in terms of speed, accuracy, ease of use, and to solve some specific problems experienced by users of search engines.

Correspondingly, this paper will focus in one of those many technique which is called Latent Semantic Analysis (LSA) and more specifically the attempt to enhance this technique using Tagging Algorithm.

2 Indexing by Term and Indexing by Concept

Indexing by term means the original item are used as a basis of the index process. Roslan [4] defines that indexing by term is when the terms of the original items are used as a basis of the index process. Indexing by term uses two major techniques for creation of the index which are statistical technique and natural languages technique. Statistical technique is uses vector models and probabilistic model. The calculation of weight in statistical technique uses statistical information such as frequency of occurrence of words and the distribution of words in document collection. Whereas, natural language technique is also uses some statistical information, but perform more complex parsing to define the final set of index [4].

However, along with the statistical technique and natural language technique, most of the retrieval method uses exact term-matching technique to discover and to rank the relevance of the documents with respect to the formulation of the query word [5].

Wang et al. [6] defines indexing by concept as the basis of the way to express the same idea that can improve the performance of retrieval. Conceptual indexing is different from term indexing which treats each of the occurrences as a different index and uses thesauri or another query expansion technique to enlarge a query to find the other ways the same thing has been represented. Indexing by concept determines a canonical set concept based on a set of terms and uses them as a basis for indexing all item. Furthermore, the determined set of concepts does not have a label associated with each concept, but is a mathematical representation [4].

2.1 Latent Semantic Indexing

Latent Semantic Indexing (LSI) is actually a part from overall Latent Semantic Analysis (LSA). LSI essentially refers to the indexing process in the LSA algorithm while LSA is the overall process including the result. LSI is categorize in the conceptual indexing techniques where it increase the performance of the similarity search [6].

One scenario in human-machine interaction in IR is by natural language queries where the user formulates a request by providing a number of keywords or some free-form text, and then request the system to provide the relevant data in some compatible representation. The documents that is returned to the users usually contain one or many query terms.

However, most of the IR method is based on simple term matching strategies to examine the ranking of relevance of document with the word in the query given from the user. Furthermore this literal matching technique suffers from synonyms in document [7]. In addition, literally term matching has seriously bad drawbacks mostly because of the ambivalence of words and their unavoidable insufficiency of precision as well as due to personal style and individual differences in word usage.

It is not uncommon to find a group of different words which express the same meaning (synonymy) [8]. Then again, a single word might have a few different meanings (polysemy) depending on the context [4, 7, 9]. This term variability may introduce ambiguity to the conceptual structure of the collection. Hence, exact term-matching method turns out to be unsuccessful and inaccurate to retrieve relevant document that do not share words with user's queries due to the problem of words usage. Therefore, exact matching strategies can't satisfy user because it treat words as if they are not dependent to each other, although it is quite obvious that they are not [4, 8].

Furthermore, Roslan [4] states that individual terms and keywords are not adequate discriminators of the semantic content of the documents and queries. Different user will illustrates the same information using different words. Hence, the performance of the conventional retrieval models often suffers from either missing relevant documents which are not indexed by the key words used in the query, but by synonymy or retrieving the not relevant documents which are indexed by unwanted words of the keywords in the query [4, 8].

3 Malay Morphology

Malay is an agglutinative dialect with rich morphology [10, 11]. There are three fundamental morphological operations are [12, 13]:

- **Affixation**—new words are produced by applying the affix to the root word. Affixation can be done in three ways [14]; infixes, prefixes, and suffixes. Infixes put the affix in between original words, prefixes put them before, and suffixes put them at the end. Example of infix would be the word *telunjuk*; which originates from the word *tunjuk* (point at); the affix is *-el-*. Example of prefixes would be the word *memukul*; originates from the word *pukul* (hit); the affix is me-. Example of suffix would be the word *fikiran*; originating from the word *fikir* (think); the affix is -an.

- **Reduplication**—new words produced by duplicating the word itself. There are three known types of reduplication in Malay language; full reduplication, partial reduplication, and rhythmic reduplication. Full reduplication repeats the exact root words and separate them with a dash (–). An example of full reduplication would be the word is *layang-layang*, which comes from the word *layang*. Partial reduplication is formed by taking the first part of the original word and combining it to the beginning of it. It's mainly used to indicate plural form of the word. Example of partial reduplication would be the word *dedaun* (plural) which comes from the word *daun* (single). Rhythmic reduplication repeats the whole word, but alters one or more of its phonemes in the second word. Examples of this are *kuih-muih*, *gunung-ganang*, *desas-desus*, and *gundah-gulana*.
- **Compounding**—new words produces by combining simplex word to another single word. One example of compound words is *air mata*; combination of the word *air* (water) and *mata* (eye) which means teardrops in English.

4   Part-of-Speech Tagging

POS tagging can be described as the process on identifying or classifying a text into their corresponding class or part of speech based on their meaning and based on the word relations. There are many types of POS tagging algorithm and those methods can be divided into three as follows:

- Rule-based POS (RPOS) tagging
- Transformation-based tagging
- Stochastic (probabilistic) tagging

Referring to Rayner [14], Table 1 listed out the Malay POS tag that will be used as a references in this paper. There are 4 main word types identified as noun, verb, adjective and function. There are a few sub types for each word type.

Table 1 also listed out the possible sub types for each type of word class but in this paper, it will only focus on the word types. So, for different sub types that relies on the same word type will be assign according their word type not their sub type.

Table 1 Malay word types and sub types [14]

Word type	Sub type (English)	Sub type (Malay)
Noun	Noun	*Kata nama*
	Proper noun	–
Verb	Verb	*Kata kerja*
Adjective	Adjective	*Kata adjektif*
Function	Conjunction	*Kata hubung*
	Interjection	*Kata seru*
	Interrogative	*Kata tanya*
	Command	*Kata perintah*
	–	*Kata pangkal ayat*
	Auxiliary (Amplifier)	*Kata bantu*
	–	*Kata penguat*
	Particles	*Kata penegas*
	Negation	*Kata naif*
	–	*Kata pemeri*
	Preposition	*Kata sendi nama*
	–	*Kata pembenar*
	Direction	*Kata arah*
	Cardinal number	*Kata bilangan*
	–	*Kata penekan*
	–	*Kata pembenda*
	Adverb	*Adverb*

4.1 Rule-Based POS Tagger

Rule-based part-of-speech tagging is a tagging algorithm that is conducted based on the rules to identify the class of the word. Table 2 shows the noun affixing identification rules, Table 3 listed out the adjective affixing identification rules and Table 4 illustrates the verb affixing identification rules.

In this paper, we applies a straightforward Malay RPOS tagger. Tables 2, 3 and 4 describes the affixing rules involve in this paper. First, the RPOS tagger starts by checking the word in the tagged word corpus. This tagged word corpus was automatically generated by applying the affixing rules on all the root words available from [4]. If the word exist and only have single tag, then the tagging process completed. But if the word have more than single tag, identification of the word type relation will be conducted on that word. Table 5 listed out the word type relation that used in this paper. Apart from that, if the word does not exists in the tagged word corpus, before going through the tagging process, affixing rules will be applied first.

Table 2 Noun affixing identification rules [14]

Rules	Prefix	Next character	Sequences of character	Suffix	
				May end with	
1a	pe	ny, ng, r, l and w	a–z	an	–
1b	pem	b and p	a–z	an	–
1c	pen	d, c, j, sy and z	a–z	an	–
1d	peng	g, kh, h, k and vowel	a–z	an	–
1e	penge	–	a–z (3–4 character)	an	–
1f	pel or ke	–	a–z	an	–
1g	juru, maha, tata, pra, swa, tuna, eka, dwi, tri, panca, pasca, pro, anti, poli, auto, sub, supra	–	a–z	–	–
1h	not started withme, meng, mem, menge, ber, be, di, diper	–	a–z	–	an, at, in, wan, wati, isme, isasi, logi, tas, man, nita, ik, is, al

Table 3 Adjective affixing identification rules [14]

Rules	Prefix	Next character	Sequences of character	Suffix	
				May end with	
2a	ter, se, bi	–	a–z	–	–
2b	ke	–	a–z	an	–
2c	not starting with di and men	–	a–z	–	in, at, ah, iah, sequences of vowels then wi and sequences of consonants end ending with i

5 Result and Discussion

In this paper, seven experiments are conducted. The first experiment conducted to compare the time taken for extracting the normal term list and the tagged term list, total number of both lists and also the time taken for the creation of term document matrix. All other experiments record all the results of the retrieval system by using different dimension and threshold value. For all the experiments, 2028 Malay translated hadith used as the test collection. Table 6 records the result for the first experiment on the test collection.

Table 4 Verb affixing identification rules [14]

Rules	Prefix	Next character	Sequences of character	Suffix	
				May end with	
3a	me	ny, ng, r, l, w, y, p, t, k, s	a–z	–	–
3b	mem	b, f, p and v	a–z	kan or i	–
3c	men	d, c, j, sy, z, t and s	a–z	kan or i	–
3d	meng	g, gh, kh, h, k and vowel	a–z	–	–
3e	menge	–	a–z (3–4 character)	–	–
3f	memper or diper	–	a–z	kan or i	–
3 g	ber	not r	a–z	kan or an	–
3 h	bel	–	a–z	–	–
3i	ter	not r	a–z	–	–
3j	ke	–	a–z	–	an
3 k	–	–	a–z	–	i or kan
3 l	di or diper	–	a–z	kan or i	–

Table 5 Word Type Relation [14]

Word types	Valid sequence of word types
Noun	Adjective, adverb, verb, noun
Verb	Adverb, noun
Adverb	Verb, adjective, noun

Table 6 Experiment 1

	Term_List	Term_ListTag
Number of terms extracted	935	1233
Time taken for extraction	2 s	18 s
Time taken for creating Term Document Matrix	38 s	3 min 40 s

From experiment two up to experiment seven, Fig. 1 summarize the average recall percentage of LSA retrieval system and LSAT retrieval system for each defined threshold values. Followed by Fig. 2 shows the graphically comparison of average percentage of precision value for both retrieval system respectively. Meanwhile, next Fig. 3 illustrates the average percentage F-Measure for both retrieval system.

Figure 4 illustrates the graph that averagely summarize all the values for recall, precision and F-Measure for each threshold value in previous figure for LSA retrieval system and LSAT retrieval system.

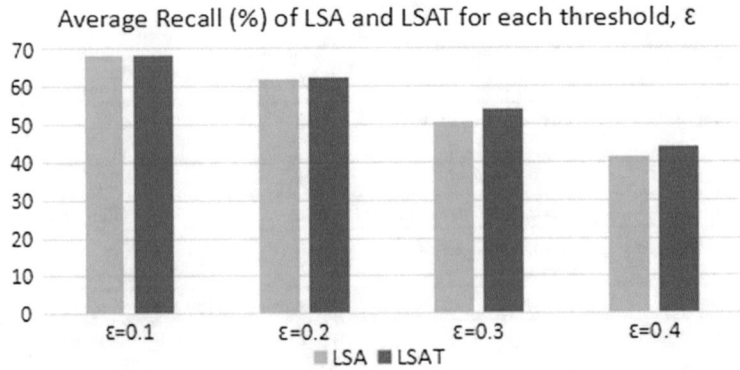

Fig. 1 Average recall percentage (%) for LSA and LSAT

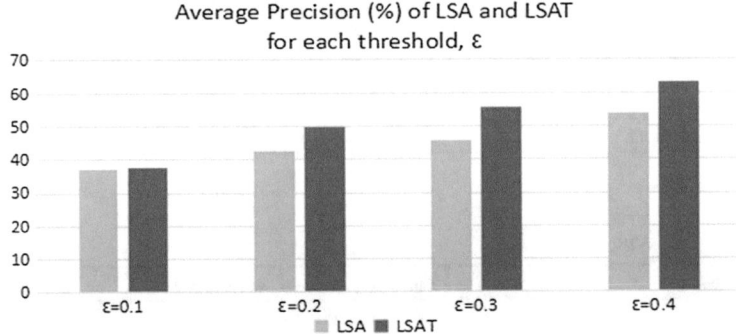

Fig. 2 Average precision percentage (%) for LSA and LSAT

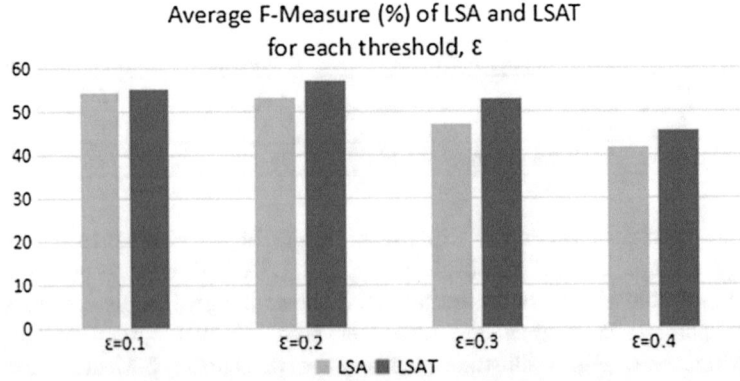

Fig. 3 Average F-measure percentage (%) for LSA and LSAT

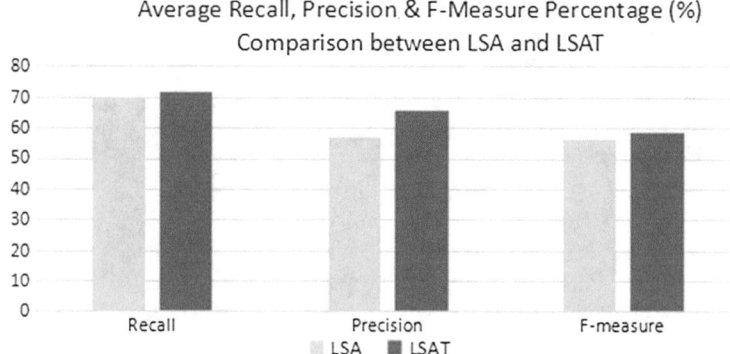

Fig. 4 Average recall, precision and F-measure percentage (%) for LSA and LSAT

From all the experiment results show that LSAT retrieval system able to enhance the performance of LSA retrieval system.

5.1 Analysis of Similarity in LSA and LSAT

LSA analyze and study the relationship among terms from documents by producing set of concepts related to the terms and documents. This is an example where by having tagging algorithm in LSA helps to reduce the number of unrelated retrieved document, in this section the k-dimensional value is set to be 220 and the threshold value is 0.2. The following Table 7 stated all the result on the experiment by using query "*hukum bernazar*". Based on relevant judgment by the experts, there are 11 relevant documents for this query.

From the table, shows that LSA method able to retrieve 10 related documents out of 11. While LSAT method able to retrieve all 11 relevant documents for this query. Referring to the table, LSA method also includes some unrelated documents as the retrieval result. Number of this unrelated documents retrieved will effect the percentage of recall and precision value. Next Fig. 5 illustrates the representation of the terms occurrence with and without additional tagging information in LSA.

From the figure, showing that majority of the unrelated document retrieved from the word "*hukum*". By applying tagging algorithm on the query "*hukum berhadas*",

Table 7 Experiment Result for LSA and LSAT on Query "hukum bernazar"

	LSA	LSAT
Total retrieved	36	11
Total unrelated retrieved	25	–
Total relevant retrieved	10	11
Recall (%)	90.91	100
Precision (%)	27.78	100

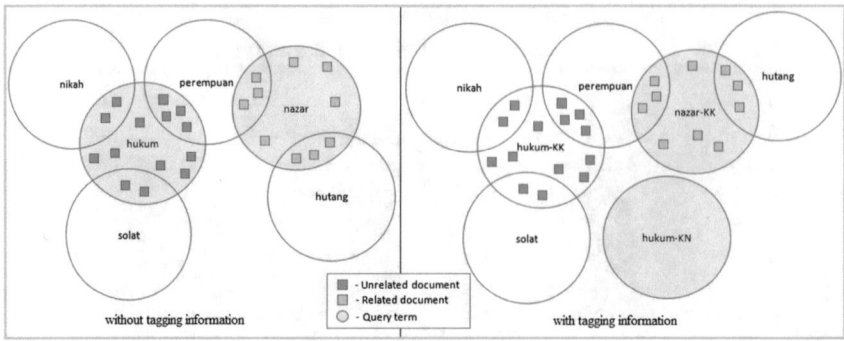

Fig. 5 Representation of terms occurrence with and without tagging information

the word "*hukum*" is actually classified as "*Kata Kerja—KN*" and the word "*bernazar*" is classified in "*Kata Kerja—KK*". Without this information, the word "*hukum*" in the query will be treated as one and because of that, a few of the unrelated documents had been retrieved. Therefore, by having that extra information, it helps to eliminate the retrieval of the unrelated documents and at the same time improve the performance of the retrieval engine.

6 Conclusion

This paper study the LSI algorithm, and then try to improve the retrieved documents by LSI algorithm by applying tagging algorithm. Through out the evaluation of both retrieval system, LSAT retrieval system able to enhance averagely approximate to 3.60% of the result performance from LSI retrieval system.

Acknowledgements This research is based upon work supported by Ministry of Higher Education (Malaysia) under Fundamental Research Grant Scheme (FRGS/1/2015/ICT01/UiTM/03/1) and Faculty of Computer and Mathematical Sciences, Universiti Teknologi MARA, Shah Alam, Selangor, Malaysia.

References

1. SivaKumar, A.P., Premchand, P., Govardhan, A.: Indian languages IR using latent semantic indexing. Int. J. Comput. Sci. Inf. Technol. (IJCSIT) 3(4) (2011)
2. Kumar, Ch.A., Radvansky, M., Annapurna, J.: Analysis of a vector space model, latent semantic indexing and formal concept analysis for information retrieval. Cybern. Inf. Technol. 12(1) (2012)
3. Baeza-Yates, R., Ribeiro-Neto, B.: Modern Information Retrieval: The Concepts and Technology Behind Search (2012)

4. Sadjirin, R., Rahman, N.A.: Efficient retrieval of Malay language documents using latent semantic indexing. In: 2010 International Symposium on Information Technology (ITSim), vol. 3. IEEE (2010)
5. Reshma, O.K., Sreejith, C., Reghu Raj, P.C.: An effective Malayalam information retrieval system using query expansion. In: International Conference on Control Communication and Computing (ICCC) (2013)
6. Wang, H., Guo, Y., Li, J., Shi, X.: Research of the conceptual representing of documents based on light ontology. In: 9th International Conference on Fuzzy Systems and Knowledge Discovery (2012)
7. Phadnis, N., Gadge, J.: Framework for document retrieval using latent semantic indexing. Int. J. Comput. Appl. (0975–8887) **94** (2014)
8. Alfathe, M., Al-Taie, S.: Document retrieval system, a case study. Int. J. Eng. Res. Appl. **6**(7), 20–22 (2016)
9. Skallman, E.: The interplay of synonymy and polysemy: the case of arrojar, echar, lanzar and tirar. Master Thesis, University of TromsØ, 2012
10. Babu, A., Sindhu, L.: A survey of information retrieval models for Malayalam language processing. Int. J. Comput. Appl. (0975–8887) **107** (2014)
11. Hasmy, H., Bakar, Z.A., Ahmad, F.: Construction of computational lexicon for Malay language. In: IVIC, LNCS, vol. 9429, pp. 257–268 (2015)
12. Kassim, M.N., Maarof, M.A., Zainal, A., Wahab, A.A.: Word stemming challenges in Malay texts: a literature review. In: Fourth International Conference on Information and Communication Technologies (ICoICT) (2016)
13. Sharum, M.Y., Abdullah, M.T., Sulaiman, M.N., Murad, M.A.A., Hamzah, Z.A.Z.: MALIM —a new computational approach of Malay morphology. In: Proceedings of 4th International Symposium on Information Technology—ITSim: KL, vol. 2, pp. 837–843 (2010)
14. Alfred, R., Mujat, A., Obit, J.H.: A ruled-based part of speech (RPOS) tagger for Malay text articles. In: ACIIDS, Part II, pp. 50–59 (2013)

Exploiting Distance Graph and Hidden Topic Models for Multi-label Text Classification

Thi-Ngan Pham, Van-Hien Tran, Tri-Thanh Nguyen
and Quang-Thuy Ha

Abstract Hidden topic models, the method to automatically detect the topics which are (hidden in a text) represented by words, have been successfully in many text mining tasks including text classification. They help to get the semantics of text by abstracting the words in text into topics. Another new method for text representation is distance graph model, which has the ability of preserving the local order of words in text, thus, enhancing the text semantics. This paper proposes a method to combine both hidden topic and distance graph models for opinion mining in hotel review domain using multi-label classification approach. Experiments show the efficiency of the proposed model provides a better performance of 4% than that of the baseline.

Keywords Distance graph model · Hidden topic model/LDA · Multi-label classification (MLC) · Text classification

T.-N. Pham · V.-H. Tran · T.-T. Nguyen · Q.-T. Ha (✉)
Vietnam National University (VNU), University of Engineering
and Technology (UET), Hanoi, Vietnam
e-mail: thuyhq@vnu.edu.vn

T.-N. Pham
e-mail: nganpt.di12@vnu.edu.vn

V.-H. Tran
e-mail: hientv@vnu.edu.vn

T.-T. Nguyen
e-mail: ntthanh@vnu.edu.vn

T.-N. Pham
The Vietnamese People's Police Academy, Hanoi, Vietnam

© Springer International Publishing AG 2017
D. Król et al. (eds.), *Advanced Topics in Intelligent Information
and Database Systems*, Studies in Computational Intelligence 710,
DOI 10.1007/978-3-319-56660-3_28

1 Introduction

The vector space representation is one of the most common representations for text due to its simplicity. In spite of its efficiency, this method treats each document as an unordered "bag of words", so it loses information about the structural ordering of the words in the document. To deal with this problem, Aggarwal and Zhao [1] defined the concept of distance graph which represents the document in terms of the distances between the distinct words. This is the kind of natural intermediate representation which makes it effective for text processing. Distance graph representation has been proven effectiveness in some text mining applications such as clustering, classification, indexing and retrieval...

Hidden topic models are useful to improve the semantics of the text representations, in which, the model of Latent Dirichlet Allocation (LDA) [2, 3] is used broadly in many works, such as [4–8]. This method can exploit the valuable knowledge which is hidden in the text to illustrate the characteristics of a text in a more extensive manner. Ramage et al. [5] proposed Labeled LDA model to learn user word-tag correspondences directly. Inkpen and Razavi [7] introduced a novel text representation method by using multiple level LDA model for different numbers of topics. Topic models have been also applied to multi-label learning problems [4, 6, 8]. Rubin et al. [6] proposed a more flexible set of LDA models for MLC including a model that takes into account prior label frequencies, and one that can additionally account for label dependencies which lead to significant improvement in classification performance. In other work [4, 8], hidden topic model was used for enriching data features and increasing the effectiveness of classifier by combining various features like traditional TF-IDF, bigrams, unigrams and LDA features into Vector Space Models.

In this paper, we combine the distance graph and hidden topic models, in which the representation of distance graphs is applied on the topic sentence to preserve the information about the distance and order between the hidden topics in the documents. We will present experimental results on MLC model classifying user reviews about one thousand Vietnamese hotels to show the effectiveness of the approach.

This paper is organized as follows. In the Sect. 2, we introduce the distance graph model and its application in text classification. In the next section, we describe the proposed model, which gives more details about the process of MLC using the combination of hidden topic model and distance graph representation. Experiments and results are presented in Sect. 4. Recent studies related to our work are introduced in Sect. 5. Conclusions are showed in the last section.

2   Distance Graph Model

A new text representation of distance graph in [1] was introduced to overcome the disadvantages of vector-space model by preserving the order (in term of distance) between the words in a document. According to [1], a distance graph of order k for a document D drawn from a corpus C is defined as a graph $G(C, D, k) = (N(C), A(D, k))$, where $N(C)$ is the set of nodes, and $A(D, k)$ is the set of edges.

In details, nodes in the set $N(C)$ are distinct words in the entire the corpus C. Therefore, the term "*node i*" and "*word i*" are used interchangeably to represent the index of the corresponding word in the corpus. Each edge in the set $A(D, k)$ is a directed arc from node i to node j if the *word i* precedes *word j* by at most k position. The frequency of the edge is the number of times that *word i* precedes *word j* by at most k position in the document. In a special case, in the distance graph of order 0, all nodes contain only self-loop edges (i.e., from a node to itself along with the frequency). Therefore, this representation is quite similar to the vector-space representation.

After constructing the distance graph, all nodes and edges (along with their frequency) are extracted as features to represent documents in vector space model.

The representation of higher orders gives insights about word ordering in the document. In [1], the authors examined the relative behavior of the distance graphs of different orders and show that distance graphs of low orders turn out to be most effective. For distance graphs with the orders higher than 5 or 10, it often does not bring much useful information.

Using the distance graph representation does not require the development of new mining and management techniques. We can take full advantages of all existing tools for text because the distance graph can be converted to a vector-space representation with "edge-augmented" feature.

3   A Combination of Hidden Topic and Distance Graph Models for Multi-label Review Classification

Our proposed model for Multi-Label Classification using the combination of the Hidden Topic Model and the Distance Graph Representation is described in Fig. 1. The dataset for training and evaluating the Multi-Label Classifier is crawled from a famous website containing hotel reviews in Vietnamese.

In common text classification, the text is represented as a set of independent units like unigrams/bag of words, bigrams, and/or multi-grams which construct the feature space, and the value of each feature is binary, frequency, or TF-IDF. This representation is simple; however, it disregards valuable knowledge inferred by considering the different types of relations between the words. In our proposed model, we first exploit the semantics of review sentence in term of hidden topics.

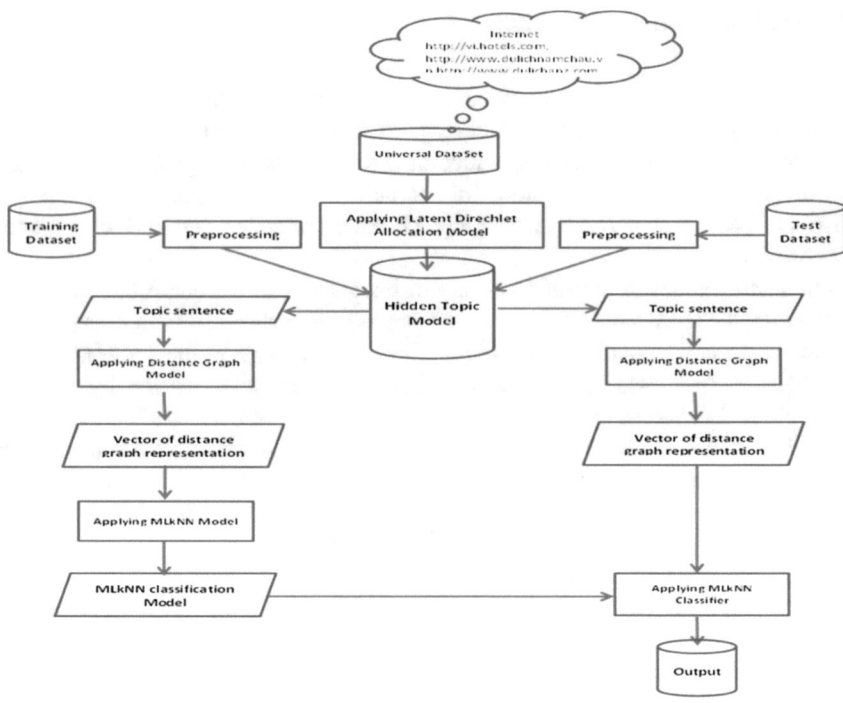

Fig. 1 Combination model of hidden topic and distance graphs for multi-label review classification

LDA [2, 3] is a generative probabilistic model for estimating the hidden topics from a corpus, in which the documents are represented as a weighted relevancy vector over latent topics, where a topic is characterized by a distribution over words. After getting the hidden topics, we convert each review text into topic sentence, in which each word is replaced with its topic having the maximum probability. Next, the distance graph model is used to build features for the reviews.

To evaluate the effect of the number of hidden topics, we build several models using different settings and varying orders for each multi-label classifier. These models will be applied for classifying newly unlabeled text reviews.

3.1 Preprocessing

Preprocessing has been implemented on reviews crawled from the website containing hotel reviews in Vietnamese. Three main sub-steps have been done in the Preprocessing step.

- Text content extraction, i.e., remove HTML tags, script, advertisements,...
- Word segmentation for enhancing the semantics (note that a Vietnamese word may contain more than one separated syllables). Tokenized segmentation is used to segment sentence. For example, the sentence "*Nhân viên phục vụ ở đây rất thân thiện.*" (Here, the waiters are very friendly) is segmented into "*Nhân_viên | phục_vụ | ở_đây | rất | thân_thiện.*"
- Stop word removal, i.e., prepositions, articles and conjunctions are removed. This considerably reduces the size of feature space and prevents the system from dealing with a large number of unrelated tokens, thus to optimize the classification performance.

3.2 Hidden Topic Model Construction and Topic Sentence Generation

In order to build Hidden topic model, we crawled data from Vietnamese websites about tourism and hotels such as http://vi.hotels.com, http://www.dulichnamchau. vn, http://www.dulichanz.com, http://bookhotel.vn, http://www.dulichvtv.com, http://chudu24.com and so on. A universal dataset with more than 24,000 documents is extracted. Several LDA models have been built on the universal dataset using the GibbsLDA++ tool [9] with different number of topics. Since we considered reviews on five aspects/class labels: Location and Price, Staff Service, Hotel Condition/Cleanliness, Room Comfort/Standard and Food/Dining, a small number of hidden topics of 10, 15, 25, 50 and 100 were selected to evaluate.

These models then will be applied on the training/testing data to generate topic sentences for all of the reviews. Table 1 lists some hidden topics along with typical words in each topic.

Each word in a sentence was replaced its topic to generate topic sentence. After this step, each review will be described as the topic sentence {(*topic(d*, 1), *topic* (*d*, 2), ..., *topic(d, n*)}, where n is the number of words in the review d (after preprocessing) and *topic(d, i)* is the topic of the word i in the review d. This representation will also reduce the size of feature space to the cardinality of the set of hidden topics.

Table 1 The hidden topic in the data set

Topic	Representative words in the review
1	*Nhân viên*/staff, *phục vụ*/service, *nhiệt tình*/fervour, *thân thiện*/friendly,...
2	*Ngon*/delicious, *món*/food, *nhiều*/much, ...
3	*Đẹp*/beautiful, *rộng rãi*/commodious, *thoáng*/well-aired...
4	*Gần*/near, *trung tâm*/central, *ví trí*/location, *đường*/road....
...	...

3.3 Distance Graph Representation on Topic Sentence

In this step, we apply the distance graph representation on topic sentences (instead of original review ones). Our purpose is to take advantages of this representation model to get the information about the orders of the hidden topics occurring in each review sentence.

Due to the fact that the size of features is equal to the number of hidden topics, it is less complicated in distance graph representation of topic sentences than that of original review sentences as in [1]. We also selected different orders in distance graph to evaluate the effectiveness of topic ordering information.

3.4 The Model of Multi-label Classifier

In this step, we use the k-nearest Neighbor based algorithm for Multi-label Classification (MLkNN) [10] on the converted features from the representation model in the previous step.

In our model, five hotel aspects were identified, i.e., Location and Price, Staff Service, Hotel Condition/Cleanliness, Room Comfort/Standard and Food/Dining as mentioned above.

4 Experiments and Results

4.1 The Datasets

We carried out experiments on the data set of customers' comments about a thousand hotels in Vietnam. About 3,700 sentences/reviews were retrieved from website http://www.chudu24.com (a famous website about Vietnam tourism) and manually annotated according to five labels. About 1,550 sentences having at least one label were extracted, of which 150 sentences were used to create the testing set, and the rest was used to create the training set. The distribution of data in 5 labels is: 1-Staff Service (645), 2-Room Comfort/Standard (552), 3-Food/Dining (372), 4-Location and Price (186), and 5-Hotel Condition/Cleanliness (240). The number of sentences having 1 label is 1183; 2 labels is 297; 3 labels is 62; 4 labels is 8.

In order to train LDA model for generating hidden topic models, the universal dataset of 24,000 articles, introductions, comments about hotels in Vietnam (from other sources as mentioned above) were also crawled. Preprocessing step is applied for all data sets, i.e., for LDA and classification.

4.2 Experimental Results

In order to evaluate the effect of distance graph representation and the Hidden topic model combination model, we took several experiments with the number of hidden topics ranging from 10 to 100 and the order of distance graph from 0 to 2. Besides our model (denoted as Distance graph + Hidden topic model), we built two baselines for comparison:

- Baseline 1 (denoted as the Term Frequency in SVM) used term frequency feature represented in the vector space model. In fact, the feature of Term Frequency is equivalent to the distance graph with the order of 0.
- Baseline 2 (denoted as Original Distance graph) used the distance graph model directly on the review sentences.

After removing stop words and low frequency ones in pre-processing step, the number of distinct words in our experiments is about 2,500. They were used to build features for experiments with both the baseline and the proposed model.

In our model, the MLkNN algorithm was implemented with default configuration, and the number of neighbor k is 1.

In traditional single label learning, the performance of the learning system is evaluated with conventional metrics such as accuracy, precision, recall, etc. However, the performance evaluation of multi-label learning algorithms is complicated as each example can be associated with multiple labels simutanously. Therefore, a number of evaluation metrics specific to multi-label learning has been proposed in the literature. In our model, we use the example-based metrics [11] which work by evaluating the performance of classification models on each test example separately and then returning the averaged value across the test set. These metrics include: *Hamming loss* (hloss in short)—the fraction of instance-label pairs which have been misclassified; *one-error*—the fraction of examples whose top-ranked predicted label is not in the ground-truth relevant label set; *coverage*—the number of steps needed, on average, to move down the ranked label list of an example so as to cover all its relevant labels; *ranking loss* (rloss in short)—the average fraction of misordered label pairs and *average precision* (AP in short)—the average fraction of relevant labels ranked higher than a particular label. Furthermore, for average precision, the larger values the better performance; whereas, for the other four metrics, the smaller values the better performance.

The results of the experiments are described in the Table 2. The up and down arrows are put on the right side of each measure to indicate the higher/lower, the better performance. We highlight the best results in bold. The results show that using the representation of distance graph gives better results than those of term frequency. According to the results in Table 2, the baseline using Term Frequency in VSM has the lowest Hamming loss which means the least misclassification error including the prediction error and the missing error, but it gets the worst (lowest value) in overall average precision which means the least average accuracy of multi-label learning. In addition, the proposed model makes an improvement in

Table 2 The experimental results

Experiment			hloss ↓	oneError ↓	Coverage ↓	rloss ↓	AP ↑
Term frequency in VSM			**0.0547 ± 0.0050**	0.6626 ± 0.0186	0.9473 ± 0.1438	0.0733 ± 0.0123	0.7487 ± 0.0201
Original distance graph			0.1401 ± 0.0148	0.2459 ± 0.0405	0.9693 ± 0.1355	0.1586 ± 0.0301	0.8250 ± 0.0296
Distance graph + hidden topic model	Order 0	10 topics	0.1369 ± 0.0136	0.2184 ± 0.0193	0.8204 ± 0.1006	0.1255 ± 0.0172	0.8542 ± 0.0136
		15 topics	0.1359 ± 0.0211	0.2136 ± 0.0341	0.8217 ± 0.0962	0.1263 ± 0.0171	0.8559 ± 0.0197
		25 topics	**0.1283 ± 0.0167**	0.2056 ± 0.0359	**0.7695 ± 0.0955**	**0.1150 ± 0.0208**	**0.8645 ± 0.0222**
		50 topics	0.1398 ± 0.0157	**0.2157 ± 0.0263**	0.8566 ± 0.1094	0.1338 ± 0.0212	0.8507 ± 0.0188
		100 topics	0.1473 ± 0.0104	0.2464 ± 0.0227	0.9215 ± 0.1063	0.1513 ± 0.0184	0.8318 ± 0.0172
	Order 1	10 topics	0.1518 ± 0.0178	0.2384 ± 0.0283	0.8567 ± 0.1227	0.1351 ± 0.0215	0.8405 ± 0.0203
		15 topics	0.1479 ± 0.0140	0.2384 ± 0.0274	0.8707 ± 0.0865	0.1390 ± 0.0160	0.8397 ± 0.0168
		25 topics	**0.1310 ± 0.0167**	**0.2056 ± 0.0305**	**0.8384 ± 0.1010**	**0.1300 ± 0.0206**	**0.8565 ± 0.0195**
		50 topics	0.1401 ± 0.0111	0.2317 ± 0.0311	0.9082 ± 0.1076	0.1458 ± 0.0234	0.8364 ± 0.0209
		100 topics	0.1642 ± 0.0146	0.2913 ± 0.0361	0.9966 ± 0.1133	0.1677 ± 0.0223	0.8041 ± 0.0213
	Order 2	10 topics	0.1412 ± 0.0113	0.2103 ± 0.0194	0.7910 ± 0.1050	**0.1163 ± 0.0151**	0.8592 ± 0.0117
		15 topics	0.1360 ± 0.0133	0.2203 ± 0.0280	0.8090 ± 0.1152	0.1238 ± 0.0210	0.8541 ± 0.0207
		25 topics	**0.1271 ± 0.0161**	**0.1995 ± 0.0286**	**0.7835 ± 0.0935**	0.1175 ± 0.0178	**0.8646 ± 0.0194**
		50 topics	0.1546 ± 0.0098	0.2719 ± 0.0315	0.9236 ± 0.0897	0.1508 ± 0.0171	0.8197 ± 0.0183
		100 topics	0.1546 ± 0.0098	0.2719 ± 0.0315	0.9236 ± 0.0897	0.1508 ± 0.0171	0.8197 ± 0.0183

performance of MLC in comparison with the original method of distance graph [1] in all five metrics. Additionally, the results indicate that the number of hidden topics is the best at 25, and the worst at 100. This may relate to the kind of short review texts in our experiments and the number of hidden topics should be reasonable to the number of predefined label set.

5   Related Work

Text classification has been regarded as a practical and effective method in text mining tasks. In order to improve the performance of such an important task, we always need an informative and expressive method to represent the texts. A number of approaches with topic modeling-based algorithms have been developed such as LDA to enhance the semantics of text. The feature of hidden topic model is used in several classification work such as [4, 7, 12, 13], in which LDA model is applied on a corpus to infer its major topics to be used for document representation. To evaluate the effectiveness of using hidden topic features, experiments were made with different representation such as: TF-IDF representation, hidden topics representation, and the combination representation of TF-IDF and hidden topics. In [7, 13], Inkpen and Razavi built a multi-level topic representation to explore the most discriminative representation for the task of the text classification. All the work showed a very good classification performance with the use of hidden topic features.

Correlations among words are chosen to improve the quality of the classification results in several below work. Aggarwal and Zhao [1] proposed the distance graph representations which preserve information about the relative ordering between the words in the graphs. This presentation is proved to be effective with a large number of different classification, clustering and similarity search applications. Rossi et al. [14] presented a textual document classification algorithm which uses the structure of a bi-particle heterogeneous network to induce a classification model. Mei et al. [15] proposed two different approaches including Block Identification and Matching (BIM) and Itinerant Dynamics with Emergent Attractors (IDEA). In which BIM model shares many similarities with text segmentation models based on lexical similarity [1]. In [16], Poyraz et al. proposed a representation of words and the higher-order paths between terms based on undirected graph exploring the latent information in higher-order paths. They used this novel semantic smoothing representation for the Naïve Bayes algorithm. Altinel et al. [17, 18] also explored the higher-order paths between document as well as the terms to incorporate semantic information into the Support Vector Machines (SVM) algorithm.

As mentioned above, both the approaches with topic modeling-based algorithms such as LDA and approach of new representation of text data based on graph for exploring latent information and relation among words are received more attention.

This paper is the first approach of combining the hidden topic model to extract topics from text data and the representation of topics based on distance graph representation.

6 Conclusions and Future Work

This paper show an experimental study on the MLC models on Vietnamese hotel reviews. We proposed a combination model of the hidden topic and the distance graph models, in which the hidden topic model identifies the topic of each word in a review sentence; then the topic sentence will be represented base on distance graph. Experiments have been implemented to show that the proposed algorithm is better than the original distance graph representation. Although our model gets some promising results, more investigations and experiments should be carried out to validate this approach, such as running this model on some widely used benchmark datasets for text classification, and using other classifiers in the model for evaluation.

Acknowledgements This work was supported in part by VNU Grant QG-15-22.

References

1. Aggarwal, C., Zhao, P.: Towards graphical models for text processing. Knowl. Inf. Syst. (KAIS) **36**(1) (2013)
2. David, M.: Blei: probabilistic topic models. Commun. ACM (CACM) **55**(4), 77–84 (2012)
3. Blei, D.M., Ng, A.Y., Jordan, M.I.: Latent Dirichlet allocation. J. Mach. Learn. Res. (JMLR) **3**, 993–1022 (2003)
4. Pham, T.-N., Phan, T.-T., Nguyen, P.-T., Ha, Q.-T.: Hidden topic models for multi-label review classification: an experimental study. In: ICCCI 2013 (2013)
5. Ramage, D., Hall, D., Nallapati, R., Manning, C.D.: Labeled LDA: a supervised topic model for credit attribution in multi-labeled corpora. In: EMNLP, pp. 248–256 (2009)
6. Rubin, T.N., Chambers, A., Smyth, P., Steyvers, M.: Statistical topic models for multi-label document classification. Mach. Learn. **88**(1–2), 157–208 (2012)
7. Inkpen, D., Razavi, A.H.: Text representation and general topic annotation based on latent Dirichlet allocation. Studia Universitatis Babes-Bolyai, Informatica **58**(2) (2013)
8. Trejo, J.V.C., Sidorov, G., Miranda-Jiménez, S., Ibarra, M.M., Martínez, R.C.: Latent Dirichlet allocation complement in the vector space model for multi-label text classification. Int. J. Comb. Optim. Prob. Inform. **6**(1) (2015)
9. Phan, X.-H., Nguyen, C.-T.: GibbsLDA++. http://gibbslda.sourceforge.net/ (2007)
10. Min-Ling, Z., Zhou, Z.-H.: ML-KNN: a lazy learning approach to multi-label learning. Pattern Recogn. **40**(7), 2038–2048 (2007)
11. Tsoumakas, G., Katakis, I., Vlahavas, I.P.: Mining multi-label data. Data Min. Knowl. Discov. Handb. **2010**, 667–685 (2010)
12. Phan, X.-H., Nguyen, C.-T., Le, D.-T., Nguyen, L.-M., Horiguchi, S., Ha, Q.-T.: Classification and contextual match on the web with hidden topics from large data collections. IEEE Trans. Knowl. Data Eng. **23**(7), 961–976 (2011)

13. Razavi, A.H., Inkpen, D.: Text representation using multi-level latent Dirichlet allocation. In: Advances in Artificial Intelligence, pp. 215–226. Springer, (2014)
14. Rossi, R.G., de Andrade Lopes, A., de Paulo Faleiros, T., Rezende, S.O.: Inductive model generation for text classification using a bipartite heterogeneous network. J. Comput. Sci. Technol. **29**(3), 361–375 (2014)
15. Mei, M., Vanarase, A., Minai, A.A.: Chunks of thought: finding salient semantic structures in texts. In: 2014 International Joint Conference on Neural Networks (IJCNN). IEEE (2014)
16. Poyraz, M., Kilimci, Z.H., Ganiz, M.C.: Higher-order smoothing: a novel semantic smoothing method for text classification. J. Comput. Sci. Technol. (2014)
17. Altinel, B., Ganiz, M.C., Diri, B.: A semantic kernel for text classification based on iterative higher–order relations between words and documents. In: Artificial Intelligence and Soft Computing. Springer (2014)
18. Altinel, B., Ganiz, M.C., Diri, B.: A simple semantic kernel approach for SVM using higher-order paths. In: 2014 IEEE International Symposium on Innovations in Intelligent Systems and Applications (INISTA) Proceedings. IEEE (2014)

Financial Reports and Financial News—An Information Content Gap Analysis

Jia-Lang Seng, Kuan-Ying Huang and Hsiao-Fang Yang

Abstract This paper hopes to find the relationships between financial news articles and the stock market and the information content gap between financial news articles and footnotes to financial statements. We use 90 listed companies in Taiwan Stock Exchange of 2013 to test stock market responses to the information content in the financial news articles and footnotes to financial statements. The content analysis technique used to find the information content to stock market reactions and compare the information content gap between financial news articles and footnotes to financial statements. Then, we find that optimistic sentiment expressed in financial news articles positively relates to the stock price movement. However, we are unable to reach a conclusion that information disclosed in the footnotes to financial statements is significantly enough to represent the existence of information content gap as compared to financial news articles.

Keywords Cumulative abnormal return · Financial news articles · Content analysis · Footnotes to financial statements

1 Introduction

Major content included in the financial reports are Statement of Financial Position, Statement of Comprehensive Income, Statement of Stockholders' equity, Statement of Retained Earnings and footnotes to financial statements. However, besides financial reports and other formal announcement issued by managers and compa-

J.-L. Seng (✉) · K.-Y. Huang · H.-F. Yang
Department of Accounting, College of Commerce,
National Chengechi University, Taipei City 11605, Taiwan ROC
e-mail: seng@nccu.edu.tw

K.-Y. Huang
e-mail: stevenhuangky@gmail.com

H.-F. Yang
e-mail: hfyang.wang@gmail.com

© Springer International Publishing AG 2017
D. Król et al. (eds.), *Advanced Topics in Intelligent Information and Database Systems*, Studies in Computational Intelligence 710,
DOI 10.1007/978-3-319-56660-3_29

333

nies, other sources of information provide not only accounting figures but also information with a different perspective to the public. Compared to periodic financial reports, these alternatives can provide information timely and help investors act instantly on receiving each piece of information. Therefore, current and prospective investors may combine financial reports provided by companies with more current financial news articles together to perform better decision making.

Online news is the easiest way one can capture what is happening and what might happen in the future in no time. Efficient Markets Hypothesis states that the market reacts instantaneously according to new incoming information [9, 10]. Livnat and Mendenhall [16] suggests that analysts' forecasts are a better proxy for the market expectation of earnings [16]. However, developments in the field of behavioral finance are increasingly spurring accounting researchers to question this view [6, 13–15]. Lee [15] challenges the traditional assumption that the price adjustment to new information is always rational and instantaneous, and calls for more work that incorporates the findings of behavioral finance into capital market research in accounting. Therefore, better understand the process by which stock prices incorporate accounting information. Mian and Sankaraguruswamy [18] examine cross-sectional differences in the stock price response to earnings. The study takes a time series approach and helps answer the call for more work that incorporates the findings of behavioral finance in capital market research in accounting [15]. Using a recent proxy of investor sentiment constructed by [2, 3], they examine how the stock price reaction to earnings news varies with investor sentiment. The results show that the stock price response to good earnings news is higher during periods of high sentiment, whereas stock price reaction to bad earnings news is higher during periods of low sentiment. There are many possible types of information can move stock prices, such as rumors, inside information, and news articles. Cecchini et al. [4] creates dictionaries of key words that can help predict fraud and bankruptcy and shows that if combine text and quantitative information together, they give the best results.

Financial news can discover various information about a company [17]. The return earnings relation can be modeled if relevant information from financial news could be captured [5]. Studies [8, 11, 19–22] have used the news to study the effect on the stock market. In addition, [12] indicate that the tone of news influence investor's reaction. Thus, news releases may influence investor's behavior and expectations and in turn affect stock market fluctuations. In the financial markets, changes in stock prices are the consequences of many unpredictable actions taken by investors and companies and even economic condition of a country. As the standardized financial reports are required for presentation, there is comparatively little room for companies and managers to express information in his favorable way. Thus, the research issues of this paper are as follows: 1. Does the language used throughout financial news articles provide a signal to the market response? 2. With the aid of detail information contained in the financial reports, does the market reaction to new information is more associated to it than merely news article alone? This paper plans to implement a dictionary-based content analysis method that is capable of analyzing the frequency of words used and build an overall sentiment of each article after proper evaluation. By evaluating the intricacies of article sentiment and words used, we seek to find out whether the role sentiment and textual information can play in the stock market. The remainder of this paper is organized as follows. Section 2 presents research method. Section 3 discusses results and Sect. 4 is the discussion and conclusion.

2 Research Method

2.1 Hypothesis Development

Asquith et al. [1] use analyst reports to explore the effect on the stock market. "In the end, stock ratings and target prices are just the skin and bones of analysts' research. The meat of such reports is in the analysis, detail, and tone" ("When a stock's rating and target collide", Business Week Online, April 25, 2002). Therefore, *the hypotheses 1: Footnotes to financial statements provide information and are associated with market returns around annual financial reports released dates.* In addition, the *hypotheses 2: Positive sentiment and information content in financial news articles combined with financial reports around annual financial reports released dates is more associated with market responses than footnotes to financial statements alone.*

2.2 Data and Sample

This paper not only collects financial news articles of selected companies but also annual financial statements as well. The detailed processes of data collection are as follows. The firm list of listed companies of 2013 is taken from the website of Taiwan Stock Exchange (TSE), for an initial sample of 882 companies. Companies in the Finance and Insurance Sector are excluded because of different capital structures and industry characteristics. The post-release dates of annual financial reports published during a six working day periods are collected, that is, [0, +5] working days, as the event day being the release day, suggesting that the week after each annual financial reports are released would have an impact on that company's stock prices. Most of the data sources used in this paper are from Taiwan Economic Journal database (TEJ), as most of the information of public companies is included in the database. For each working days' window, which depends on the publishing date for each company, this paper collects a number of accounting and financial market variables. For each annual financial reports announcement dates, the regression coefficients estimated over the trading day window [−200, −21] relative to an earnings announcement day were used to compute abnormal returns during the [0, +5] window. Excluding 20 trading days before financial report announcements avoids the confounding of the estimated coefficient with annual financial report announcements. Because of the adoption of International Financial Reporting Standards (IFRS) since 2013, the reporting of nonrecurring earnings components is no longer required for disclosure. Therefore, we do not include the reporting of nonrecurring earnings components. In order to identify the presence of detailed financial statements, this paper counts words and phrases in financial news articles. After deleting missing data, the numbers of remaining companies and news are 90 and 212, respectively. The number of industry breakdown are: manufacture of beverages (1), manufacture of wearing apparel and clothing accessories (2), manufacture of fabricated metal products (1), manufacture of food products (2), manufacture of plastics products (1), manufacture of machinery and equipment (4), waste collection, treatment and disposal activities; materials recovery (1), manufacture of chemical material (7), manufacture of pharmaceuticals and medicinal chemical products (1), manufacture of rubber products (1), manufacture of other non-metallic mineral products (2), manufacture of electronic parts and components (27), manufacture of computers, electronic and optical products (16), manufacture of motor vehicles and parts (1), construction of buildings (1), retail trade (7), water transportation (2), air transport (1), accommodation (1), food and beverage service activities (1), telecommunications (1), computer systems design services (1), financial intermediation (1), real estate development activities (6), and wholesale trade (1).

2.3   Content Analysis Method

2.3.1   To Measure Notes to Financial Reports and Scoring

Notes to the annual financial reports are also scored by seven criteria pointed out by the Financial Supervisory Commission (FSC), as the competent authority responsible for the development, supervision, regulation, and examination of financial markets and financial service enterprises in Taiwan. This paper defines a variable NOTESCORE to represent the scoring of notes to financial statements, ranging from 1 to 7. This scoring is constructed manually because the format and ways of presentation in the annual financial reports are not machine readable without specific modifications. The manual scoring criteria of annual financial reports are summarized as follows: (a) Valuation of Investment Property; (b) Material Component of Property, Plant, and Equipment; (c) Benefits to the Management; (d) The Impact of IFRSs Implementation; (e) The Impact of the New IFRSs Standard; (f) Three level of fair value measurement of financial instruments; (g) Disclosure of Material Accounting Measurement.

2.3.2 To Measure News Articles and Scoring

This paper collects news articles from Knowledge Management Winner (KMW) database. This research searches financial news articles using company names and codes (stock ID) as two key words and further screens out the news articles unrelated to the financial reports. Only news articles related to financial information, industry analysis or stock performances and forecasts are chosen. The team of professors and three graduate students read the textual content of financial news articles and score each news report based on the amount of information disclosed in the financial news. They score each news report based on scoring criteria. The building processes of the word lists are manually select positive, negative and scoring-related words. To increase the validity of the word lists, they repeat the building processes many times to reduce subjectivity. The scoring of financial news articles is between 1 and 5. The scoring criteria are summarized as follows: (a) Financial news corresponds to the financial reports and extends the reports; (b) Financial news corresponds to the financial reports and provides the cross-year or cross-firm or cross-industry analysis; (c) Financial news corresponds to the financial reports and provides further information to explain the notes in the reports; (d) Financial news corresponds to the financial reports and provides opinions on the effect stock trends; (e) Financial news corresponds to the financial reports and provides further information about earnings, investments, merger and acquisition, management, headquarters changes, and accounting method changes. In addition, we use TONE and TOTALSCORE to measure the financial news articles.

2.4 Regression Model

Hypothesis 1 (as Eq. 1) predicts that notes to the financial statements are associated with market returns. Hypothesis 2 (as Eq. 2) predicts that positive sentiment in the news after annual financial reports released is associated with market returns around the post-released dates.

$$
\begin{aligned}
CAR_i = {} & \beta_0 + \beta_1 SURP_i + \beta_2 BEAT_i + \beta_3 DIV\_INC_i \\
& + \beta_4 BM_i + \beta_5 LOGREV_i + \beta_6 FILELAG_i + \beta_7 NOTESCORE_i + \varepsilon_i
\end{aligned}
\tag{1}
$$

$$
\begin{aligned}
CAR_i = {} & \beta_0 + \beta_1 SURP_i + \beta_2 BEAT_i + \beta_3 DIV\_INC_i + \beta_4 BM_i \\
& + \beta_5 LOGREV_i + \beta_6 FILELAG_i + \beta_7 NOTESCORE_i + \beta_8 DET\_FS_i \\
& + \beta_9 TONE_i + \beta_{10} TOTALSCORE_i + \varepsilon_i
\end{aligned}
\tag{2}
$$

where i indexes the company observation; CAR is the cumulated abnormal return over the six days' working day window starting on the published date of financial reports; SURP is the difference between TEJ annual actual earnings and the most recent consensus analyst earnings forecast made prior to the earnings announcement, scaled by stock price measured at the beginning of the year; BEAT equal to one if announced earnings for the year of 2013 meet or exceeded analysts' expectations (i.e., when SURP $\geq$ 0) and zero otherwise; DIV_INC representing a firm that dividend declared in 2014 greater than prior year, DIV_INC is equal to one if the dividend change is positive, compared to prior year, and is 0 if the dividend change is less or equal to zero or there was no dividend announcement made in 2014; BM is Book to market ratio as the book value of equity scaled by the market value of equity (both measured at the end of the current year); LOGREV is current year sales and use its natural logarithm; FILELAG is number of days between the annual financial reports release date of each company's and average companies release date; NOTESCORE is the scoring of financial news articles, depending on the abundance of information discussed in the news articles; DET_FS is equal to one if the sum of BS_D and SCF_D to be equal to or greater than one and is zero otherwise (BS_D and SCF_D to be one word identified are greater than one for balance sheet related words and sentences greater than one for statement of cash flow related words and sentences, respectively); TONE is the differences between frequency counts of optimistic words and negative words separately and divided by the sum of frequency count of positive and negative words; TOTAL-SCORE is the sum of the scores of footnotes to financial statements and financial news articles.

3 Research Results and Analysis

Table 1 represents descriptive statistics for all accounting, financial market, and con-tent analysis variables.

Table 1 Descriptive statistics

Variable	Obs.	Mean	Median	Min	Max	Std. dev.
CAR	90	0.012	0.012	−0.268	0.148	0.051
SURP	90	0.006	0.001	−0.834	0.991	0.157
BEAT	90	0.611	1.000	0.000	1.000	0.490
DIV_INC	90	0.644	1.000	0.000	1.000	0.481
BM	90	0.670	0.619	0.094	1.961	0.373
LOGREV	90	17.036	16.862	13.827	22.098	1.678
FILELAG	90	4.169	−0.020	−4.020	62.980	10.377
DET_FS	90	0.667	1.000	0.000	2.000	0.519
NOTESCORE	90	5.700	6.000	4.000	7.000	0.771
TONE	90	0.684	0.834	−1.000	1.000	0.387
TOTALSCORE	90	10.211	10.000	8.000	12.000	1.008

Table 2 The regression coefficients of Eq. 1

Variables	Expected sign	Coefficient	t-stat
Intercept		−0.0314	−0.50
SURP	+	0.0425	1.19
BEAT	+	0.0187	1.61
DIV_INC	+	−0.0229*	−1.99
BM	+	0.0157	1.02
LOGREV	+	0.0048	1.41
FILELAG	+	0.0010	1.87
NOTESCORE	+	−0.0087	−1.25
Adjusted R2			0.0877
Sample size			90

Notes * denote statistical significance at the 10% level, based on a two-tailed t-test

The results of Table 2 show the different from prior research, the coefficient on BEAT is not statistically significant. The coefficient on SURP is not significant when BEAT is included in the regression. DIV_INC, BM and LOGREV are not statistically significant. The coefficient on FILELAG is not statistically significant. NOTESCORE is not statistically significant to the market response. Based on the results, this paper cannot reject Hypothesis 1.

The results of Table 3 show the coefficient on DIV_INC is statistically significant but with opposite direction. BM and LOGREV are positive and statistically significant, suggesting that the market responds positively to higher BM and higher revenue companies. The coefficient on FILELAG is statistically significant, indicating that the later companies release their financial reports than other companies, the higher the market response. The coefficient on DET_FS is not statistically significant, which is different from prior research [7]. The coefficient on TONE is

Table 3 The regression coefficients of Eq. 2

Variables	Expected sign	Coefficient	t-stat
Intercept		−0.1616	−2.09
SURP	+	0.0142	0.42
BEAT	+	0.0109	1.00
DIV_INC	+	−0.0344**	−3.08
BM	+	0.0328*	2.16
LOGREV	+	0.0084*	2.52
FILELAG	+	0.0012*	2.31
DET_FS	+	0.0183	1.85
TONE	+	0.0588***	3.79
TOTALCORE	+	−0.0032	−0.65
Adjusted R2			0.2166
Sample size			90

Notes *, ** and *** denote statistical significance at the 10%, 5%, and 1% levels, respectively, based on a two-tailed t-test positive and statistically significant. However, the coefficient on TOTALSCORE is not statistically significant. From these results, that is to say, the results do support Hypothesis 2.

4 Discussion and Conclusion

This paper aims to find possible connections between financial news articles and the stock price movement through content analysis methods. In additional to news articles alone, financial reports are the most formal source of public information companies can provide to the public. This paper hopes to find that the additional information disclosed in the footnotes to a financial statement does provide information that is more abundant and the stock price movement is more associated with it. As a result, this paper examines whether the language, sentiment and information content in the news articles and footnotes to financial statements can generate a market response.

This paper uses 90 listed companies in TSE of 2013 to test whether the stock market responds to the information content in the financial news articles and notes to the annual financial statements. By implementing content analysis methods to analyze the news articles and financial reports. This paper finds that optimistic sentiment expressed in financial news articles are significantly enough to be a credible signal to the stock price movement to a certain level. Moreover, based on the scoring criteria of this paper, financial news articles about basic company's fundamentals are associated with the stock price movement. However, this paper is unable to reach a conclusion that information disclosed in the footnotes to financial statements can be a source of explanation for market responses and that the footnotes to financial statements may not provide more decision-relevant information to investors and prospects, as the results show.

This paper uses 90 samples to evaluate the market price movement and there is relatively fewer sample selected compared to other research. Due to the limitation of the period and computerized tools to help analyzed data, this paper is unable to find all variables statistically significant. Consequently, if the number of companies and textual analysis tools can be expanded and specified, the market reaction to the financial news articles and footnotes to financial statements can be identified further. When it comes to textual analysis in the beginning phase, it requires lots of manual work to help better train data to retrieve features to be included for machine learning. Therefore, if financial news articles being evaluated as well as footnotes to financial statements are more thoroughly read and if one can include third parties, such as other professionals or experts, to help analyze the information content in both the news articles and annual financial statements, the findings and conclusion may be elaborated further.

In addition, this paper does not separate footnotes to financial statements into different segments, only scores them into a single figure. The scoring criteria brought out by the FSC may be biased to companies without some kind of assets and will later become requirements instead of suggestions, making these unsuitable for scoring. If one can utilize other content analysis methods especially suitable for processing footnotes to financial statements, he may be able to decide which part of the footnotes to financial statements will generate public interests and therefore combine his findings into future research.

The CAR calculation focuses on the stock price movement compared to the mar-ket. In addition to stock price changes, the trading volume may also be affected. One can include the changes in trading volume as well in future research. This paper selects [0, +5] time window to calculate CAR because of the focus on the post-release dates of the annual financial statements. For future research, one can expand and compare the pre- and post-release effects of the annual financial statements by using [−1, +1] or [−3, +3] time window and further uses the built model to testify and compare results of a different period.

Acknowledgements This research is supported by NSC 102-2627-E-004 -001, MOST 103-2627-E-004 -001, MOST 104-2627-E-004-001, MOST 105-2811-H-004-035.

References

1. Asquith, P., Mikhail, M.B., Au, A.S.: Information content of equity analyst reports. J. Financ. Econ. **75**(2), 245–282 (2005)
2. Baker, M., Wurgler, J.: Investor sentiment and the cross-section of stock returns. J. Financ. **61**(4), 1645–1680 (2006)
3. Baker, M., Wurgler, J.: Investor sentiment in the stock market. J. Econ. Perspect. **21**(2), 129–152 (2007)
4. Cecchini, M., Aytug, H., Koehler, G.J., Pathak, P.: Making words work: using financial text as a predictor of financial events. Decis. Support Syst. **50**(1), 164–175 (2010)
5. Chen, K.T., Lu, H.M., Chen, T.J., Li, S.H., Lian, J.S., Chen, H.: Giving context to ac-counting numbers: the role of news coverage. Decis. Support Syst. **50**(4), 673–679 (2011)
6. Daniel, K.: Discussion of: "testing behavioral finance theories using trends and sequences in financial performance", (by Wesley Chan, Richard Frankel, and S.P. Kothari). J. Account. Econ. **38**, 51–64 (2004)
7. Davis, A.K., Piger, J.M., Sedor, L.M.: Beyond the numbers: measuring the information content of earnings press release language. Contemp. Account. Res. **29**(3), 845–868 (2012)
8. Engelberg, J.: Costly information processing: evidence from earnings announcements. http://papers.ssrn.com/sol3/papers.cfm?abstract_id=1107998 (2008)
9. Fama, E.F.: The behavior of stock-market prices. J. Bus. **38**(1), 34–105 (1965)
10. Fama, E.F.: Efficient capital markets: a review of theory and empirical work. J. Financ. **25**(2), 383–417 (1970)
11. Gidófalvi, G.: Using news articles to predict stock price movements. http://cseweb.ucsd.edu/~elkan/254spring01/gidofalvirep.pdf (2001)
12. Henry, E.: Are investors influenced by how earnings press releases are written? J. Bus. Commun. **45**(4), 363–407 (2008)
13. Hirshleifer, D., Teoh, S.H.: Limited attention, information disclosure, and financial reporting. J. Account. Econ. **36**(1), 337–386 (2003)
14. Hirshleifer, D., Teoh, S.H.: The psychological attraction approach to accounting and disclosure policy. http://papers.ssrn.com/sol3/papers.cfm?abstract_id=1359967 (2009)

15. Lee, C.M.: Market efficiency and accounting research: a discussion of 'capital market research in accounting' by S.P. Kothari. J. Account. Econ. **31**(1–3), 233–253 (2001)
16. Livnat, J., Mendenhall, R.R.: Comparing the post-earnings announcement drift for surprises calculated from analyst and time series forecasts. J. Account. Res. **44**(1), 177–205 (2006)
17. Ma, Z., Sheng, O.R., Pant, G.: Discovering company revenue relations from news: a network approach. Decis. Support Syst. **47**(4), 408–414 (2009)
18. Mian, G.M., Sankaraguruswamy, S.: Investor sentiment and stock market response to earnings news. Account. Rev. **87**(4), 1357–1384 (2012)
19. Mittermayer, M.A.: Forecasting intraday stock price trends with text mining techniques. In: Proceedings of the 37th Annual Hawaii International Conference on System Sciences, Kauai, HI, USA (2004)
20. Sadique, S., In, F.H., Veeraraghavan, M.: The impact of spin and tone on stock returns and volatility: evidence from firm-issued earnings announcements and the related press coverage. http://papers.ssrn.com/sol3/papers.cfm?abstract_id=1121231 (2008)
21. Tetlock, P.C., Saar-Tsechansky, M., Macskassy, S.: More than words: quantifying language to measure firms' fundamentals. J. Financ. **63**(3), 1437–1467 (2008)
22. Wuthrich, B., Cho, V., Leung, S., Sankaran, K., Zhang, J.: Daily stock market forecast from textual web data. In: 1998 IEEE International Conference on Systems, Man, and Cybernetics, San Diego, CA, USA (1998)

The Great National Photocorpus of 20th-Century Vietnamese. Origins, Assumptions and Goals

Piotr Wierzchoń

Abstract Lexicography is the science and practice of making dictionaries. Its development has led to new techniques for the visual presentation of lexicographic entries. This article focuses on the technique of photodocumentation, which enables a textual quotation to be shown in its natural context. We aim to present a technological system which will make it possible, relatively cheaply, to produce a monolingual dictionary together with quotations and chronologisation—that is, the date at which a given word first appears. We consider the example of Vietnamese. As a preliminary database of material we selected just over 100 books, which we scanned and from which we excerpted quotations to illustrate the natural use of the headwords.

Keywords Linguochronologisation · Photodocumentation · Polish language · Vietnamese · 20th-century texts · Corpus · OCR

1 Introduction

The political and social upheavals which took place on the Indochinese peninsula in the 20th century have created a unique opportunity to construct, based on the example of that area, a special type of lexicographical collection. This will contain a vast amount of lexical material (several tens of thousands of headwords) taken from real printed texts of the 20th century. We intend here to discuss the stages and parameters of that process. Material will be collected in accordance with the photodocumentation formula, where each excerpt will be illustrated by means of an

Research reported in this paper was supported by the Polish Ministry of Education under Grant no. 0014/NPRH3/H11/82/2014, *Narodowy Fotokorpus Języka Polskiego.* 🄷

P. Wierzchoń (✉)
Institute of Linguistics, Adam Mickiewicz University, Poznań, Poland
e-mail: wierzch@amu.edu.pl

© Springer International Publishing AG 2017
D. Król et al. (eds.), *Advanced Topics in Intelligent Information and Database Systems*, Studies in Computational Intelligence 710,
DOI 10.1007/978-3-319-56660-3_30

original photograph of a fragment of text. Discussion and testing of the excerptology concept for Vietnamese will make it possible to apply that experience also to European languages, including Polish. In other words, we plan to use the example of Vietnamese to address the general problem of the collection of lexical material in photodocumentary form.

The photocorpus has a relatively short history. The idea was born in March 2016, during the *8th Asian Conference on Intelligent Information and Database Systems* held in the Vietnamese city of Da Nang. However, the preparations which made it possible to begin work on the project had lasted for more than a decade, and required the application of knowledge from several different disciplines. The procedures were tested on material from other languages [2, 7], mainly Polish (see http://www.nfjp.pl [8, 9]).

In short, in this article we shall present methods for the collection of Vietnamese words discovered in texts of the 20th century. This is an extremely varied period, rich historically, reflecting the impact of colonialism, the war years, the partition era and national unification, and we are therefore dealing with an extremely rich set of material whose value cannot be overestimated: it may be used for many types of analysis in linguistics and in the humanities in general.

For valuable remarks concerning the Vietnamese language, as well as Vietnamese examples, I would like to thank Michał Dziopak, a student of Vietnamese at the Institute of Linguistics of Adam Mickiewicz University, Poznań.

2   Vietnamese

Vietnamese is a language spoken by more than 78 million people worldwide [1], chiefly in Vietnam itself. Consideration may also be given to ethnic minorities living in Vietnam and to Vietnamese living in other countries, including the United States, France, Poland, Laos and Cambodia. Vietnamese is a member of the **Austroasiatic** language group. It is an **isolating** language, meaning that utterances in the language are made up of words that are invariant in form. There are three main dialects of Vietnamese: (a) the **northern dialect**, spoken by Vietnamese in the north of the country, as far as the province of Thanh Hóa, (b) the **central dialect**, spoken in the central part of the country, especially the ancient capital and its surroundings—Huế, (c) the **southern dialect**, spoken in the south of the country and in Hồ Chí Minh City.

While many Asian countries have developed their own writing systems (Korea, Japan, Thailand, etc.), Vietnam has based its system on those used in foreign countries (the Latin alphabet) and has never developed its own system independently. In the early period of the country's existence, Vietnamese writing was influenced by China, which invaded the state on multiple occasions. Under Chinese rule, Chinese writing was used, and many centuries later a writing system developed based on the Chinese—**Chữ Nôm**. The era of **Chữ Nôm** was also the time when the first Christian missionaries reached Vietnam. A missionary who made a significant contribution to the creation of the modern Vietnamese alphabet was Alexandre de Rhodes. He created the foundations of a system for writing Vietnamese using the Latin alphabet. He was the author of a Vietnamese–Portuguese–Latin dictionary (từ điển Việt—Bồ Đào Nha—Latin). The writing system using that alphabet was adopted officially for the Vietnamese language in the early 20th century [1]. Nôm continued to be used until the second half of the 19th century, but only by the educated and upper classes; it later died out. Poorer people were illiterate up to the time of the French invasion.

A separate problem in the language is the existence of Sino-Vietnamese words. This results from the fact that, firstly, the Vietnamese majority (Kinh) migrated to the area of modern-day northern Vietnam from southern parts of China, and secondly, northern Vietnam was invaded many times by China and remained under occupation for hundreds of years. This Chinese occupation led to the appearance of **Sino-Vietnamese (Hán)** words. It is estimated that today **more than half** of Vietnamese words are of Chinese origin, although in many cases these are archaisms.

3 Vietnamese Lexicography in the Context of Chronologisation

Present-day Vietnamese lexicography is dominated by bilingual dictionaries, such as Việt-Anh (Vietnamese–English), Việt-Nhật (Vietnamese–Japanese), Việt-Pháp (Vietnamese–French), and Việt-Trung (Vietnamese–Chinese). Alongside general bilingual dictionaries there are also specialist dictionaries (such as từ điển chuyên ngành kinh tế Việt-Anh—a Vietnamese–English economic dictionary), grammatical dictionaries (such as từ điển ngữ pháp tiếng Anh), dictionaries of pronunciation (such as Từ điển phát âm tiếng Anh), and so on.

Monolingual dictionaries include phraseological dictionaries (từ điển nhóm từ và thành ngữ), dictionaries of synonyms and antonyms (từ điển từ đồng nghĩa, từ điển từ trái nghĩa), etymological dictionaries (từ điển từ nguyên), terminological dictionaries (từ điển thuật ngữ), dictionaries of archaisms (tx điển từ Việt cổ), and grammatical dictionaries (từ điển ngữ pháp tiếng Việt).

Some specific dictionaries may be mentioned: Văn Tu Nguyễn, *Từ điển từ đồng nghĩa tiếng Việt*, Nhà xuất bản Giáo dục, 2001 (Dictionary of Synonyms in Vietnamese), *Từ điển từ đồng nghĩa trái nghĩa tiếng Việt*, Nhà xuất bản Hồng Đức, 2016 (Dictionary of Synonyms and Antonyms in Vietnamese), *Từ Điển Tiếng Việt (Ngôn Ngữ Việt Nam)*, Nhà xuất bản Từ Điển Bách Khoa, 2014 (Dictionary of Vietnamese), Nguyễn Lân, *Từ Điển Thành Ngữ Và Tục Ngữ Việt Nam*, 2014 (Vietnamese Phraseological Dictionary), *Từ Điển Hán-Việt*, Nhà xuất bản Văn Hoá Thông Tin, 2012 (the Han-Viet dictionary), and Ngọc San Nguyễn, Văn Thiện Đinh, *Từ điển từ Việt cổ*, Nhà xuất bản Văn hóa thông tin, 2001 (Dictionary of Archaisms of Vietnamese). From our historical standpoint, the most important dictionary is Nguyễn Thị Huệ, *Từ điển từ nguyên giải nghĩa*, Nhà xuất bản Văn hóa thông tin, 2002.

A source which can be used to look up original quotations for headwords is the online *Southeast Asian Languages Library* (http://www.sealang.net/library/), which states:

> The **Southeast Asian Languages Library** is an ambitious, technically innovative plan to create essential **digital resources** for all national and major minority Southeast Asian languages. We will build on existing US/ED initiatives to add **new facilities** for seventeen languages, including Acehnese, Balinese, *Cebuano*, Hiligaynon, Ilocano, *Indo-nesian*, *Javanese*, Maguin-danao, *Malay*, Mien, *Tagalog*, Tetum, Wa, and Waray, and extend existing resources for five more: *Burmese, Cambodian, Lao, Thai*, and *Vietnamese* (ten "critical" LCTLs are in *italic*).

It is clear that there are few Vietnamese sources in which the reader can see the original sentence contexts to illustrate particular headwords. The concept presented here will answer that need.

4   Etymology and Chronologisation

The distinction between etymology and chronologisation is fairly clear [3, 4, 11]. In etymology we seek explanations or hypotheses on how a given word came into existence, where it originated, from what language it might have been borrowed into another. For example, Vietnamese includes the following words:

1. bi-a—from English *billiards*;
2. phô mai—from French *fromage* 'cheese';
3. Hải—from the Chinese for 'sea'; it also appears as a morpheme in certain words, such as hải sản 'seafood';

   - bánh pizza—from Italian *pizza*;
   - mì spaghetti—from Italian *spaghetti*.

As the above examples show, the giving of **etymologies** involves finding the language from which a given word originates. However, the work of an etymologist is not limited to this. For example, the Vietnamese counting game "oẳn tù tì" comes from the English "one, two, three" (the name is used in southern Vietnam for the popular game of rock, paper, scissors), a phrase which was adopted during the period when American troops were stationed there.

In turn, **chronologisation** involves proposing a specific date for the appearance of a given word or expression. Naturally, this may be done with different degrees of precision (for example, to the nearest day, month, year, decade, or even century). Below we give examples of our hypotheses concerning the dating of words and phrases in Vietnamese: **1800:** Ấp cây đợi thỏ (To hug a tree, waiting for a hare), Cuộc bể dâu (A mulberry field has turned into a sea, the sea has turned into a mulberry field; Great change), **1850:** chạnh lòng cố quốc (longing for one's homeland), phế hưng (fall and rise), **1900:** chủ nghĩa cá nhân (individualism), cừu dịch (a hated person), **1950:** trường kỳ kháng chiến (long battle, long resistance), **1986–2000:** xóa bỏ tập trung (decentralisation), bỏ tem phiếu (abolition of food rationing), xã hội đặt hàng (social procurement), khoán sản phẩm (piecework), chuyển sang kinh tế thị trường định hướng xã hội chủ nghĩa, (transformation to a market economy with socialist orientations), kinh tế vĩ mô (macroeconomy), kinh tế vi mô (microeconomy), tư nhân hóa (privatisation), hiện đại hóa nền kinh tế (economic modernisation), hội nhập kinh tế thế giới (integration into the world economy), toàn cầu hóa (globalisation).

Naturally, the dates given above represent hypotheses as regards chronologisation. The question arises of whether it is possible to develop a more reliable system which would enable the obtaining of this type of chronological information—that is, which would provide a given linguistic unit together with the date of its first appearance. The system of lexicography and chronologisation that has been developed by us will be discussed further in this article.

5 Diachronic Trends

In practice, there do not yet exist tools to determine trends in historical word frequencies for 20th-century Vietnamese. As regards recent years, such a tool is provided by Google Trends, in which users can trace the popularity of a given word or phrase in Vietnamese texts; for example, *vi rút Zika* (August 2016) (Fig. 1).

Of course, there does not exist at present a system enabling verification of the chronologisation of Vietnamese words across the whole of the 20th century (for Polish see http://www.nfjp.pl [8, 9]), although to a certain very limited extent such ad hoc verifications can be performed using Google Books.

6 The Great National Photocorpus of 20th-Century Vietnamese

Below we shall describe the operational steps contained in a certain optimum lexicographic programme leading to the obtaining of a vast number (tens or hundreds of thousands) of Vietnamese headwords, together with attestation in the form

Fig. 1 *Vi rút Zika* (Google trends)

of quotations, and their chronologisation. This is a concept that has already been tested for Polish (the *Polish National Photocorpus* under the direction of Jan Wawrzyńczyk, available at http://www.nfjp.pl); see also: [10, 11].

6.1 The Database

The first step to be taken is the collection of a database of material. For the purposes of a prototype we collected more than 100 books written in Vietnamese. We made efforts to ensure that they were as representative as possible of the second half of the 20th century (access to works from the first half of the century currently involves high costs, as such books are very expensive). Listed below are some examples of the books included (Table 1).

As the list shows, the collection includes some books translated into Vietnamese (Gobineau Marcel, Gaines J. Ernest, etc.). This is not a problem, because our aim is to provide a description from the perspective of chronologisation of Vietnamese words, namely those that have been written and published in the Vietnamese language. From each book we took a photograph (excerpt) of the title and publication date, for example (Table 2).

6.2 OCR

All of the texts underwent an OCR procedure using the ABBYY FineReader program (ver. 11), to obtain textual layers [5, 7]. The quality of the results of this process is illustrated by the following example (Fig. 2).

Table 1 Title, author, and year of publication

Title	Author	Year of publication
Bản tự truyện của bà Giêng Pítman	Gaines J. Ernest	1986
Bầu Trời Chia Cắt	Vonphơ Crixta	1985
Bí Thư Cấp Huyện	Đào Vũ	1983
Bốn Mùa Yêu	Gobineau Marcel	1994
Cái Sân Gạch Và Vụ Lúa Chiêm	Đào Vũ	1972
Cây Sồi Mùa Đong	Alecxanđrơ Grin	1982
Chê Ghê-Va-Ra	La-Vơ-Rét-Xki	1977
Chế Rượu Vang Trái Cây Trong Gia Đình	Vũ Công Hậu	1983
Chiếc Mũ Kêpi	Côlet S.G.	1990
Chuyện Vui của Bili Booccơ	Phơrăng Hácđi	1983
Cổ Học Tinh Hoa	Nguyễn Văn Ngọc, Trần Lê Nhân	1988
Con Đường Đau Khổ tập 2	Tôlxtôi Alêkxêy	1976

Table 2 Title, place, and year of publication in photodocumentation

W_46	NHỮNG NĂM ẢO MỘNG	1988
W_47	HAI NGƯỜI ĐÀN BÀ	HÀ NỘI 1982
W_48	người đi xuyên tường	1983
W_49	TRĂN TRỞ	HÀ NỘI — 1986
W_5	NHÀ BÚP BÊ	Hà-nội — 1970

It can be observed that the text obtained as a result of OCR may contain certain inaccuracies (for example, *xảy ra*).

6.3 Tokenisation

For Vietnamese, OCR errors are not the only problem in the automatic processing of texts. A far greater challenge is the division of text into segments, namely words or phrases. In Vietnamese not every space is a word separator, as there exist words that consist of two or more separate syllables. Of course, this task of segmentation may be performed manually, for example for the fragment given above: *Đúng*—It is true, *chính thế đấy*—it is exactly thus, *rõ ràng*—clearly, *là* thế—so, *Mấy*—several etc.

In the case of large data collections such a manual procedure is not feasible (it is too costly and would take too much time). For this reason, in the procedure of

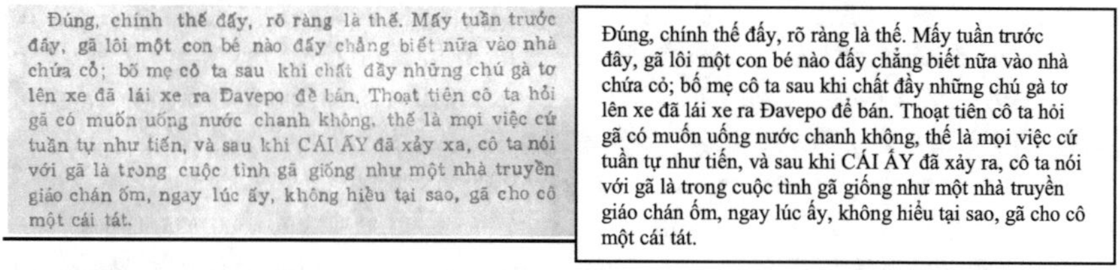

Đúng, chính thế đấy, rõ ràng là thế. Mấy tuần trước đấy, gã lôi một con bé nào đấy chẳng biết nữa vào nhà chứa cô; bố mẹ cô ta sau khi chất đầy những chú gà tơ lên xe đã lái xe ra Đavepo để bán. Thoạt tiên cô ta hỏi gã có muốn uống nước chanh không, thế là mọi việc cứ tuần tự như tiến, và sau khi CÁI ẤY đã xảy xa, cô ta nói với gã là trong cuộc tình gã giống như một nhà truyền giáo chán ốm, ngay lúc ấy, không hiểu tại sao, gã cho cô một cái tát.

Đúng, chính thế đấy, rõ ràng là thế. Mấy tuần trước đây, gã lôi một con bé nào đấy chẳng biết nữa vào nhà chứa cô; bố mẹ cô ta sau khi chất đầy những chú gà tơ lên xe đã lái xe ra Đavepo để bán. Thoạt tiên cô ta hỏi gã có muốn uống nước chanh không, thế là mọi việc cứ tuần tự như tiến, và sau khi CÁI ẤY đã xảy ra, cô ta nói với gã là trong cuộc tình gã giống như một nhà truyền giáo chán ốm, ngay lúc ấy, không hiểu tại sao, gã cho cô một cái tát.

Fig. 2 A paragraph from scanned book

creating a list of words for the Vietnamese corpus, we used a Vietnamese tokenising tool [6]. This produced the following output:

```
<s>                                      <w t="word">là</w>
    <w t="name1">Đúng</w>                <w t="word">thế</w>
    <w t="punctuation">,</w>             <w t="punctuation">.</w>
    <w t="word">chính</w>                <w t="word">Mấy</w>
    <w t="word">thế</w>                  <w t="word">tuần</w>
    <w t="word">đấy</w>                   <w t="word">trước</w>
    <w t="punctuation">,</w>           </s>
    <w t="word">rõ ràng</w>        etc.
```

Comparing these results, we can observe the advantages and defects of the tokenising tool: segments such as *rõ ràng*, *con bé* and *xe* were identified both by the tool and by a native Vietnamese speaker, but in several cases the tool divided the material differently.

6.4   Creation of a Word List

The next operational step was the identification of words and phrases required to be illustrated by a quotation. We collected all of the results marked <w t="word">, and next filtered them using the list at https://www.vi.wiktionary.org/wiki/Trang_Chính. From this dictionary we selected all headwords within the tag <h1 id="firstHeading" "class="firstHeading" lang="vi">từ điển</h1> (for the example of the headword từ điển). A part of this list is given below:

ân ái, án Anh, ăn ảnh, An ấp, An Bá, an bài, ăn bám, ăn bận, an biên, an bình etc.

This list was used to find words and phrases in the file obtained from the tokenising tool vnTokenizer. A common file was created (i.e. segments from vnTokenizer and wiktionary.org), namely:

ác báo, ác cảm, ác chiến, ác giả ác báo, ác mộng, ác nghiệp, ác nghiệt, ác nhân, ác ôn, ác quả, ác tâm, ác thú etc.

Of course, the common file also contained larger units (phrases), such as *bách chiến bách thắng, công thành danh toại, địa phương chủ nghĩa, không tiền khoáng hậu, khủng hoảng chính trị, thanh thiên bạch nhật, thao thao bất tuyệt, thiên hình vạn trạng, thuận buồm xuôi gió, thủy quân lục chiến, tiến thoái lưỡng nan, tràng giang đại hải, trùng trùng điệp điệp*, etc. vnTokenizer also accepted French and English units such as *concept, concierge, concours, concubine, condamné, condition, conditions, conduire, conduite*.

vnTokenizer selected from the texts some words of Chinese origin which are very rarely used. This is probably linked to the time of publication of some of the books used by the program. Some words of Chinese origin have entered the everyday language and are regarded as normal Vietnamese words, but there are also some which are not in use and continue to be considered Chinese words. The analysed texts did not contain many words of the latter type.

6.5 Acceptance of Photodocumentation

In this step we focused on checking whether or not the OCR process had read the texts correctly. This was done using a form with three columns. The first contained the word being searched for, and the second an image of the word in its photographic citation. The third column is used by the human annotator to record a decision on whether or not the word is a correctly recognised string of characters. It may be that the OCR system recognises a word incorrectly, leading to a word being included erroneously in the dictionary. Since we wish to avoid such situations, every quotation needs to be manually verified. This verification is performed using the third column, in which three decision fields are provided: WORD (correctly recognised), PHRASE (correctly recognised), and WRONG (meaning that some character was read incorrectly by the OCR system, and hence the illustration in the photograph is not an illustration of the headword currently being considered). Segments of the last type were rejected (Table 3).

6.6 www.corodo.vn

Accepted headwords were sent to the website. For each headword we excerpted a textual illustrative paragraph to serve as a quotation. The final dictionary object consists of the following objects:

Table 3 Interface for OCR acceptance

(a) headword, (b) quotation, (c) location, (d) date of publication of text.

We plan eventually to include more than 100,000 headwords from Vietnamese. It will be possible to browse the entries both *a fronte* and *a tergo*.

7   Conclusions

In relation to what is a huge research undertaking, requiring expertise on many fronts (library science, linguistics, applied information science, etc.), we have formulated a theory of excerption and chronologisation, designed especially for this purpose, regulated by concrete directives oriented towards the chronologisation of material from the 20th century. Its chief tasks are:

(a) excerption, namely discovery (searching and finding) in a mass of texts; and
(b) chronologisation, with a precision of at least 1 year of Vietnamese headwords together with attestation in the form of quotations.

We used for that purpose Vietnamese books, which we scanned and subjected to OCR, and from which we selected headwords for further illustration in context. It was our aim to present an automatic procedure for revealing to readers new and possibly forgotten worlds. We have demonstrated how to restore memory of pre-war and post-war Vietnam. The result of the project as a whole will be a collection of documents in the form of excerpted and scanned paragraphs relating to the everyday life of the Vietnamese nation. The project presented here, titled Great National Photocorpus of 20th-Century Vietnamese, is unprecedented in the history of Vietnamese scholarship. Never before has lexical material for the Vietnamese language been presented in such quantity (we plan to include more than 100,000 headwords) and in such a form (faithful photodocumentation of each headword).

References

1. Austin, P. (ed).: 1000 Languages. The Worldwide History of Living and Lost Tongues. Thames & Hudson, London (2008)
2. Dzienisiewicz, D., Wierzchoń, P.: On the Japaneseness of Polish: A Linguochronological Approach, vol, 3, pp. 53–76. Opuscula Iaponica & Slavica, (2016)
3. Graliński, F., Wierzchoń, P.: RetroC—a corpus for evaluating temporal classifiers. In: Vetulani, Z., Mariani, J. (eds.) Proceedings of 7th Language and Technology Conference, pp 245–249. Poznań (2015)
4. Graliński F.: Polish digital libraries as a text corpus. In: Vetulani, Z., Uszkoreit, H. (eds.) Proceedings of 6th Language and Technology Conference, pp. 509–513. Poznań (2013)
5. Iwanowski, M.: Fotoaddenda do leksykografii polskiej, Warszawa (2009)
6. Le-Hong, P., Nguyen, T.M.H., Roussanaly, A., Vinh, H.T.: A hybrid approach to word segmentation of Vietnamese texts. In: Proceedings of the 2nd International Conference on

Language and Automata Theory and Applications, pp. 240–249. Tarragona, Spain, Springer, LNCS 5196 (2008)

7. Smith R.: An Overview of the Tesseract OCR Engine. In: ICDAR '07 Proceedings of the Ninth International Conference on Document Analysis and Recognition, vol. 2. Washington, DC (2007)

8. Wawrzyńczyk, J.: 1000 słów zadośćuczynienia. (Wypiski ze strony http://www.nfjp.pl). Wawrzyn, Warszawa (2016)

9. Wawrzyńczyk, J., Nasze Drobne Kompensacje Leksykograficzne, czyli jak wzbogacamy zasoby strony http://www.nfjp.pl (2016)

10. Wierzchoń P., Fotodokumentacja 3.0, „Język. Komunikacja. Informacja" 2009, t. 4, pp. 63–80 (2009)

11. Wierzchoń P., Graliński F., Z kart historii „parcia na" neologizmy, „Poradnik Językowy" 2016, z. 4, pp. 110–129 (2016)

The Impact of User Sentiment Aroused by The-Day-of-the-Week on the Recommendation Effectiveness in Microblog

Shih Yun Weng, Ping Yu Hsu, Ming Shien Cheng and Phan-Anh-Huy Nguyen

Abstract Microblogging has become an increasingly popular platform for users to post their views and comments online due to the ease of posting and replying. This generated an abundance of sentiment database which could be used to study the-day-of-the-week sentiment patterns of users. We adopted sentiment database to extract sentiment expressions from the posted Plurk messages to investigate whether there are sentiment fluctuations in the days of a week and if there are opportunities to use the-day-of-the-week sentiment patterns to maximize the effectiveness of sentiment-based product recommendation. The experimental results showed that users' posts are significantly strong negative on Monday and strong positive on Friday, Saturday, and Sunday. We speculate that the recommended products with positive sentiment words were more effective during Monday and Saturday because of the approach-avoidance motivation. People would usually have negative sentiment after 5 days continuous work during Friday working hours, so that recommended products with negative sentiment words are more effective. Whereas on Friday night the positive sentiment increased after off duty, and caused the recommendation products with positive sentiment words more effective. Negative sentiment on Sunday due to the coming blue Monday may cause negative sentiment recommendations more effective.

S.Y. Weng (✉) · P.Y. Hsu · P.-A.-H. Nguyen
Department of Business Administration, National Central University,
No.300, Jhongda Rd., Jhongli City 32001, Taoyuan County, Taiwan, ROC
e-mail: 984401019@cc.ncu.edu.tw

P.Y. Hsu
e-mail: 984401019@cc.ncu.edu.tw

P.-A.-H. Nguyen
e-mail: 984401019@cc.ncu.edu.tw

M.S. Cheng (✉)
Department of Industrial Engineering and Management, Ming Chi University
of Technology, No.84, Gongzhuan Rd., New Taipei City 24301,
Taishan District, Taiwan, ROC
e-mail: mscheng@mail.mcut.edu.tw

© Springer International Publishing AG 2017
D. Król et al. (eds.), *Advanced Topics in Intelligent Information
and Database Systems*, Studies in Computational Intelligence 710,
DOI 10.1007/978-3-319-56660-3_31

355

Keywords Microblog · Sentiment word · Plurk · The-day-of-the-week

1 Introduction

Microbolog users' posts are often with emotions. Many scholars have also used microblog users' emotions to predict the stock market [19], national election results, etc. [23]. Lazarus in 1975 [12] pointed out that a person's emotions are not permanent; they change over time. In addition, a person's emotions also directly affect the behavior of another [2]. Rystrom in 1989 [18] also pointed out that the inner emotions of investors have an impact on financial market investment decisions. The emotional changes of investors also lead to different investment returns during the-day-of-the-week. Since people's emotions are not static and that they change with time [12], people experience varying degrees of pessimism or optimism during the day of the week.

Many scholars also pointed out that the day of the week effect exists in the personal emotions of university students, office workers, married men, etc. [3, 7, 21]. Many scholars also found from the study on the-day-of-the-week changes in stock returns that the earnings on Mondays are significantly lower than earnings in other days of the week [1, 5, 8, 11, 17]. Rystrom in 1989 [18] also pointed out that most people do not like Mondays, considering it a bad day with negative emotions. By Friday, people start looking forward to the next 2 days off, thus the emergence of positive emotions.

This study adopted the emotion thesaurus provided by research of Cho Shu-Ling et al. [4] to extract every article with emotions posted by Plurk users and gain an insight into whether users' posting emotion dimensions during different days of the week draw the same group of followers to recommended products embedded in discussion threads by the robot and further enhance the product recommendation result. At the same time, recommended products with positive and negative emotional contents were also embedded to further explore whether recommend products with different emotions affect the product recommendation result. Furthermore, gain an insight into the relationship between individuals' emotional state and product recommendation during individualized marketing through social networking site information.

This paper is organized as follow: (1) Literature review: this part contains a review of emotional analysis and applications, the day-of-the-week effect; (2) Research methodology: this part describes the content of research hypothesis in this study; (3) Result discussion and management implications: this part describes the result of research hypothesis test and discusses management implication; (5) Conclusion and future research: The contribution of the hypothesis developed in this study is proposed, and possible future research direction is discussed.

2   Literature Review

2.1   Emotional Analysis and Applications

With the surge of analyses on emotions in recent years, emotion related information can be used to observe many important events. For example, Schumaker et al. [19] pointed out that emotions can predict stock market rises and falls. With the rapid rise of social networking sites and the increasing availability of information, emotions can be extensively used on social networking sites. For instance, O'Connor et al. [15] used posted emotions on Twitter to study the opinion poll from 008 to 2009 during the US presidential election period. Tumasjan et al. [23] used Twitter users' political emotions to predict the 2009 election results. Mishne and Glance in 2006 [14] also used emotions expressed in posts of bloggers on Blogpulse to predict the movie box office [10].

Many scholars found in their study that when individuals are in a positive mood, they will attempt to protect their positive emotions to avoid situations that may generate negative emotions [20, 24]. When people are in a negative mood, they tend to change the current situation to change the negative emotions [20]. The Westbrook's research in 1987 [25] showed that people are less willing to process messages that undermine joy. In other words, people with positive emotions will reinforce their interpretation of positive emotions and reduce the importance of negative attributes. Although many scholars use the emotions contained in online users' posts in various studies, no researcher has researched and recommended products with positive and negative emotion words in order to understand different emotional states or whether their recommendation of products with positive and emotion words will affect the product recommendation result embedded in discussion threads to obtain practical implications.

2.2   The Day-of-the-Week Effect

A Monday is often regarded as a "bad day", because it marks the end of rest and the first day of work. Many people therefore experience negative emotions on a Monday. Most people start showing positive emotions on a Friday, as holidays are coming. Rystom in 1989 [18] believes that investors express different emotions in different days of the week and generate varying degrees of optimism and pessimism, leading to different investment returns each day. Many psychologists believe that emotions affect personal behavior. Rystom [18] also believes that investors' decisions are also affected by personal emotions. Other than the investment market, some scholars also conducted research targeting the emotional changes of university students, office workers, and married men during different days of the week. For example, The research of Stone et al. [21] targeting married men shows that the men showed more positive emotions on weekends, while they were more depressed and down compared to Fridays. The research of Farber in 1953 [7] targeting university students shows that Mondays were the least pleasant day for all, while Fridays and Saturdays were everyone's favorite days. Pecjak in 1970 [16] conducted a research on university students' color, emotion, and day association and found that Mondays strongly associated with gray and vitality. Therefore, it is deemed that Mondays are associated with negative emotions. Saturdays are associated with red and laughter, and Sundays are associated with white and happiness. Therefore, it is deemed that Saturdays and Sundays are associated with positive emotions. Froggatt in 1970 [9] targeting office workers show that short-term unexcused absences had the highest incidences on Mondays and the lowest incidences on Fridays.

Many scholars conducted research on emotional changes of market investors, office workers, and students. However, few scholars applied social networking platforms, especially microblog users, to analyze the emotional changes of users during different days of the week. For example, Logunov and Panchenko in 2011 [13] applied emoticon picture used by users on Twitter to predict emotional cycles. It was found by surprise that happiness emotion use increased on Fridays, while sadness emoticon use decreased. They more frequently used sadness emoticons on Mondays, but used them less on weekends when happiness emotions usage increased. In view of this, the Plurk users' posts with emotions were applied to analyze the emotional changes of users during the-day-of-the-week and embed products in discussion threads in order to gain an insight into whether the emotional changes of users during the-day-of-the-week affect product recommendation results.

3   Research Methodology

On Plurk, there is an indicator used to reflect plurkers' degree of activeness on Plurk called Karma. The higher the Karma value, the more Plurk use authorization received (such as setting timeline titles, using special emoticons, etc.). The degree of plurker activeness (such as frequently publishing posts), number of friends and fans, and interactions among plurkers all affect the Karma value. Many plurkers make friends with robots to increase the Karma value and obtain more information. Plurk robots are virtual accounts created by third-party users.

A robot intended for recommending books was created in this study, through which book-loving plurkers can be attracted to make friends. http://www.books.com.tw that offers a variety of products such as books, magazines, CD, DVD, and so on is the most famous online bookstore in Taiwan. It is the source of the recommended books in this study, and all the recommended books are bestselling books chart. When plurkers publish posts, the system will automatically embed books that may interest users in

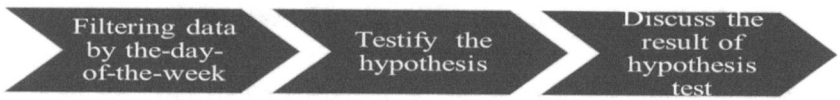

Fig. 1 Research process

discussion threads through the recommendation engine, through which the number of clicks can be observed, will help determine the product recommendation results.

Figure 1 shows the research process. Data is first selected based on the-day-of-the-week. Then, data hypothetical verification is carried out. Finally, the verification results are discussed.

3.1 Data Collection

The posts published by every plurker who made friends with robot accounts from November 9th, 2015 to January 21st, 2016 were first collected. The recommendation system then randomly recommended books on http://www.books.com.tw based on the emotion thesaurus of Chuo Shu-Ling [4]. The contents of the recommended products are with both positive and negative emotions. When plurkers publish new posts the recommendation system will first match the published post contents with the corresponding thesaurus and send the matched emotion words to http://www.books.com.tw through the Google's own search engine API and then search books containing the emotion words in the book summary. The first page result of a book randomly searched and selected serves as the commended product as a reply to the published post. The recommended contents are then embedded in discussion threads, including book title, brief introduction, and website hyperlink.

All interested plurkers can click the links after seeing the reply. The recommendation system will then record the number of clicks as important information for further analysis. Figure 2 shows the data collection processes in this study.

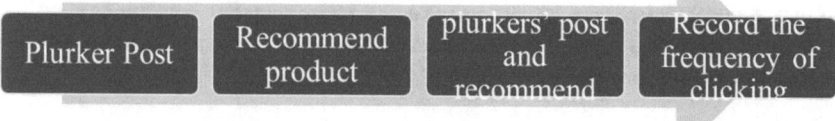

Fig. 2 Data collection processes

3.2   Variable Definition and Operational Definition

Emotions of posts: According to the 613 emotion words proposed by Chuo Shu-Ling et al. [4], there are 218 emotion description words and 395 emotion-inducing words. Emotions of Plurk posts can be divided into the following two polarity types: (1) Posts with Positive Emotions: this type of posts refers to posts whose emotion word dimension of 5–9. (2) Posts with Negative Emotions: this type of posts refers to posts whose emotion word dimension of 1–5.

Posts may contain mixed emotion words (i.e. both positive and negative emotion words). In this study, based on the positive and negative dimensions of each emotion word proposed by Chuo Shu-Ling et al. [4], the weighted average calculation was adopted to calculate whether posts fall under the attributes of positive or negatives. The determination equation is as follows:

$$\overline{V} = \frac{\sum_{i=1}^{n} W_i V_i}{\sum_{i=1}^{n} W_i}$$

W_i Representing the frequency of No. i of emotion words that appear in a post
V_i Representing the positive and negative dimension of No. i of emotion words
n   The number of emotion words

The above equation derives at, which represents the positivity or negativity of each post. According to the positivity and negativity assessment standard of Chuo Shu-Ling et al. [4], posts with positivity and negativity of less than 5 points are classified under negative attributes; those with five points are classified as neutral posts; those not within the research scope are deleted.

Emotions of product recommendations: Based on the 613 emotion words proposed by Chuo Shu-Ling et al. [4], the product recommendations are classified into the following two types: (1) Product Recommendations with Positive Emotions: this type of product recommendation refers to recommended products whose linked contents contain positive emotion words. (2) Product Recommendations with Negative Emotions: this type of product recommendation refers to recommended products whose linked contents contain negative emotion words.

Clicks: The number of clicks for a product recommendation is the variable for observing product recommendation results. The higher the number of clicks on a product recommended, the greater the product recommendation result.

Intensity of positive and negative emotions in posts: The calculation equation is as follows:

$$D = |V - N|$$

V   represents the positive and negative dimensions of emotion words.
N   is the reference point of the positive and negative dimensions of emotion words.
   According to the emotion thesaurus of Chuo Shu-Ling [4], the positive and

negative dimensions (V) of emotion words in this study are set as 1–9 points, with the reference point of five. Based on this equation, the higher the D value, the greater the intensity of posts with positive and negative emotions.

3.3   Research Hypotheses

In view of the past studies, it can be inferred that the contents of plurkers' posts are affected by the "blue Monday" phenomenon, making the negative emotions in posts seem "stronger". Therefore, the first hypothesis proposed states:

H1: Plurkers' posts on Mondays have stronger negative emotions than positive emotions.

If everyone in anticipation of the coming 2-day weekend shows positive emotions, stronger positive emotions' will be seen in posts of plurkers. Therefore, the second hypothesis proposed states:

H2: Plurkers' posts on Fridays have stronger positive emotions than negative emotions.

People tend to have positive emotions on Saturdays. Therefore, it is inferred in this study that plurkers' posts contain stronger positive emotions. Therefore, the third hypothesis proposed states:

H3: Plurkers' posts on Saturdays have stronger positive emotions than negative emotions.

People tend to show positive emotions on Sundays. It is therefore inferred in this study that plurkers' posts have stronger positive emotions. Therefore, the fourth hypothesis proposed states:

H4: Plurkers' posts on Sundays have stronger positive emotions than negative emotions.

It is inferred in this study that under circumstances where plurkers' posts on Mondays are with strong emotions, product recommendations with positive emotion words can better enhance the number of clicks compared to product recommendations with negative emotion words. Therefore, the fifth hypothesis proposed states:

H5: Under the situation that posts on Mondays have stronger negative emotions, product recommendations with positive emotion words receive more clicks than product recommendations with negative emotion words.

It is inferred in this study that under circumstances where posts are with strong positive emotions on Fridays, product recommendations with positive emotion words can better enhance the number of clicks compared to product recommendations with negative emotion words. Therefore, the sixth hypothesis proposed states:

H6: Under the situation that posts on Fridays have stronger positive emotions, product recommendations with positive emotion words receive more clicks than product recommendations with negative emotion words.

It is deemed in this study that under circumstances of posts with strong positive emotions on Saturdays, product recommendations with positive emotion words can better enhance the number of clicks compared to product recommendations with negative emotions. Hence, the seventh hypothesis proposed states:

H7: Under the situation that posts on Saturdays have stronger positive emotions, product recommendations with positive emotion words receive more clicks than product recommendations with negative emotion words.

Under circumstances of posts with strong positive emotions on Sundays, product recommendations with positive emotions can better enhance the number of clicks compared to product recommendations with negative emotion words. Therefore, the eighth hypothesis proposed states:

H8: Under the situation that posts on Sundays have stronger positive emotions, product recommendations with positive emotion words receive more clicks than product recommendations with negative emotion words.

4   Result Discussion and Management Implications

Table 1 shows the hypothesis test result and the discussion covers the following:
   Possible Reasons for Higher Clicks for Product Recommendations with Negative Emotion Words than Product Recommendations with Positive Emotion Words.
   The experimental results for H6 and H8 do not coincide with the hypotheses. First, Friday evening is the time people feel the happiest. Further analysis was conducted targeting posts with positive emotions on Fridays. It was found that

Table 1 Result of hypothesis test

Hypothesis	Test method	P value	Significant
H1	t-test	0.002	Yes
H2	t-test	0.000	Yes
H3	t-test	0.000	Yes
H4	t-test	0.000	Yes
H5	t-test	0.004	Yes
H6	t-test	0.127	No
H7	t-test	0.01	Yes
H8	t-test	0.116	No

during daytime on Fridays, product recommendations with negative emotion words have a higher average number of clicks compared to product recommendations with positive emotions. However, on Friday evenings, product recommendations with positive emotions have a higher average number of clicks compared to product recommendations with negative emotions. This finding conflicts with the avoidance theory previously proposed by Elliot and Thrash [6]. However, the research of Tomkins [22] shows that people that have long suppressed their negative emotions tend to seek opportunities to express their negative emotions. It is therefore inferred in this study during daytime on Fridays, people may be fully suppressed after working or attending class for 5 days consecutively. The pursuit of balance in emotions expressed therefore resulted in the higher number of clicks for product recommendations with negative emotions than product recommendations with negative emotions.

Targeting the experimental results for H8, Farber in 1953 [7] sees Sunday as a special day. Someone believe that since the next day is a workday, Sunday should be the day to enjoy to the fullest; others start feeling anxious, depressed, and stressful, as the 2-day weekend is about to come to an end. According to the research of Tomkins [22] it has been found that people under suppressed negative emotions tend to pursue balance in expressed emotions. Therefore, this study deems that such "mood" may have affected the experimental results and may be the reason for the higher clicks for product recommendations with negative emotions than clicks for product recommendations with positive emotions.

Although this research has confirmed microblog users on Fridays have the tendency to post messages with positive emotions, they enjoy clicking recommended products with negative emotion words. Therefore, after further dividing the day and night, it was found that users enjoy clicking recommended products with negative emotions during day time, and the opposite is true at night when users start to feel truly relaxed. This situation continues until Saturday when users are in a good mood all day. Based on this, there search on only the "day-of-the-week" is rather inadequate. That is, emotional changes not only take place during different days of the week, emotions may change on the same day as time goes by, This perspective is worthy of exploration by research workers. Furthermore, although people tend to have positive emotions on Sundays, scholars have pointed out in their study that Sunday is a special day. While people are enjoying themselves, they are shadowed by "This is the last day of the weekend. There is work tomorrow." [7], this research has confirmed this point. Under the situation that users' posts on Sundays are with strong emotions, product recommendations with negative emotion words have higher clicks than product recommendations with positive emotions. Hence, as far as research workers are concerned, instead of classifying a single day under positive emotions or negative emotions, it is the people's inner delicate feelings that require more consideration.

However, as far as practical workers are concerned, the experimental results in this study show that besides recommending products on industry related websites, recommendations on microblog social networking sites may also be considered. Based on the emotional rise and fall of a vast majority of potential clients that have not shown interest in a particular industry or product, products that cater to their needs may be recommended. To utilize emotions expressed in posts by social networking sites to enhance product recommendation results, the research results in this study may be considered, specifically designing robots on social networking sites that automatically publish posts and recommend products. This recommendation approach not only attracts users but also draws the attention of online users with the same interests, thereby broadening the product recommendation exposure and enhancing product recommendation clicks.

5 Conclusion and Future Research

This study confirmed that emotions indeed exist during the-day-of-the-week and affect users' posts. The research results show that plurkers' posts on Mondays show strong negative emotions, and their posts on Fridays, Saturdays, and Sundays show strong positive emotions. The research results also show that under circumstances where plurkers' posts on Mondays are with strong negative emotions and their posts on Saturdays are with strong positive emotions, recommended products with positive emotion words had higher clicks than products with negative emotion words. Although the plurkers' posts throughout Fridays show strong positive emotions, recommended products with negative emotion words produced better results perhaps due to the depressed emotions during the day or after working for five consecutive days. It was by evenings, in a truly relaxing manner, did the recommendation results of products with positive emotion words get better. Sundays were special days. Although the plurkers' posts showed strong positive emotions, the recommendation results of products with negative emotion words were better, perhaps due to the fact that there was work the next day. This finding also shows that research workers should not classify a certain day under "positive emotions" or "negative emotions"; instead, they should take into account the inner delicate feelings of people. For distributors that wish to use emotions expressed by users through posts on social networking sites to enhance product recommendation results, they may design robots to automatically send posts and carry out product recommendation on social networking sites in reference to the research results herein.

Emotion words were adopted in this study to determine the emotions of posts. However the sole use of emotion words to analyze emotions may lead to misjudgment of emotions. For instance, when numerous negative words appear in posts, such as "no", "without", etc., emotions expressed may be opposite from predicted emotions. Therefore, in addition to emotion words, the use of negative thesauruses, emoticons, or "grammatical analysis" as references for determining emotions in posts can effectively reduce errors of this type. The sources of Chinese emotion words are not as complete as English emotion thesauruses. The emotion thesaurus proposed by Chuo Shu-Ling et al. [4] was adopted for all the analyses in this study. The thesaurus only contains 218 emotion description words and 395 emotion inducing words, 613 emotion words in all. If more emotion thesauruses can be created, the accuracy of emotion word analyses can be more effectively enhanced. This study also unexpectedly found that as far as Fridays and Sundays are concerned, the "avoidance theory" previously mention by Elliot and Thrash [6] does not apply, especially for Sundays. Fridays and Sundays are therefore deemed worthy of further investigation and research by relevant disciplines.

References

1. Aggarwal, R., Rivoli, P.: Seasonal and day-of-the-week effects in four emerging stock markets. Financ. Rev. **24**(4), 541–550 (1989)
2. Baumeister, R.F., Vohs, K.D., DeWall, C.N., Zhang, L.: How emotion shapes behavior: feedback, anticipation, and reflection, rather than direct causation. Pers. Soc. Psychol. Rev. **11**(2), 167–203 (2007)
3. Christie, M.J., Venables, P.H.: Mood changes in relation to age, EPI scores, time and day. Br. J. Soc. Clin. Psychol. **12**(1), 61–72 (1973)
4. Chuo, S.-L., Chen, C.-C., Tseng, J.-Y.: Chinese emotions and related psychophysiological data in Taiwan. Chin. J. Psychol. **55**(4), 439–454 (2013)
5. Cross, F.: The behavior of stock prices on Fridays and Mondays. Financ. Anal. J. **29**(6), 67–69 (1973)
6. Elliot, A.J., Thrash, T.M.: Approach-avoidance motivation in personality: approach and avoidance temperaments and goals. J. Pers. Soc. Psychol. **82**(5), 804 (2002)
7. Farber, M.L.: Time-perspective and feeling-tone: a study in the perception of the days. J. Psychol. **35**(2), 253–257 (1953)
8. French, K.R.: Stock returns and the weekend effect. J. Financ. Econ. **8**(1), 55–69 (1980)
9. Froggatt, P.: Short-term absence from industry: I Literature, definitions, data, and the effect of age and length of service. Br. J. Ind. Med. **27**(3), 199–210 (1970)
10. Glance, N., Hurst, M., Tomokiyo, T.: Blogpulse: automated trend discovery for weblogs. In: WWW 2004 workshop on the weblogging ecosystem: aggregation, analysis and dynamics, Vol. 2004 (2004)
11. Keim, D.B., Stambaugh, R.F.: A further investigation of the weekend effect in stock returns. J. Finance **39**(3), 819–835 (1984)
12. Lazarus, R.S.: A cognitively oriented psychologist looks at biofeedback. Am. Psychol. **30**(5), 553 (1975)
13. Logunov, A., Panchenko, V.: Characteristics and predictability of Twitter sentiment series. In: 19th International Congress on Modelling and Simulation, pp. 1617–1623 (2011)
14. Mishne, G., Glance, N.S.: Predicting movie sales from blogger sentiment. In: AAAI Spring Symposium, Computational Approaches to Analyzing Weblogs, pp. 155–158 (2006)
15. O'Connor, B., Balasubramanyan, R., Routledge, B.R., Smith, N.A. (2010). From tweets to polls: linking text sentiment to public opinion time series. ICWSM, 11(122–129), pp. 1–2
16. Pecjak, V.: Verbal synesthesiae of colors, emotions, and days of the week. J. Verbal Learn. Verbal Behav. **9**(6), 623–626 (1970)
17. Rogalski, R.J.: New findings regarding day-of-the-week returns over trading and non-trading periods: a note. J. Finance **39**(5), 1603–1614 (1984)
18. Rystrom, D.S., Benson, E.D.: Investor psychology and the day-of-the-week effect. Financ. Anal. J. **45**(5), 75–78 (1989)
19. Schumaker, R.P., Zhang, Y., Huang, C.N., Chen, H.: Evaluating sentiment in financial news articles. Decis. Support Syst. **53**(3), 458–464 (2012)

20. Schwarz, N., Clore, G.L.: Feelings and phenomenal experiences. Social Psychol.: Handb. Basic Principles **2**, 385–407 (1996)
21. Stone, A.A., Hedges, S.M., Neale, J.M., Satin, M.S.: Prospective and cross-sectional mood reports offer no evidence of a "blue Monday" phenomenon. J. Pers. Soc. Psychol. **49**(1), 129 (1985)
22. Tomkins, S.S.: Affect, imagery, consciousness: Vol. I: The positive affects (1962)
23. Tumasjan, A., Sprenger, T.O., Sandner, P.G., Welpe, I.M.: Predicting elections with twitter: what 140 characters reveal about political sentiment. ICWSM **10**, 178–185 (2010)
24. Wegener, D.T., Petty, R.E.: Mood management across affective states: the hedonic contingency hypothesis. J. Pers. Soc. Psychol. **66**(6), 1034 (1994)
25. Westbrook, R.A.: Product/consumption-based affective responses and post-purchase processes. J. Mark. Res., pp. 258–270 (1987)

Multi-sentence Compression Using Word Graph and Integer Linear Programming

Dung Tran Tuan, Nam Van Chi and Minh-Quoc Nghiem

Abstract Multi-sentence compression is the task of generating a single sentence from a set of sentences about the same topic. In this work, we explore the use of syntax factor in combination with informativeness and linguistic quality in an Integer Linear Programming framework. Compression candidate paths generated by a word graph are re-ranked using frequent words. Then top k-shortest paths are used as the variables for Integer Linear Programming formulation. Our system improves over state of the art in both English and Vietnamese datasets.

Keywords Multi-sentence compression · Integer linear programing · Sentence fusion

1 Introduction

In the context of information explosion, the need to synthesize and distil information from multiple sources is increased. A lot of people prefer reading a short news highlight instead of long paragraphs, especially those who do not have much time. Finding an effective summary method is the inevitable solution to this problem. There are many ways to classify an automatic summarization system [1]. In this paper, we focus on the problem of multi-document summarization using multi-sentence compression technique.

Multi-sentence compression is a useful task in natural language processing and can apply to multi-document summarization task. It takes a group of sentences about

D.T. Tuan (✉) · N. Van Chi · M.-Q. Nghiem
Faculty of Information Technology, University of Science,
VNUHCM 227 Nguyen Van Cu, Dist. 5, Ho Chi Minh City, Vietnam
e-mail: ttdung@student.hcmus.edu.vn

N. Van Chi
e-mail: vcnam@fit.hcmus.edu.vn

M.-Q. Nghiem
e-mail: nqminh@fit.hcmus.edu.vn

© Springer International Publishing AG 2017   367
D. Król et al. (eds.), *Advanced Topics in Intelligent Information
and Database Systems*, Studies in Computational Intelligence 710,
DOI 10.1007/978-3-319-56660-3_32

the same topic as input, then produces an output through a compression system, which was a short summary and often accompanied by a length limitation.

The majority of current summary systems are based on extractive approach, which generates output by extracting verbatim sentences or phrases from the original sentences. The key point of this approach is to identify the most important information and put them into the summary. By contrast, abstractive summary systems could generate new sentences or phrases by combining existing information. Generally, extractive systems achieve higher results than abstractive systems; this has been proven through previous works. This is understandable since extracting something out is always easier than creating new ones. Genest et al. [2] showed that existing extractive approach are very close to the way of human-written summary. Therefore, we need to promote deeper research in abstractive approaches.

Filippova [3] proposed a word graph-based approach for multi-sentence compression based on sentence fusion. This approach made a series of pioneering works on multi-document summarization and multi-sentence compression task. Her approach often leads to shorter alternatives of the original sentences, therefore may generate incomplete sentences.

Based on Filippova's work, Banerjee et al. [4] proposed an abstractive multi-document summarization method based on multi-sentence compression which uses ILP for sentence selection. The combination of word-graph and ILP sentence selection has shown promising results. In this approach, they put the TextRank to the ILP formulation as a part of the variable weights, offs with the random selection of 200 candidate paths from word-graph. This might lead to good path missing for the next steps as well as instability result. Moreover, this method did not ensure the syntax correctness of the sentences. The 3-gram language model was insufficient while dealing with long phrases or long segmentations.

Our research focuses on the legacy of the previous achievements of sentence compression and fusion approach that would solve the previous limitations. In this paper, we proposed a new compression method, which included the syntax factor combining with informativeness and language model factors to optimize ILP formulation. Our system could apply to address both multi-sentence compression and multi-document summarization problem to produce a short, concise and readable compressive summary.

2 Related Works

Sentence compression was developed in many previous works using both rule-based and statistical-based approaches, including the noisy-channel model and decision tree method [5–7]. These methods transform the input parse tree into the compressed tree by deleting some elements in the original tree. However, most of the sentences are ambiguous and have multiple possible parse trees. Then deleting a node in a parse tree might lead to ungrammatical output sentence.

Statistical approaches to sentence compression have also been explored. In this paradigm, rules about which syntactic constituents can be merged were introduced [8]. Thadani and McKeown [9] also present an approach that uses discourse information as well as a statistical approach to compression task using Integer Linear Programming (ILP). A scoring function is computed to determine which word is kept or removed. ILP is a state-of-the-art sentence selection method which finds the output that maximizes an objective function subject to certain constraints. Their approach has an advantage over rule-based and tree-based approaches because it is unsupervised thus does not require labeled training data. However ILP is an NP-hard problem, and they did not use an approximation function.

Filippova [3] introduced a graph-based technique for sentence compression that constructs a word-graph for multiple relevant sentences in a group, then fuses information and generates a new sentence. Her method requires only Part-Of-Speech (POS) tags of words and the output sentences are ranked to calculate edge weights. The path's grammar is learned from vertex word forms including POS tags and its adjacent nodes; therefore this method provides a reliable way of generating informative and grammatical sentences. Her approach includes the following two main steps: (1) same words with same POS tags are mapped to the same node; (2) a weight ranking function is used to select the best output sentences.

Boudin and Morin [10] improved Filippova's approach by re-ranking the fusion candidate paths according to the number of relevance key phrases they contain to generate more informative output sentences. Words and part-of-speech tags extracted from every candidate are used to build the second word-graph, in which each word is denoted by a node. TextRank [11] algorithm has been applied to calculate edge weights between nodes. KeyPhrases were specified by setting a threshold for edge weights between adjacent words and a syntactic pattern. These key phrases were used to re-rank candidate paths, and the system chose the best paths for the final compression.

Similar to key phrases approach, Luong et al. [12] proposed a re-ranking candidates method using frequent words to reduce incorrect mapping (**MSC-FRank**). Before constructing word graph, the system sorts all source sentences by their duplicated words with each other. If a sentence has several duplicated words inside it, and those words also appear in other sentences, that sentence will be sorted with higher priority for mapping because these sentences could create more context than other sentences. Their word-graph building strategy also has some minor changes, they immediately add a directed edge from the recently mapped node to the previously mapped node instead of adding edges after all nodes are mapped into the graph. This allows the system to check the path from a recent node to the new candidate.

Banerjee [4] proposed an abstractive multi-document summarization based on multi-sentence compression combined with Integer Linear Programing for sentence selection (**ILP-Summ**). After pre-processing and clustering the document, the remain clusters are used as the input to build the word-graphs. Each word-graph is created by one cluster containing the relevant sentences. The word-graph construction strategy is similar to the previous works. There is no re-ranking algorithm applied to the fused sentences. To optimize the system performance, they chose random k-shortest paths and fed them into the ILP objective function. They used a 3-gram language model to ensure the linguistic quality of the sentences.

3 Proposed Method

Our proposed multi-sentence compression approach consists of two main steps: *Sentence Fusion* and *Sentence Selection*.

3.1 Sentence Fusion

In the sentence fusion step, our system generates a directed word-graph from the input sentences within a cluster following the procedure of Luong et al. [12]. In the word-graph, a node represents a word or a punctuation and its POS tags. Two nodes are linked by an edge if they are adjacent words or punctuation in a sentence. The edges are added immediately after a new node is added to the graph. Two dummy nodes (start node and end node) are automatically generated. The first word of each sentence connects with the start node, and the last word of each sentence connects with the end node.

Figure 1 illustrates an example word-graph. The input contains two sentences:

S1: In Asia Japan Nikkei lost 9.6% while Hong Kong's Hang Seng index fell 8.3%.

S2: Elsewhere in Asia Hong Kong's Hang Seng index fell 8.3% to 12,618.

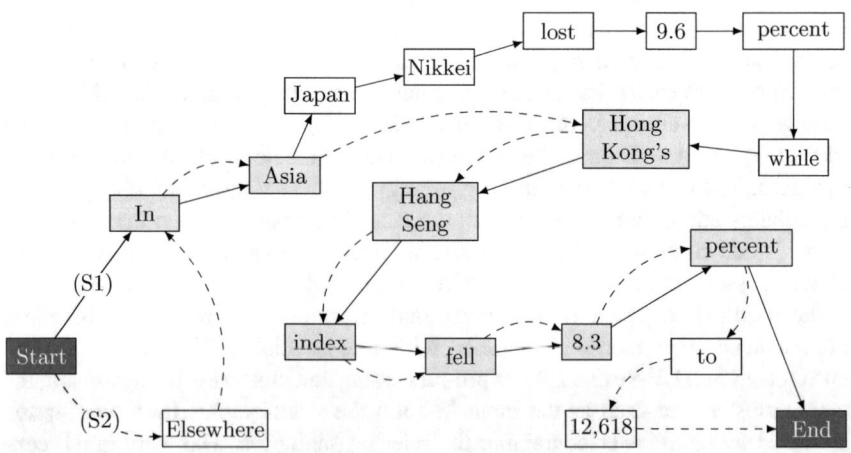

Fig. 1 Word graph and a possible compression path (*gray nodes*)

As we can see, the two input sentences refer to the same topic, but differs in sentence length, syntax, and the detail of information. The solid directed arrows connect the words in the first sentence $S1$, while the dotted arrows join the words in the second sentence $S2$.

We have several candidate paths between the dummy start and end nodes. For example, we can generate these paths:

- *In Asia Hong Kong's Hang Seng index fell 8.3%.*
- *Elsewhere in Asia Hong Kong's Hang Seng index fell 8.3%.*
- *Elsewhere in Asia Japan Nikkei lost 9.6% while Hong Kong's Hang Seng index fell 8.3%.*

To make the generated sentences original, we do not consider paths that too similar to any of the input sentences. We set a threshold d to remove all paths that have cosine similarity (with the original input sentences) equal or greater than d. To avoid too long or too short sentences, we set the threshold n and l to filter out the inappropriate paths. If a length of a path is less than n or larger than l, we remove that path. We also use the frequent words re-ranking technique [12] to reduce incorrect mapping node to word-graph and get a top k-shortest paths from the graph as the input for the next step - sentence selection. If k is too small, we could not get as much as possible variable paths. If k is too big, we could overload the ILP formulation.

3.2 Sentence Selection

In the sentence selection step, we select the best path in the word-graph as our fusion sentence. Following Banerjee et al. [4], we formulate the best path selection as an ILP problem. Banerjee et al. choose the path that maximizes important information and linguistic quality. Different from Banerjee et al., we also incorporate a syntax score to the ILP formulation to ensure the syntax rightness of the output sentence. Each path (from the generated k-shortest paths) is considered a variable with the corresponding coefficients for the ILP objective function. Solving the objective function would get the best paths that maximizes important information, readability, and syntax rightness. We made these constraints to get concise summary and avoid similar information from different sentences included in it.

Informative Score We use TextRank scores [11] to generate importance content of a sentence. TextRank algorithm is an unsupervised method to build a graph for each sentence, each node corresponds to one word, two related words will be connected by an edge. The score of each node in the graph is calculated as the Eq. 1:

$$I(V_i) = (1 - \delta) + \delta \times \sum_{V_j \in adj(V_i)} \frac{w_{ji}}{\sum_{V_k \in adj(V_j)} w_{jk}} I(V_j) \tag{1}$$

where V_i represents the word, $adj(V_i)$ denotes the adjacent nodes of V_i and δ is the damping threshold. The computation converges to return final word importance scores. The informative score of a path is obtained by summing the scores of each word in a single path. For example: $I(p_i)$ is the informative score of path p_i. TextRank has some advantages as it is an unsupervised method, so we do not need a gold corpus for training. Furthermore, the TextRank algorithm is based on word concurrence and does not require any grammar input. Thus it is language independent and we can apply this ranking method for many different languages.

Linguistic Score To compute Linguistic Quality score, we use a language model to assign language scores for every candidate paths. More specifically, we use a 3-gram language model. Supposed a candidate path p_i has q words, the score assigned for this path is defined as Eq. 2:

$$LQ(p_i) = \frac{1}{1 - LL(w_1, w_2, ..., w_q)} \tag{2}$$

where:

$$LL(w_1, w_2, ..., w_q) = \frac{\log_2 \prod_{t=3}^{q} P(w|w_{t-1}w_{t-2})}{L} \tag{3}$$

As we can see in Eq. 3, we compute the probability of different sets of 3-gram in each path then averaged them by the total number of the conditional probabilities computed L.

Syntax Score To calculate the syntax score, we parse the input sentence to get a syntax tree, then we calculate the probability of the subtrees using the Probabilistic Context-Free Grammars (PCFG). We then apply Inside Algorithm [13] to the model. Let $S = x_1, x_2, \ldots, x_n$ be a set of words in a sentence, where x_i is the ith word in the sentence. Let $\pi(i, j, X)$ for any $X \in N$ (non-terminal set), for any i, j with $1 \leq i \leq j \leq n$, to be the set of all parse trees for words $x_i \ldots x_j$ such that non-terminal X was the root of the tree.

$$\pi(i, j, X) = \begin{cases} p(X \rightarrow x_i) & \text{if } X \rightarrow x_i \in R \\ 0 & \text{otherwise} \end{cases} \tag{4}$$

where X is a non-terminal and $p(X \rightarrow x_i)$ is the probability of terminal x_i by X in the rule set R. The sum of all available subtree scores is the syntax score of the input sentence. Supposed that path p_i has n words, the syntax score is calculated as following equation:

$$S(p_i) = \sum_{t \in T(1, n, S)} p(t) \tag{5}$$

where S is the start symbol in PCFG quintuple, and $p(t)$ is the probability of every non-terminal X in the corpus that X is at the root of the tree.

To select the best paths from the cluster, we combine three factors include informative score $I(p_i)$, linguistic quality score $LQ(p_i)$ and syntax score $S(p_i)$ in an optimization framework. We maximize the following objective function:

$$F(p_1, ..., p_k) = \sum_{i=1}^{k} \frac{I(p_i).LQ(p_i).S(p_i).p_i}{T(p_i)} \tag{6}$$

Each p_i represents a binary variable, that can take a binary value depending on whether the corresponding path is selected in the final summary or not.

$$p_i = \begin{cases} 1 & selected \\ 0 & otherwise \end{cases} \tag{7}$$

If $p_i = 1$, the path has been selected to summary and vice versa if $p_i = 0$. $T(p_i)$ is the number of words in a path. This index is used to normalize the weights in favor of the shorter paths. We use several constraints to solve the ILP problem and retain the best paths which maximize the objective function that contains important content, new information without redundant and duplication.

Supposed that S is the set of all candidates paths, C is the set of all paths which are selected into the final summary.

$$\forall p_i \in S, \forall p_j \in C$$
$$p_i + p_j \leq 1 \ if \ sim(p_i, p_j) \geq \theta \tag{8}$$

The new sentence is added to the summary if and only if it has cosine similarity with the existing one in the summary $sim(p_i, p_j)$ lower than or equal a similar threshold θ.

4 Experimental Results

4.1 Settings

Data We evaluate our approach on English and Vietnamese datasets. For English, we use 300 pairs of English sentences from Amazone's Mechanical Turk services collected by McKeown et al. [14]. For Vietnamese, we use the Multi-Sentence Compression data of Luong et al. [12].

Stanford Parser[1] is used to get the parse tree of input sentences. We use the Stanford Treebank and VLSP Treebank[2] to train the parsers for English and Vietnamese, respectively. The SRI Language Modeling Toolkit[3] is used to train our language

---

[1] http://nlp.stanford.edu/software/lex-parser.shtml.
[2] http://vlsp.hpda.vn/.
[3] http://www.speech.sri.com/projects/srilm/.

Table 1 Comparison of English and Vietnamese MSC systems

	English			Vietnamese		
Metric	MSC-FRank	ILP-Summ	ILP-SLT	MSC-FRank	ILP-Summ	ILP-SLT
ROUGE-1	55.654	67.813	**68.039**	76.679	78.179	**80.232**
ROUGE-2	43.981	56.232	**56.584**	65.715	70.872	**70.899**
ROUGE-SU4	38.360	53.064	**53.196**	60.975	70.175	**70.203**
BLEU	30.551	33.144	**34.203**	58.159	52.979	**59.054**

models. For English, we use the English Gigaword corpus.[4] For Vietnamese, we collect the training data from Vietnamese news websites. This data contains 95,143 documents and 2,175,962 sentences. We use the Python LP Modeler (PuLP)[5] as an ILP solver.

Metrics We use the two traditional metrics: Recall-Oriented Understudy of Gisting Evaluation (*ROUGE*) and Bilingual Evaluation Understudy (*BLEU*) for evaluation. While ROUGE is recall-oriented, BLEU is precision-oriented. We use $F1$ scores from *ROUGE*–1, *ROUGE*–2, *ROUGE*–SU4 and 4-grams for *BLEU* measurement.

Parameter Settings We choose $n = 5$ (included start node and end node) since we noticed that the incomplete sentences often have less than three words. k is set to 200 to reduce the computational overload of the ILP. d is set to 0.8 to drop out the candidate paths which are too similar to the source sentences. The threshold θ is set to 0.5 to ensuring the selection of novelty information. We set the length limit for compressed sentences to 20 words and the length of the summary to 100 words. We compare our system (ILP-SLT) with two state-of-the-art systems: MSC-FRank and ILP-Summ. All parameters are the same as previous work settings for fair comparison.

4.2 Results

Multi-sentence Compression Task As can be seen from Table 1, all of the *ROUGE* and *BLEU* scores obtained by our systems outperforms other systems on both English and Vietnamese datasets. ROUGE scores increased slightly while BLEU scores showed significant increments. As Table 1 shows, it is not sufficient if we only use the language model to compute the linguistic quality of the compressed sentences. Integrating syntax scoring function to filter out grammatically incoherent sentences has shown significant improvements.

In Table 1 we can see that the *ROUGE*–2 scores in every system are always higher than the corresponding *ROUGE*–SU4 scores. It is not common in Multi-Document Summarization task but reasonable in the Multi-Sentence Compression task. Since

---

[4]http://www.keithv.com/software/giga/.

[5]https://www.coin-or.org/PuLP/.

Table 2 Comparison of MDS English systems (DUC 2004 dataset)

Metric	MSC-FRank	ILP-Summ	ILP-SLT
ROUGE-1	0.36454	0.38179	**0.38659**
ROUGE–2	0.09792	0.11752	**0.11943**
ROUGE–*SU*4	0.11417	**0.13467**	0.13374
BLEU	0.09413	0.11232	**0.12096**

the output is a short sentence, the number of 4-gram matching is reduced as the length of the sentence decreased. Our experiment shows that if a summary has less than 30 words, its *ROUGE*–2 score is higher than the respective *ROUGE*–*SU*4 score. In the experimental datasets, the average lengths of the compressed sentences for English and Vietnamese are 25.45 and 29.86 words, respectively.

Another remarkable result is that the system scores using Vietnamese dataset are better than using English dataset. It is understandable since the English dataset is created in an abstractive manner while it is extractive in Vietnamese dataset. Both *ROUGE* and *BLEU* scores are based on the overlapping of n-grams. This could lead to the better results for Vietnamese dataset over the English dataset.

Multi-document Summarization Task Our second task is to evaluate the contribution of multi-sentence compression to the problem of multi-document summarization.

The input of each system is a set of documents on the same topic; the output is a 100-word length summary. We used Banerjee's sentence clustering method [4] to cluster similar sentences from all documents into groups. In each group, we applied our method including sentence fusion and sentence selection to generate one compressed sentence. By adding the output sentences of each group, we obtained the final summary.

Table 2 shows the results of multi-document summarization systems on DUC 2004 dataset.[6] Our proposed system achieved higher results in *BLEU*, *ROUGE*–1, and *ROUGE*–2 scores than other systems. There was no significant differences between *ROUGE*–*SU*4 scores of our system and the state-of-the-art system (ILP-Summ).

4.3 Analysis

Consider the following two example sentences:

E1 *The sanctions imposed to force Gadhafi to turn over the two Libyans wanted in the 1988 bombing over Lockerbie, Scotland, that killed 270 people.*

---

[6]http://duc.nist.gov/duc2004/.

E2 *The U.N. imposed sanctions since 1992 for its refusal to hand over the two Libyans wanted in the 1988 bombing that killed 270 people killed.*

The first one is generated by ILP-SLT and the second one is generated by ILP-Summ. **E1** is an unambiguous statement while the phrase *"killed 270 people killed"* in **E2** is difficult to understand and confusing. The 3-gram language model failed to identify this case because both *"killed 270 people"* and *"270 people killed"* have high probabilities. By using a syntax parser that assigns low probabilities to the paths which have long and redundant information, our proposed method overcomes the restrictions of the n-gram language model and achieved better results. The trade-off, however, is the executing time. Our system runs slower than the systems that do not use a syntax parser.

5   Conclusion and Future Work

In this paper, we have explored the use of syntax factor in combination with informativeness and linguistic quality in an ILP framework for multi-sentence compression task. Experimental results on English and Vietnamese datasets showed that using syntactic information improves the performance of the system in both multi-sentence compression and multi-document summarization tasks. Future work will address the processing time problem for this task. Dynamic programming could be a potential technique.

Acknowledgements This work was supported by the Ho Chi Minh City Department of Science and Technology, Grant Numbers 15/2016/HD-SKHCN.

References

1. Nenkova, A., Maskey, S., Liu, Y.: Automatic summarization. In: Proceedings of the 49th Annual Meeting of the Association for Computational Linguistics: Tutorial Abstracts of ACL 2011, pp. 3:1–3:86 (2011)
2. Genest, P.E., Lapalme, G., Yousfi-Monod, M.: Hextac: the creation of a manual extractive run. Génération de résumés par abstraction (2013)
3. Filippova, K.: Multi-sentence compression: finding shortest paths in word graphs. In: COLING 2010, 23rd International Conference on Computational Linguistics, Proceedings of the Conference, 23–27 August 2010, pp. 322–330. Beijing, China (2010)
4. Banerjee, S., Mitra, P., Sugiyama, K.: Multi-document abstractive summarization using ILP based multi-sentence compression. In: Proceedings of the 24th International Conference on Artificial Intelligence, pp. 1208–1214 (2015)
5. Daumé, III, H., Marcu, D.: A noisy-channel model for document compression. In: Proceedings of the 40th Annual Meeting on Association for Computational Linguistics, pp. 449–456 (2002)
6. Turner, J., Charniak, E.: Supervised and unsupervised learning for sentence compression. In: Proceedings of the 43rd Annual Meeting on Association for Computational Linguistics, pp. 290–297 (2005)

7. Galley, M., McKeown, K.: Lexicalized markov grammars for sentence compression. In: HLT-NAACL, pp. 180–187 (2007)
8. Clarke, J., Lapata, M.: Global inference for sentence compression: an integer linear programming approach. J. Artif. Intell. Res. **31**, 399–429 (2008)
9. Thadani, K., McKeown, K.: Sentence compression with joint structural inference. In: Proceedings of the Seventeenth Conference on Computational Natural Language Learning, CoNLL 2013, 8–9 August 2013, pp. 65–74. Sofia, Bulgaria (2013)
10. Boudin, F., Morin, E.: Keyphrase extraction for n-best reranking in multi-sentence compression. In: Human Language Technologies: Conference of the North American Chapter of the Association of Computational Linguistics, Proceedings, June 9–14, 2013, pp. 298–305. Westin Peachtree Plaza Hotel, Atlanta, Georgia, USA (2013)
11. Mihalcea, R., Tarau, P.: Proceedings of EMNLP 2004, pp. 404–411 (2004)
12. Luong, A., Tran, N., Ung, V., Nghiem, M.: Word graph-based multi-sentence compression: Re-ranking candidates using frequent words. In: 7th International Conference on Knowledge and Systems Engineering, pp. 55–60 (2015)
13. Collins, M.: Probabilistic context-free grammars (PCFGs). Lecture Notes (2013)
14. McKeown, K., Rosenthal, S., Thadani, K., Moore, C.: Time-efficient creation of an accurate sentence fusion corpus. In: Human Language Technologies, pp. 317–320 (2010)

A Performance Comparison of Feature Extraction Methods for Sentiment Analysis

Lai Po Hung and Rayner Alfred

Abstract Sentiment analysis is the task of classifying documents according to their sentiment polarity. Before classification of sentiment documents, plain text documents need to be transformed into workable data for the system. This step is known as feature extraction. Feature extraction produces text representations that are enriched with information in order to have better classification results. The experiment in this work aims to investigate the effects of applying different sets of features extracted and to discuss the behavior of the features in sentiment analysis. These features extraction methods include unigrams, bigrams, trigrams, Part-Of-Speech (POS) and Sentiwordnet methods. The unigrams, part-of-speech and Sentiwordnet features are word based features, whereas bigrams and trigrams are phrase-based features. From the results of the experiment obtained, phrase based features are more effective for sentiment analysis as the accuracies produced are much higher than word based features. This might be due to the fact that word based features disregards the sentence structure and sequence of original text and thus distorting the original meaning of the text. Bigrams and trigrams features retain some sequence of the sentences thus contributing to better representations of the text.

Keywords Sentiment analysis · Feature extraction · Text representation · N-grams · Part-of-speech

L.P. Hung (✉) · R. Alfred
Faculty of Computing and Informatics, Universiti Malaysia Sabah,
Kota Kinabalu, Sabah, Malaysia
e-mail: janelai627@gmail.com

R. Alfred
e-mail: ralfred@ums.edu.my

© Springer International Publishing AG 2017
D. Król et al. (eds.), *Advanced Topics in Intelligent Information
and Database Systems*, Studies in Computational Intelligence 710,
DOI 10.1007/978-3-319-56660-3_33

1 Introduction

Sentiment Analysis is the task of classifying opinioned words, sentences or documents into groups of positive or negative sentiment [14]. Public opinions are important as it shapes the perception, reputation and image of products or social matters. Harvesting and analyzing opinions are commonly done to discover public opinions regarding a matter. Given the large volumes of opinions available in the internet, automated opinion analysis is easier via sentiment analysis. Sentiment analysis has three tasks, feature extraction, feature selection and classification [8]. Feature extraction produces different representation of plain text documents called features. Feature selection will then select and filter those features for relevant features to the topic and lastly a machine learning classifier will use the features to classify documents.

Plain text are usually broken down and represented in a form that the computer can process during text analysis. "Feature extraction" in this context is the process that extracts features of the text, not to be confused with extraction of product features. Breaking down the sentences of words may distort the original meaning in the document. Changing the sequence of words or taking out some words may even make the document meaningless. When a text is broken down to form other representations, some of the information it carries will be lost. In an effort to compensate for the lost, many kinds of features are extracted to recover as much information as possible, such as single words or word phrases. The classifier uses the features for analysis. The simplest features are each single word in the text, this is a common representation called the Bag-Of-Words representation. However, a lot of information is lost as all structures and connection between words are broken down. In consideration of this, phrases are also used as features as they preserve some structure and relations. There are many other different features explored for the purpose of capturing the most features out of the original text, such as Part of Speech tags, Sentiment score tags and Synonym tags. Effective feature extractions can retain more information contained in the text, and helps the classifier make the right judgement with regards to the sentiment of the document. Thus, effective feature extraction methods are required in order to have higher accuracy results in analyzing sentiment of a set of documents. The aim of this paper is to investigate and compare the performances of word features and phrasal features used in sentiment analysis of reviews. The result will be a discussion of the performance of different features and what contributes to the performance.

The rest of the paper follows, where Sect. 2 will highlight some related works on feature extraction. Section 3 will describe the experimental setup of this work. Section 4 discusses and analyses the results obtained. Section 5 will conclude the paper.

2   Related Work

There are some related works which compares the performance of Sentiment analysis when different kinds of feature types are used. They are shown in Table 1.

Some examples of feature types are mentioned in the table above, N-gram features such as unigrams and bigrams are more common than tag-based features. N-gram features are words or characters, the N represents how many words or character make up for a feature. Unigram is a single word, bigram is a two word pair, trigram is a 3 word phrase and so on [8]. Tag-based features are features of words which have been tagged like with their Part-of-Speech (POS) [14], Senti-wordnet [4] scores or semantic groups [8]. The tags can either serve as just markers for picking out word features or the tags can serve as part of the features in analysis. Related works revealed that N-grams tend to be more effective widely used as primary features [6, 15–17, 20].

Pang, Li and Vaithyanathan were among the first to introduce machine learning sentiment analysis, they experimented with different machine learning techniques and feature types. They found that unigrams consistently provide the best accuracy when used with different machine learners, however they pointed out that bigrams maybe useful for word sense disambiguation but not in their experiment setting. The authors in [13] and [10] compared unigrams with other features to enrich text representation but unigrams still showed better performance. The work in [19] however found that word-relation features (bigram) performed better than word features (unigram) when they are applied with an ensemble classifier and attributed the good performance to word relation information in the word pairs. Another comparison using Chinese documents found that bigrams in terms of word and Chinese characters attained the best results and that unigrams worked well with less documents [9]. The work in [5] focuses on ensembles of different feature types and classifier, their work emphasizes on the combination of word features and various information tags to produce best results. Many of the works related to sentiment analysis have used unigrams as features because it is easy and effective, however the word relations in phrases are undeniably important. Phrasal features such as bigram and trigram preserves some word relation however they require more computation load and thus are not common when dealing with simple sentiment analysis work. In this work, the paper compares the performance of the sentiment analysis system that applies different types of features extracted using unigrams, bigrams and trigrams in English reviews.

3   Experimental Setup

Figure 1 shows the experiment framework used in this work. In this experiment, the main objective is to investigate the performance of a sentiment analysis system based on different types of features used in constructing the text representation of

Table 1 Related works of feature extraction comparison

Author	Features	Classifier	Dataset	Best feature
Pang et al. [14]	Unigram, bigram, POS	Naïve Bayes (NB), Support vector machines (SVM), Maximum entropy (ME)	Cornell movie review	Unigram
Li and Sun [9]	Unigram, bigram, trigram	NB, SVM, ME, Artificial neural network (ANN)	Hotel reviews	Bigram
O'Keefe and Koprinska [13]	Unigram, Sentiwordnet	NB, SVM	Cornell movie review	Unigram
Li et al. [10]	Unigram, part-of-speech	SVM	Cornell movie review	Unigram
Xia et al. [19]	Part-of-speech, unigram, bigram	NB, ME, SVM	Multi-domain sentiment dataset	Bigram
Hassan et al. [5]	Words, Sentiwordnet, part-of-speech, semantic	SVM, NB, ANN, decision tree, logical regression	Telecommunications, technology firms, pharmaceutical drugs	Words + Sentiwordnet + semantic

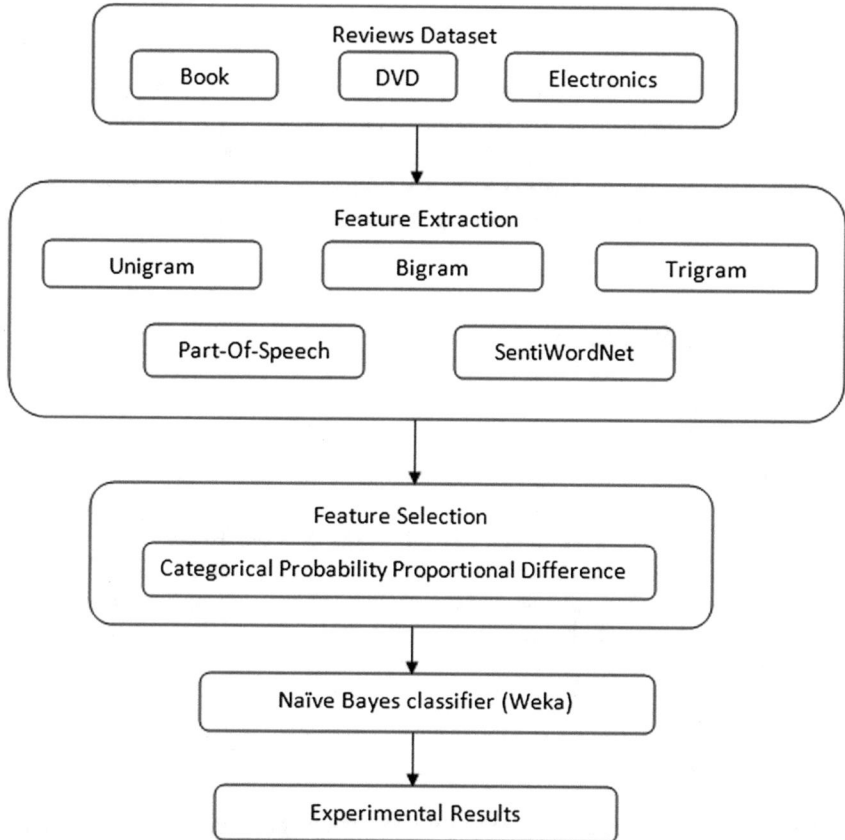

Fig. 1 Experimental setup

the reviews. The sentiment analysis system will apply a single feature selection technique (Categorical Probability Proportional Difference, CPPD) and a single classifier technique (Naïve Bayes) but different feature types will be applied and tested.

The plain text reviews will be represented as multiple feature types. Each feature type is a feature set which will have its features selected and classified and the accuracy of sentiment analysis with each feature set will be recorded. This experimental setup is applied so that the results will reflect the effectiveness of the feature type for sentiment analysis. This is to fulfill the objective of this work where the sentiment analysis accuracies of different types of features can reflect the performance of the features and provide the comparison of results for analysis. The feature types that will be featured in this experiment are unigrams, bigrams, trigrams, part-of-speech tagged features and Sentiwordnet features. The feature selection method used will be CPPD [1] and the classifier implemented is the Naïve Bayes classifier [8, 14].

3.1   Dataset

The dataset used in this work is the Multi-Domain Sentiment Dataset [2]. The dataset contains various English language product reviews. The reviews chosen in this work are of the topics Books, DVD and Electronics. 1000 reviews are chosen for each topic to form initial datasets. These datasets are then preprocessed and represented as different feature types for further classification.

3.2   Feature Extraction

3.2.1   Unigram

Unigrams are features of single words similar to the Bag-of-words feature (e.g. good, bad) [7]. Unigrams are simple to create and has been used in most sentiment analysis task as mentioned in the related works section. Every word in a text documents is considered as a feature. Unigrams can be easily formed by tokenization where a text documents is read as a string of words with spaces in between. The tokenizer will separate the each word in the string based on the blank spaces [10, 13, 14].

3.2.2   Bigram

Bigrams are part of N-gram features where a feature is made up of two words in this work (e.g.: good_part, bad_thing) [12]. Unigrams disregard word structure and sequence, causing the concern that semantics of the sentence is distorted. Bigram is a way of preserving a little of the sentence structure. They are formed by pairing off each word with its left and right neighbor. So each word forms two bigrams [9, 19].

3.2.3   Trigram

Trigrams work similarly as bigrams but they are made up of three words instead (e.g.: very_good_part, not_bad_thinking) [12]. Each word is paired off with both neighboring word to form a trigram. Trigrams are phrase features that preserve even more of the sentence structure and sequence. Collocations and phrases that express stronger sentiment can be easily captured with trigram. Although trigrams are less popular than unigrams and bigrams, it does show potential from some literatures [3, 12].

3.2.4   Part-of-Speech (POS)

Part-of-Speech (POS) refers to the grammatical category a word belongs to [8]. Using the Stanford POS tagger, each word in a document is tagged with its corresponding POS (e.g. good_adj, day_noun). Then, the words that belong to a chosen POS are extracted as one of the features. The types of POS include Adjectives, Verbs and Adverbs. These types are chosen because they are able to describe the behaviors or characteristics of an object and they are likely to hold sentiment words that indicate polarity. In this work, the types of POS that will be included are Adjectives, Verbs, Adverbs and Nouns as some nouns can also hold descriptive abilities [10, 14].

3.2.5   Sentiwordnet

Sentiwordnet is a sentiment lexicon which holds words and its corresponding sentiment scores [4]. Words can be tagged with the sentiment scores where negative score indicates the word has negative sentiment and positive score indicates the word has positive sentiment. However for this work, instead of tagging the words with sentiment score, each word in the document is compared with the words in the lexicon. Assuming most words in the sentiment lexicon has sentiment polarity, only words which are found in the lexicon are kept as features [5, 13].

3.3   Feature Selection

For feature selection, the Categorical Probability Proportional Difference (CPPD) [1] will be used in this work, it is a fairly new method with promising results. CPPD uses Categorical Proportional Difference (CPD) [18] and Probability Proportional Difference (PPD) [1], two different values to select features. CPD scores rare features higher, whereas PPD scores features with higher frequency better. So CPPD is effective as combining threshold will help eliminate features extremely rare or too common features. The formula of CPD and PPD in relation to feature and class are as shown:

$$CPD(f,c) = \frac{A - B}{A + B} \tag{1}$$

$$PPD(f) = \frac{nPos(f)}{sPos + S} - \frac{nNeg(f)}{sNeg + S} \tag{2}$$

where A is the number of time word and category appear together and B is the number of times the word occurs without the category, nPos(f) refers to the number of positive documents which feature term occurs in, nNeg(f) refers to the number of negative documents in which feature term occurs in, sPos refers to the total number of feature terms in positive class, sNeg refers to the total number of feature terms in negative class and finally S refers to the total number of feature terms.

In the experiments conducted, ten CPPD joint thresholds are automatically generated for each feature selection method of each dataset experiment with values in increasing order. As the threshold value selected gets higher the features selected will be reduced. CPD has a maximum value of 1.0, thus for this work the CPD threshold will be set as 1.0 throughout the experiments. Only PPD thresholds are selected. The ten PPD thresholds are generated by ordering all the unique PPD values of features in increasing order and selecting the value encountered within equal intervals in position as the thresholds. Ten sets of sentiment analysis will be run with each feature set using the joint thresholds of CPD and PPD. Only features with values fulfilling both thresholds in CPPD are selected for classification in each set. Results for the set of features selected with CPPD thresholds that have the best sentiment analysis accuracy of each method will be shown in the next section of results from this work.

3.4   Classifier

The classifier used in this work is the Naïve Bayes classifier. Built based on the Bayes Theorem, it works out the possibility of a feature belonging to a class by assuming the variables have conditional independence [8]. The Naïve Bayes classifier is a popular classifier in sentiment analysis as it can be seen based on the related works highlighted in Sect. 2. This is because it is easy to implement just by calculation.

The Naïve Bayes classifier in this work is implemented through WEKA (Waikato Environment for Knowledge Analysis), a data mining tool developed by the University of Waikato. Weka contains various text processing functions and machine learning algorithms which can be readily applied.

4   Results and Discussion

The results of the experiment are shown in the following tables. Table 2 shows the initial feature set size of each feature type. The results shown in the Table 3 are the results of the best performing set with the features selected by CPPD thresholds.

The table shows the results of sentiment analysis after using the different extracted features and the experiment was repeated on three different datasets. CPPD is made up of two thresholds, CPD and PPD. CPD has a maximum threshold of 1.0 and was the threshold value maintained throughout this experiment because CPD's ability in retaining rare features produced a large feature set although the threshold value is high. Only the best performing set out of the ten sets of

Table 2 Initial feature set size

	Book	DVD	Electronics
Unigram	18786	17540	7953
Bigram	92806	93788	49210
Trigram	145995	149533	80953
Part-of-speech	17149	17266	7725
Sentiwordnet	9708	9342	4542

Table 3 Results of sentiment analysis with different features

CPD threshold: 1.0

Feature set	PPD threshold	Feature count	Reduction (%)	Accuracy (%)
Book dataset				
Unigram	0.00006	1983	89.44	90.6
Bigram	0.00002	1451	98.44	96.1
Trigram	0.00001	5227	96.42	**97.2**
Part-of-speech	0.00006	1854	89.19	91.3
Sentiwordnet	0.00012	1279	86.83	87.7
DVD dataset				
Unigram	0.00006	2041	88.36	83.1
Bigram	0.00002	1818	98.06	91.9
Trigram	0.00001	7973	94.67	**95.5**
Part-of-speech	0.00006	2026	88.27	83.1
Sentiwordnet	0.00012	1305	86.03	78.5
Electronics dataset				
Unigram	0.00015	1013	87.26	77.4
Bigram	0.00003	4862	90.12	**94.8**
Trigram	0.00002	4981	93.85	94.3
Part-of-speech	0.00015	992	87.16	77.2
Sentiwordnet	0.00024	623	86.28	71.9

experiments are shown in the table above. The details of the experiment are listed in columns such as the PPD threshold used, feature count for the set, reduction in features of the set compared to the initial full dataset and the accuracy for the experiment set. The accuracy numbers in **bold** shows the best results out of the feature types for the dataset.

From the observations of results above, trigrams produced the best results in sentiment analysis accuracy for the Book and DVD dataset. Bigrams produced the best results for the Electronics dataset, however in this case trigrams only lacked 0.5% accuracy in comparison. This would mean that generally the best performing feature is trigram. The best set results of Unigrams and Part-of-Speech features in each dataset were almost similar with differences of less than 1% accuracy.

Sentiwordnet features achieved the lowest accuracy in comparison with the other methods. Through CPPD feature selection the best performing sets all managed to achieve their respective best results after their feature sets have been reduced to more than 85%. This means that only the selected 15% of features were sufficient to bring about the best result for each feature set.

The results of sentiment analysis shows that phrase based features performed better than word based features. Bigrams and trigrams are considered phrase based features as each feature has more than one term. Unigrams, Part-of-Speech and Sentiwordnet features are considered word features as each feature is a word only. Agreeing with literature [3, 11, 19] which states that sentence structure and sequence in text processing tasks, the results of this experiment shows that the more longer the phrase the more accurate the classifier performs, seeing that in our experiment trigrams with majority outperforms bigrams. The semantics of a word and a phrase which contains the word might be different, therefore it might not be accurate to take a single word and judge the whole document's sentiment with it. Neighboring words to a sentiment word has the power to influence its sentiment polarity, therefore retaining more neighbor words assures that the semantics and sentiment interpreted is as exact as possible to the original sentence. The good performances of trigrams indicate that the presence of a sentiment phrase in a document is a better indicator of document sentiment. However, the longer the phrase the more features tend to be generated, such that can be seen in the original size of the feature sets, the number of features of trigram and bigrams are much more than unigrams and word based features. Besides that, looking at the best performing sets trigrams also require more training features to produce better results. Bigrams and trigrams produce more features because as there are more there are more unique phrases formed with the pairing and grouping of words, thus there are less redundant features. In this case there might be also a lot of rare features which only occur once in the dataset, so feature selection is important in such cases to weed out those features and prevent feature sets from being too big.

The reason of for the popularity of unigrams with sentiment analysis tasks is also shown through the results. Unigrams are easy to generate and the full feature set size is small compared to generating phrases. Given that the ease to generate unigrams and the small feature set, the performance of unigrams is still quite satisfactory, scoring accuracies of more than 75%. Therefore unigram is an adequate feature to use for simple sentiment analysis considering it has less processing load, the tradeoff is a little less in accuracy compared to using bigrams and trigrams. From the results, a comparison between variants of word based features can also be assessed. It also shows that the results of word-based feature; unigram, part-of-speech and sentiwordnet have accuracies which are quite close, especially the results for unigrams and part of speech are quite similar. This would indicate that the part-of-speech groups Adjectives, Verbs, Adverbs and Nouns are the actual key players in pointing out sentiment words which are useful for classification. Especially since the size of the initial feature sets of unigrams and Part-Of-Speech features and also their best performing sets have similar feature count. Sentiwordnet features performed the poorest among the features assessed in this work, all though the accuracies exceeded 70%. The feature count of Sentiwordnet features extracted is also the lowest, implicating that the feature count might be too low and inability to provide as much information as the other features for the classifier to make its decisions. However, given that the word based features accuracies are not as high as bigrams and trigrams, relying on the sentiment of words only are not enough to conclude the sentiment polarity of documents with high accuracy. The word relation between phrases captures the distinguishing information of sentiment polarity of the document, however more features are required in the process than other word based features.

5 Conclusion

In this work, various feature extraction techniques are tested to discover the most effective feature type which could improve the task of analyzing sentiments of the text corpus. The types of features extracted in this paper include unigrams, bigrams, trigrams, part-of-speech and Sentiwordnet features. The various feature types are explored to provide classifier with the features that contain the most information to aid with classifying sentiment polarities. Based on the obtained results, phrase based features (e.g., bigrams and trigrams) or features which are made up of more than one word, provide better representation of the text documents as the results of sentiment analysis were much more accurate in comparison with word based features (e.g., unigram, part-of-speech and Sentiwordnet). This is because phrases capture better meanings and expressions of ideas in the overall document. However, phrases produce more features and more phrasal features are also required for the training of the classifier. Therefore the processing load of using phrases is heavier on the system. For word based features instead, the feature generated and used are less and thus processing load are lighter. However, the accuracy of word based features is not as accurate as phrasal features. As future work, besides using single feature types as a feature set for classification, a combination of feature types as a single feature set can be tested. For weaker feature types such as Sentiwordnet, their features can serve as reinforcements to stronger feature types by adding more information into the feature sets.

References

1. Agarwal, B., Mittal, N.: Categorical probability proportion difference (CPPD): a feature selection method for sentiment classification. In: Proceedings of the 2nd Workshop on Sentiment Analysis where AI Meets Psychology (SAAIP 2012), pp. 17–26. Mumbai (2012)
2. Blitzer, J., Dredze, M., Pereira, F.: Biographies, bollywood, boom-boxes and blenders: domain adaptation for sentiment classification. In: ACL (2007)
3. Dave, K., Lawrence, S., Pennock, D.M.: Mining the peanut gallery: opinion extraction and semantic classification of product reviews. In: Proceedings of WWW, pp. 519–528 (2003)

4. Esuli, A., Sebastiani, F.: SentiWordNet: a publicly available lexical resource for opinion mining. In: Proceedings of Language Resources and Evaluation (LREC) (2006)
5. Hassan, A., Abbasi, A., Zeng, D.: Twitter sentiment analysis: a bootstrap ensemble framework. In: International Conference of Social Computing (SocialCom), pp. 357–364 (2013)
6. Joachims, T.: Text categorization with support vector machines: learning with many relevant features. In: Proceedings of the 10th European Conference on Machine Learning, pp. 137–142 (1998)
7. Lai, P.H., Alfred, R., Hijazi, M.H.: A review on feature selection methods for sentiment analysis. Adv. Sci. Lett. **21**, 2952–2956 (2015)
8. Lai, P.H., et al.: A review on the ensemble framework for sentiment analysis. Adv. Sci. Lett. **21**, 2957–2962 (2015)
9. Li, J., Sun, M.: Experimental study on sentiment classification of Chinese review using machine learning techniques. In: Proceedings of International Conference of Natural Language Processing and Knowledge Engineering, NLP-KE (2007)
10. Li, S., Zong, C., Wang, X.: Sentiment classification through combining classifiers with multiple feature sets. In: NLPKE, pp. 135–140 (2007)
11. Liao, C., Alpha, S., Dixon, P.: Feature preparation in text categorization. ADM03 workshop (2003)
12. Narayanan, V., Arora, I., Bhatia, A.: Fast and accurate sentiment classification using an enhanced naive bayes model. In: Intelligent Data Engineering and Automated Learning–IDEAL 2013, pp. 194–201. Springer (2013)
13. O'Keefe, T., Koprinska, I.: Feature selection and weighting methods in sentiment analysis. In: Proceedings of the 14th Australasian Document Computing Symposium, Sydney, Australia, 4 Dec 2009
14. Pang, B., Lee, L., Vaithyanathan, S.: Thumbs up? sentiment classification using machine learning techniques. In: Proceedings of the Conference on Empirical Methods in Natural Language Processing (EMNLP), Philadelphia, Association of Computational Linguistics, pp. 79–86, July 2002
15. Sarkar, S.D., Goswami, S.: Empirical study on filter based feature selection methods for text classification. Int. J. Comput. Appl. **81**(6), 0975–8887 (2013)
16. Sharma, A., Dey, S.: Performance investigation of feature selection methods and sentiment lexicons for sentiment analysis. In: IJCA Special Issue on Advanced Computing and Comm Technologies for HPC Applications, vol. 3, pp. 15–20 (2012)
17. Sharma, A., Dey, S.: A boosted svm based ensemble classifier for sentiment analysis of online reviews. SIGAPP Appl. Comput. Rev. **13** (2013)
18. Simeon, M., Hilderman, R.: Categorical proportional difference: a feature selection method for text categorization. In: Proceedings of the 7th Australasian Data Mining Conference, vol. 87, pp. 201–208 (2008)
19. Xia, R., Zong, C., Li, S.: Ensemble of feature sets and classification algorithms for sentiment classification. Information Sciences, vol. 181, pp. 1138–1152 (2011)
20. Zheng, Z., Wu, X., Srihari, R.: Feature selection for text categorization on imbalanced data. SIGKDD Explor. **6**(1), 80–89 (2004)

A Study of Plagiarism Checker System Based on Chinese Word Segmentation and SQL Intersection Operation Technique

Ming-Hsiung Ying, Jui-Wen Fan and Hsiu-Min Yu

Abstract The issue of article plagiarism has become increasingly important due to research paper plagiarism that have been frequently proclaimed. Previous studies on plagiarism detection have been conducted mainly by using vectors, statistics and matrices methods. However, the plagiarism detection system has been found to fail to correctly locate the plagiarized paragraphs as a result of sentence modification or superfluous words insertion by plagiarists, thereby leading to imprecise detection and ineffective plagiarism prevention. This study used the Chinese word segmentation and SQL intersection operation technique to build a plagiarism checker system to enhance the plagiarism detection precision. For the purpose of the study, an SQL Intersection Operation was introduced for the plagiarism detection, which was proven to be more effective than previous studies which read a massive quantity of words into program variables. This study's results show higher efficiency and more precision of the plagiarism detection method proposed than the commercial plagiarism-detection software available on the market.

Keywords Plagiarism checker system · Chinese word segmentation · SQL intersection operation · Plagiarism detection

1 Introduction

The widespread Internet allows people to easily access articles produced by others. Some people acquire these articles and turn them into their own using plagiarism or modification of wording. Consequently, article plagiarisms are frequent.

M.-H. Ying (✉) · J.-W. Fan
Department of Information Management, Chung Hua University, HsinChu, Taiwan
e-mail: ms0389729@gmail.com; mhying@chu.edu.tw

H.-M. Yu
Language Center, Chung Hua University, HsinChu, Taiwan
e-mail: kuo@chu.edu.tw

© Springer International Publishing AG 2017
D. Król et al. (eds.), *Advanced Topics in Intelligent Information and Database Systems*, Studies in Computational Intelligence 710,
DOI 10.1007/978-3-319-56660-3_34

391

Previous studies in the field of plagiarism detection were focused mostly on the use of vectors, statistics or matrices. However, the detection could be bypassed by modifying the wording or adding superfluous words in the key sentences. For this, a prototype article Plagiarism Checker System was proposed in this study based on **Chinese word segmentation** and **SQL Intersection Operation** as an attempt to address the issue of reduced plagiarism detection accuracy due to the addition of superfluous words and modification of wording. Finally, the prototype was compared with the Plagiarism Checker Systems available in the market for efficiency and accuracy as the evaluation of the study results. From the above, the purposes of this study are to:

- Propose a new approach to tackle the poor accuracy of plagiarism detection at present; and
- Provide the study results for the review of industrial sector and researchers.

2 Literature Review

2.1 Chinese Word Segmentation

Lin et al. (2010) indicated that Chinese word segmentation lays the foundation of Chinese text processing [1]. Chen (1992) also pointed out that Chinese differs from English to a large extent in terms of structure and semantics, and in analyzing Chinese sentences a single Chinese character, a graphic unit, does not necessarily constitute the smallest semantic unit, which often involves a sequence of characters [2]. Therefore, unlike English, which uses alphabetic writing system where the boundary between words is clearly defined, the boundary between Chinese words is vague. As a result, before we analyze and check the Chinese words and their corresponding meaning in sentences, we need to parse sequences of Chinese characters into meaningful units.

For the word segmentation analysis on Chinese as a language, the Chinese Knowledge Information Processing (CKIP) group of Academia Sinica developed a Chinese word segmentation system (the CKIP system). The system is based on the principle of Chinese structure tree, and extracts sentences from the Academia Sinica Balanced Corpus of Modern Chinese and modifies them for the semantic knowledge in the analysis system, which makes the system more accurate.

2.2 Plagiarism Detection Method

Astronomical quantity of information exists in the cyber space, which suggests a daunting task just to identify the source of article plagiarism in a sea of online documents and eventually the paragraphs or sentences of the plagiarism [3]. Wu et al. (2008) reported the comparison of keywords used in the past. However, the accuracy of keyword comparison is easily compromised by using words of the same pronunciation, replacing some words with others or adding punctuation marks. A number of researchers proposed several quick search methods that move a string back and forth to identify identical ones, such as violence search and quick search [4, 5].

3 Research Methods and Design

3.1 Superfluous Words Analysis

Previous study on plagiarism detection mainly by using vectors, statistics, matrices, shift-position methods. However, addition of superfluous words, word modification or word order inversion made deliberately by the plagiarists all lead to imprecise plagiarism detection. To solve this problem, this study takes superfluous words and word order into account, and proposes new method of plagiarism detection by adopting Chinese parsing techniques and SQL intersection operation.

3.2 Plagiarism Detection Procedure

Figure 1 shows the plagiarism detection process, the process is described as follows

- In this study, the prototype system will send some words to CKIP Chinese Word Segmentation System. Sequences of words are segmented first by the Chinese parser developed by CKIP group, Academic Sinica.
- The output generated by the CKIP parser is online repaired through Google XML to increase the precision rate of the later detecting process.
- The vocabularies are saved in a database after the vocabulary patching, and the positions of vocabulary in the strings are marked for later plagiarism detection.
- The set algorithm provided by the database is used to calculate the similarity between word strings. Superfluous words and punctuation marks are not included in the calculation.
- Plagiarism detection is performed if the vocabulary overlapping rate exceeds the threshold defined in the system. No comparison is made if the threshold is not exceeded. The comparison is performed by using the set algorithm provided by the database. For the purpose of the study, the threshold was defined at 20%. This threshold is user-definable.

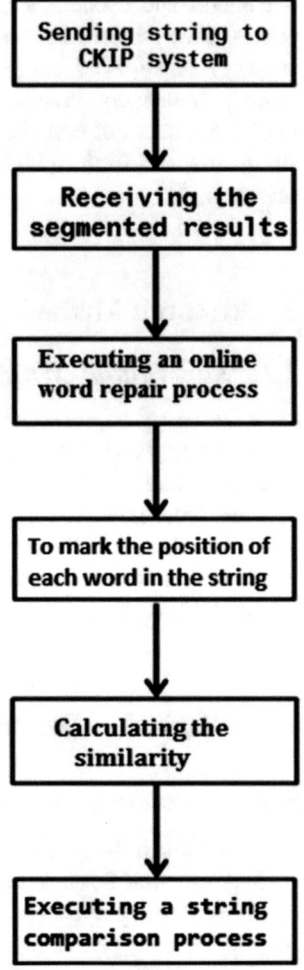

3.3 *Chinese Parsing Procedure*

In the present study, sequences of words are preprocessed by the CKIP parser,
which, however, puts a limit on the length of sequences of words as input. This
study therefore has articles demarcated into units of no more than 250 words and
batch transmit articles unit by unit to the CKIP parser for parsing analyses. Figure 2
shows the batch transmission procedure, and following is the procedure of batch
transmission

- Decide if a sequence is longer than 250 Chinese characters. If yes, then batch
  transmit the sequences to the CKIP parser; if not, transmit them directly to the
  parser.

Fig. 2 The Chinese word
segmentation procedure

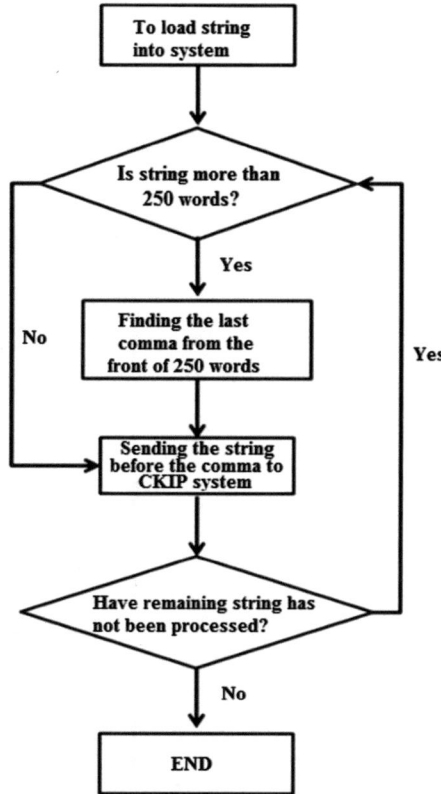

- For a sequence longer than 250 Chinese characters, the system will automatically locate the last comma within the first 250 Chinese character, and transmit the word sequence before the comma to the CKIP parser.
- After completing the above two steps, our system will decide if there exist any word sequences left unparsed. If not, then this is the end of parsing; if yes, the system will continue the first two parsing steps, including parsing the sequences after the comma located in previous step. This process continues until all the words are parsed.

3.4 Words Repair Procedure

Past research has used a built-in dictionary to repair words; however, as new words are introduced almost every day in the modern society, it is time-consuming and effort-intensive to manually repair words. This study therefore intends to adopt

Fig. 3 The words repair
procedure

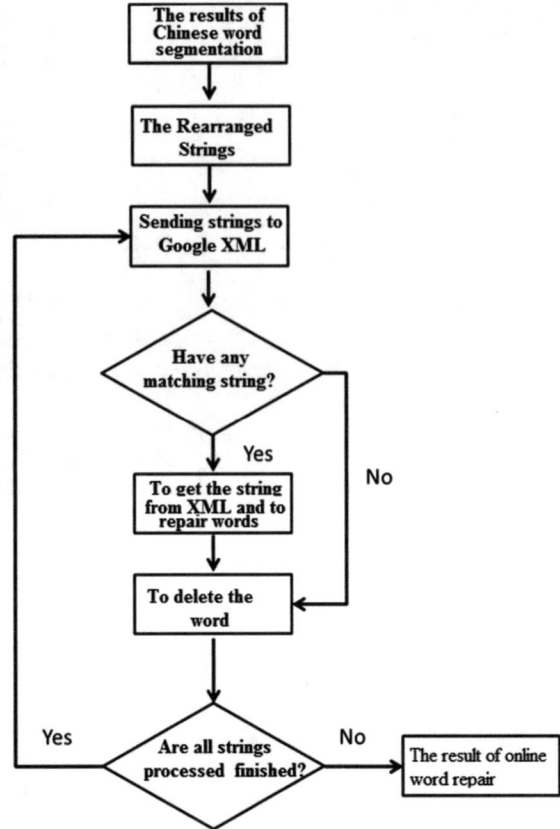

online XML provided by Google to repair words to reduce manual effort and to automatically update the dictionary when new words are identified.

To speed up the rate of online word repair, we first have all the sequences of words parsed by the Chinese parser developed by CKIP, Academia Sinica, and then online repaired by XML offered by Google to increase the precision rate of comparison. Figure 3 shows the online word repair process, and the Fig. 4 is a result of online word repair.

Fig. 4 The result of online
word repair

> Before the word repair process:
> ARCS¦動機¦模式¦整合¦了¦許多¦學習¦動機¦理論¦
>
> After the word repair process:
> ARCS動機模式¦整合¦了¦許多¦學習動機理論¦

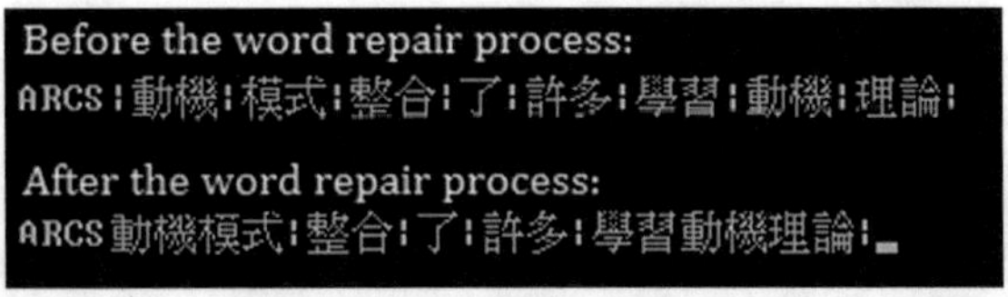

3.5   Establishment of Vocabulary Sequence and Calculation of the Overlap and Similarity Rates of Vocabularies

The sequence of phrases is important in Chinese word strings. A different sequence of words may deliver a completely different meaning. For this, the phrases contained in every string was sequentially numbered and saved in the database. The sequence of words was considered for the study is because every vocabulary found is saved in a database. Without a proper sequence, errors may occur in the database.

Figure 5 shows the establishment of vocabulary sequence and the calculation of vocabulary overlapping. The following are the process and description of the establishment of vocabulary sequence and the calculation of vocabulary overlapping:

- The segmented and patched word strings are sequentially numbered in phrase and saved in database. Superfluous words and punctuation marks are marked in the NotNeed column as 1 and 2, respectively, as shown in Fig. 5:
- As the vocabulary sequencing is done, the set algorithm provided by the database is used to determine the vocabulary overlapping rate and eliminate superfluous words and punctuation marks, which serves to prevent errors in overlapping rate calculation due to excessive superfluous words and punctuation marks. For the determination of overlapping rate, users are allowed to define their own overlapping rate threshold and comparison is performed when the defined threshold is exceeded. It is the belief of author that vocabulary overlapping more than 20% indicates a good probability of plagiarism. As such, the 20% of overlapping was defined as the comparison threshold for the study. The comparison is performed when the overlapping is greater than the user-defined threshold, and no comparison is performed if not. Equation 1 provides the overlapping rate.

$$M = \frac{COUNT(A \cap B)}{COUNT(A)} * 100\% \tag{1}$$

FileNO	FileParagraph	Number	ParagraphText...	NotNeed
45	3	3	整合	0
45	3	4	了	1
45	3	5	許多	0
45	3	6	學習動機理論	0
45	3	7	，	2

Fig. 5 Establishment of vocabulary sequence

M   word overlap rate
COUNT(A ∩ B)   A number of the intersection with the B words
COUNT(A)   The number of all vocabulary in A

The similar rate refers to the same text number ratio that some text strings A is found in the string Y, the similarity rate is calculated as follows:

$$SR = \frac{COUNT(CPsc)}{COUNT(CPs)} * 100\% \tag{2}$$

SR   similar rate
COUNT(CPs)   the total number of words comparing strings
COUNT(CPsc)   the number of labeled words of comparing strings

For example, string A contains 50 vocabularies and, therefore, COUNT(CPs) is 50. After comparison, 30 vocabularies in A are marked and, therefore, COUNT(CPsc) is 30. Equation 2 produces SR = 60%.

3.6 Intersection Operator

Use the intersection operator provided by database, the comparing strings intersect the being compared string, in the intersection operation is not the punctuation into them, the sort order is by comparison string (Fig. 6).

3.7 Plagiarism Test

The steps employed in this study for the detection of article plagiarism are the use of database intersection algorithm, vocabulary sequencing, processing of repeated

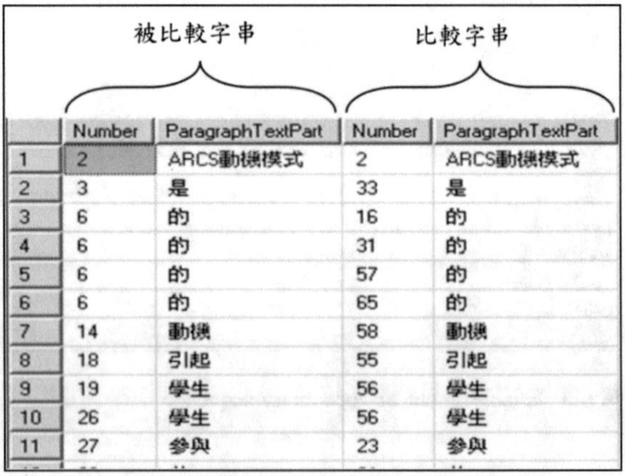

Fig. 6 Intersection and ordering

	Number	ParagraphTextPart	Number	ParagraphTextPart
1	2	ARCS動機模式	2	ARCS動機模式
2	3	是	33	是
3	6	的	16	的
4	6	的	31	的
5	6	的	57	的
6	6	的	65	的
7	14	動機	58	動機
8	18	引起	55	引起
9	19	學生	56	學生
10	26	學生	56	學生
11	27	參與	23	參與

被比較字串   比較字串

vocabularies, confirmation of vocabulary sequence, and marking. The following are the steps for the method proposed in this study for plagiarism:

- The SQL Intersection Operation provided by the database is used to intersect the comparing and compared strings and sequencing is performed based on the compared string. Punctuation marks are not included during the comparison. The sequencing is based on the compared string.
- At the end of sequencing, an automatic check is performed for any repeated values in the sequenced contents. The repetition processing is conducted if yes, and vocabulary sequencing and distance confirmation are conducted if not. The repetition processing and vocabulary sequencing and distance confirmation are described as follows:

  A. For a repetition, the number of the repeated value and that of the previous value are marked as OBU and BU, respectively. During the process, OBU remains unchanged, but BU will move from one piece of data to the next.
  B. The repeated values are captured through the database and sequenced using the number of the compared string. A value is obtained by subtracting the number to which OBU corresponds in the comparing string from the one to which BU corresponds in the comparing string, and is called the vocabulary interval for this study. This interval is considered between 0 and 3. The vocabulary interval is considered between 0 and 3 for this study because, if a string or redundant word is inserted between phrases, the string can be marked as an identical string. Therefore, the vocabulary interval is added for this study. This parameter is user0definable.
  C. If no repetition is found, a check is performed to see if the gap between the number of the comparing string and the previous number falls between 0 and 3. A record entry is made if yes.

- Once the repetition and vocabulary sequence are confirmed, the vocabularies that meet the criteria are marked.

4 Verify the Effectiveness of the System

In an era of booming information technology like today, many documents are digitized. Some people make minor modifications to these documents by adding superfluous words in several strings or inserting a string between phrases and then turn them into their works. This scenario was simulated in the study using the prototype developed and the quick search in the past for a comparison, as described below:

- The contents of two word strings, CP and BP, are compared for identical parts, which are the red parts in the following figures. Figure 7 shows the plagiarism detected successfully by the quick search approach, while Fig. 8 provides the

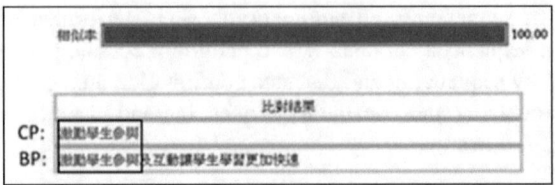

Fig. 7 Results of quick search methods without the addition of superfluous words

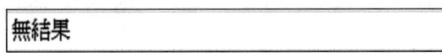

Fig. 8 Comparison results of the proposed prototype without the additional of superfluous words

Fig. 9 Result of quick search approach after the addition of superfluous words

Fig. 10 Comparison result of the prototype proposed after the addition of superfluous words

plagiarism identified effective by the prototype proposed and the indication of plagiarism similarity rate.
- With CP unchanged, several superfluous words are inserted into BP, and the results are presented in Figs. 9 and 10. Figure 9 is the result of quick search, which fails to make a positive detection of plagiarism. Figure 10 indicates that the prototype proposed correctly identify the plagiarism of article.

Tables 1 and 2 are the comparisons without and with the addition of superfluous words, respectively, in terms of analysis time, comparison time and accuracy rate.

Table 1 Analysis and comparison results without superfluous words

System performance	Prototype proposed	Quick search method
Analysis time (s)	10	0.134
Comparison time (s)	0.20	0.11
Accuracy (%)	100	100

Table 2 Analysis and comparison results with superfluous words

System performance	Prototype proposed	Quick search method
Analysis time (s)	13	0.15
Comparison time (s)	0.23	0.12
Accuracy (%)	100	0

The analysis time was long because the vocabulary patching and Chinese word segmentation were performed online. For comparison time the difference was no significant. However, the accuracy rate remained at 100%. It is clear that, despite more time spent on analysis, the prototype proposed was more effective in identifying the identical parts in the strings to be compared.

5 Conclusions and Suggestions for Future Studies

5.1 Conclusions

In a society pervasive of the well-developed internet, people tend to surf the Internet to access data and information which are already digitized. This convenience makes it easy to rewrite other authors' articles simply by using the mouse and keyboards to modify the contents of files, thereby resulting in problems of plagiarism. Previous research employed methods of statistics, vectors, and displacement to postulate mappings between sequences of words. A drawback of these methods is that the plagiarized texts are made not simply by quoting but by inserting some superfluous and sequences of words, a plagiarist technique hard to detect and to achieve checking precision. To better detect plagiarism, this study has proposed a new plagiarism detection method based on a Chinese parsing system and SQL intersection operation along with a technique of online word repair.

Previous works used to repair words to facilitate word analysis by establishing artificial corpus. However, this method is time-consuming and effort-intensive given the high cost of artificially expanding vocabulary due to constant addition of new words coined in our changing society. This study adopts online word repair via Google XML for word corpus construction and real-time new word addition.

This study has reported a new prototype of the plagiarism detection system whose functions include membership registration, membership management, file management, article plagiarism detection, word managements, etc. To verify the performance of our system, we have compared the outcome of our detection system and that of the quick searching method. The results have indicated that both systems can locate the same portion between two documents by a 100% rate of similarity identification, but when superfluous words are inserted into word sequences, the quick searching method fails to identify the same portion between two documents whereas our system can still reach a 100% rate of similarity identification yet with an efficiency lower than expected.

5.2   Contributions

- A prototype system was proposed based on Chinese word segmentation and SQL Intersection Operation and featured highly accurate plagiarism detection.
- The prototype proposed was able to identify accurately the identical parts and mark them, and the accuracy was not compromised when superfluous words or word strings were added or deleted.
- It was proposed to perform the comparison by "phrase" and bring in Google XML for online vocabulary patching in order to eliminate the trouble to establish new vocabularies manually.

5.3   Suggestions for Future Studies

- The study was focused on the comparison of the same strings. If some of the words or phrases in the strings were replaced with synonyms, it is likely to see compromised accuracy of the prototype. An idea for future studies is to incorporate semantic analysis for improved accuracy.
- The vocabulary patching was performed online for this study, and an extended period of time was spent on the online vocabulary patching and analysis. For future studies, the online vocabulary patching may be combined with vocabulary database developed by researchers for accelerated analysis efficiency.
- Files were uploaded in this study only for string analysis. An idea for future study is to analyze the structure of the files. For example, it is possible to identify whether the content of a file is a study paper or an operation document in order to distinguish the comparison methods.
- User interface was not a consideration in the prototype. Only string comparison and plagiarism detection were included. An idea for future studies is to take the user interface into consideration, as to enhance the completeness of the interface.
- In the study only the files uploaded by users were compared to the database contents. A possible direction for future studies is to include online files in the scope of comparison; i.e. comparing similar webpages and online data using proxies.

References

1. Lin, C.X., Chen, Z.J., Ling, C.C.: combined with a long term priority sequence labeled with Chinese word segmentation research. Inf. Secur. Commun. **15**(3–4), 161–179 (2010)
2. Chen, W.H.: Using short string indexing in Chinese and English full-text retrieval. DA-Yeh J. **1**(1), 161–173 (1992)
3. Horspool, R.N.: Practical fast searching in strings. Softw. Pract. Exp. **10**, 501–506 (1980)
4. Wu, D., Zhou, X., Zhang, H.: The pattern matching algorithms formalized analyze in chinese strings. Intell. Inf. Technol. Appl. **1**, 403–407 (2008)
5. Saini, A., Bahl, A., Kumari, S., Singh, M.: Plagiarism checker: text mining. Int. J. Comput. Appl. **134**(3), 8–11 (2016)

Text Summarization Based on Classification Using ANFIS

Yogan Jaya Kumar, Fong Jia Kang, Ong Sing Goh and Atif Khan

Abstract The information overload faced by today's society has created a big challenge for people who want to look for relevant information from the internet. There are a lot of online documents available and digesting such large texts collection is not an easy task. Hence, automatic text summarization is required to automate the process of summarizing text by extracting only the salient information from the documents. In this paper, we propose a text summarization model based on classification using Adaptive Neuro-Fuzzy Inference System (ANFIS). The model can learn to filter high quality summary sentences. We then compare the performance of our proposed model with the existing approaches which are based on neural network and fuzzy logic techniques. ANFIS was able to alleviate the limitations in the existing approaches and the experimental finding of this study shows that the proposed model yields better results in terms of precision, recall and F-measure on the Document Understanding Conference (DUC) data corpus.

Keywords Text summarization · Neural network · Fuzzy logic · ANFIS

Y.J. Kumar (✉) · F.J. Kang · O.S. Goh
Faculty of Information and Communication Technology,
Universiti Teknikal Malaysia Melaka, 76100 Durian Tunggal, Melaka, Malaysia
e-mail: yogan@utem.edu.my

F.J. Kang
e-mail: jkfong93@gmail.com

O.S. Goh
e-mail: goh@utem.edu.my

A. Khan
Department of Computer Science, Islamia College Peshawar, Peshawar 25120,
Khyber Pakhtunkhwa, Pakistan
e-mail: atifkhan@icp.edu.pk

© Springer International Publishing AG 2017
D. Król et al. (eds.), *Advanced Topics in Intelligent Information
and Database Systems*, Studies in Computational Intelligence 710,
DOI 10.1007/978-3-319-56660-3_35

405

1   Introduction

Text summarization has grown into a field of interest to explore, as information from online sources has become the current trend for everyone to search for information. A brief summary of a text document is useful for humans to quickly extract the important information in the text. For example, when a user looks for a news topic on the web, the search engine would retrieve many articles related to that news. It would be helpful to provide the summary of these articles to the readers instead of having them going through those lengthy texts.

Thus, a summary can aid humans in obtaining and understanding the main idea discussed in the text provided. Automatic summarization was introduced so that computers can be programmed to select and extract key points from the relevant text.

Generally, there are two types of summary which can be generated i.e. extractive and abstractive based summaries [1]. An extractive summary comprises of the original sentences which are selected from the input document. Such summaries can be obtained using methods of sentence extraction, statistical analysis and machine learning techniques. On the other hand, an abstractive summary contains sentences that have to be reconstructed using deep natural language analysis [2]. Most studies in text summarization resolves around extractive based approaches.

In this paper, we propose an extractive text summarization model based on classification using a hybrid soft computing approach known as Adaptive Neuro-Fuzzy Inference System (ANFIS). The proposed model will be used to classify sentences as summary sentence and non-summary sentence. We also compare our proposed model with existing approaches which are based on neural network and fuzzy logic technique.

The rest of this paper is organized as follows: Sect. 2 discusses on the related works concerning this study. Section 3 outlines the proposed model. The experimental results and discussion are given in Sect. 4. Finally, we end with conclusion in Sect. 5.

2   Related Works

In the recent past, soft computing based approaches have gained popularity in its ability to determine important information across documents [3–7]. For instance, a number of studies have modelled summarization systems based on fuzzy logic reasoning in order to select important sentences to be included in the summary [4, 8]. First, the features influencing the importance of a sentence are determined, such as, title word, sentence position, thematic word, etc. Then selected sentence features are used as the input to the fuzzy system. The scores for each sentence are then derived using fuzzy rules scoring. The sentences with high fuzzy score will be selected to be included in the summary until the desired summary length is obtained.

Apart from sentence scoring, fuzzy logic has also been used for semantic analysis to produce text summary. For example, Kumar et al. [9] investigated the cross document relations that exist between sentences and used fuzzy logic to rank sentences based on the type of cross document relations. The authors in [10] extracted the semantic relations between concepts using fuzzy reasoning to select summary sentences. This method which is based on latent sematic analysis improves the quality of summary.

Although all the above works support the benefits of employing fuzzy based reasoning for extracting important sentences from the document, there is a limitation concerning this method. Human or linguistic experts are required to determine the rules for the fuzzy system. Furthermore, the membership functions need to be manually tuned. These can be a very tedious and time consuming process. Moreover, the performance of the fuzzy system can be affected by the choice of rules and parameters of membership function [11].

Besides fuzzy logic, neural network models have also been employed in text summarization studies whereby its learning capabilities are used to identify summary sentences from the input text document [3]. Megala et al. [6] used a three layered feedforward network model to learn the patterns in summary sentences. The resulting trained network is then applied to new input documents to determine if a sentence should be included in the summary.

In another related work, Sarda and Kulkarni [7] used a similar neural network model with the combination of Rhetorical Structure Theory (RST). The RST relations that exist in the sentences which are picked out by their neural network model are used to form high quality summaries. Fattah and Ren [12] in their study proposed an improved content selection approach using probabilistic neural network. They used probability function to better estimate the weights of their neural network model.

Although neural network model has been useful in term of its learning capabilities, the model provides little information about the relationship between the input and output.

3 Proposed ANFIS Based Text Summarization

From the related works discussed above, it can be observed that among the two soft computing techniques that have been associated with text summarization, fuzzy logic implementation is based on knowledge–driven reasoning whereas neural network is based on data–driven approximation. Taking these observations into consideration, a better summarization system can be modeled by considering the advantages of both approaches and avoiding their drawbacks.

ANFIS (Adaptive Neuro-Fuzzy Inference System) is one example of such hybridization [13]. It combines the explicit knowledge reasoning of fuzzy logic

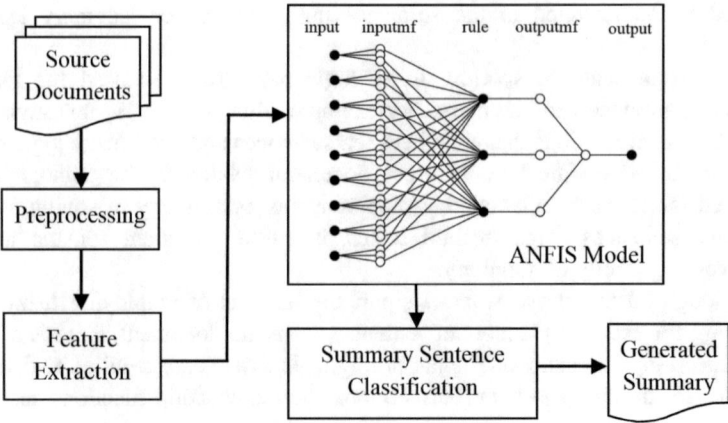

Fig. 1 Text summarization based on ANFIS model

system, which can explain input output relationship and the implicit knowledge of neural networks, which can be learnt. Past studies have shown that there are limitations with regards to fuzzy based approaches as human experts are required to determine the rules and tune the membership functions for the fuzzy system. Our hypothesis is this study is that a better fuzzy system can be produced by integrating the learning and adaptive capabilities of neural network to improve the identification of summary sentences.

The architecture of our proposed model is shown in Fig. 1. The ANFIS model is trained to classify document sentences as summary and non-summary sentence. Based on the available Document Understanding Conference (DUC) 2002 dataset, we prepared our training set which comprises of the features representing sentences with its corresponding output type i.e. summary or non-summary sentence. After computing the feature values for every sentence from the training set, we input them for the training of ANFIS. Once the training is completed, the resulting classifier model will be able predict the output score of a new document sentence to determine its class i.e. as summary or non-summary sentence.

3.1 Extraction of Features

The input documents which have been preprocessed are represented as vector of features. These features are the attributes that are used to represent summary sentences. We extract nine features from each sentence. These features have been extensively used in text summarization studies. Each feature is given a value between 0 and 1 where values around 0 indicate a low presence of the feature in the sentence while values around 1 indicate a strong presence of the feature in the sentence. The nine features that were selected as the input for ANFIS include title feature, sentence similarity, sentence location, numerical data, temporal feature, sentence length, proper noun, nouns and verbs, and frequent semantic term.

3.1.1 Title Feature

The sentence that contains the word(s) in the document title will be given high score. This can be computed by counting the number of matching characters between the words in sentence and the words in the document title.

$$f_1 = \frac{number\ of\ title\ words\ in\ sentence}{number\ of\ words\ in\ document\ title} \tag{1}$$

3.1.2 Sentence to Sentence Similarity

This feature will measure the similarity between sentence S and other sentences in the document set. The similarity is computed by using the Jiang's semantic similarity measure [14].

$$f_2 = \frac{\sum sim(S_i, S_j)}{\max\left(\sum sim(S_i, S_j)\right)} \tag{2}$$

3.1.3 Sentence Location

The placement of the sentence in the text is important where the beginning sentences in the document will be given high scores using the equation below.

$$f_3 = \frac{length\ of\ document - sentence\ position + 1}{length\ of\ document} \tag{3}$$

3.1.4 Numerical Data

Sentence containing statistical data is important for summary where the feature can be computed using the equation below.

$$f_4 = \frac{number\ of\ numerical\ data\ in\ the\ sentence}{length\ of\ sentence} \tag{4}$$

3.1.5 Temporal Feature

Temporal extraction is used to extract explicit event containing date or time expressions in the article and is measured based on the equation below.

$$f_5 = \frac{number\ of\ temporal\ information\ in\ the\ sentence}{length\ of\ sentence} \tag{5}$$

3.1.6 Length of Sentence

Sentence that is long is considered to inherit important information. Hence, the length of the sentence is computed using the equation below.

$$f_6 = \frac{number\ of\ words\ occuring\ in\ the\ sentence}{number\ of\ words\ occuring\ in\ the\ longest\ sentence} \tag{6}$$

3.1.7 Proper Noun

Sentence containing proper noun is considered to be an important sentence. The scores of sentence that contains proper noun are computed with the equation below.

$$f_7 = \frac{number\ of\ proper\ nouns\ in\ the\ sentence}{length\ of\ sentence} \tag{7}$$

3.1.8 Number of Nouns and Verbs

The sentences that contains more nouns and verbs are considered to inherit important information. The score can be computed with the equation below.

$$f_8 = \frac{number\ of\ nouns\ and\ verbs\ in\ the\ sentence}{length\ of\ sentence} \tag{8}$$

3.1.9 Frequent Semantic Term

This feature is used to determine the commonness of a term. A term that is used frequently is probably related to the topic of the document. We consider the top 10 words as the maximum number of frequent semantic terms.

$$f_9 = \frac{number\ of\ frequent\ terms\ in\ the\ sentence}{\max\left(number\ of\ frequent\ terms\right)} \tag{9}$$

3.2 ANFIS Model

ANFIS is introduced for text summarization in this paper. The nine sentence features which have been described in the previous section will become the input to our ANFIS model. Each crisp input will be transformed into fuzzy value using a membership function. Parameters in this layer are generally referred to as premise parameters and are used to adjust the shape of the membership function. The fuzzy values will be used as incoming signals to compute the firing strength of the corresponding rule. The output of each rule is combined with the linear combination of input variables. Parameters in this layer are referred to as the consequent parameters. The final output is then computed by measuring the aggregation of all incoming signals. Figure 2 depicts our ANFIS model structure. The detailed description of the basic ANFIS architecture is not presented in paper, however, it can be found in our past paper [15].

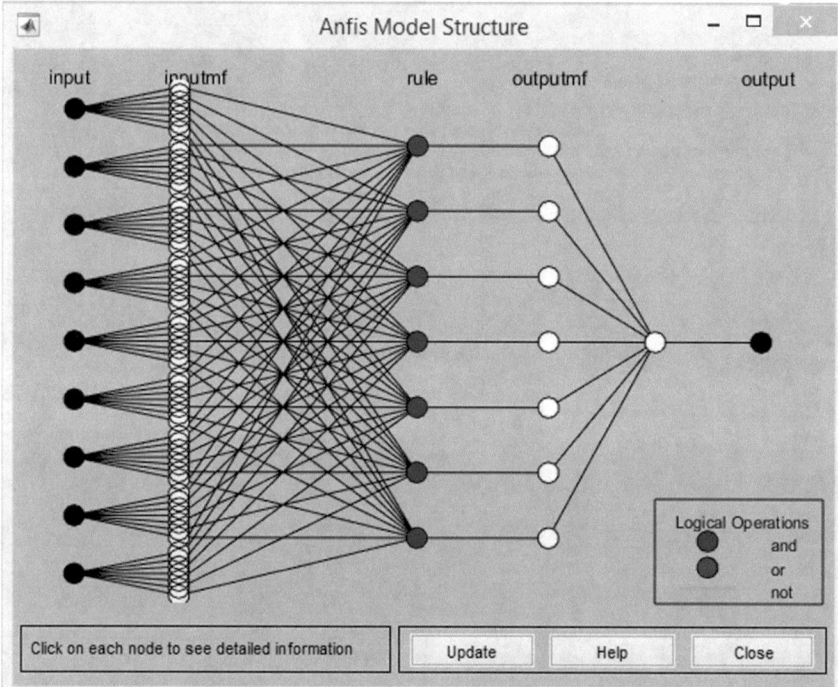

Fig. 2 ANFIS model structure

3.2.1 ANFIS Learning Method

In conventional fuzzy reasoning based text summarization, the rules were to be decided by an expert; which is the limitation concerning fuzzy inference system (FIS) in text summarization. However, in ANFIS model, no expert is required to manipulate the rules as the rules can be generated automatically by using subtractive clustering method.

Subtractive clustering algorithm estimates the cluster number and cluster centers automatically by mapping the input-output training data. Each instance is seen as a potential cluster center and the instances that have a value that is in the range of the first cluster will be included as the first cluster. Else, the instance will form a new cluster. The process will repeat until all instances are included in the clusters. An important advantage of using a clustering method to find rules is that the resultant rules are more tailored to the input data than they are in an FIS generated without clustering [16]. This reduces the problem of combinatorial explosion of rules when the input data has a high dimension (the curse of dimensionality). Figure 3 shows the fuzzy rules obtained based on the created data clusters.

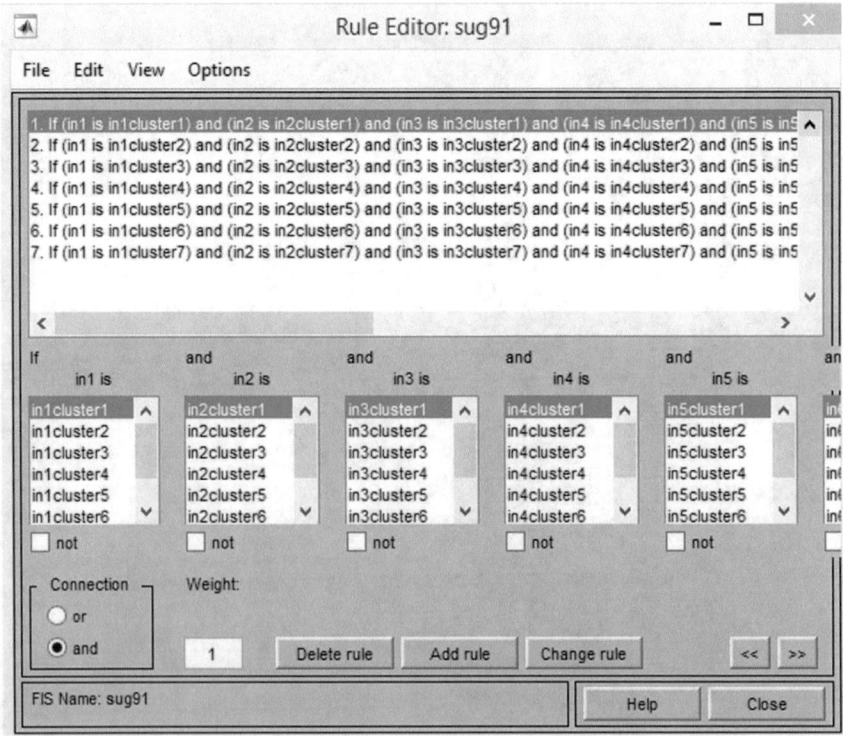

Fig. 3 Rules generated by ANFIS using subtractive clustering method

To train the ANFIS model, a hybrid method that is a combination of least-square estimation and backpropagation gradient descent method is used. Least-squares Estimate (LSE) is used to minimize the squared error of the actual output and the target output. The backpropagation method is combined with LSE to update the parameters of the membership functions. Backpropagation method originates from multilayer feedforward neural networks where the network is computed by using the gradient descent method to minimize the sum of squared errors. Backpropagation works by each input weights having their own learning rate, where the learning rate will change over time for each iteration.

A forward pass and a backward pass are included in the hybrid optimization method where forward pass is used to calculate the error measure. The error rates are propagated from the output end towards the input end in the backward pass, where all parameters are updated. The combination of fuzzy inference to represent knowledge in the form of fuzzy rules and membership functions with the learning ability of neural network enables the membership functions parameters to be adjusted directly from the output data. Figure 4 shows the membership function of our input data after training.

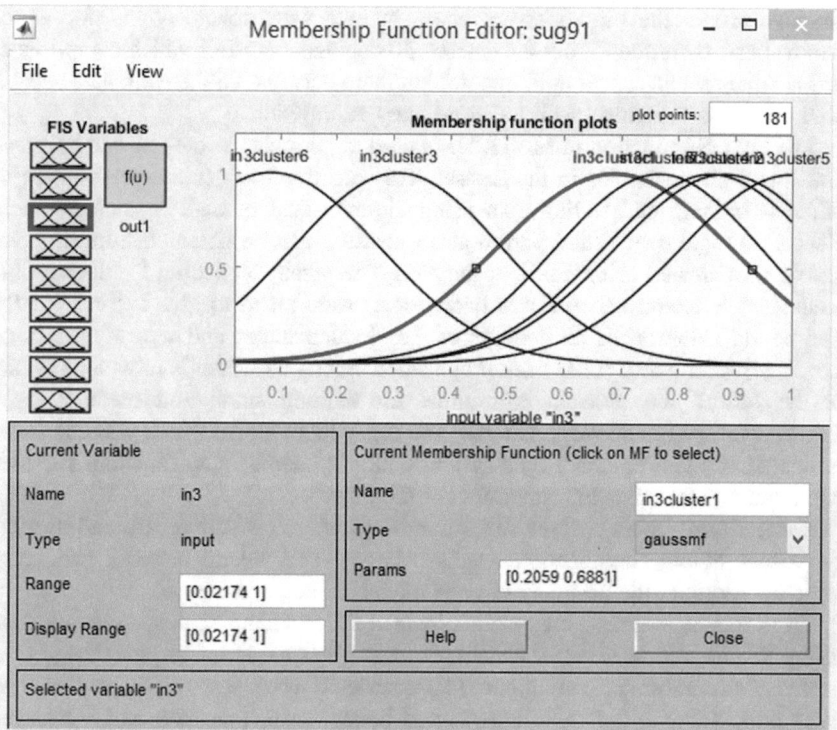

Fig. 4 ANFIS membership functions for the third input feature

3.2.2 Sentence Classification

The trained ANFIS model is then used to classify new input sentences to one of its class i.e. summary or non-summary sentence. The ANFIS model output, which is the predicted sentence score is used to set the classification rule for ANFIS to classify the sentence into binary value (1 or 0). Sentences which are classified to class '1' represents summary sentence while sentences which are classified to class '0' represents non-summary sentence. The threshold value used to classify the predicted output to one of these two classes were selected based on experimental observation which gave us the least root mean square error (RMSE).

4 Experimental Results and Discussion

For this study, we used the DUC (Document Understanding Conference) 2002 dataset and run the experiment using MATLAB R2015a. The dataset contains articles with sample human summary extracts. Based on this dataset, we prepared our training and testing data by labelling each sentence with its class (i.e. 1 or 0). The dataset was preprocessed first before extracting the sentence features. Steps that are involved in the preprocessing phase include word tokenization, stop-words removal and stemming. From the dataset, 30 document clusters which comprises of 362 sample sentences was split into 75% training data and 25% testing data using 5 hold out cross validation with balanced class distribution.

The hold out function in MATLAB is used to permute the dataset into balanced amount of class samples in the dataset. This function enables the permutation of different training and testing data using different fold of data in each iteration. Hence, multiple results and performance measure can be obtained from different testing data created using hold out function. The parameter setting for the baseline methods (i.e. neural network and fuzzy logic) were tuned to give optimal results. The neural network model consists of four hidden nodes and was trained using Levenberg-Marquardt (LM) back propagation technique which is proven to be one of the fastest and efficient algorithms for training small and medium sized feed-forward neural network patterns. The fuzzy logic model which was compared in this study uses Mamdani-type system with two membership functions for each input and five membership functions for the output.

Table 1 and Figure 5 show the average precision, average recall, and average F-measure of the classification results using ANFIS, neural network and fuzzy logic. In addition, the accuracy of each model is shown in Fig. 6.

From this experiment, the results obtained for summary sentence classification using ANFIS, neural network and fuzzy logic gives us an insight into the performance of these soft computing based approaches towards text summarization. It is clear from Table 1 that ANFIS produces better recall, precision and F-measure when compared to neural network and fuzzy logic. Based on literature, the

Table 1 Classification results using average precision, average recall, and average F-measure

Model	Average F-measure	Average precision	Average recall
ANFIS	0.7054	0.7128	0.6982
Neural network	0.5437	0.5272	0.5426
Fuzzy logic	0.5584	0.6538	0.5148

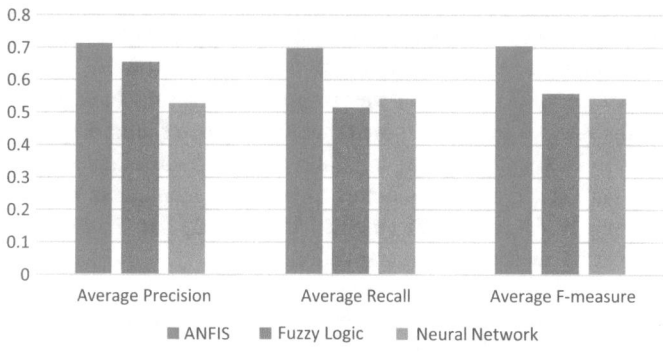

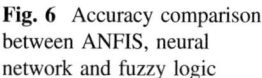

Fig. 5 Classification performance using average precision, average recall, and average F-measure

Fig. 6 Accuracy comparison
between ANFIS, neural
network and fuzzy logic

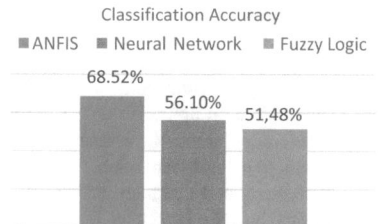

performance of the fuzzy logic based approach is often affected by the selection of
fuzzy rules and membership functions.

However, ANFIS, a hybrid approach which takes the advantages from both
neural network and fuzzy logic has improved the accuracy of the summarization
model in determining the sentences which should be included in the summary. With
the aid of a training algorithm, it enables the process of tuning of the parameters of
membership functions for each sentence feature. These results support our
hypothesis is this study that a better fuzzy system can be produced by integrating
the learning and adaptive capabilities of neural network to improve the identification of summary sentences using optimized fuzzy membership functions and rules.
However, in order to affirm the effectiveness of the classification results, performance evaluation is needed to evaluate the final summary generated for a document. This can be achieved using standard evaluation metric for summarization.

5 Conclusion

In this paper, a study on hybrid soft computing based approach to improve summary sentence selection is investigated. The key motivation of this paper is the growing number of research studies in text summarization based on soft computing approaches. In order to compensate the disadvantages of one approach with the advantages of another approach, a hybrid method called ANFIS is proposed to better identify summary sentences. The proposed approach was able to alleviate some of the limitations in the current text summarization models. The experimental results show that ANFIS achieved better classification results in terms of precision, recall and F-measure. It should be noted that the quality of summary sentences was not evaluated in this study. In our ongoing work, we are attempting to train the ANFIS classifier on larger data samples to generate summaries and evaluate the summaries using ROUGE (Recall Oriented Understudy for Gisting Evaluation)–a standard evaluation metric for summarization.

Acknowledgements This research work supported by Universiti Teknikal Malaysia Melaka and Ministry of Education, Malaysia under the Research Acculturation Grant Scheme (RAGS) No. RAGS/1/2015/ICT02/FTMK/02/B00124.

References

1. Kumar, Y.J., Goh, O.S., Halizah, B., Ngo, H.C., Puspalata, C.: A review on automatic text summarization approaches. J. Comput. Sci. **12**(4), 178–190 (2016)
2. Khan, A., Salim, N., Kumar, Y.J.: Genetic semantic graph approach for multi-document abstractive summarization. In: Fifth International Conference on Digital Information Processing and Communications (ICDIPC), pp. 173–181 (2015)
3. Keyan, M., Srinivasagan, K.: Multi-document and multi-lingual summarization using neural networks. In: International Conference on Recent Trends in Computational Methods, Communication and Controls (ICON3C), pp. 11–14 (2012)
4. Patil, M.P.D., Kulkarni, N.J.: Text summarization using fuzzy logic. Int. J. Innov. Res. Adv. Eng. (IJIRAE) **1**(3), 42–45 (2014)
5. Rucha, S., Apte, S.: Improvement of text summarization using fuzzy logic based method. OSR J. Comput. Eng. (IOSRJCE). **5**(6), 5–10 (2012)
6. Megala, S.S., Kavitha, A., Marimuthu, A.: Enriching text summarization using fuzzy logic. Int. J. Comput. Sci. Inf. Technol. **5**, 863–867 (2014)
7. Sarda, A., Kulkarni, A.: Text summarization using neural network and rhetorical structure theory. Int. J. Adv. Res. Comput. Commun. Engineering, IJARCCE **4**(6), 49–52 (2015)
8. Suanmali, L., Salim, N., Binwahlan, M.S.: Fuzzy logic based method for improving text summarization. Int. J. Comput. Sci. Inf. Secur. **2**(1) (2009)
9. Kumar, Y.J., Salim, N., Abuobieda, A., Albaham, A.T.: Multi document summarization based on news components using fuzzy cross-document relations. Appl. Soft Comput. **21**, 265–279 (2014)
10. Babar, S.A., Patil, P.D.: Improving performance of text summarization. Proc. Comput. Sci. **46**, 354–363 (2015)
11. Albertos, P., Sala, A.: Fuzzy logic controllers. Advantages and drawbacks. In: VIII International Congress of Automatic Control, vol. 3, pp. 833–844 (1998)

12. Fattah, M.A., Ren, F.: Automatic text summarization. World Acad. Sci. Eng. Technol. **13**, 192–195 (2008)
13. Loganathan, C., Girija, V.: Investigations on hybrid learning in ANFIS. Int. J. Eng. Res. Appl. **4**(10), 31–37 (2014)
14. Jiang, J.J., Conrath, D.W.: Semantic similarity based on corpus statistics and lexical taxonomy. In: Proceedings of International Conference Research on Computational Linguistics, pp. 1–15 (1997)
15. Lim, E.A., Jayakumar, Y.: A study of neuro-fuzzy system in approximation-based problems. Matematika **24**, 113–130 (2008)
16. Moh'd Arikat, Y.: Subtractive neuro-fuzzy modeling techniques applied to short essay auto-grading problem. In: International Conference on Sciences of Electronics, Technologies of Information and Telecommunications (SETIT), pp. 889–895 (2012)

Part V
Innovations in Web and Internet Technologies

Analysis of Indoor Positioning Based on BLE

Wenjun Zhu, Shinheon Kim, Jaemin Hong and Chonggun Kim

Abstract Indoor positioning systems based on BLE (Bluetooth Low Energy) beacon signal strength is studied. Student positioning in a classroom is a target application of this study. BLE is low energy consuming and one directional low energy consuming communications. A suggested layout of multiple BLE for deciding student positions is proposed and the experimental results based on the environment are analysed. The proposed position assumption method based on three signal strengths are useful for indoor positioning.

Keywords BLE · Signal strength · Indoor positioning · Classroom environment

1 Introduction

In recent years, a lot of work which uses global positioning system (GPS) and wireless network technology to overcome the constraints of time and place using a variety of information have been studied. At the same time, the needs for indoor positioning of mobile objects are increasing. For example, indoor positioning in a shopping mall or take attendance check in classroom. Wide area positioning systems (e.g. satellite positioning system) do not work correctly in indoor environments. Major methods of indoor positioning methods are Cell-ID, Trilateration and

W. Zhu · S. Kim · J. Hong · C. Kim (✉)
Department of Computer Engineering, Yeungnam University,
280 Daehak-Ro, Gyeongsan, Gyeongbuk 38541, Republic of Korea
e-mail: cgkim@yu.ac.kr

W. Zhu
e-mail: ccxr@outlook.com

S. Kim
e-mail: rubymix80@ynu.ac.kr

J. Hong
e-mail: hjm4606@naver.com

© Springer International Publishing AG 2017
D. Król et al. (eds.), *Advanced Topics in Intelligent Information
and Database Systems*, Studies in Computational Intelligence 710,
DOI 10.1007/978-3-319-56660-3_36

421

Fingerprint [1]. In this study, the trilateration method for indoor positioning based on Bluetooth low energy beacons signal strength. For calculating the distance from the mobile object to each beacon transmitter using RSSI (Received Signal Strength Indicator) values, the Chipcon formula [2] was used.

2 Previous and Related Studies

2.1 Other Methods for Bluetooth Indoor Positioning System

A time based Bluetooth indoor positioning system showed in [3], that system is intended to have an accuracy of 1 m to locate any mobile Bluetooth device without additional hardware in the mobile and without any changes in its software.

On the other hand, Bluetooth signal strength based indoor positioning methods in, an MNN (multiple neural networks) system is able to localize correctly the user 89% of the times. System accuracy is 0.5 m on the average. In, that system has a fairly good precision with 50% probability of being within about 1.5 m of the actual when the mobile phone on which the system is running does not move.

2.2 A Method for Positioning

At present, most of indoor positioning algorithm is based on distance measuring which is measure the distance from the mobile object to anchor nodes, then calculate the coordination of the mobile object. At least three specific points could be used by the mobile object for deciding position by using trilateration. Suppose A, B, C are three points of known location, D is the point of unknown location. (x_a, y_a), (x_b, y_b), (x_c, y_c), (x, y) are the coordinates of point A, B, C, D. d_a, d_b, d_c are the distances from D to A, B, C (Fig. 1).

According to the Pythagorean theorem, we can draw Eq. (1)

$$\begin{cases} \sqrt{(x-x_a)^2 + (y-y_a)^2} = d_a \\ \sqrt{(x-x_b)^2 + (y-y_b)^2} = d_b \\ \sqrt{(x-x_c)^2 + (y-y_c)^2} = d_c \end{cases} \tag{1}$$

By solving Eq. (1), we can get the coordinate of point D (x, y) [4].

$$\begin{bmatrix} x \\ y \end{bmatrix} = \begin{bmatrix} 2(x_a - x_c) & 2(y_a - x_c) \\ 2(x_b - x_c) & 2(y_b - x_c) \end{bmatrix}^{-1} \begin{bmatrix} x_a^2 - x_c^2 + y_a^2 - y_c^2 + d_c^2 - d_a^2 \\ x_b^2 - x_c^2 + y_b^2 - y_c^2 + d_c^2 - d_b^2 \end{bmatrix}$$

Fig. 1 Trilateration

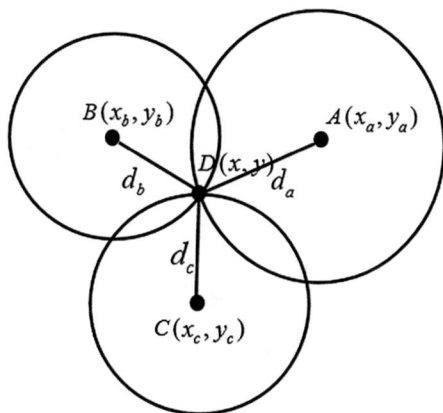

Therefore, the key of positioning by trilateration is to get the distances from the mobile object to three known location points.

2.3 A Method for Getting Distance by Radio Signal Strength

Friis formula [5] and Chipcon formula are used for calculating the distance between two objects (http://www.ti.com). Although prediction and calculation of the distance based on radio signal strength are very complex depend on environments. Calculation error on distance decision is a major problem which has to be solved. In this study, the Chipcon formula (2) is employed to simplicity for distance calculation.

$$RSSI = -(10n \log_{10} d + A). \tag{2}$$

RSSI is signal strength, n is the propagation loss, d is the distance, A is the TX Power of transmitter. The distance between the receiver and transmitter can be calculated by using the Chipcon formula: d = 10^((abs(RSSI) − A)/(10 × n)) (3). Signal strengths affected by the surrounding environment may be varied by time changed [6, 7].

$$d = 10^{\frac{|RSSI| - A}{10*n}}. \tag{3}$$

2.4   Feature of BLE

As a low energy device, BLE had these features: low operation cost and low power consumption, BLE is 60–80% cheaper than traditional Bluetooth. Transmitter can last up to 3 years on a single coin cell battery [8]. As a receiver, almost all of the smart devices running different operating system can support BLE.

3   Proposed Classroom Positioning Model

3.1   Positioning for BLE

Figure 2 is the schematic target environment for using BLE indoor positioning system. The proposed system can be used for analyzing preference positions of participated students in the classroom.

The working process of this positioning system is as follows:

1. Getting the position of each mobile object by using BLE RSSI signal strength.
2. Send the position information to server by using Wi-Fi or cellular data.
3. If the position changes of mobile object more than a specified range, then send and update the newest position information of the mobile object on the server.

3.2   RSSI Signal Characteristics of BLE Beacon

According to the different environment to decide the propagation loss n and transmitter TX Power A in Chipcon formula. The TX Power A is always set as

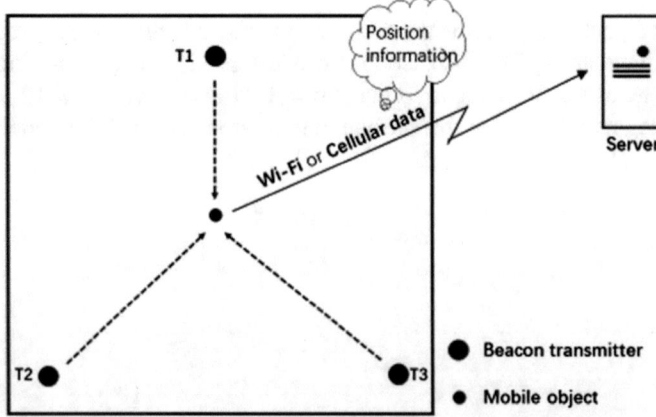

Fig. 2 The schematic of BLE indoor positioning system

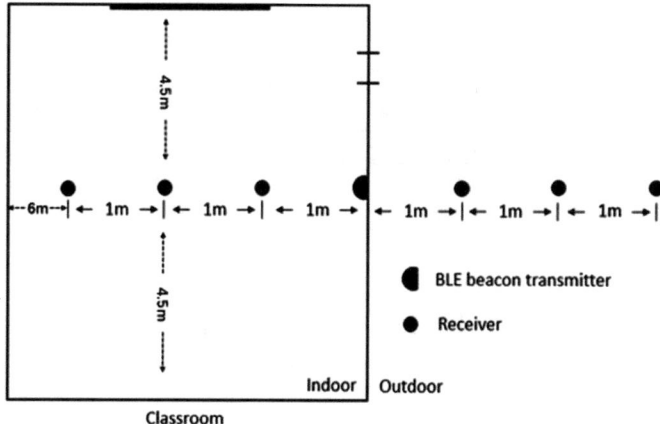

Fig. 3 Experimental environment for sampling

−20 dBm in Wi-Fi environments [9, 10]. In this experiment, the n was set as 2.0 and the A was set as −59 dBm. Because it has the minimum error in our experimental environment.

In the experimental environment showed at Fig. 3, the distance from the transmitter to the mobile object measured at 1–3 m, the calculated distance using RSSI signal strength is similar to the actual distance was showed at Table 1 [11].

In a normal indoor radio environment, the calculated distance is similar to the actual distance. When the receiver distance from the transmitter over 3 m, the RSSI almost has no change, similar to [9], in [9], 5 m is set as coverage in Wi-Fi environment.

4 Positioning in a Classroom

As mentioned, based on the result of RSSI signal characteristics test, thus we define the meaningful coverage of each BLE transmitter is 3 m, then set 3 BLE transmitters as an equilateral triangle, the height of transmitters is 1.2 m from floor [12], and the length of triangle sides is 3 m that can make sure any point in the equilateral triangle can be covered by 3 transmitters in meaningful range [10, 13]. We use three Raspberry Pi 3 as transmitters and Samsung Galaxy S3 as receiver. The height of transmitters and receiver is 1.2 m.

The RSSI was measured for 50 times at each test point. Median filter was selected for remove noise. Table 2 shows the RSSI values of each test point. Figure 4 is the schematic experimental environment for positioning test and the actual position and calculated position of each test point.

We try to find three probable locations of the mobile object. The probable location is solved by the distance to each two transmitters (ex. T1&T2, T1&T3,

Table 1 RSSI values are corresponding to the distance

Indoor			Outdoor		
Actual distance (m)	Average RSSI (dbm)	Calculated distance (m)	Actual distance (m)	Average RSSI (dbm)	Calculated distance (m)
1	−60	1.0393	1	−69.3	1.8335
2	−72.2	1.9613	2	−70.8	2.1548
3	−74.9	2.4368	3	−78.6	3.4972

Table 2 RSSI (−dBm) of each test point

Test point	RSSI (−dBm)		
	T1	T2	T3
P1	45.4	61	69.4
P2	46.2	63.2	68.8
P3	49.6	71	66.6
P4	55	57.4	59
P5	61	61.8	60.8
P6	60.8	65.6	57.8
P7	59.2	54.4	55.2
P8	58.8	59.8	53.8
P9	62	63.4	45

T2&T3). For every transmitter group. Two circles with center at each transmitter and radius equal to the calculated distance from the mobile object to transmitters are drawn. Theoretically, the only intersection of two circles is the mobile object position. But in the vast majority of cases, that only intersection remained impossibly hard to solve. The 5 cases are shown on Fig. 5:

Case 1. The ideal case, two circles have only intersection point
Case 2. Two circles have two intersection points, two circles' radius both shorter than the interval of two transmitters. In this case, a line to connect the two intersection points is drawn, and the middle point of this line is decided as assumed probable position of mobile object.
Case 3. Two circles have two intersection points, two circles' radius both longer than the interval of two transmitters. In this case, the same method with case 2 was used for deciding assumed probable position of mobile object.
Case 4. Two circles have no intersection point because of the sum of two circles' radius is short than the interval of two transmitters. In our experiment, probability of this case appears is highest. When the sum of R1 and R2 is shorter than interval of transmitters, an amplification x is set, then suppose the result of $(R1 + R2) \times x$ equal interval of transmitters and find the solution of this equation. We can know that VR1 is $x \times R1$ and VR2 is $x \times R2$. The intersection point of VR1 and VR2 is decided as assumed probable position of the mobile object.

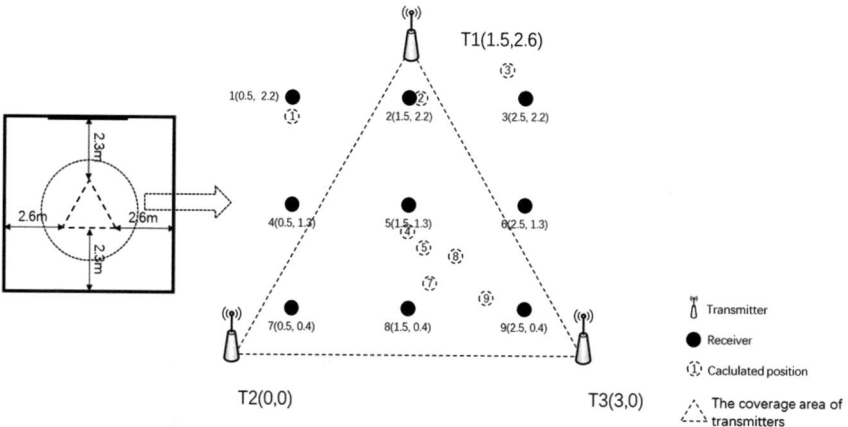

Fig. 4 Actual position and calculated position of each test point

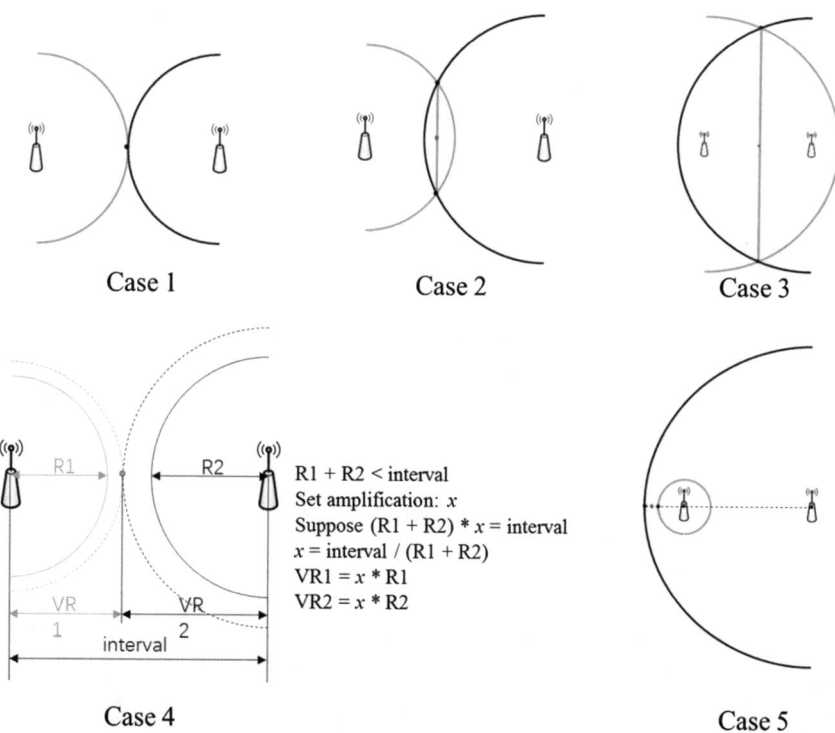

Fig. 5 5 cases of the mobile object probable location

Fig. 6 Position of the mobile
object

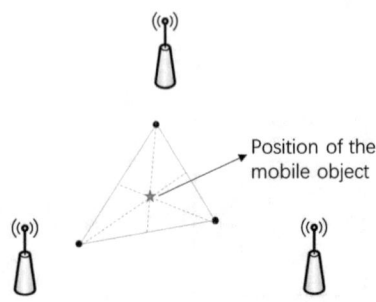

Position of the
mobile object

Case 5. Two circles have no intersection point because of one circle is include another one. In this case, a line to connect two transmitters is drawn, then extend the line until it has intersection points with both circles, the middle point of two intersection point is decided as assumed probable position of the mobile object (Fig. 6).

Finally, we use three probable points as vertex and draw a triangle, then find the mass center of the triangle as the mobile object position.

Table 3 shows the actual coordinates, calculated coordinates by using trilateration method and the error of each test point.

In this experiment, −59 dBm was set as the TX Power of transmitters, 2.0 was set as the propagation loss n in Chipcon formula. All the three transmitters covered a 7.8 m^2 area. The coordinate of each tested point showed in Table 3 is calculated by trilateration. For calculating the distance from the receiver to each transmitter, the median RSSI value was used. The three test points of P1, P2 and P3 nearby transmitter 1 shows an amazing accuracy of less than 0.3 m. The worst case appears at P4 and P7. Plainly, the distance from receiver to transmitter 2 is nearer than transmitter 3 at P4 and P7. But the RSSI of transmitter 3 is similar to the transmitter 2, it could lead to making the calculated coordinate in the middle of transmitter 1 and transmitter 3, then cause a position error more than 1 m. In most cases, the

Table 3 Actual coordinates, calculated coordinates and error

Test point	Actual coordinate	Calculated coordinate	Error(m)
P1	(0.5, 2.2)	(0.5, 2.0)	0.2
P2	(1.5, 2.2)	(1.6, 2.2)	0.1
P3	(2.5, 2.2)	(2.3, 2.4)	0.3
P4	(0.5, 1.3)	(1.5, 1.1)	1.0
P5	(1.5, 1.3)	(1.6, 0.9)	0.4
P6	(2.5, 1.3)	(2.1, 0.9)	0.6
P7	(0.5, 0.4)	(1.6, 0.7)	1.1
P8	(1.5, 0.4)	(1.9, 0.8)	0.5
P9	(2.5, 0.4)	(2.2, 0.5)	0.4
Average error (m)		0.51	

method gave an accuracy of 0.3–0.6 m. Compare to other signal strength based methods. This method shows a higher accuracy of 0.3–0.6 m. But the coverage of transmitters in this experiment is smaller than others. A larger coverage experiment needs to test in the future while keeping the accuracy at the same level.

5 Conclusions

BLE RSSI signal strengths are studied for indoor location tracking.

Although the indoor positioning method based on RSSI signal strength has less cost than some kinds of other indoor positioning method, because it does not require any other extend devices [4]. But the performance is barely satisfied. Because of a complex radio environment can make a huge error of location; a cleaner radio environment can improve the precision of indoor positioning based on RSSI. BLE Advertising is a one-way communication method [8]. It is hard to use a time-based positioning method on BLE devices.

However, precision of indoor positioning based on RSSI signal strength is not high enough in some case in this study. In the future, more studies of radio propagation model can help to improve the precision of indoor positioning method based on RSSI.

Acknowledgements This research was supported by the Leading Human Resource Training Program of the Regional Neo industry through the National Research Foundation of Korea (NRF) funded by the Ministry of Science, ICT and Future Planning (216C000360). This work was also funded by the BK21 + program of the National Research Foundation of Korea (NRF).

References

1. Kim H (2006) Wireless LAN-based LBS services
2. Chipcon formula. http://www.ti.com
3. Fischer G, Dietrich B, Winkler F (2004) Bluetooth indoor localization system
4. Wang G., Liu H., Hu H., Shen S.: Indoor localization algorithm: a survey (2011)
5. Friis, H.T.: Friis transmission equation (1946)
6. Machaj, J., Brida, P., Piché, R.: Rank based fingerprinting algorithm for indoor positioning (Sept 2011)
7. Dong, Q., Dargi, W.: Evaluation of the reliability of RSSI for indoor localization (Aug 2012)
8. What is iBeacon? A Guide to Beacons. http://www.ibeacon.com/what-is-ibeacon-a-guide-to-beacons/
9. Park, H.J., Hong, J.M., Zhu, W.J., Lee, D.H., Zin, H., Kim, Y.R., Kim, C.G.: Analysis of selected indoor location APs in Wi-Fi environments (2016)
10. Park, H.J., Min, G.W., Zin, H.C., Kim, Y.R., Kim, C.G.: An efficient AP allocation for position tracking lased on Wi-Fi (Oct 2015)

11. Bahl, P., Padmanabhan, V.N.: Enhancements to the RADAR user location and tracking system. Technical Report MSR-TR-2000-12, Microsoft Research (2000)
12. Liu, H.-H., Yang, Y.-N.: Wi-Fi based indoor positioning for multi-floor environment (Nov 2011)
13. Sohn, J.-H., Hwang, G.-H.: Development of position awareness algorithm using improved trilateration measurement method (Mar 2013)

A Fusion Technique of Schema and Syntax Rules for Validating Open Data

Shin'ya Yamaguchi and Kimio Kuramitsu

Abstract Schema validation is an important technique that achieves reliable data exchange on the Web. Historically, schema validation has been developed intensively in XML. However, open data, which receive much attention recently, are exchanged in various formats such as JSON and CSV. As XML schema validators are not available in other different formats, most of open data are exchanged without schema validation. In this paper, we present a new approach that provides a schema constrained syntax definition by creating a fusion of a schema definition and a syntax definition of a data format described in PEG. By applying our approach, we can obtain a generated parser serving as a schema validator that validates open data.

Keywords Schema validation · Open data · Schema language · Parsing expression grammar · Parser generator · Parsing

1 Introduction

Schema validation is a classic problem with data exchange, but a new challenge rises from time to time. The problem was first focused in the context of EDI (Electronic Data Interchange), which requires a strong consistency of data exchange. As EDI was developed before the Internet age, the interoperability between senders and receivers is not open, which leads to easier agreements on the syntax and the semantics.

As data exchange on the Web evolves, the relation between senders and receivers becomes more complicated and open. Under such situations, XML has been proposed as a standard data format, and schema validation techniques on XML have been developed intensively in early 2000s. As a result, schema languages such as

S. Yamaguchi (✉) · K. Kuramitsu
Yokohama National University, Yokohama, Japan
e-mail: yamaguchi-shinya-tm@ynu.jp; y.shinya.kml@gamil.com

K. Kuramitsu
e-mail: kimio@ynu.ac.jp

© Springer International Publishing AG 2017
D. Król et al. (eds.), *Advanced Topics in Intelligent Information and Database Systems*, Studies in Computational Intelligence 710,
DOI 10.1007/978-3-319-56660-3_37

431

XML/DTD, Relax NG, XML Schema have been accepted to industry standards on the Web.

In terms of data exchange, the significant gap between the earlier Web and the present Web exists in the diversity of data formats. Typically, *open data* has been receiving much attention recently. Open data are exchanged in various formats. Whereas XML format was mainly used on the Web in past days, we get to use various data formats that are represented by lightweight formats such as JSON and CSV.

Despite the intensive development of XML schema validators, many of open data are exchanged without schema validations. In other words, there is no way to ensure the consistency of data exchange in most formats except XML. Schema validations for various data formats are clearly necessary to achieve more reliable open data exchange.

We focus on *parsing expression grammar* (PEG), which forms a basis for the modern parser generator. We use PEG to achieve a schema validation for open data in which various data formats are used. In this paper, we present a novel approach called *schema fusion*, which provides a schema constrained syntax definition by fusing a schema definition and a syntax definition of a data format described in PEG. By applying our approach, we can obtain a generated parser that serves as a schema validator for open data.

The rest of this paper is structured as follows. In Sect. 2, we analyze current states and problems associated with data exchange on the Web. In Sect. 3, we describes the requirements and the approach for open data schema validators, and in Sect. 4, we illustrates details of schema fusion. In Sect. 5, we discuss related works and Sect. 6 provides the conclusion to this paper.

2 Motivation

In this section, we analyze current problem states of data exchange on the Web, especially the tendency of open data. As results of our analysis, we confirm the following three points.

- **The importance of open data is growing.**
- **Schema definitions are toward the standardization.**
- **Formats exchanged on the Web are diversified.**

2.1 Data Application

We start to analyze the difference in data exchange between past and present from the view of data applications.

On the early Web, EDI (Electronic Data Interchange) was mainly used in data exchange. The data exchange using EDI is in accordance with the protocols that are promised between the sending side and the receiving side. That is, the contents of exchanged data are restricted by the protocols. The protocols also restrict the receiver and the domain of the data applications. Although EDI allows us to automate a part of commercial transactions, we are forced to rebuild the systems when changing the protocols.

On the other hand, open data, which has become a trend on the present Web, are provided in standard open text data formats. In other words, open data are exchanged in the format that can be used by anyone without any restriction. In comparison to EDI, because the number of open data users are unspecified, the receiver can determine the domains of data. By using open data freely, new value and business are created continuously. For example, Climate Field View [1], which has been developed by The Climate Corporation, is a realtime information distributing service for agriculture workers. As the open data is spread in future, it is considerable that application area of open data will be expanded.

Additionally, the importance of open data as a social infrastructure is growing [2]. Open data include traffic, water and sewerage, electricity and other public services information. It has attracted attention for use in the applications and the services based on these social open data for maintaining public infrastructures.

2.2 Schema Vocabulary Standard

Schema validation has been used for a long time as a technique for enhancing the reliability of data exchange on the Web.

Schema definitions are described in schema languages. Since we can freely choose vocabularies used in schema definitions, the relationships between instance data and schema definitions can be ambiguous. As a simple example, we assume the case of using vocabulary Name in a schema definition. In this case, except the author of the schema definition, we can not judge whether the Name indicates names of persons or names of any other kind. This is ambiguous because we have no other choice but to infer the author's intention from the context of the data.

To avoid these ambiguities, several vocabulary catalogs for schema definition have been standardized [3–5]. Using standardized vocabularies, we can define schema definitions clearly and judge meanings of data automatically. In this paper, we call such standardized vocabularies *schema vocabulary standard*.

2.3 Format Diversity

There is a lot of format using data exchange on the present Web. Not only traditional text data formats such as XML and CSV but also new formats including RDFa and microdata are widely spread. While text data formats are used mainly, image data formats and the formats depending on specific softwares are still exchanged. These

Table 1 Format-specific number of data in the open data catalogs

Format	DATA.GOV(US)	DATA.GOV.UK	DATA.GO.JP
HTML	74825 (22.0%)	11433 (37.3%)	6286 (29.7%)
XML	42632 (12.6%)	389 (2.7%)	151 (0.7%)
PDF	35082 (10.3%)	1074 (3.5%)	8733 (41.3%)
CSV	12472 (3.7%)	5376 (17.5%)	720 (3.4%)
JSON	10859 (3.2%)	196 (0.6%)	0 (0%)

situations are same in the field of open data. DATA.GOV, which is the governmental open data catalog of US, deals with 49 formats.

Table 1 shows the format-specific number of data in the open data catalogs in US, UK(DATA.GOV.UK) and Japan(DATA.GO.JP) as on February 2016. The formats on the table are Top 5 formats appearing in DATA.GOV. Through the three catalogs, HTML format has a large share of the total. This means that XML format, which has established schema validation techniques, is not mainstream of the data formats. Rather, XML format is not preferred, and instead, CSV and JSON formats are often used. Accordingly, we must support the data format diversity in order to validate open data on schema definitions.

3   Schema Technique for Open Data

In this section, we will show the requirements of schema validation techniques for open data and present our key idea.

3.1   Requirement

The requirements of schema validation techniques for open data are summarized as follows:

1. Conforming to schema vocabulary standard
2. Format independent

"Conforming to schema vocabulary standard" means that all of the vocabularies appearing in schema definitions are defined by using a schema vocabulary standard. This allows us to automatically process the data conforming to the schema definition because there are no ambiguities in the schema definition.

"Format independent" means that schema definitions and these schema validation are not dependent on specific formats. For example, DTD and XML Schema depend on the XML format. As we shown in Sect. 2.3, open data are provided in a variety of formats. As various formats are flooding the present Web, it is not practical to integrate all of the formats into one specific format for both the data providers and the receivers. By using a schema definition that is independent of the format, we can validate the date described in multiple formats for same one schema definition. As a result, the reliability of open data exchange can be improved.

Schema validators for open data are also required its practical performance. Following are the requirements for the implementation.

3. Linear time performance
4. Performing in multiple programming language environments

3.2 Strategy

In order to fulfill the above requirements, we focus on a parser generator. A parser generator is a program that generate a parser from syntax definitions described in formal grammar such as context free grammar (CFG) and parsing expression grammar (PEG) [6]. We particularly focus on PEG because the parser generator based on PEG can satisfy the above requirements 2, 3, 4. PEG has the expressiveness for describing most of the syntax of text data formats appearing in open data. Table 2 summarizes the text formats that can be described in PEG and its number of the productions. Another good point of PEG based parser generators is its linear time performance based on packrat parsing algorithm [7]. Additionally, since there are many implementations in various language environments, we can perform PEG parsers in multiple language environments supported by the PEG based parser generators.

As a simple example, we describe a syntax definition of JSON format in Fig. 1. The productions of PEG are denoted as form $A = e$, where A is a nonterminal and e is a parsing expression. In this example, we can obtain the generated parser accepting JSON data from the syntax composed of 13 productions.

Our key idea is simple. That is, we create a fusion of a syntax definition described in PEG and a schema definition expressed by using schema vocabulary standard. We call this idea *schema fusion*. Concretely, we synthesize a syntax definition embedded schema constraints, and then, we perform a parser generated from the syntax definition as a schema validator. As a good representation of this idea, we can use

Table 2 Syntax definitions of data formats described in PEGs

Format	# of Productions
HTML	28
XML	16
CSV	5
JSON	13
RDFa	21
Microdata	19

```
File      = _ Value _ !.
Value     = String / Number / Object / Array
            / Null / True / False
Object    = '{' Member (',' Member )* '}'
Member    = String _ ':' _ Value
Array     = '[' Value (',' Value )*  ']'
String    = '"' (!'"' .)*  '"' _
True      = 'true' _
False     = 'false _
Null      = 'null' _
Number    = '-'? INT (FRAC EXP? / '' ) _
INT       = '0' / [1-9] [0-9]*
FRAC      = '.' [0-9]+
EXP       = [Ee] ( '-' / '+')? [0-9]+
_         = [ \t\r\n]*
```

Fig. 1 A JSON syntax definition in PEG

existing PEG parser generator implementations. This means that we can reduce the cost of developing schema validators for each data format. Moreover, the PEG based schema validators can support format diversity by fusing one schema definition and multiple syntax definitions in PEG. Additionally, to operate schema validation and parsing at the same time is also a good approach because we can expect reducing the complexity.

4   Schema Fusion

Schema fusion is a technique for rewriting a syntax definition of data formats into a schema constrained syntax definition. We create a fusion of a syntax definition described in Nez, which is one of PEG families formal grammar, and a schema definition described in Celery, which is a syntax independent schema language. We call this output syntax definition *schema constrained syntax definition*. Then, we can do schema validation to parse input text data by using a parser generated from a schema constrained syntax definition. That is, the input data is valid if the parsing is successful, and invalid if the parsing is failed. In the rest of this section, we present the components of schema fusion: schema vocabulary standard Schema.org, schema language Celery, PEG based formal grammar Nez and the algorithm which create a fusion of Celery and Nez.

4.1   Schema.org: Schema Vocabulary Standard

We use Schema.org as a schema vocabulary standard of schema fusion. Schema.org [5] is a representative schema vocabulary standard established by Google, Microsoft and Yahoo in 2011. Schema.org has been adopted by the search engines of Google and Yahoo.

The schema definitions of Schema.org are represented as sets of `type`. A `type` is constructed based on multiple `property`. A `property` has other `type` or primitive data type in Schema.org such as `Boolean` and `Number`. In addition, a `type` can be defined as a union type, which comprises multiple `type`. In Schema.org, these types are defined hierarchically as a tree structure.

4.2   Celery: Schema Definition

In schema fusion algorithm, we introduce Celery, which is a schema language conforming to Schema.org, as the language for describing schema definitions. Celery is a schema language that is independent from data format syntax. In Celery, a schema definition is called **type**. The syntax of Celery is defined in Fig. 2.

A **type** is constructed of multiple **type** declarations. A **type** declaration nominates a type of Schema.org and specifies the properties. Where `TypeName` and `propertyName` correspond to a type and a property name defined in Schema.org. A type definition for a property is given by `Type`. Schema type applies other types defined by **type** declarations and primitive type. `#Primitive` includes the basic data types of Schema.org (http://schema.org/DataType) such as Integer, String, etc.

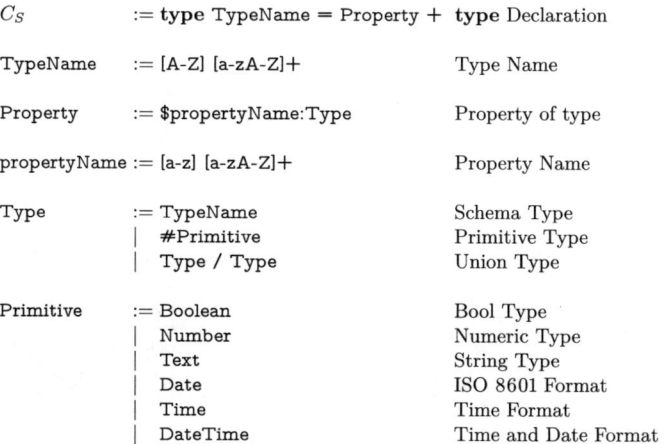

Fig. 2 Syntax definition of Celery

A union type matches either of the specified two types. Celery has equivalent expressiveness to Schema.org by describing schema definition in these syntaxes.

A **type** declaration only constrains specified properties. That is, the properties that are not specified will be ignored when the schema validation. If a type definition of a property conflicts with the definition in Schema.org, the properties in the **type** declaration take priority.

Following is an example of a schema definition denoted Person type (http://schema.org/Person) that is described in Celery syntax.

```
type Person {
   $name:#Text
   $birthDate:#Date
   $adderess:PostalAdress / #Text
}
```

4.3   Nez: Syntax Definition

We describe a syntax definition of a data format by using Nez, which is a PEG-based pure declarative syntax foundation [8]. One of the reason is that Nez has the extended operators for constructing *abstract syntax tree* (AST). We map a schema definition defined by Celery into syntax definition described in Nez by corresponding the AST constructions of Nez and the tree structure of Schema.org. As a result, we can obtain not only the validation results but the AST typed by the schema definition. Moreover, Nez satisfies the requirements denoted in Sect. 3.1 because Nez provides the parser generators for multiple language environments, the grammar translators for other PEG-based tools and the parser optimizer.

Tables 3, 4 shows a part of the Nez operators [8]. Note that the PEG operators shown in Table 3 are equivalent to Ford's definitions [6].

Table 3 PEG Operators

PEG	Description
'*abc*'	String
[*abc*]	Character class
.	Any character
A	Nonterminal application
(p)	Grouping
$p?$	Option
$p*$	Zero or more repetition
$p+$	One or more repetition
$\&p$	And predicate
$!p$	Negation
$p_1 p_2$	Sequence
p_1/p_2	Prioritized choice

Table 4 Nez extended operator

AST	Description
{e}	Tree constructor
$*lable*(e)	Subtree constructor
{$ e}	Folding constructor
#t	Tagging
' '	Replaces a string

Annotations for Syntax Definitions In schema fusion algorithm, we annotate the productions of data formats by using Nez AST operators. Based on these annotations, we map a **type** declaration of Celery into the productions of a syntax definition. Note that while **type** declarations construct tree data structures, syntax definitions of data formats do not necessarily have a tree data model. For example, the CSV format represents data as relational model. CSV data are only composed of values and delimiters. Thus, we cannot map a **type** declaration directly into the syntax definition of CSV because the properties of **type** declaration have a key and a value. Hence, we introduce the following two types of annotations into a syntax definition.

- #object: Annotated syntax is composed of key and value
- #list: Annotated syntax is only composed of values.

These annotations correspond to the tagging of tree constructors in Nez. For instance, when a production A is represented in A= {e #object}, the production A is judged as an #object type syntax. An #object annotated production has $prop labeled expressions that represent properties. Moreover, a $prop labeled expression has expressions that correspond to $key and $value. On the other hand, the productions that are annotated #list only have $value labeled expressions. In this method, we only have to annotate into existing syntax definitions. That is, we don't need to modify the component of the syntax definitions. By this way, we can map schema definitions into syntax definitions of any data formats.

Table 5 shows results of attaching #object and #list annotation into the syntax definitions appearing in Table 2. There is no change on the number of productions in every syntax definition.

Finally, we describe the syntax definition of JSON format redefined by AST operators and annotations in Fig. 3. In comparison to Fig. 1, we can add AST constructions to the productions by merely adding slight changes.

4.4 Fusion Algorithm

The fusion algorithm that creates a fusion of Nez and Celery operates in the following three steps.

Table 5 Syntax definitions in Nez grammar

Format	# of Productions	# of $object	# of $list
HTML	28	2	1
XML	16	2	1
CSV	5	0	1
JSON	13	1	1
RDFa	21	1	0
Microdata	19	1	0

```
File     = _ Value _ !.
Value    = String / Number / Object / Array
         / Null / True / False
Object   = { '{' $prop(Member) (',' $prop(Member) )* '}' #object}
Member   = { $key(String) _ ':' _ $value(Value) #Member}
Array    = '[' { $value(Value) (',' $value(Value) )*  #list} ']'
String   = '"' { (!'"' .)*  #String }'"' _
True     = { 'true' #True} _
False    = { 'false #False } _
Null     = { 'null' #Null } _
Number   = { '-'? INT (FRAC EXP? #Float / " #Integer) } _
INT      = '0' / [1-9] [0-9]*
FRAC     = '.' [0-9]+
EXP      = [Ee] ( '-' / '+')? [0-9]+
_        = [ \t\r\n]*
```

Fig. 3 A JSON syntax definition in Nez

1. *Specifying*: Specifying productions that embed schema definitions by analyzing a syntax definition of target data format.
2. *Deploying*: Deploying properties defined in schema definitions into specified productions.
3. *Merging*: Merging deployed productions into specified syntax definition and generating new productions for each Celery **type**.

In this subsection, we illustrate each step of the algorithm by example to create a fusion JSON syntax and the following schema definition.

```
type Person =  $name:#Text $birthDate:#Date
```

Specifying We start to specify target productions from annotations attached in a syntax definition. That is, we find productions that have #object or #list tag in the syntax definition. In JSON syntax definition, the production Object is specified because it has #object tag.

```
Object = {'{' $prop(Member) (',' $prop(Member)*)'}' #Object}
```

Deploying Next, we deploy properties defined in schema definitions into `$prop` labeled parsing expressions in the specified production. In the above example, the parsing expressions mapped by nonterminal symbol `Member` are deployed.

```
Member = {  $key(String)":" $value(Value) #Member}
```

The property deployment is executed by *deploy* functions of each annotations. In the case of `#object`, the data are composed of key and value. Thus the function is executed to rewrite parsing expressions has labels `$key` and `$value`. In an expression labeled `$key`, a name of a property is embedded directly. A type of property is deployed recursively into an expression labeled `$value`. If the type is a schema type or union type, the function continues to apply the fusion algorithm as long as the type is not a primitive type. The following represent examples of rewritten productions by embedding `Person` type into `Member`.

```
Member#Person$name =  '"name"' ":" $name(Text)
Member#Person$age =  '"birthDate"' ":" $birthDate(Date)
```

Note that `Text` and `Date` productions representing the primitive type of Celery are added into the syntax definition since these types don't exist in the original JSON syntax.

Additionally, in the case of `#list` annotation, the function only deploys `$value` because there are no expressions representing `$key`.

Merging Lastly, we merge productions generated in previous step into the production specified in *Specifying* step. That is to say, in this step, we generate a new production that accepts only input constrained by a schema definition.

```
Member#Person = Member#Person$name
        / Member#Person$birthDate
```

Since a syntax annotated by `#object` has a key, an occurrence of properties is unordered. Indeed, the properties such as XML tags and JSON object have no particular order except that there are restrictions of schema definitions. Therefore, we represent `Member` production as a choice (`'/'`) of productions deployed schema definitions. The above production can accept any order of `Member#Person$name` and `Member#Person$birthDate`. On the other hand, in the case of `#list`, we can not specify each property without order restriction. Hence, productions of `#list` are represented by sequence, instead of choice.

Finally, fusion algorithm halt after merging `Object` production and `Member #Person` is as follows:

```
Object#Person =
   { '{' Member#Person (',' (Member#Person)* '}' #Person}
```

5   Related Work

Not only Nez, there are other practical parser generators including ANTLR [9, 10] and Yacc [11]. As these tools have been developed for use in constructing programming language processors, there is no instance to use the parser generator as a schema validator generator. Our schema fusion technique is a first attempt to a apply parser generator technique into schema validation. The extensions of Nez are considered to suit schema validation. In future work, we will support additional semantic schema constraints such as *uniqueness* constraint and *n* times repetition of a property that are supported in common schema languages of XML.

6 Conclusion

In this paper, we present *schema fusion*, which is a schema validation technique for open data. Schema fusion supports diversity of data format by combining a schema definition and a syntax definition represented in PEG. In future work, we will show the extensions for PEG that are required to represent semantic schema constraints such as a uniqueness constraint.

References

1. Government, U.: Climate Field View—DATA.GOV. https://www.data.gov/applications/climate/ (2010)
2. Jaakkola, H., Mäkinen, T., Eteläaho, A.: Open data: Opportunities and challenges. In: Proceedings of the 15th International Conference on Computer Systems and Technologies, pp. 25–39. CompSysTech '14, ACM, NY, USA (2014). doi:10.1145/2659532.2659594
3. Dan Brickley, L.M.: Foaf vocabulary specification 0.99. http://xmlns.com/foaf/spec/ (2014)
4. IPTC: Rnews Embedding Metadata in Online News. https://iptc.org/standards/rnews/ (2013)
5. Schema.org Community Group: Schema.org. https://schema.org (2012)
6. Ford, B.: Parsing expression grammars: a recognition-based syntactic foundation. In: Proceedings of the 31st ACM SIGPLAN-SIGACT Symposium on Principles of Programming Languages, pp. 111–122. POPL '04, ACM, NY, USA (2004). doi:10.1145/964001.964011
7. Ford, B.: Packrat parsing:: Simple, powerful, lazy, linear time, functional pearl. In: Proceedings of the Seventh ACM SIGPLAN International Conference on Functional Programming, pp. 36–47. ICFP '02, ACM, NY, USA (2002). doi:10.1145/581478.581483
8. Kuramitsu, K.: Nez: practical open grammar language. Tech. Rep. (2015). arXiv:1511.08307
9. Parr, T., Fisher, K.: Ll(*): The foundation of the antlr parser generator. In: Proceedings of the 32Nd ACM SIGPLAN Conference on Programming Language Design and Implementation, pp. 425–436. PLDI '11, ACM, NY, USA (2011). doi:10.1145/1993498.1993548
10. Parr, T., Harwell, S., Fisher, K.: Adaptive ll(*) parsing: The power of dynamic analysis. In: Proceedings of the 2014 ACM International Conference on Object Oriented Programming Systems Languages & Applications, pp. 579–598. OOPSLA '14, ACM, NY, USA (2014). doi:10.1145/2660193.2660202
11. Johnson, S.C.: Yacc: Yet Another Compiler-Compiler, vol. 32. Bell Laboratories Murray Hill, NJ (1975)

Job Description Language for a Browser-Based Computing Platform—A Preliminary Report

Arkadiusz Danilecki, Tomasz Fabisiak and Maciej Kaszubowski

Abstract In this paper we report on our work-in-progress on a new job description language intended for use in a browser-based voluntary computing platform. The language has workflow-control features, will enable the automatic data distribution and allows job creators to react to special events and failures.

Keywords Volunteer computing · Browser-based computing

1 Introduction

The total computing power of machines in private possession of regular users is hard to estimate, but in Poland alone it must be in range of tens of petaflops. Most of the time this immense computing power is unused. Voluntary computing make it possible for users to share their resources with scientists. Unfortunately, projects such as BOINC [2] require users to download and install a specialized software. As a result, projects based on those two frameworks attracts almost exclusively males, most of which have higher-than-average technical expertise [6].

An alternative approach to voluntary computing takes advantage of the popularity of World Wide Web [1, 3, 5, 21]. The basic idea is to employ user's browsers to execute scripts solving scientific problems. Such scripts could be injected to pages served by a website. The users browsing the site may be not even aware that they participate in a computation, effectively trading their computing resources in exchange for the site's contents. This approach has the potential to reach to vastly more diverse group of users.

A. Danilecki (✉) · T. Fabisiak · M. Kaszubowski
Institute of Computing Science, Poznań University of Technology, Poznań, Poland
e-mail: adanilecki@cs.put.poznan.pl

T. Fabisiak
e-mail: fabisiak.tomek@gmail.com

M. Kaszubowski
e-mail: maciej.kaszubowski94@gmail.com

© Springer International Publishing AG 2017
D. Król et al. (eds.), *Advanced Topics in Intelligent Information and Database Systems*, Studies in Computational Intelligence 710,
DOI 10.1007/978-3-319-56660-3_38

443

We have already created a prototype of the browser-based computing environment [9]. This paper contributions consists of design of the job description language which will allow the environment to automatically distribute the input data. We have developed the parser for our language and we are in process of implementing the missing server-side functionality.

The rest of paper is structured as follows: in Sect. 2 we briefly describe the envisaged architecture for our platform. Our new contribution is contained in Sect. 3, where we describe our new job description language. The related work on browser-based and voluntary computing is outlined in Sect. 4. Section 5 concludes the paper.

2   PovoCop Architecture

In this section we describe the architecture of Poznań Open Voluntary Computing Platform (PovoCop) (Fig. 1 "PovoCop envisaged architecture"). We provide only the basic outline needed for understanding our main contribution presented in Sect. 3; more details are in [9]. The subset of the architecture is implemented in a prototype,[1] written in JavaScript and using node.js [19]. The final version will be implemented in Elixir using Phoenix framework [16].

We start by introducing the *job template* and *instance* concepts. A job template is a passive object, a program, a description of a problem. A job instance is an active object, a running distributed computation, solving a problem. Jobs must be divisible into *tasks*: computationally independent units of execution.

We differentiate five roles: *Job owners* are responsible for job creation and submission. They are interested in job status and results. *Plantations* are worker nodes—browsers on user machines. Within each Plantation, many *Worker* threads may run, each executing a single task. *Recruiters* are HTTP servers visited by users. A Recruiter should inject a simple bootstrap JavaScript code into the pages it serves. *Schedulers* maintain jobs: distribute tasks to Plantations, monitor job status etc. Finally, *Directories* function as brokers between Plantations and Schedulers. The first prototype implements only a single Scheduler, communicating directly with browsers.

We will now describe a typical workflow in our envisaged architecture. First, a Job Owner submits a job to *Schedulers* (see (1) at Fig. 1 "PovoCop envisaged Architecture"). Job is packed into a *jobundle*, a .zip archive with a particular structure. A Scheduler registers at Directory (step 2 at Fig. 1). and stores there a set of initial tasks (one for each job). Later, a user visits one of the Recruiters' sites. The browser loads a page and executes bootstrap code (3), which sends a request for a JavaScript code from a Directory (4)—at this moment user's browser becomes a Plantation node. The directory sends back a task, and address of Scheduler responsible for the job (5). After finishing the task, Plantation sends the results to the Scheduler, and receives a new task to execute (6).

---

[1] available at https://github.com/Vatras/BrowserComputing.

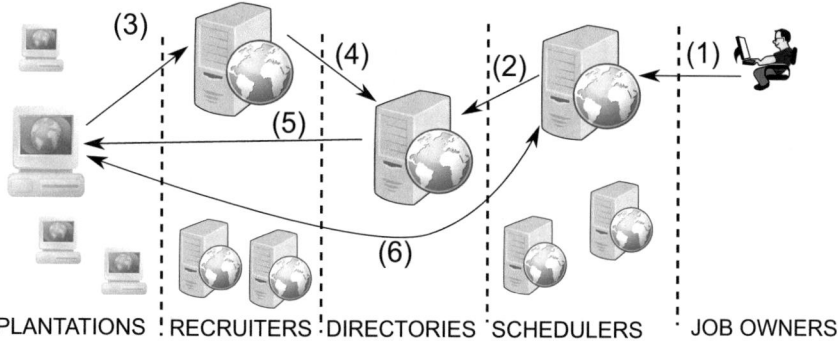

Fig. 1 PovoCop envisaged architecture (uses images from openclipart.org)

3 PovoCop Job Description Language

In this section we will describe the details of the PJDL. We will start by motivating creation of the language. We will proceed by introducing the basic syntax. Then we will explain how to use PJDL to describe the resources, scripts used by jobs, event handling and data distribution.

3.1 The Introduction

In our first prototype a job creator had to explicitly implement both the data distribution and the result merging, writing the scripts on both the server- and the client-side. Moreover, when discussing the design of PovoCop we realised that there may be many commonly shared, simple and repetetive tasks. Therefore, we decided to create a simple language to separate exception handling from application logic, help the data distribution, automate repetetive tasks, while still allowing the job creators to implement required functionality themselves.

In the final version, the application logic (set of algorithms solving a particular problem) will be implemented in JavaScript, while PJDL will be used to describe (1) from where the input data should be taken, (2) how should the data be distributed, (3) when to run which script or another action (such as sending HTTP request to a specified URL), (4) how to react to failures, and (5) how to combine the results.

With this goals in mind we designed PovoCop Job Description Language (PJDL). We started writing our first skeleton parser in C++, using civetweb framework.[2] Then we rewrote the skeleton code in Elixir functional language (see Sect. 3.7 for implementation details). The skeleton used for comparison allowed users to post .zip archive at a specified address. The server then unzipped the archive and parsed the

---

[2] Available at https://github.com/szopeno/skeleton-parser-civetweb/.

`main.pjdl` file (at that point only few statements were supported by the skeleton parser). The C++ code had 1590 lines of code, of which parser+lexer had 1331 lines. A comparable Elixir code had 850 lines (not counting page templates, and most of them autogenerated), of which parser+lexer had 429 lines of code. The Elixir skeleton parser used in comparison was written by one of us the very first day he started learning the language. We tested both parsers using the following command line:

```
time for i in 'seq 1 100'; do
    curl --form file[title]="simple" \
    --form file[archive]=@simple_jobundle.zip \
    http://localhost:4000/file;
done
```

The Elixir version was more than 13 times faster (8 s vs. 106^3). We were impressed enough to complete the parser in Elixir.

We stress once again that what follows is a report on a work-in-progress, different from our node.js-based JavaScript prototypes. So far, we have implemented the parser for our proposed language. Unlike in previous prototypes, none of the server functionality (e.g. task distribution, event handling etc.) is completed in an Elixir-based version.

3.2 Jobundles

Jobs will be submitted to PovoCop as *jobundles* (job bundles): .zip archives with a particular structure. A jobundle should contain a `main.pjdl` file with a job description, optional `index_template.html` to be used as a part of a job description on a website, and two folders: `resources` and `scripts`. The former with input data for a job, the latter with scripts used by a job. After a jobundle is submitted, a *job template* is created. In order for a job to become available for distribution in PovoCop, a *job instance* have to be created. Job instances may be created manually using Restful interface, or automatically by an environment, depending on the `main.pjdl` contents.

3.3 The Basic Syntax

Job description should be written into `main.pjdl` file. Within the file, each line should have the following format:

```
keyword [<list-of-attributes>] [<list-of-names>]
```

---

[3]Probably result of our lack of familiarity with civetweb. Unzipping the archive and parsing took less than 0.003 s per request, the bottleneck was *mg_handle_form_request* function.

```
job
    short "Example job description"
    results_visibility .public
    results "http://www.cs.put.poznan.pl/adanilecki/results"

    resources
        instance
            vector "input" embedded "input.dat"

    activity
        condition
            completed "goal"
        http put "http://site.com/file"
            from value "started"
```

Fig. 2 An example PJDL file

A list of attributes may contain one or more string *attributes*. Each *name* is a string within paranthes. A line in this format will be called a *node*.

Each *node* has associated indentation level, i.e. number of whitespaces used to indent the line. A node with indentation level n becomes a parent to following nodes with higher indentation level. A parent node **NamedNode** with its children form an *NamedNode section*.

Sections may be written more concisely by joining the lines in form of:

```
<parent node> = <node 1>, <node 2>
```

This is equivalent to:

```
<parent node>
        <node 1>
        <node 2>
```

A top-level node must consist in a single keyword **job**, with no attributes or names. All other nodes are children of a **job** node. The **job** children nodes are divided into three groups: basic job description, resource description, and workflow description.

The Fig. 2 "Example PJDL file" provides a taste of PJDL syntax. In the following sections we will describe syntax in more detail, explaining the parts of the PJDL files, meaning of the keywords and their possible attributes.

3.4 Basic Job Description

The job creator should provide a short job description in form of a **short** node, with no attributes and with a single *name*. Two more lines are optional: **results_visibility** allows job creator, by using a proper attribute, to declare whether results should be public (**.public**), visible to registered user contributors (**.contributors**), visible to the particular user group (**.team**), or visible only to job creator (**.owner**). By default, public visibility is assumed, as we think that by making results publicly accessible a job is more likely to attract volunteers willing to share their resources. A job owner may further declare where the final result should be stored, using **results** keyword with a name containing the destination URL for the final results. Finally, by using **instantiation** with attribute **.instant** job creator can declare job to be instantiated immediately after submission.

3.5   Resource Description

Resources. A **resources** node describes the input data for the job, with either **instance** or *resource* children nodes. The latter should start with one of the following keywords: **vector**, **array**, **cube**, **raw**, **folder**. The keyword declares the resource type. The first three are, respectively, one-, two- and three-dimensional arrays. The **raw** declares the resource will be handled by the scripts provided by the job creator. A **cube** is a three-dimensional array. A **folder** describes a resource consisting of files. Each of the keywords may have additional attributes. First two attributes describe the resource location: **embedded** indicates that resource is included within jobundle, a **remote** attribute states that resource should be downloaded from remote url on instantiation. Four additional attributes declare behavior on job instantiation: **.copy** indicates the resource should be copied from job template to job instance, **.shared_rw** indicates the resource will be shared read-write between the job instances, **.shared_ro** declares the resource will be shared read-only between job instances, and finally .**ask** indicates that a job creator will be presented with a form to upload a resource on instantiation (a resource will become private to a job instance).

Each resource must have a single *name*. It may have (depending on the attributes) the children nodes describing the dimensions of the resource and its location.

The **instance** node will be used during instantiation. It may have many *resource* children nodes. On instantiation, system will automatically create more than one job instances, depending on the number of **instance** nodes, by automatically combining the input data from different **instance** sections. With two **instance** nodes, one with n resource nodes and second with m resource nodes, system will first create m instances each with first resource from the first section and one of the resources from the second section, then it will create m job instances, each with second resource from the first section and one of the resource from the second section—repeating the process until it will create $n \times m$ job instances. If no **instance** nodes are present, system assumes it should create only one job instance on instantiation.

Scripts. A **scripts** node declares the scripts to be used. Each node starts with a **script** keyword and may have either **.embedded** or **.remote** attributes. The former indicates the script is included within a jobundle, while the latter states that script must be copied from a remote location on instantiation. We envisage that scripts could be written in any language which could be compiled into LLVM bytecodes. Scripts will have to follow several general guidelines, to be specified in near future.

3.6 Workflow Description—Activities

Each activity starts with **activity** keyword, with one of the three possible attributes indicating when activity should be run. Activity is executed when a condition specified within a child **condition** node is true. A **condition** node can have **and**, **or**, **xor** or **not** subsections (which can be nested) and can test whether a particular goal is met (**completed** "named_goal"), whether an url exists (**exists** "url"), and whether two resources match or contain a particular value (**equals**). Activity with no condition is run on init.

An activity may contain three possible actions: **http**, **run** and **distribute**. The **http** action will allow to run http requests (with attributes **put**, **get**, **post** and **delete**) on specified "url address" *name*, possibly with data fetched from **from** url, with result posted to url specified in **result** node. The **run** will run a specified JavaScript script, either in asynchronous mode, or waiting for its completion. The most important action will be **distribute** "task.js", which will describe the tasks to be sent to browsers connecting to PovoCop platform.

Activities will allow job creators to design complicated workflows, executing remote scripts, handling the failures, waiting for a new input data and so on. As such, the PJDL could be also thought as a simple workflow language.

Data Distribution. The **distribute** section describes how the input data will be divided and distributed between the tasks. A **distribute** should have an **instance** child nodes, each with the **resource** children nodes. A **resource** can have attributes indicating that the data should be injected into scripts sent to browser (**inject**), sent via WebSockets (**.websockets**, the default), the script will get the data from a specified url (**get**). The **raw** resources distribution must be manually handled by job creators. In case of vectors, the system will distribute the data by dividing vector into disjoint ranges, in arrays it will divide the data into rows, in cubes it will divide it into slices. The folder resources will be distributed based on files.

An **instance** section is used for a task distribution, used analogously to **instance** sections in job instantiation described above. For example, in the following **distribute** section:

```
distribute
        instance
                resource"A"
                resource  "B"
```

Assume "A" and "B" are arrays declared earlier in **resources** section and both have *n* rows. In that case system will create at least *n* tasks (the first with the first rows from "A" and "B", the second with second rows from "A" and "B" and so on). If, however, a job creator would put those resources into two separate **instance** sections, system would create at least $n \times n$ tasks instead; first n tasks with the first row of the A array and subsequent rows of the "B" array, then n tasks with the second row of the "A" array and so on. The system should create more tasks than needed, introducing redundancy; that this, for each chunk of input data, many tasks should be created acting on the same data. This will be used in result verification and to increase the fault tolerance.

Results Handling. A **distribute** node should have **results** section. Within the section three children nodes may be present. The **url** specifies the address at which partial results of the activity will be stored. The **combines** allows the job creator to specify how to handle partial results. The **.self** attribute with "script" *name* states that the result merging will be handled by the specified script. Other possible attributes are **.min**, **.max**, **.avg**, **.sum**, **.nomerge**, **.vector**, **.array**. The **.nomerge** means that results are left as they are. The **.vector** joins the results into a single vector. The **.array** puts each result into a separate row of result array.

Finally, a job creator can declare the method of the result verification using attributes **.none**, **.majority_voting**, **.weighted_voting**, **.spot_checking**, **.self**. In case of **.self** a *name* of script used for result verification should be given. For majority voting, results are deemed correct if majority of redundant tasks (with same input data) returns the same result. In weighted majority, the *plantations* which returned correct results in past will be given more weight during voting. In spot checking, the system will occasionally sent tasks for which the result is already known.

Event Handling and Fault Tolerance. The **distribute** node may have **events** section. Within this section, job creator may specify reactions to particular events: **task_completed** (a single task has successfully ended), **job_completed** (all tasks have successfully ended), **task_failed**, **verification_failed**, **job_failed**, **task_started**, **job_started**. For each event, a reaction may be one of **run** a script, sending a **http** request, **complete** a named goal, **retry** (the task), **fail** (stop the job), **wait_and_retry** (retry the task after a delay).

3.7 The Implementation

The elixir code of the parser is provided at github.[4] It has roughly 1700 lines (not counting the dependencies and page templates). To test it, one should first get the dependencies (including elixir, erlang packets, node.js). Note node.js is not actually run, it is used by Phoenix only for asset management. Next, one should run *mix deps.get* and *npm install* commands. The server is run by executing *mix phoenix.server*. With a parser standalone, one should run *mix escript.build* and then

---

[4]parser+http server is at https://github.com/mkaszubowski/pjdl_parser- parser standalone is at https://github.com/szopeno/pjdl_standalone_elixir.

run *./lexer ex.pjdl*, or, preferably, *iex -S mix* and then within iex console issue the command *MyApp.parse "ex.pjdl"*.

Jobundles should be uploaded at `http://localhost:4000/file`. The submitted jobs can be inspected at `http://localhost:4000/uploads`. At the time of submission, using web interface one can inspect and delete the uploaded jobs, with a possibility of editing the simple fields like job description.

4 Related Work

The first successful attempt at voluntary computing was GIMPS (Great Internet Mersenne Prime Search) started in January 1996 [22]. The most known, however, was Seti@Home, officially started in 1999. Based on experiences gathered during Seti@Home project, a team from Berkeley university created BOINC (Berkeley Open Infrastructure for Network Computing) [2]. BOINC is a common infrastructure for distributed voluntary computing. Currently there are at least 57 projects based on BOINC, with close to half million volunteers and more than nine hundred thousands computers, contributing on average 10.5 PetaFLOPS.

A concept related to voluntary computing is so called "citizen science" or "crowd computing". In Zooniverse [18] humans provide their unique skills (such as pattern recognition) to solve problems for which there is no known fast algorithmic approach (e.g. classifying wild bees captured by motion cameras jungle near Serengeti). Similar approach is *Games With a Purpose* (GWAP) such as FoldIt [10], where scientific problems are presented to users as puzzles.

Using browsers to enable users' participation in voluntary computing is not a novel idea. To our best knowledge it was first proposed in 1996 in a Charlotte system [4]. The attempts at utilizing browsers in voluntary computing may be roughly divided into the three generations: the first was based on Java applets, with proposals such as DAMPP [20] or Bayanihan [17]; the second, starting in 2007, includes approaches based on single-threaded JavaScript [5, 11, 12]; and finally, within last few years many advanced solutions and proposals appeared, usually employing Web-Workers standard, exposing Restful interface and often based on node.js. (e.g. [7, 8, 13–15], see our survey [9] for more examples). Most of those proposals share many similarities. They usually distinguish three roles in the architecture: the creators of the original code to be executed, solving a problem, the user machines voluntary computing resources, and the servers acting as brokers between job creators and volunteers.

5 Conclusions

In this paper we have presented a snapshot of our ongoing work on PovoCop platform, which will be used to facilitate users' participation in voluntary computing by employing the common browsers. We have described the design of the job description language and reported on the current state of its implementation. As far as we know, we are the first to propose a separate job description language for a browser-based computing environments with workflow-control features. Our initial skeleton prototype is written in elixir and is freely available on github.

In future we will continue the work on our prototype, both based on node.js and on elixir/phoenix.

References

1. Alexandrov, A.D., Ibel, M., Schauser, K.E., Scheiman, C.J.: SuperWeb: towards a global web-based parallel computing infrastructure. In: 11th International Parallel Processing Symposium, pp. 100–106. IEEE Press (1997)
2. Anderson, D.P.: BOINC: a system for public-resource computing and storage. In: 5th IEEE/ACM International Workshop on Grid Computing, pp. 4–10. IEEE Press (2004)
3. Baratloo, A., Karaul, M., Kedem, Z.M., Wijckoff, P.: Charlotte: Metacomputing on the web. Future Gener. Comput. Syst. **15**(5–6), 559–570 (1999). Oct
4. Baratloo, A., Karaul, M., Karl, H., Kedem, Z.M.: An infrastructure for network computing with java applets. Concurr. Pract. Exp. **10**(11–13), 1029–1041 (1998)
5. Boldrin, F., Taddia, C., Mazzini, G.: Distributed computing through web browser. In: 2007 IEEE 66th Vehicular Technology Conference, pp. 2020–2024. IEEE Press (2007)
6. Cusack, C., Martens, C., Mutreja, P.: Volunteer computing using casual games. In: Future Play 2006 International Conference on the Future of Game Design and Technology, pp. 1–8 (2006)
7. Czarnul, P., Kuchta, J., Matuszek, M.: Parallel computations in the volunteer-based comcute system. In: Wyrzykowski, R., Dongarra, J., Karczewski, K., Waśniewski, J. (eds.) PPAM. LNCS, vol. 8384, pp. 261–271. Springer, Heidelberg (2013)
8. Duda, J., Dłubacz, W.: Distributed evolutionary computing system based on web browsers with JavaScript. In: Manninen, P., Öster, P. (eds.) PARA. LNCS, vol. 7782, pp. 183–191. Springer, Berlin/Heidelberg (2013)
9. Fabisiak, T., Danilecki, A.: Browser-based harnessing of voluntary computational power. Technical Report RA-11/16, Poznań University of Technology, Poznań (2016)
10. Khatib, F., Cooper, S., Tyka, M.D., Xu, K., Makedon, I., Popović, Z., Baker, D., Players, F.: Algorithm discovery by protein folding game players. Proc. Natl. Acad. Sci. **108**(47), 18949–18953 (2011)
11. Klein, J., Spector, L.: Unwitting distributed genetic programming via asynchronous JavaScript and XML. In: Proceedings of the 9th Annual Conference on Genetic and Evolutionary Computation. pp. 1628–1635. GECCO'07, ACM, New York, NY, USA (2007)
12. Konishi, F., Ohki, S., Konagaya, A., Umestu, R., Ishii, M.: RABC: A conceptual design of pervasive infrastructure for browser computing based on ajax technologies. In: 7th IEEE International Symposium on Cluster Computing and the Grid, pp. 661–672. IEEE Press (2007)
13. Martínez, G.J., Val, L.: Capataz: a framework for distributing algorithms via the world wide web. CLEI Electron. J. **18**(2) (2015)
14. Meeds, E., Hendriks, R., Al Faraby, S., Bruntink, M., Welling, M.: Mlitb: machine learning in the browser. PeerJ Comput. Sci. **1**, e11 (2015)
15. Merelo-Guervos, J.J., Castillo, P.A., Laredo, J.L.J., Garcia, A.M., Prieto, A.: Asynchronous distributed genetic algorithms with javascript and json. In: 2008 IEEE Congress on Evolutionary Computation (IEEE World Congress on Computational Intelligence), pp. 1372–1379. IEEE Press (2008)
16. Pereira, P.A.: Elixir Cookbook. Packt Publishing Ltd (2015)
17. Sarmenta, L.F.: Bayanihan: Web-based volunteer computing using java. In: Masunaga Y., Katayama T., Tsukamoto M. (eds.) Worldwide Computing and Its Applications, WWCA'98, LNCS, vol. 1368, pp. 444–461. Springer (1998)

18. Simpson, R., Page, K.R., De Roure, D.: Zooniverse: Observing the world's largest citizen science platform. In: 23rd International Conference on World Wide Web, pp. 1049–1054. WWW'14 Companion, ACM, New York, NY, USA (2014)
19. Tilkov, S., Vinoski, S.: Node.js: using javascript to build high-performance network programs. IEEE Internet Comput. **14**(6), 80–83 (2010)
20. Vanhelsuwe, L.: Create your own supercomputer with Java. JavaWorld (January 1997). http://www.javaworld.com/jw-01-1997/jw-01-dampp.ibd.html
21. Wilkinson, S.R., Almeida, J.S.: QMachine: commodity supercomputing in web browsers. BMC Bioinform **15**(1), 1 (2014)
22. Ziegler, G.M.: The great prime number record races. Not. AMS **51**(4), 414–416 (2004)

Measuring the Effectiveness of Knowledge Driven Web Applications

Supriya Chakraborty, Novarun Deb and Nabendu Chaki

Abstract Today, web applications are often considered as real-life reflections of enterprises. Patrons do invest large finances. Besides, lot of time and effort are put in towards development and maintenance of the web applications. The effectiveness of such web applications is an indicator for both further investment and refinement of existing content, services and human resources. In this work, a new approach is proposed to quantify the effectiveness of knowledge driven web application. This is named as Unique Track Measure Orientation (UTMO). The proposed UTMO computes the effectiveness from navigation of different types of information in accordance with the business value. The business value could be anything-the brand value, turnover, or cumulative profit.

Keywords Effectiveness · Metric · Quantification · Business value · Trial and error · Visibility

1 Introduction

In recent past, significant investment in terms of time, effort and money is quite a common phenomenon towards development and maintenance of web application. Consequently, works for measuring effectiveness of these web applications using quantifiable metrics has gained attention. The web applications may be categorized as "transaction oriented" *e.g., Amazon, dishtv, ticketbooking* and as "driven by knowledge" *e.g., institute web application, social networking site*. The measurement of

S. Chakraborty
Greater Kolkata College of Engineering and Management, Dudhnai, India
e-mail: supriyachakraborty@acm.org

N. Deb · N. Chaki (✉)
University of Calcutta, Kolkata, India
e-mail: nabendu@ieee.org

N. Deb
e-mail: novarun@acm.org

© Springer International Publishing AG 2017
D. Król et al. (eds.), *Advanced Topics in Intelligent Information and Database Systems*, Studies in Computational Intelligence 710,
DOI 10.1007/978-3-319-56660-3_39

455

effectiveness in accordance with business value of transaction oriented web application is relatively simple. The same is not true for knowledge driven web applications. The momentum of social networking sites, content network in cloud, and recent globalization of educational institutes adds to the challenges.

A total of thirteen (13) propositions including age, orientation of interpretation, psychology, religion, sex, age and politics for designing text and graphics of web application are considered in [1]. The study is quite exhaustive and presents deep insights from diverse perspectives that could be used towards assessing web applications. However, no experimental validation has been reported in [1]. In the recent past, web analytics has been performed to provide better insight of the web applications. A comprehensive content has been observed in [2] that has described contemporary technology issues of web analytics. However, metrics of measuring such effectiveness has been found in [3–6].

Broadly, such attempts have three different steps. The first step is regarding elicitation of data which is captured from log server, client side, and other channels (sales data). The second step is all about determination of data or grouping of data that influence the objective of the web application. The last step is proposing methodologies, formulations, preparing data and experiments that measure the effectiveness of the web applications.

In [4], a metric called, *guidance performance indicator* (GPI) has been proposed by considering transition type, weight, and efficiency. Advancement by *Kano's domain quality index* has been proposed over GPI with five (5) quantifiable quality parameters in [3]. Besides, [5] and [6], the buying processes have been described. As per significance, pages have been considered as *Action Page*, and *Target Page* (Payment). The successful execution from action page to target page has been called *conversion*. Metrics including *contact efficiency, relative contact efficiency and conversion efficiency* have been defined and exemplified. Besides, the navigation from one page to another page has been formulated by *confidence* metric.

In this context, few tools are studied such as EtailQ [7], Webqual-4 [8], Sitequal [9], and NetQual [10] which have been used for evaluation of websites effectiveness. The WebQual-4 has used metrics as *mean, standard deviation, standard error from mean, skewness, and kurtosis*.

The above study reveals that a web page may have multi-variant data. The navigation of users is measured for different types of pages. Therefore, existing approaches measures the effectiveness with higher abstraction. Thus, the significance of navigations for clients in accordance with each type of information is often overlooked. Besides, varied types of clients typically navigate generic information as well as information on their specific interest. Existing works consider either each type of information or group of different types of information to measure the objectives of web application. However, grouping of information by intuition involve personal bias and often dilutes understandings.

The proposed work in this paper aims to address the followings:

- A systematic method to classify different types of information navigated by clients in accordance with the business value is to be formulated.

- The effectiveness of the web application is to be measured in accordance with this classification.
- A customized framework is to be proposed to classify multi-variant data. This distinguishes each type of information with a unique header and provides functions to measure the effectiveness. A different scheme is proposed to elicit client navigation information for other web applications.
- Significant association, visibility and benefit of repetitive navigation for each type of information are described and formulated.

This paper is organized as follows. This introduction section presents identification and analysis of existing gaps in measuring effectiveness. In the next section, a brief description of related works is discussed. Section 3 illustrates the metrics and formulations. The work is supported by experimental results. The paper ends with concluding comments.

2 Preliminaries and Related Work

In following sub-sections, some of the recent and good works from the existing literature are discussed in the context of the three major functionalities of the proposed work.

2.1 Collection of Data

Broadly two different technologies are practiced to track web data, either by using *the HTTP cookie*, and/or using *server side log*. The former is used to track client side whereas the latter is used for server side tracking. A cookie is used to record browsing activities of the client *e.g. clicking specific buttons, authentication, which pages were visited in the past, arbitrary pieces of information (name, address, etc.) that the client previously entered into the form.*

Numerous works are available on sharing of client side tracking data with the outer world that aims to handle the issue of privacy. Some consider that analytics software is nothing but a client side JavaScript code treated as a third party [11]. When a client visits a website, the analytics code is loaded up and run in the browser's engine. Such analytics code then communicates with the analytics servers with *client data like location, type of operating system, browser and information collected from browser.* In a completely different approach, Vela´squez used digital camera to capture the reflection light of eyes to extract the eye rotation through variations in the corneal reflection. Based on such analytics on log data, the key objects of web page have been identified. Another interesting approach in which the crawler has loaded the requested page when client types the starting URL as reported in [12].

2.2 Grouping of Navigation Data on Different Types of Information

The session data of clients have been clustered using hierarchical agglomerative technique in [13]. Clusters are formed based on navigation of pages. A unique characteristic is selected for a cluster, and remaining characteristics are used to build other clusters. Reichle, Perner et al. have prepared the data from the server side log with examples.

2.3 Measuring Effectiveness

The design of web application structure plays a crucial role in the effectiveness of web application. In this context, graph based model, tree based model, Markov chains, ontological paradigm have been briefly described by [14]. A minimum spanning tree is derived from graphs towards navigating website structure as a tree. In [15], an observation based experiment for executing a series of tasks for admission on an educational institute has been reported. Two more distinct approaches also have been observed in [16] and [17]. Both syntactic and session levels similarities have been considered by Storm, Kraemer et al. to form the cluster [16]. The method has been titled as semi-supervised competitive fuzzy clustering. Next, the site features and user behaviors have been determined based on specific interest by selecting the window length. A time series analysis, called selected episode graph, is used to measure the impact of behavior of user group on changing sessions. Li, Sun, and others have used temporal information as duration of visit of client to pages to assess the relevance of pages [17]. The effectiveness is also explored from different perspectives. In [18], the efficiency of content has been measured with the optimized location so that transportation is minimized. Wang, Zhu et al. claimed the comprehensive performance improvement by designing infrastructure for content delivery network in cloud [19].

3 Framework to Measure Effectiveness of Web Applications

In this work, a framework is proposed to measure the effectiveness based on the navigation of information by the clients. The framework has three components: *Unique orientation, track orientation, and measure orientation.*

Unique Orientation: Each type of information is uniquely identified by a distinguished header. In the implementation level, the term *label* is more accepted over the term header. No information is published without any header. The information may or may not be heterogeneous.

Track Orientation: The proposed framework provides a default layout that contains each type of information in a single page. Therefore, web server log is customized to elicit the tracking information. However, for other web sites, functionalities are provided using hover [20] to track the navigation of clients. The text and images are hovered and when the client navigates information under any header, the event is tracked and forward to the analytics server.

Measurement Orientation: The time that a client spent for a specific type of information is subject to the length of the text. However, an image may take less time to interpret. We have no room for such uncertainty. The importance of each type of navigated information is subject to the duration of total time spent by the client in all the sessions; this is called *importance of navigated information* (I_{Hj}). Again, within this duration, a client may encounter a specific type or group of different types of information more than once and may even spend different amount of times. The performance indicator δ_{word_ψ} in $_\psi GPI$ [3, 4] have considered duration and length of web page for all users.

In this paper, we assume that the same information may be retrieved in different sessions for different durations by the same client. Thus, durations of all sessions are included to compute the importance in the following equation, where *i is the client index, H_j are for each type of header information, j is the header index, n is the total number of H_j, s is the session index, D_t is a duration in each session a user i have spent, t is the duration index, and l is the index of the text length.*

$$I_{H_{j,i}} = \frac{\sum_{s=1}(H_{j,i,s} \cdot D_{j,i,s,t})}{\sum H_{j,i,t}} \tag{1}$$

$I_{H_{j,i}}$ implies importance of header H_j for each client i. $H_{j,i,s}$, $D_{j,i,s,t}$, and $\Sigma H_{j,i,t}$ signify number of sessions starting from 1 to n, duration of navigated time, and length of the corresponding header information respectively. Summation of $I_{H_{j,i}}$ for all clients is computed as:

$$I_{H_j} = \sum I_{H_{j,i}} \tag{2}$$

Co-relating with the business value: The importance of each type of information (I_{Hj}) is correlated with the business value (v) using the following well known equation where n is the number of pairs of data:

$$r_{v.I_{H_j}} = \frac{n \sum v I_{H_j} - (\sum v)(\sum I_{H_j})}{\sqrt{n((v^2) - (\sum v)^2)} - \sqrt{n(\sum I_{H_j}^2) - (\sum I_{H_j})^2}} \tag{3}$$

The higher value of $r_{v.I_{Hi}}$ implies the high association between each type of important information and the business value.

Significant Association: The importance of each type of information is influenced when a client navigates different types of information. The association of different types of information with the business value is crucial. More specifically how $r_{v.I_{Hi}}$ is associated with $r_{v.I_{H_j}}$; where $j' \in r_{v.I_{H_j}}$ and $j' \cap j'' \neq NULL$.

The different information types having strong linear association with the business value, is called *significant association* (SA). It is computed as:

$$SA_{v,H1,H2,\ldots Hn} = \left\{ \frac{\sqrt{(r_{v,H1})^2 + (r_{v,H2})^2 + \cdots + (r_{v,Hn})^2 - (2r_{v,H1}r_{v,H2}, \ldots, r_{v,Hn}, r_{H1,H2,\ldots r_{Hn-1,Hn}})}}{1 - \left\{(r_{H1,H2})^2 + (r_{H2,H3})^2 + \cdots + (r_{Hn-1,Hn})^2\right\}} \right\} \quad (4)$$

A trial and error method is used to choose the Hj' and to check values of SA. The higher values of SA is considered and lower values of SA are discarded. A value close to +1 could be considered as higher association. A domain knowledge is necessary to intuitively choose the Sj' for trial.

Visibility of SA: It is already stated that SA implies the linear associations ranging from 0 to +1 between business value and different types of information thst are navigated by clients. The visibility ratio between clients of $SA_{v,j}$ and all existing users of $SA_{v,j}$ signify a very important characteristic. To be more precise, the duration of navigation of each session is considered in the visibility ratio. If the visibility ratio is low, there is a chance to promote the business for H_j of $SA_{v,j}$. The determination of the threshold value to judge the low and high values of visibility ratio may depend on management policies. Obviously, the high visibility ratio implies the low scope for endorsement among existing clients. The equation of visibility ratio is shown below:

$$vis_j = \frac{\sum_{i=1}^{n} \wedge i \to each H_j \in SA_{v,j}, s=1, t \in s, i \to s \, for \, each \, H_j \in each \, SA_{v,j} \, i*s*t}{\sum_{i=1}^{n} \wedge i \to all H_j \in all \, SA_{v,j}, s=1, t \in s, i \to s \, for \, all \, H_j \in all \, SA_{v,j} \, i*s*t} \quad (5)$$

$$symbol' \to 'represents \, navigation$$

Benefit of Repetitive Navigation: The navigation of same types of information by a client is significant. The interpretation is manifold. For educational site, the interpretation is referring the content to the students; for e-commerce site, the interpretation is to check the availability and comparing with others etc. The repetition of navigation of same information could be either visiting the same type of information in the same session or in the different sessions. Indeed, there could be a chance of increasing visibility by other users with the effect of repetitive navigations of similar information of one user. The following equation is used to measure the benefit of repetitive navigation of same information:

$$Benfit_{H_j} = \left\{ \sum_{s=1}^{n} \left\{ s*(1-2)^{1-s} \right\} \cdot t \right\} * r_{v,H_j} \quad (6)$$

4 Experiments and Result Analysis

The experiment is performed on knowledge driven web site of self-financing institutes. It is assumed that the financial transaction is not part of such web applications and also, admissions of students are done through central counseling. A customized web layout, as shown in Fig. 1, is designed to track individual information for the client. The customized design ensures one header with its content be displayed on the monitor at a time. Long multi-variant content would scroll down on the monitor. Both horizontal and vertical menus are part of the home page. The home page is referred in each page. The layout would not permit publication of a web page until and unless, the header text is specified in the header section of the layout. One auto-generated function, called *header_distinctive() is used* to associate a unique number with each header. This *generates finite number of distinctive header numbers and associate with specific header text*. Template of the page layout is generated by a function *default_layout()*. Only one layout is designed in the miniature version of the experiment.

Collection of navigation information for web application: The PHP based application is used for experiments. In the server side, Apache HTTP logs are configured and accessed for elicitation of experimental data. Log Modules *mod_log_config* and *mod_setenvif* are configured and log related directives *CustomLog, LogFormat, and SetEnvIf* are used of apache web server. The format of the access log is configured using *mod_log_config* string. The *CustomLog* defines the access log relative to the ServerRoot. Some related entries are:

- The IP address of the client which has made the request to the server is logged. A log post-processor logResolve is used to determine the hostname. However, for proxy based clients, IP of the proxy is logged.
- The time format is: [day/month/year:hour:minute:second zone].

The proposed layout is page wise container of each type of information. Thus, the server side log is enough to collect the navigation information of clients. However, for other web applications that contain more than one types of information in a single page, a different scheme is adopted. This is described below.

A custom listener for hover [20] intents for each type of information using the *Google Tag Manager* (*gtm*) is implemented. The listeners are assumed to be limited

Fig. 1 Layout of the web page

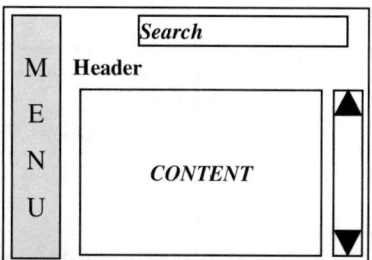

to *Form submit*, *Link Clicks*, and *Clicks* for any *DOM* object. A *gtm.hover* event is pushed to the data layer. A new tag is defined, *Hover_Intenets_Listner* and used *gtm.element*, *gtm.elementClasses*, and *gtm.elementID* as available with the other listeners of gtm. The duration of the mouse over on each type of information is computed using *mouseenter* and *mouseleave* javascript events.

One of the implementation level limitations is that if the mouse pointer is not over the specific information type which is observed by the client, nothing will be sent to the *gtm*. Therefore, web layouts that are single column are often preferred, and no advertisement is reflected at both the left and right side of the page layout.

Data Design and Results: During the experiment, total 39 headers and one home page (H_1) are considered as headers. This counts to total 40 *different types of information*. The importance of headers navigated by five most frequent clients and average importance of all headers navigated by all clients are shown in Figs. 2 and 3 respectively.

From Figs. 2 and 3, the followings observations are noted:

- Most frequent users navigated more from home page (H_1) to location (H_{26}), faculty (H_{17}), course (H_{18}), placement (H_{11}), and admission (H_{24}) information.
- The average importance of H_1 is maximum (86% approximately) whereas 12 among 40 different types of information have importance below 10%.
- Another important observation is that often user exits from the *home page* itself. This is called *bounce back* from the *home page*. The average bounce back from

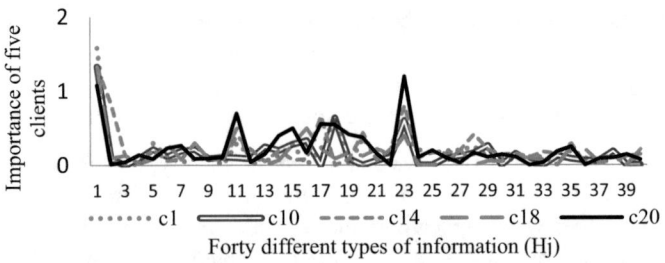

Fig. 2 Importance of forty different types of information for five most frequent clients

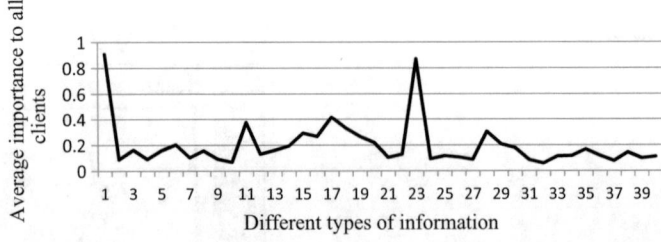

Fig. 3 Average importance of forty different types of information

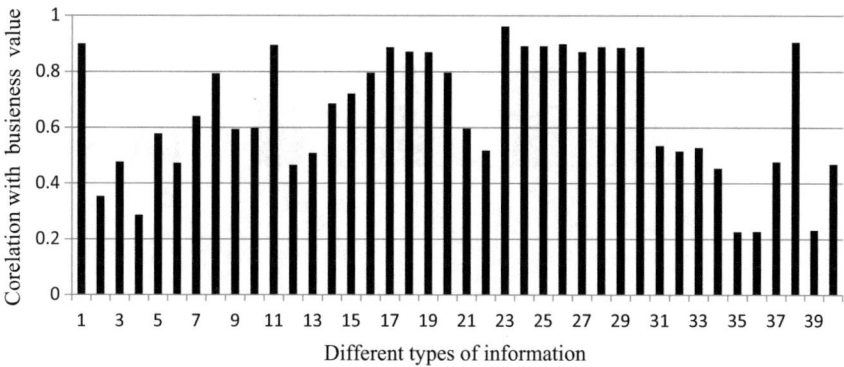

Fig. 4 Correlation between business value and each type of navigated information

home pages is approximately 68%. The expert clients use direct address to reach on other information bypassing the home page data. In this work, the ratio between number of home page navigation and total number of other types of information navigation is computed to measure the bounce back. In comparison, the Google Analytics shows that the bounce back is 98.91%. The Google Analytics computes the bounce back with the exits from the home page without traversing other pages or headers or information.

The correlation between business value and each type of navigated information by clients is shown in Fig. 4. Here, a result is summarized by applying trial and error method for determining association between business value and different types of navigated information using Eq. (4). The *Number of Admission* (v) is considered as business value. The summarized result is shown in Table 1.

In Fig. 4, the strongest corelation of business value is observed with the admission information (H_{24}). Since, the navigated data is collected during the admission process, clients mostly navigates admission related information. However, H_{24} has been avoided in further analysis for SA. The 4718 hits by clients in H1 has occured in first counselling during admission. However, account section reports that only 60 candidates took admission during first counselling.

Table 1 Significant association (SA) excluding home page navigation

V	Hj'	SA
Business value is considered as number of admissions in the academic year 2016	H_{11} (placement), H_{26} (location)	0.85
	H_{17} (faculty), H_{18} (course)	0.87
	H_{17}, H_{18}, H_{19} (accreditation)	0.98
	H_{31} (campus life), H_{32} (industry visit)	0.56
	H_{12} (research projects), H_{13} (conference), H_{14} (workshop, seminar), H_{15} (industry visit)	0.52
	H_{35} (sports), H_{37} (cultural events)	0.27

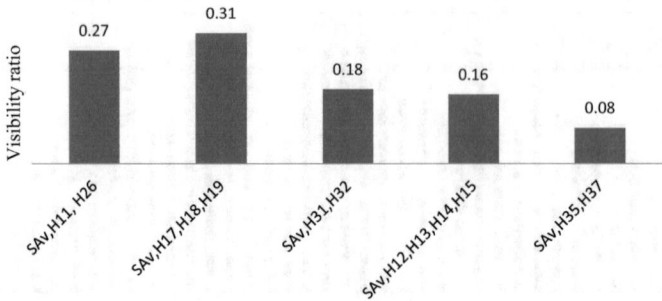

Fig. 5 Visibility ratio of significant information (SA)

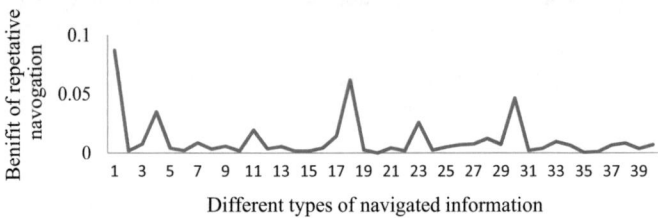

Fig. 6 Benefit of repetitive navigated information by all clients

From Table 1, students consider the reputation of the institute before they enroll their own candidature. The college for which this experiment is performed ranks below the average level of the state. So, students mostly choose this college during the first counselling as a backup option in case their applications in higher ranking colleges are rejected. Again, the maximum association with business value is found for H_{17}, H_{18}, and H_{19}. These are even higher than H_{11}, or H_{26}. The visibility ratio of SA in Fig. 5 suggests that scopes of more promotions exist for H_{12}, H_{13}, H_{14}, H_{15}, H_{35}, and H_{37}.

The benefit of total repetitive navigation for all clients on each type of information is shown in Fig. 6. Here, H_1, and H_{23} has greater than 50% benefits, while its lower for others. Page load time is suggested to be less for higher repetitive information.

5 Conclusion

The proposed work quantifies the effectiveness of knowledge driven web application in accordance with the business value and navigated information. The high bounce back from web applications is considered while measuring the effectiveness. In our experiment, bounce back is 67.96%. We do suggest a search option in the home page. It's found that average navigation of different types of information for the most frequent clients as well as others are approximately same. This implies the standardization, quick interpretation and ease of links for web applications. The lower visibility ratio of significant association of different types of information suggests a scope for promotional activities among existing clients.

References

1. Zahed, F., Pelt, W.V., Song, J.: A Conceptual framework for international web design. IEEE Trans. Prof. Commun. **44**(2), 83–103 (2001)
2. Constantine, J.A., Boucouvalas A.C.: Future proof analytics techniques for web 2.0 applications. In: International Conference on Telecommunications and Multimedia, Greece, pp. 214–219, (2014)
3. Ali S.R., Khan A., Baig M.M.F.: Implementation of Kano's model in web metrics for information driven websites—KDQI. In: International Conference on Information and Communication Technologies (ICICT), Indonesia, pp. 1–6, (2015)
4. Stolz C., Viermetz M., Skubacz M., Neuneier R.: Guidance performance indicator—web metrics for information driven web sites. In: IEEE/WIC/ACM International Conference on Web Intelligence (WI), France, pp. 186–192. IEEE Xplore (2005)
5. Spiliopoulou, M., Pohle, C.: Data mining for measuring and improving the success of web sites. Data Min. Knowl. Disc. **5**(1), 85–114 (2001)
6. Carneiro A.R., Jorge A.M., Brito P.Q., Domingues M.A.: Measuring the Effectiveness of an e-commerce site through web and sales activity. In: Springer Proceedings in Mathematics and Statistics, vol. 73, pp. 149–162, (2001)
7. Etail technology. http://etailsolutions.com/technology/
8. Elangovan N.: Evaluating perceived quality of b-school websites. J. Bus. Manag. vol. **12**(1), 92–102 (2013)
9. SiteQual measure. http://www.emeraldinsight.com/doi/abs/10.1108/174103904105724
10. Process of Netquall. https://www.netquall.com/process/
11. January E., Breeanne P., Martin M.: Misuse, play, and disuse: technical and professional communication's role in understanding and supporting website owners' engagement with Google Analytics. In: International Professional Communication Conference, pp. 80–84 (2015)
12. Demarty G., Maronnaud F., Breton G., Hallé S.: SiteHopper: abstracting navigation state machines for the efficient verification of web applications In: 9th International Workshop on Web Service and Formal Methods (WS-FM), pp. 103–117. Springer (2013)
13. Anupama D.S., Sahana D., Gowda B.: Clustering of Web User Sessions to Maintain Occurrence of Sequence in Navigation Pattern. In: 2nd Symposium on Computer Vision and the Internet, Elsevier Procedia Computer Science, pp. 558–564, India, (2015)
14. Żatuchin D.: Problem of website structure discovery and quality valuation. In: Conference on Computer Science and Information Technology, pp. 117–122. IEEE (2011)
15. Nagpal R., Mehrotra D., Bhatia P.K.: Task based effectiveness evaluation of educational institute websites. In: International Conference on Computational Techniques in Information and Communication Technology, India, pp. 315–319. IEEE (2016)
16. Storm K., Kraemer E., Aurrecoeche C.: Website evolution: usability evaluation using time series analysis of selected episode graphs. In: International Symposium on Web Systems Evolution (WSE), USA, pp. 27–36. IEEE (2009)
17. Li Z., Sun M.T., Dunham M.H., Xiao Y.: Improving the web site's effectiveness by considering each page's temporal information. In: 4th International Conference on WAIM, China, pp. 47–54. Springer (2003)

18. Jin, Y., Wen, Y., Guan, K.: Toward cost-efficient content placement in media cloud: modeling and analysis. IEEE Trans. Multimedia **18**(5), 807–819 (2016)
19. Wang Z., Zhu W., et. al.: CPCDN: content delivery powered by context and user intelligence. IEEE Trans. Multimedia **17**(1), 92–103 (2014)
20. Code my views posting: designing hover styles and the future of the technique. https://codemyviews.com/blog

Responsive Data Table Solution with New Scrolling Control Gesture for Better User Experience

Lukáš Čegan

Abstract According to available statistics, nowadays more user browse websites on mobile devices than on a desktop computer. This brings new challenges for web developers on how to deal with such a heterogeneous environment of devices. Today, several different solutions are on the market for building responsive data tables. However, these solutions have their limits in the event that the table contains a large number of records. In this paper a new solution is proposed based on a new scroll control gesture and elimination of superfluous loading records. The impacts of the proposed solution to the user experience were investigated through an experiment whose results are presented and discussed in this paper.

Keywords Data table · User experience · Responsive design

1 Introduction

Dynamic development in the mobile technology area, cheaper and more sophisticated devices on the market, faster and achievable communications network, all bring an increasing number of users who access the Internet via mobile devices. According to related statistics, the number of mobile users was greater than desktop users. It brings to web developers a variety of new challenges which deal with design and programming web applications. These applications must be designed for such a large group of diverse devices, while providing a good user experience. Currently, one of the most used approaches to deal with this fact is the application of responsive design.

L. Čegan (✉)
Faculty of Electrical Engineering and Informatics, Department of Information Technology, University of Pardubice, Pardubice, Czech Republic
e-mail: lukas.cegan@upce.cz

© Springer International Publishing AG 2017
D. Król et al. (eds.), *Advanced Topics in Intelligent Information and Database Systems*, Studies in Computational Intelligence 710,
DOI 10.1007/978-3-319-56660-3_40

467

The term "Responsive web design" was coined by a leading web developer Ethan Marcotte in his book [1]. He describes responsive design as a new way of designing the ever-changing web. The goal of responsive design is to create easy web reading and content navigation with minimal resizing, scrolling and panning, across a wide range of different devices from smart phones to desktop computer monitors. Also, the responsive web design is a subject of numerous scientific studies today. A variety of approaches have been proposed to implement responsive design in mobile e-learning systems [2, 3] or support better mobile access to virtual and remote laboratories [4]. An additional published solution is focused on adaptation of usability principles in responsive web design technique in the e-commerce area [5]. Also, some authors present a solution for responsive web design based on the resolution [6]. All of these studies have only one goal to achieve a better user experience which is nowadays the holy grail of all web developers. Responsive design is just one of the techniques to achieve a better user experience. In addition, there are many other techniques to improve user experience. For example, some authors present in their papers solutions how to improve the user experience on mobile apps through data mining, based on collecting user context data by Google Services API [7]. Or, the authors try to enrich user experience by visualizing video sounds with sound word animation [8].

This paper is focused on improving the user experience with a web data table by combine responsive design and a new way of controlling and loading table records. The data table is often the dominant component of websites because it offers the ability to display a lot of information in a standard structured way, which every user understands well. Unfortunately, making data tables mobile friendly, is in a large number of cases like trying to cram an entire basketball team into a Volkswagen Beetle car. Very often it is impossible. If the developer does not apply any responsive design technique, on the small screen, the data table may be displayed in two ways. First, the table can be reduced to the size of the screen (see Fig. 1a). This

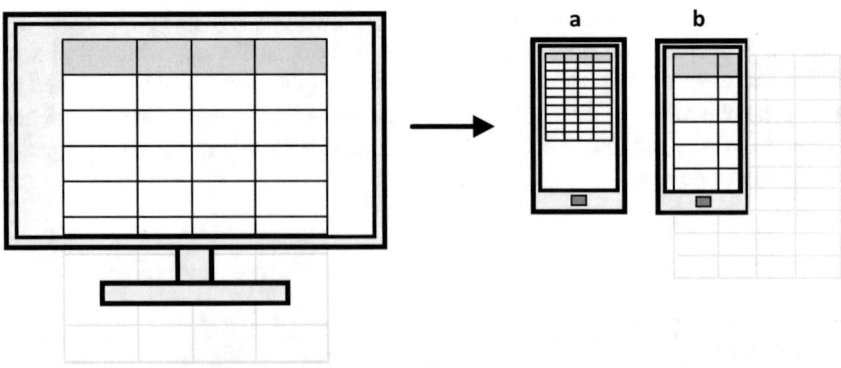

Fig. 1 Data table—desktop computer/mobile device

leads to the fact that the user must use the microscope to be able to read the table values. Second, the table can remain in regular size but at a cost that the user sees on the screen, only a small portion of the table at a time (see Fig. 1b). In this case, the user is able to read the values of the table, but to get an overview of the entire table, the user must repeatedly move the contents of a table on the screen, which is very time-consuming and annoying.

The web developers used many different approaches which are used in the practical world for solving this problem. Every approach has a positive and negative impact for user experiences. Unfortunately, there is no single universal solution. It is very important to rigorously assess all factors of table (like the number of rows and columns, format of entries, editable and sortable records, etc.) for selecting an appropriate approach for concrete data table. In this paper a new solution is proposed based on an intelligent data table control and elimination of superfluous loading table records which save the transmitted data and allow the users to more quickly find the records of interest in the data table.

The paper is organized as follows. After introducing the objective of this paper, the responsive design is presented in Sect. 2. The Sect. 3 described the proposed solution for viewing multicolumn data table. The Sect. 4 presented the practical experiments. The results of the experimental analysis are discussed in Sect. 5. Finally, the last section gives conclusions and future research opportunities followed by references at the end.

2   Responsive Data Table

Today, web developers use several different approaches which have been developed for establishing tables that can scale well in different viewport sizes. Common solutions are based on pure HTML and CSS. A suitable example can be a solution implemented in one of the most widespread CSS framework Bootstrap. To create a responsive table, Boostrap applies the overflow-x property: auto to wrap the table and overflow-y property: hidden for a window less than 768 px. This causes, on the small screen, the content of table to be available by scrolling horizontally (see next source code).

Example of a Twitter Boostrap 3.0 responsive table solution [9]

```
program Inflation (Output)
<style>
.table-responsive {
  min-height: .01%;
  overflow-x: auto;
}
@media screen and (max-width: 767px) {
  .table-responsive {
    width: 100%;
    margin-bottom: 15px;
    overflow-y: hidden;
    -ms-overflow-style: -ms-autohiding-scrollbar;
    border: 1px solid #ddd;
  }
}
</style>
<div style="table-responsive">
  <table>
    ...
  </table>
</div>
```

Besides pure HTML and CSS solutions, advanced approaches based on Java-Script exist also. The example of this approach is responsive a data table from the Foundation Zurb framework. Their solution consists in pinning the first column and making the rest of the table horizontally scrollable. To achieve this functionality, Zurb uses jQuery library to manipulate the DOM and CSS [10]. However, this is not the only solution on the market. Generally, current responsive data table solutions based on JavaScript can be divided to three basic groups; key/value table, table with hiding less important columns, table with inverted axes.

2.1 Key/Value Data Table

A large data table with many columns cannot be reduced to the size of a small smartphone screen. An approach called Key/Value table consists of replacing each row of the table by its own new table. This new table contains only two columns. The first column contains the "keys", which are the headers of the original table. The second column contains the "values", which are the records of the original row (see Fig. 2). In other words, each row of the source table is displayed in the new small table, which is formed by the first row (header table) plus the actual row of the source table. Then, this table is displayed vertically instead of horizontally. This solution eliminates the need for horizontal scrolling. On the other hand, this solution requires more vertical scrolling. Another disadvantage is the lost possibility to simply compare each value between rows (records no longer positioned in one column).

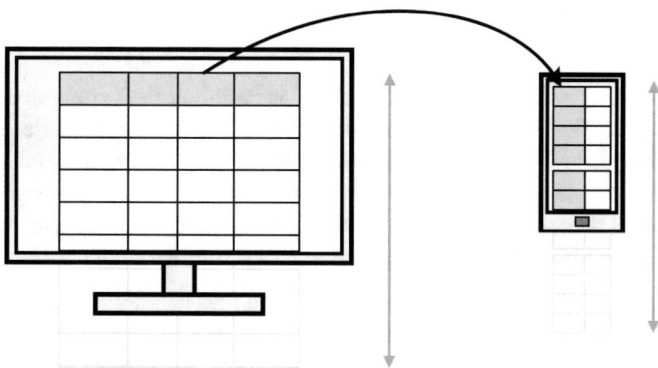

Fig. 2 Key/Value data table

2.2 Table with Hiding Less Important Columns

A solution called "Table with hiding less important columns" is based on prede-termined rules that determine which columns will be displayed in the available width of the screen. Each column of the table has a defined priority level that determines the rendering order of the columns. If the columns of the table have no determined level priority, then the columns with the smallest width are rendered as first (see Fig. 3). For the user perspective, the biggest disadvantage is loss of control over what part of the content will be displayed. Some of the solutions solve this problem by adding a drop-down menu through which the user can select columns to display.

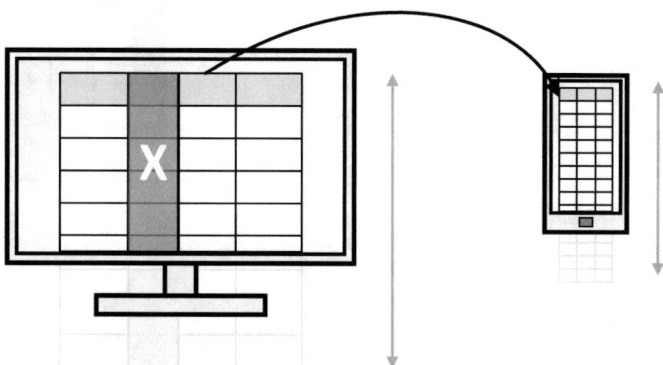

Fig. 3 Table with hiding less important columns

2.3 Table with Inverted Axes

The last solution "Table with inverted axes" is based on the assumption that users used a mobile device to read more in portrait than landscape orientation (see Fig. 4). Therefore, if we flip the table axes, we can view more columns at once. In this solution the data table is scrollable in the direction from left to right and vice versa. Due to the user losing an overview of the table, it is necessary to fixed a header of table and enable scrolling only the table entries exclude table header.

2.4 Limitation of Responsive Data Table Solutions

Approaches outlined above can help improve user experience in a mobile environment greatly. However, these approaches have their limits in the event that the table contains a large number of records. In this case the records delivered to a user has to be limited. A Commonly used approach is pagination. This means that the control bar is attached to the table from which the users can select a page of records. This method of data control has been successfully used on desktops but in a mobile environment, leads to a worse user experience, because web users have to work intensively with the control buttons. A newer method is sequential loading of records, automatically triggered by scrolling the contents of the table. The disadvantage of this method is a large amount of data that is unnecessarily loaded in the case that the user requests the records at the end of the table.

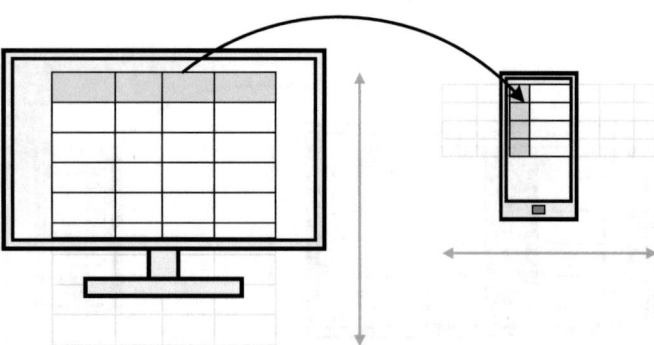

Fig. 4 Table with inverted axes

3 Proposed Solution

Based on the identification of the main limitations of current responsive data table solutions, which are listed in Chap. 2, we proposed a new robust solution that significantly speeds up the web users' ability to achieve the required records, and it eliminates an unnecessary load of records. The proposed solution aims to improve the user experience, which is essential for the successful website.

3.1 Touch Gesture

The core of the proposed solution is based on the new touch gestures through which a mobile user calls up a hint window with the table rows index (see box with letter F in Fig. 5). The gesture is detected by a proprietary JavaScript library when the user scrolls the content of the table very quickly (on the background is permanently calculated current speed scrolling). When web browser captures this gesture, normal scrolling of table content is interrupted and instead of it, the user scrolls the index value in a hint window. If the user is satisfied with the selected value, the user closes the selection of the index value by tapping on the hint window. Then web browser hides the hint window, creates an AJAX request and sends it to the web server. Web server processes the request and sends the requested records back to the web browser. Then the user can continue browsing the table from a place that is very close to the target record which the user wants to display.

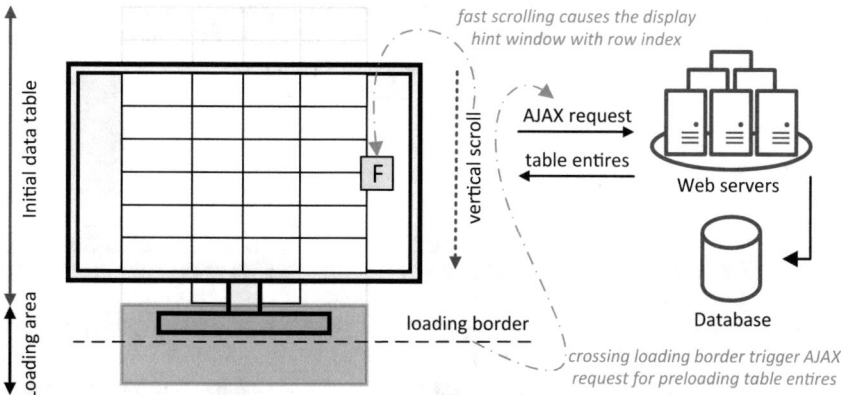

Fig. 5 Architecture of proposed solution

3.2 Loading and Rendering Table Records

The loading of table records is, in the proposed solution, triggered in two different ways. First, the data loading is triggered when the user confirms the selection of an index value as described above. Second, when a user scrolls the table at a normal speed and at the same time, the loading area crossed the loading border (see Fig. 5). This event also triggers an AJAX request to the web server as in the first case. Because the records loading consumes a certain amount of time, it is recommended to set the loading border at a distance corresponding to half the height of the device screen in order to achieve continuous data rendering.

The records are included to the table by JavaScript immediately after the web browser receives them from web server. In this case, "include" means to create for each row of records a new DOM elements TR, and TD, and inject them into the right place in the table.

4 Experiment

The effectiveness of the proposed solution has been investigated in three test scenarios. For each scenario a special website was created that has been tested on twenty users of different age and gender. For each scenario, respondents were given a different task. The data for each website comes from OpenFlights service [11]. Furthermore, each scenario was tested in various network environments that simulate different bandwidth and latency. To simulate the network environment Linux tools Netem (Network Emulator) and TBF (Token Bucket Filter) was used. The parameters for each scenario was shown in Table 1.

The performance metrics in this experiment were the total time required to complete the task (TTT), overall size of the transferred data from the server to the client (TDS) and total number of AJAX requests (TNR).

Table 1 Parameters for each experiment scenario

	Scenario A	Scenario B	Scenario C
Dataset	Airports.dat—12 columns, 8107 rows	Airlines.dat—8 columns, 6048 rows	Airlines.dat—8 columns, 6048 rows
Index column	Name of airport—first letter	Name of airline—first letter	Airline ID—multiples of 500
User task	Find row for "Ozamis" airport	Find row for "Tran-silvania" airline	Find first row for airline ID "11726"
Network	3G—1 Mbit/s bandwidth, 300 ms latency	LTE—10 Mbit/s bandwidth, 50 ms latency	FIBER—unlimited Mbit/s bandwidth, 50 ms latency

Table 2 Results of experiment (medians of all measurement)

	Scenario A		Scenario B		Scenario C	
	A	B	A	B	A	B
Total task time (s)	12	52	14	83	28	325
Transferred data size (kb)	13	196	21	209	19	305
Total number of requests	5	59	7	71	6	129

5 Experimental Results and Discussion

For evaluation of efficiency of the proposed solution the method of A/B testing has been used. This method consists in comparing two variants, A and B, which are examined separately in a controlled experiment. Experiment results are statistically analyzed to determine which variation performs better for a given conversion goal.

In our experiment, variant B is a pure HTML table with responsive design "Table with hiding less important columns" (see Sect. 2.2) and sequential loading of records. In the variant A same table is used as in variant B, but for records loading the proposed solution is used. The next Table 2 shows the results of an experiment for each test scenario. Values represent medians of all measurements.

As the results show, from the user's perspective the proposed solution is significantly better tool than traditional solutions, because it enables users to more quickly reach the searched table records. Especially in networks with high latency is the difference substantial. The proposed solution also significantly reduces the amount of transferred data which could be reflected in a reduction of costs paid to the ISP. Both of these characteristics are certainly a positive impact on the user experience. For a better comprehensibility the experiment results are also shown in the graphs (see Fig. 6). In each graph both variant A/B are shown. Each variant is composed of three bars expressing each scenario.

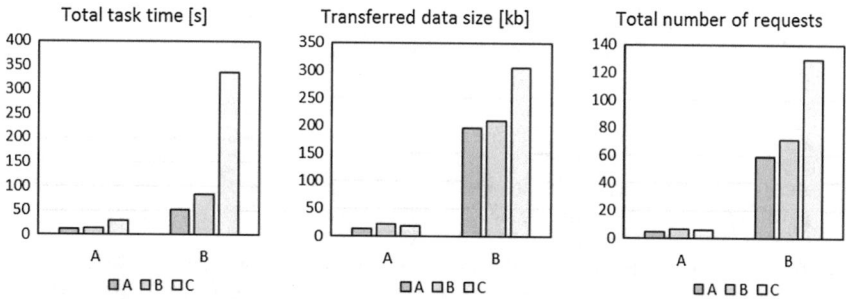

Fig. 6 Comparison of experiment results—**1** Total task time. **2** Transferred data size. **3** Total number of requests

6 Conclusion

This paper presents a new approach to display and control a data table in a web environment. The proposed solution is based on the combination of responsive design and a new control gesture through which a user can work faster with table records. Moreover, the proposed solution also significantly reduces amount of transferred data. Both of these improvements lead to better user experience. The effectiveness of the proposed solution has been investigated in a controlled experiment on the three test scenarios. The experiment results are discussed in this paper and confirm the initial research assumptions.

As future work, we are planning to implement new module to our solution which allows the web user to simply select a column of a table under which will be realized, fast scrolling of the records. In the current version of the solution is the selection of a column only in the hands of developers, which is a significant limitation of functionality from the user's perspective.

Acknowledgments This work is published thanks to the financial support Faculty of Electrical Engineering and Informatics, University of Pardubice under grant TG02010058 "Podpora aktivit proof-of-concept na Univerzitě Pardubice".

References

1. Marcotte, E.: Responsive web design. Second Ed. A book apart (2014)
2. Zhu, B.: Responsive Design: e-learning site transformation. In: Fourth International Conference on Networking and Distributed Computing, pp. 126–130. IEEE (2013)
3. Peng, W., Zhou, Y.: The design and research of responsive web supporting mobile learning devices: e-learning site transformation. In: International Symposium on Educational Technology, pp. 163–167. IEEE (2015)
4. Zervas, P., Trichos, A., Sampson, D.G., Li, N.: A responsive design approach for supporting mobile access to virtual and remote laboratories: e-learning site transformation. In: 14th International Conference on Advanced Learning Technologies, pp. 11–13. IEEE (2014)
5. Majid, E.S.A., Kamaruddin, N., Mansor, Z.: Adaptation of usability principles in responsive web design technique for e-commerce development. In: International Conference on Electrical Engineering and Informatics, pp. 726–729. IEEE (2015)
6. Lee, J., Lee, I., Kwon, I., Yun, H., Lee, J., Jung, M., Kim, H.: Responsive web design according to the resolution. In: 8th International Conference on u- and e-Service, Science and Technology, pp. 1–5. IEEE (2015)
7. Auad, T.O.S., Mendes, L.F.C., Stroele, V., David, J.M.N.: Improving the user experience on mobile apps through data mining. In: 20th International Conference on Computer Supported Cooperative Work in Design, pp. 158–163. IEEE (2016)
8. Wang, F., Nagano, H., Kashino, K., Igarashi, K.: Visualizing video sounds with sound word animation to enrich user experience. IEEE Trans. Multimedia, IEEE (2016) (Issue 99)
9. Twitter Boostrap—Overview. http://getbootstrap.com/css/#tables-responsive
10. Responsive Tables. ZURB Foundation, http://foundation.zurb.com/responsive-tables.html
11. Airport, airline and route data. http://openflights.org/data.html

Part VI
New Methods and Applications in Software Engineering

Adaptation of an ANN-Based Air Quality Forecasting Model to a New Application Area

Cezary Orłowski, Arkadiusz Sarzyński, Kostas Karatzas, Nikos Katsifarakis and Joanicjusz Nazarko

Abstract The paper presents the adaptation procedure concerning the application of an ANN-based Air Quality forecasting model that was already developed for the city of Gdańsk, Poland, and is now tested for the city of Thessaloniki, Greece. For Gdańsk the model has taken into account the city's meteorological parameters, which have been implemented using a one-way neural network for ease of learning, as well as the concentration levels of the pollutant of interest, PM10. In the process of teaching the network, four methods of propagation have been used (Back Propagation, Resilient Propagation, Manhattan Propagation, and Scaled Conjugate Gradient) for the purpose of choosing the best method. Results were then compared with real values which define the full network configuration (minimizing the forecast error). The model was then subjected to a process of adaptation for Thessaloniki. The data acquired through the process of adaptation regarding the PM10 levels used for both model training and testing purposes.

Keywords Computational intelligence · Artificial neural networks · Air pollution · Forecasting

C. Orłowski
Institute of Management and Finance, WSB University in Gdańsk, Gdańsk, Poland
e-mail: corlowski@wsb.gda.pl

A. Sarzyński (✉)
Faculty of Management and Economics, Department of Applied Business Informatics,
Gdansk University of Technology, Gdańsk, Poland
e-mail: arek3108@gmail.com

K. Karatzas · N. Katsifarakis
Department of Mechanical Engineering Informatics Systems
and Applications—Environmental Informatics Research Group,
Aristotle University, Thessaloniki, Greece

J. Nazarko
Faculty of Management, Bialystok University of Technology, Bialystok, Poland

© Springer International Publishing AG 2017
D. Król et al. (eds.), *Advanced Topics in Intelligent Information
and Database Systems*, Studies in Computational Intelligence 710,
DOI 10.1007/978-3-319-56660-3_41

1 Introduction

The continuous development of industrial technology has a strong impact on the quality of the environment [1]. This is particularly evident in urban areas where industrial and city processes have a significant impact on the quality of life of citizens [1]. The concept of Smart Cities concerning urban development and environmental management take into account these factors and suggest the need for more advanced studies concerning urban environmental pressures. Air pollution is of particular importance in this respect, and Particulate Matter of mean aerodynamic diameter up to 10 μm (PM10) is among the most important pollutants that greatly affect quality of life in Europe [2]. For several decades, governments worldwide have tried to control and limit the level of particulate matter in the air, via AQ monitoring and modelling. The state-of-play for air pollution forecasting systems has been shown in previous works by the authors [3, 4]. Currently, we present an attempt to adapt the system developed for PM10 forecasting, prepared for Gdańsk, Poland, for the city of Thessaloniki, Greece. The process of adaptation does not only have a practical dimension (contribute to the AQ modelling for Thessaloniki) but also a scientific one (the creation of procedures to adapt systems to various geographic domains of interest).

2 The Area of Interest: The City of Thessaloniki

Thessaloniki—the second most populous city in Greece is located in the northern part of the country. The city is surrounded by the Mount Hortiatis from the northeast, having the sea front of the Thermaikos Gulf at the southwest. The population of the Greater Thessaloniki Area is estimated at around one million. Industrial zones are located in the northwestern part with residential areas located in the south (Fig. 1). The climate of the agglomeration is moderate, with warm, dry summers and wet, mild winters. Average temperatures are respectively: 25.3 °C in summer and 7.0 °C in winter. There is a predominance of strong northern winds, resulting from the Eastern Mediterranean location of Thessaloniki. All these features (number of inhabitants, location, climate, and wind factors) have a significant impact on the quality of the environment. Focusing on PM10, air quality limit values have been frequently exceeded especially in the city center [5], making air pollution forecasting an important factor for improving the quality of life of the inhabitants.

3 Prediction Model of Air Pollution by PM10 Dust

The current study made use of the PM10 forecasting models already developed for Gdańsk [7–9]: Concerning the parameters affecting the future (24 h ahead) level of PM10, the current PM10 levels as well as meteorological conditions were taken

Fig. 1 Map of Thessaloniki divided into industrial zone and residential area [6]. The black rectangle corresponds to the location of the "Kordelio" AQ monitoring station

Table 1 The impact of meteorological parameters on the future levels of PM10 concentrations

Parameter	Impact
Temperature	The level of PM10 concentration decreases with increasing temperature
Dew point	The level of PM10 concentration decreases as the dew point increases
Wind direction	Depending on the existence of emission sources in the air path as well as of already polluted air masses in the area of interest
Wind speed	The level of PM10 concentration increases with decreasing wind speed
Atm. pressure	The level of PM10 concentration varies with increasing atmospheric pressure
Relative humidity	The level of PM10 concentration varies with increasing relative humidity

into account. The influence of these parameters was examined with the aid of the Pearson's correlation coefficient. Basic findings are summarized in Table 1.

On the basis of this analysis, all available parameters were selected as inputs, thus including: 6 meteorological parameters (temperature, dew point, wind direction, wind speed, atmospheric pressure, and relative humidity) and the current level of PM10 concentrations.

The algorithm selected for the development of the AQ forecasting model had to be able to deal with the non-linear relationship between inputs and the target parameter. For this purpose, Artificial Neural Networks (ANN) were selected for

Fig. 2 The basic architecture
of the ANN model: n—the
number of neurons in the
hidden layer

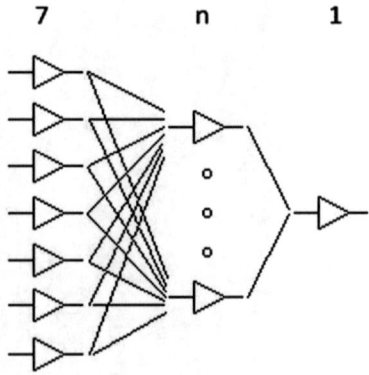

the model development [10]. A significant advantage of ANN is the ability to automatically reorganize internal structures within the problem domain in order to lead to a better and more accurate result. The architecture of the ANN developed in this study is presented in Fig. 2.

The basic model architecture had to be complemented with the specified network type (unidirectional/recursive), the number of hidden layers, the number of neurons in the hidden layer, the activation function used, and the learning method as described below:

- Network type: The vast majority of related problems can be solved using one-directional ANNs while a recursive network is typically used for complex optimization of problems. For this reason a one-directional ANN was selected in this case
- Number of hidden layers: The problem of forecasting usually requires the use of a multi-layered network (at least one more layer in addition to the input and output layers). Most problems can be solved by using a network with one hidden layer, including the problems of AQ forecasting, this being the choice in the case studied here.
- Activation function: The function to be chosen depends on the problem being solved. In multilayer networks, non-linear activation functions are commonly used. Neurons placed on the network layer are then characterized by their ability to learn and therefore a bipolar sigmoid function is selected, which is one of the main non-linear activation functions.
- Number of neurons in the hidden layer: A proper amount of neurons in the hidden layer leads to better forecasting accuracy. This is one of the most time-consuming processes because there is no definite pattern to determine the exact number of neurons. Research should be conducted to verify network performance at different configurations. Simulations were carried out for the network, comprising from 5 up to 31 neurons in the hidden layer. These values were chosen experimentally. If the results indicate the need to extend this range, then the range would be extended.

- The learning method employed: The most popular methods of ANN learning (Back Propagation, Resilient Propagation, Manhattan Propagation, Scaled Conjugate Gradient) were tested in the frame of this study, differing in the mathematical way, the speed and accuracy of teaching neurons in order to reduce the overran ANN error.

Concerning the ANN model training, meteorological data as well as air quality concentration levels, spanning two calendar years, available at 1 h intervals, were divided into two sets: the training set and the test set. Both sets included different dates for different seasons of the year, in order for the results to be reliable.

After the simulation of different network configurations it was made evident that for the city of Gdańsk, the best AQ forecasting results were generated by a network that used Back Propagation with 17 neurons in the hidden layer: the absolute average forecasting error was 7.82 $\mu g/m^3$, the maximum error was 27.44 $\mu g/m^3$ and the minimum error was 0.12 $\mu g/m^3$ (Table 2).

Table 2 The results of the ANN training method using Back Propagation and a number of 5 up to 31 neurons in the hidden layer, for Gdańsk

Back propagation			
Number of neurons	Aver. error ($\mu g/m^3$)	Max. error ($\mu g/m^3$)	Min. error ($\mu g/m^3$)
5	8.39246	30.60900	0.13300
6	8.19068	30.60900	0.13300
7	8.20002	30.60900	0.08550
8	8.06906	29.18400	0.28500
9	8.16485	33.15500	0.28500
10	8.22733	33.87700	0.06650
11	8.02354	31.06500	0.72200
12	8.19316	31.06500	0.65550
13	8.09064	31.06500	0.65550
14	8.00355	30.40000	0.19950
15	7.97970	30.40000	0.19950
16	8.02156	30.40000	0.12350
17	7.82266	27.44550	0.12350
18	7.96971	32.70850	0.12350
19	7.93870	32.70850	0.00000
20	7.78723	29.26950	0.08550
21	8.05046	29.26950	0.08550
22	8.07071	31.31200	0.08550
23	7.90875	32.84150	0.34200
24	7.97535	32.84150	0.08550
25	7.96133	32.84150	0.08550
26	7.78406	31.62550	0.11400
27	7.99385	32.96500	0.11400
28	8.04175	32.96500	0.11400
29	8.11973	29.97250	0.02850
30	8.00504	29.97250	0.02850
31	8.04030	29.97250	0.02850

4 Adaptation Process Model for Thessaloniki

Artificial neural network is a structure composed of neurons. Each neuron in a layer is connected to other neurons in adjusted layers. Each connection has a weight—numeric value—which determines how the signal flows through the network. The ANN training process is based on the recalculation of the weights of the neurons in order to minimize the overall error. As a next step, the testing process checks how thoroughly the data has been "learned" and thus the model can be used to predict the parameter of interest for new data entries. This process is very similar to the training phase. Input data are fed into the ANN in order for the network to generate the result. The difference between the ANN-based result and the actual value defines the overall error which is generated by the network.

In our study we made use of the ANN developed for Gdansk, and provided as inputs data coming from Thessaloniki, made available as on-hour measurements for two calendar years: 2004–2005. Data included meteorological parameters (temperature, dew point, humidity, atmospheric pressure, wind speed, and direction) and the levels of PM10 concentration. Of the seven PM10 monitoring stations, one was selected (station "kord" corresponding to Kordelio), for which data was the most complete.

The data was additionally cleaned to avoid any empty values, which could adversely affect the model quality during the training phase, and then were subjected to processes of standardisation and normalisation.

As a next step, the processed data were divided into a training and a testing set. The latter consisted of the 10% of the overall available data, randomly selected.

5 Research of the Degree of Generality of the AQ Forecasting Model

As a next step in the adaptation process, we studied the potential of generalization of the forecasting model developed for Gdańsk. On this basis, the first step was to test the configuration that was verified in Gdańsk, for the Thessaloniki data. The processes of training and testing the network for Gdańsk and Thessaloniki was conducted with the aid of a special platform—PLGrid. This is a computational platform that allows applications to run on high-performance clusters using a graphical user interface. The training process network ended when the difference between successive iterations of learning had fallen below the value 0.000001 (while to computational time reached 14.400 min). The average absolute error for the Gdansk AQ forecasting model applied with Thessaloniki data reached 23.68 $\mu g/m^3$. This was considered to be a high error value and thus re-testing on a modified configuration of the ANN was deemed as necessary, in order to arrive to a more efficient configuration that would lead to a smaller forecasting error.

It was therefore decided to carry out computational experiments to identify the optimal number of neurons in the hidden layer, and the best method of network training. The study once again examined all combinations of the following:

- ANN training methods: Back Propagation, Resilient Propagation, Manhattan Propagation, Scaled Conjugate Gradient
- The number of neurons in the hidden layer: 5–31.

The results of the computational study is presented in Table 3, Figs. 3 and 4.

Based on the results presented in Table 3, we can conclude that the smallest forecasting error is achieved using the Resilient Propagation method of training with 25 and 28 neurons in the hidden layer. The average forecasting error reached 9.80 $\mu g/m^3$ and 9.75 $\mu g/m^3$ respectively.

Another conclusion that can be drawn from the results is that not only the number of neurons in the hidden layer has changed, but also the method of training. For Gdańsk, the Back Propagation method proved to be the most effective, while for Thessaloniki the Resilient Propagation method achieved the best result, with a prevalence of about 2.5 $\mu g/m^3$. The network trained via the Manhattan Propagation method reached an average forecasting error slightly worse than the network taught using the Back Propagation method.

Overall it is evident that when adapting an already developed and tested ANN model for air pollution forecasting, it is necessary to carry out a computational study involving all network configurations, parameters and training methods. This is due to different external influences affecting the behavior of the network during the forecasting and teaching process—factors specific to the area where the AQ measurement stations are located.

6   Conclusions

In the present paper we discuss the problem of data-driven air pollution forecasting with emphasis on PM10. Efforts were made to adapt a model previously developed and tested for the city of Gdańsk, to be used for the city of Thessaloniki. The results of the research are as follows:

(a) The ANN model developed and tested for Gdansk did not perform satisfactory in Thessaloniki.
(b) The parameters affecting the ANN characteristics and training procedures had to be re-estimated, keeping the same basic ANN architecture (one input layer of 7 neurons, one output layer with one neuron, one hidden layer with a number of neurons ranging from 5 to 31).
(c) Computational experiments were selected as the appropriate method to resolve the problem.
(d) The new ANN characteristics resulting from the computational experiments identified the most effective ANN configuration for Thessaloniki as a 28

Table 3 Computational results for the AQ forecasting results in Thessaloniki with the aid of an ANN: column 'N'—the number of neurons in the hidden layer, 'PM10'—the average absolute error PM10, 'PM10%'—the average relative error of PM10, 'MaxP10'—the maximum absolute error PM10, 'MinP10'—the minimum absolute error PM10

N	Back Propagation				Resilient Propagation				Manhattan Propagation				Scaled Conjugate Gradient			
	PM10	PM10%	MaxP10	MinP10	PM10	PM10%	MaxP10	MinP10	PM10	PM10%	MaxP10	MinP10	PM10	PM10%	MaxP10	MinP10
5	21.4595	44.3523	55.479	0.779	21.6809	44.8099	59.279	2.4795	19.6501	40.6126	56.106	1.9285	16.2433	33.5714	42.0935	0.304
6	14.6159	30.2079	55.4125	4.7975	20.1185	41.5808	56.9705	1.9665	17.5755	36.3248	46.1215	3.192	16.746	34.6105	48.7815	0.247
7	15.5096	32.055	50.71	1.444	20.8588	43.1108	67.7435	1.7385	17.1582	35.4624	57.4265	1.1875	19.9475	41.2274	75.011	6.574
8	17.6142	36.4048	54.5575	0.019	18.8821	39.0253	44.0695	7.182	14.3301	29.6173	41.172	3.4295	20.356	42.0716	64.8935	0.0855
9	23.9609	49.5222	54.643	0.9025	20.5957	42.567	87.9975	0.7885	18.2814	37.7838	64.96	1.8905	26.8212	55.4337	66.803	10.754
10	17.6719	36.5241	37.3055	12.0745	12.7115	26.2719	52.477	3.401	16.5188	34.1408	53.4365	3.971	20.4985	42.3662	66.7935	3.648
11	28.4595	58.8199	67.5915	6.669	18.6877	38.6235	73.32	0.7885	23.9865	49.575	67.924	3.705	18.7754	38.8048	66.252	1.14
12	17.7706	36.728	58.69	7.106	15.7179	32.4855	37.2865	3.8285	21.7577	44.9685	51.7645	9.1865	20.7265	42.8374	68.399	0.2755
13	16.0591	33.1908	33.7335	8.8255	20.435	42.2348	59.4975	0.4275	18.3976	38.0239	51.4035	1.3775	17.5243	36.2191	66.252	6.5265
14	13.8836	28.6945	49.6745	2.0235	20.3524	42.0641	43.68	2.546	19.2248	39.7336	62.224	1.311	18.1952	37.6056	53.712	2.6125
15	27.4292	56.6903	59.184	2.8215	18.0439	37.2929	88.6815	0.6935	23.9463	49.492	57.398	6.4315	22.0587	45.5907	78.032	5.776
16	19.3067	39.9028	54.9945	13.395	21.8022	45.0606	52.3535	3.287	17.4395	36.0439	55.498	2.5555	21.7533	44.9594	62.566	6.0515
17	23.6847	48.9513	65.264	11.9605	18.323	37.8699	64.0765	2.394	14.9213	30.8392	52.724	1.254	20.9794	43.36	51.717	2.964
18	17.6347	36.4471	70.1755	4.807	22.1808	45.843	60.3145	2.584	22.8173	47.1585	51.09	2.375	17.4688	36.1043	51.8215	4.7975
19	20.0491	41.4373	78.9915	1.9095	28.0956	58.0677	104.566	3.5815	15.9327	32.9295	54.2535	2.3465	15.6214	32.2861	70.3845	0.494
20	25.2076	52.0988	66.043	2.5555	14.298	29.5509	37.0395	0.9785	18.1491	37.5104	61.521	2.0045	22.0536	45.5802	78.507	8.6355
21	21.7116	44.8733	102.048	0.1805	16.6262	34.3628	47.233	0.6745	17.9365	37.0709	51.1565	0.513	19.2994	39.8877	57.9775	1.9475
22	22.0536	45.5802	71.762	8.4075	18.4677	38.1689	57.056	4.332	**12.5967**	**26.0348**	**59.3835**	**2.6885**	**12.3497**	**25.5243**	**55.745**	**0.7695**
23	**12.2511**	**25.3204**	**57.854**	**0.2185**	16.6481	34.4081	53.0185	1.5865	17.1801	35.5077	42.825	0.703	17.8122	36.8141	67.297	0.6745
24	20.1646	41.6759	56.049	0.152	29.1326	60.2109	62.642	2.527	20.796	42.9809	61.939	3.477	25.6125	52.9356	97.583	1.843
25	18.6848	38.6175	39.424	0.095	9.80738	20.2698	49.513	4.712	17.1041	35.3506	51.1565	0.5985	17.2992	35.7539	63.1265	2.375
26	24.1414	49.8952	87.703	7.809	22.9985	47.5331	65.5775	6.1085	21.3857	44.1997	78.0795	7.676	23.2725	48.0994	81.585	0.9785
27	22.6609	46.8353	79.153	1.178	24.3154	50.2547	71.002	0.57	23.1739	47.8955	68.2945	1.273	21.2081	43.8327	63.877	1.406
28	16.9492	35.0304	37.429	10.507	**9.75112**	**20.1535**	**37.7045**	**0.5415**	16.5385	34.1816	61.806	2.3845	18.3209	37.8653	49.171	3.1255
29	19.1152	39.5071	64.4945	1.0735	18.8879	39.0374	59.7065	0.3895	12.843	26.5438	89.736	0.2945	28.4859	58.8742	91.066	3.933
30	21.2052	43.8267	61.1505	7.3815	18.8258	38.909	62.832	1.4915	32.5986	67.3745	78.241	3.5815	22.457	46.4139	109.734	3.306
31	15.4299	31.8904	52.7335	3.2015	27.3846	56.5981	81.8605	0.285	20.2698	41.8934	69.121	5.0445	23.1542	47.8548	101.982	4.066

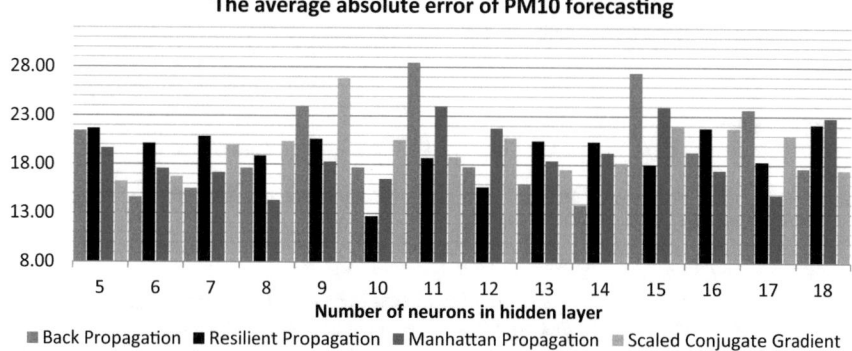

Fig. 3 The average absolute error of PM10 forecasting (5–18 neurons in hidden layer)

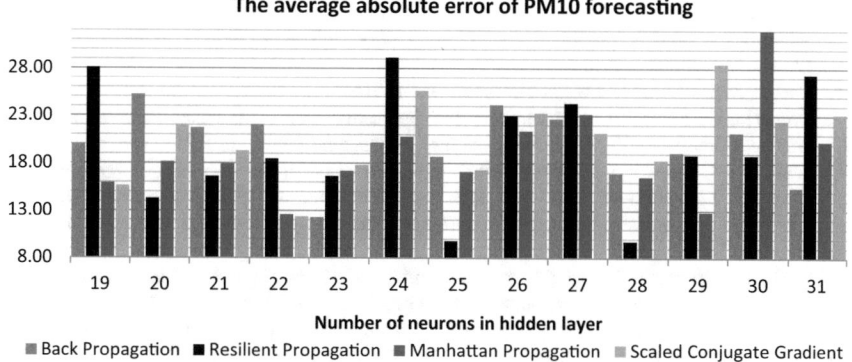

Fig. 4 The average absolute error of PM10 forecasting (19–31 neurons in hidden layer)

neurons in the hidden layer making use of the Resilient Propagation method (average forecast error was 9.80 $\mu g/m^3$).

It was also confirmed that it is possible to use an ANN to predict the hourly PM10 concentration levels (24 h in advance) with relatively good accuracy (as in Gdańsk). As a next step, further development of the model is planned to take into account weather and air quality prediction models, in order to generate a more rich meteorological set of parameters as well as AQ information for the ANN. Adjusting the artificial neural network to the new configuration of input parameters should reduce forecasting error.

Based on the current study, it was found that the process of adaptation of the network required a re-analysis of the network structure and its training process. It turned out that the models developed for Gdańsk require changes to both the structure of the model and methods of teaching. This means that the processes of adapting model predictions with the use of an artificial neural network require that each model is dedicated to an individual city.

Acknowledgment The research were conducted partly within S/WZ/1/2014 project financed from Ministry of Science and Higher Education funds.

References

1. Howes, R., Skea, J., Whelan, B.: Clean and Competitive: Motivating Environmental Performance in Industry (2013)
2. Guerreiro, C., Foltescu, V., de Leew, F.: Air quality status and trends in Europe. Atmos. Environ. **98**, 376–384 (2014)
3. Orłowski, C., Sarzyński, A.: A model for forecasting pm10 levels with the use of artificial neural networks. In: Information Systems Architecture and Technology—the use of IT Technologies to Support Organizational Management in Risky Environment, Wrocław (2014)
4. Orłowski, C., Sarzyńska, M., Sarzyński, A.: Prototyp modelu systemu samouczącego do prognozowania stężenia pyłu pm10 w powietrzu atmosferycznym. In: Innowacje w zarządzaniu i inżynierii produkcji, Opole (2014)
5. Moussiopoulos, N., Vlachokostas, Ch., Tsilingiridis, G., Douros, I., Hourdakis, C., Naneris, C., Sidiropoulos, C.: Air quality status in Greater Thessaloniki Area and the emission reductions needed for attaining the EU air quality legislation. Sci. Total Environ. **407**(4), 1268–1285 (2009)
6. Google Maps. https://www.google.com/maps/ (2016)
7. Orłowski, C., Kowalczuk, Z.: Modelowanie procesów zarządzania technologiami informatycznymi. Pomorskie Wydawnictwo Naukowo-Techniczne PWNT (2012)
8. Sitek, T., Orłowski, C.: Ocena technologii informatycznych—Koncepcja wykorzystania systemów inteligentnych, [w:] R. Knosala (red.), Komputerowo zintegrowane zarządzanie, T. 2, Oficyna Wydawnicza Polskiego Towarzystwa Zarządzania Produkcją, Opole, pp. 153–159 (2007)
9. Sitek, T., Orłowski, C.: Model of Management of Knowledge Bases in the Information Technology Evaluation Environment. In: Grzech, A., Borzemski, L., Świątek, J., Wilimowska, Z. (eds.) Information Systems Architecture and Technology: Models of the Organisation's Risk Management, Oficyna Wydawnicza Politechniki Wrocławskiej, pp. 221–231, Wrocław (2008)
10. Kosiński, R.: Sztuczne sieci neuronowe Dynamika nieliniowa i chaos (2005)

Checking Compliance of Program with SecureUML Model

Thanh-Nhan Luong, Van-Khanh To and Ninh-Thuan Truong

Abstract Access control is one of the most efficient ways to restrict resource access violations. However, in software development process, programs may not be fully complied with access policies represented in specifications. In this paper, we present an approach to verify the access policies compliance between a SecureUML model and its software program. We extract access control rules specified in SecureUML model, analyze source code of its program and propose an algorithm to check the compliance between two paradigms. Our approach can help programmers to detect some resource access violations and to improve the quality of software systems.

Keywords Specification · Compliance · Checking · SecureUML

1 Introduction

Security is always an important criterion in assessing quality of software in current stage of technology. Each software field focuses on specific security properties. However, the basic security properties of a secure software are confidentiality, integrity, availability, accountability, and non-repudiation [1]. If a software system needs to be secure, then it is important to be sure that all of the security policy is enforced by sufficiently strong mechanisms. There are organized methodologies and risk assessment strategies to assure the completeness of security policies and assure that they

T.-N. Luong (✉)
Department of Informatics, Hai Phong University of Medicine and Pharmacy,
72A Nguyen Binh Khiem, Ngo Quyen, Hai Phong, Vietnam
e-mail: ltnhan@hpmu.edu.vn; ltnhan1982@gmail.com

V.-K. To · N.-T. Truong
VNU University of Engineering and Technology,
144 Xuan Thuy, Cau Giay, Hanoi, Vietnam
e-mail: khanhtv@vnu.edu.vn

N.-T. Truong
e-mail: thuantn@vnu.edu.vn

© Springer International Publishing AG 2017   489
D. Król et al. (eds.), *Advanced Topics in Intelligent Information
and Database Systems*, Studies in Computational Intelligence 710,
DOI 10.1007/978-3-319-56660-3_42

are completely enforced. In practice, however, a model of a software system and its implementation may be created by different persons and at different phases. As a consequence, it raises the problem of incompliance between the specification and implementation of security policy in software development.

Role-based access control (RBAC) [2] is one of the methods to make confidentiality, integrity of a software system and limit access breaches. The basic principle of RBAC is the least privilege. SecureUML [3] is a modeling language based on Unified Modeling Language (UML) and RBAC model with additional support for specifying authorization constraints. SecureUML defines a vocabulary for expressing different aspects of access control, like roles, role permissions and user-role assignments. Obviously, system designers use SecureUML for modeling access policies of software systems. Thus, finding a way to prove that the software program complied with all the specification in its SecureUML model increases the reliability of software.

In this paper, we propose an approach to check the compliance of the object oriented programs against the predefined specification based on the SecureUML. The contribution of the approach includes:

- Firstly, we analyze the main function of an object oriented program as a four-level tree of authorization elements
- After that, we propose an algorithm to check whether the program is conformed to the SecureUML specification
- Finally, we propose a platform to implement the approach.

The rest of the paper is organized as follows. Section 2 presents some basic knowledge about RBAC and SecureUML metamodel. The next section discusses studies related to our work. Section 4 describes a case study of the healthcare software system. Our approach of compliance checking is introduced in Sect. 5. The last section we draw some conclusions and future works.

2   Background

The main elements of SecureUML model is represented in SecureUML metamodel which is based on RBAC model. Therefore, we recall the basic knowledge of role-based access control and briefly represent SecureUML metamodel in this section.

2.1   Role-Based Access Control

The RBAC model is depicted in Fig. 1, it includes five main elements which are *Users*, *Roles*, *Permissions*, *Objects* and *Operations* [2, 4]. A *User* can be simply understood as a person who participates in the system. A *Role* represents a work position in an organization which reflects responsibility and duty of the user. *Permissions* show legal *Operations* (sequence of actions) on *Objects* (protected resources).

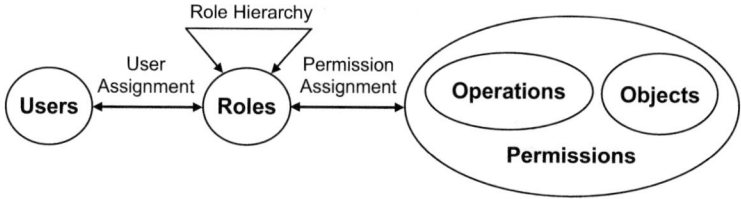

Fig. 1 Role-based access control

The center concept of RBAC is the role and the basic relationships are *Permission Assignment*, *User Assignment* and *Role Hierarchy* [4]. Specifically, an user can be assigned to many roles and a role can have several users. Each role is assigned with some permissions, is defined depending on the policies of organization. The principle of least privilege helps users in RBAC model only has sufficiently roles and permissions to carry out their duties. The inheritances between roles are described in *Role Hierarchy*. Role $r1$ inherits role $r2$ if role $r1$ has all permissions of $r2$.

In the RBAC model, users do not perform actions directly but through the roles. Therefore, if an user's job position or policy of the organization alters then the administrator of system can easily change the roles, permissions by granting or revoking them. This increases the flexibility in controlling access to system resources [2, 5].

2.2 SecureUML Metamodel

The combination of UML metamodel and the concepts of RBAC forms SecureUML metamodel, is shown in Fig. 2 [3, 6, 7]. *User*, *Role*, *Permission* and relationship

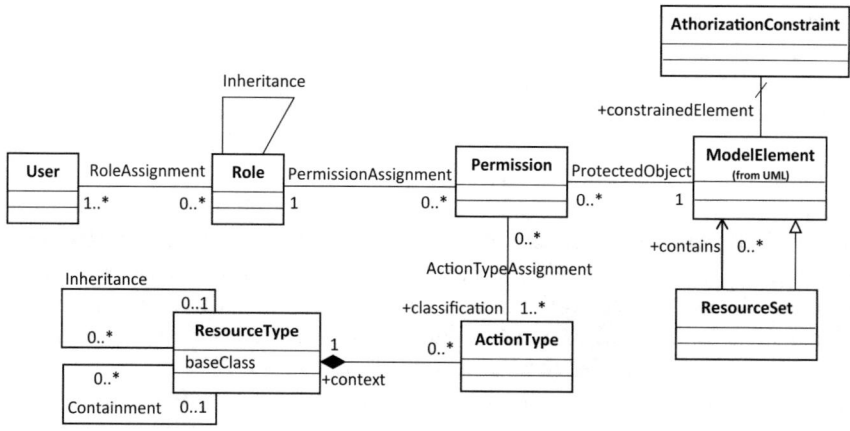

Fig. 2 SecureUML metamodel

between them are similar to *Users*, *Roles*, *Permissions* in RBAC model. *Resource Type* defined all available action types for a particularly metamodel type. *ActionType* is used to describe the semantics and classify *Permission*. A user defined set of model elements is represented by *ResourceSet* which uses to define permissions or authorization constraints. The protected resources is generalized to *ModelElement* and *AuthorizationConstraint* expresses a precondition imposed on every call to an operation of a particular resource.

3 Related Work

There have been many research works related to the verification of the software program's access policies. In this section, we summarize some studies similar to our work.

In [8], Edwin Okoampa Boadu and Gabriel Kofi Armah presented a RBAC based design architecture. They applied RBAC concepts in a Hospital management system to execute security policies and protect resources of system. Therefore, the violations relevant unauthorized accessing to medical records of patients or other resources are eliminated. Although they proposed RBAC model for information management system of hospital and a implementation program, they did not verify the compliance between model specification and its program.

Fisler [9] proposed Margrave tool which is used to analyze RBAC policies (written in the XACML language) and then translates them into a form of decision-diagram. This tool checks satisfaction of policy and property. In additional, the tool also analyzes change-impact of policies. However, this approach has not mentioned to program which implements RBAC policy.

Clarke [10] used Bounded Model Checking (BMC) to build a tool for formal verification of ANSI-C programs. The tool can apply to check safety properties and compare an ANSI-C program with another design. The approach checked pointer safety, array bounds, and user-provided assertions properties, but it does not solve resource access permissions or users.

Qamar et al. [11] specified and validated SecureUML model of a medical information system. Although authors used formal methods but they did not solve the verification work. The study of Qamar [12] proposed a formal approach to answer the question of who is accessing What in Electronic Health Record (EHR) via the form of feedback-based queries. Security policy specifications are represented by Z notation and SecureUML model. In addition, authors introduce the Jaza tool which is used to validate that specifications. Their study automatically detect authorized user of patient medical record and exposed information with that system user. However, the approach is only applied to validate SecureUML model.

All of above studies did not analyze the resource access policies in software program's source code. In our study, we extract access control rules specified in SecureUML model, analyze source code of its program and propose an algorithm to check the compliance between two paradigms.

4   A Case Study

In the healthcare software system, protecting patient informations from unauthorized access users or unexpected modifications are very important. In fact, if this system do not warrant security standards, patients cannot give private information which is significant for diagnostic. EHR [13, 14] has been used widely in health care facilities (hospitals). Beside it has brought many benefits (effectiveness, convenience,...), protecting information of EHR from unauthorized disclosure has been attending. In regulations like HIPAA (USA's Health Care Insurance Portability and Accountability Act), patients' EHRs should be confidentially kept because they contain private information. The patients and their doctors are used EHRs to monitor and treatment diseases.

Figure 3 describes a small part of EHR management system which is used to illustrate our approach. In this SecureUML model, each EHR is protected resources and contain the information about profile code, patient code, doctor code and other contents. There are three roles (Doctor, Patient and AdministrativeStaff) in system. The users need some privileges to operate their tasks in a hospital via roles. The patient can only read the information in his/her medical record, the doctors can read the information, update the contents in their patients' records and an administrative staff can read all medical records. Each user of the system have to comply above access policies.

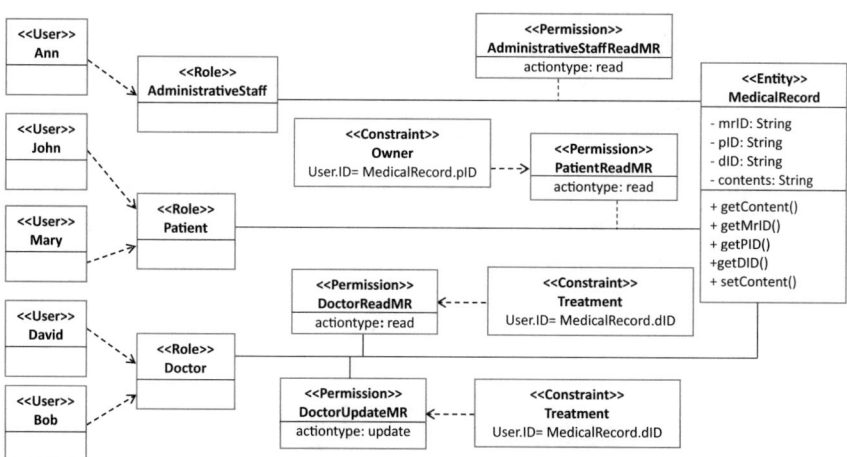

Fig. 3   A part of EHR management system using SecureUML

5   Approach

In this section, we give an approach to analyze a program as an abstract tree of authorization elements and then we propose an algorithm to check the compliance of the program with its SecureUML model.

5.1   Program Analysis

To facilitate the process of extracting information from the source code, we assume that the software system presented in Fig. 3 written by Java. We choose Java to illustrate our approach because it is a popular object-oriented programming language and there are many tools supported for analyzing source code.

We use a completely four-level tree to represent the main function of a program. The tree is built as follows:

- Root of the tree is called the node *main* and it has n children (n is the number of users in the program).
- Each node of level 1 represents one user in the program, namely user's name. The number children of each node on this level is equal the number assigned roles for that user.
- Node named r on level 2 has m children if user u (parent of r) with role r make m resources exploitations in the program.
- Each resource exploitation can be understood as a permission of user. It has three children which are action type, protected resource and condition. They can be created from method's specification or annotations of methods in the program. If an exploitation without any condition then condition's value is an empty string.

A path from a node on level 1 to a node on level 3 and children of it construct a tuple $<user, role, cond, actiontype, resource>$ to describe resources exploitation of user according to the user's role.

According to the tree building rule, we build the four-level tree of the source code of sample program (Appendix) presented in Fig. 4.

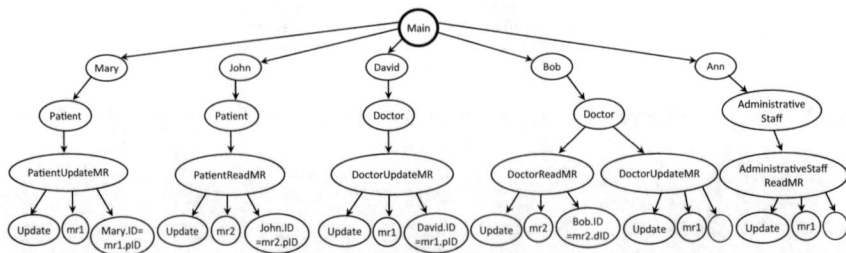

Fig. 4   Analyzing the main function as a tree

5.2 Algorithm for Checking Compliance

The input of algorithm is specification file of SecureUML model (SFile) and source code of program (PFile). Violation reports will be created in case of representing resources exploitation in the program but lack of it's specifications in the model.

From the description file of SecureUML model (such as *.XMI), we extract resource access rules to the set S of tuples $<r, at, rs, ct>$ where an any user in role r is executed action type at on resource rs if constraint ct to be satisfied.

In the our algorithm, the function $GetChildren(T, N)$ gets all children of node N in order from left to right and N is any branch node of tree T. The function $CheckExist(S, t)$ returns value *true* or *false*. It receives false value if t is not described in S and gets true value in the opposite case.

Algorithm 1 Algorithm for checking compliance

Status ⟵ **Compliance Verification** (*SFile, PFile*) {
1: Build the set $S = \{< r, at, rs, ct >\}$ from *SFile* //File describes SecureUML model
2: Build tree T from *PFile* //See Sect. 5.1
3: For each *user* ← $GetChildren(T, main)$
4:   For each *role* ← $GetChildren(T, user)$
5:     For each *prm* ← $GetChildren(T, role)$ {
6:       *act* ← $GetChildren(T, prm)$
7:       *res* ← $GetChildren(T, prm)$
8:       *cond* ← $GetChildren(T, prm)$
9:       $t =< user, role, act, res, cond >$ //a resource exploitation.
10:       If ($CheckExist(S, t) == false$) //t is not described in S
11:         Status := "Violation"
12:     }
13: Return
14: }

According to the Algorithm 1, there are two violations in sample code: (1) user *Mary* with the role *Patient* performed permission *PatientUpdateMR* and (2) user *Bob* with the role *Doctor* performed the permission *DoctorUpdateMR*. The violation is declared in the first case because the role *Patient* in the SecureUML model does not include the action type *update*. In the second case, the role *Doctor* has the permission *update* on the *MedicalRecord* resource if the constraint *User.ID = MedicalRecord.dID* is satisfied. However, in the program, doctor *Bob* performed *update* medical record without any condition. Therefore, this is also an access violation. In short the exploitations of users in the sample program are listed in the Table 1 (Appendix part).

Table 1 The exploitation of users in the sample program

User	Role	ActionType	Resource	Condition
Mary	*Patient*	*update*	*mr1*	*Mary.ID=mr1.pID*
John	Patient	Read	mr2	John.ID=mr2.pID
David	Doctor	update	mr1	David.ID=mr1.dID
Bob	Doctor	Read	mr2	Bob.ID=mr2.dID
Bob	*Doctor*	*update*	*mr1*	
Ann	AdministrativeStaff	Read	mr1	

6 Conclusion and Future Work

We have presented an approach to check the compliance of a software program with its SecureUML model. In which, the access policies specified in the model is extracted to the set of tuples to describe permissions of roles. While, the program's source code is analyzed into the four-level tree which shows user's resource exploitations. The algorithm allows us to check all the exploitation in the program and gives resource access violation reports. In addition, our work is illustrated on a case study of the EHR management system.

We used Soot library in Eclipse [15] to analyze source code of a program written in Java. From the analysis results of Soot and the programmer's description file about methods of source code, we extracted all information about users, roles, action types, resources and conditions. However, we will use Eclipse to design the SecureUML model and export it to file *.XMI so we just used a list to store resource access rules of model. Our tool detects illegal resource exploitation of users. We will also extend our work and finish the support tool to solve larger case studies with inheritance role in the future.

Acknowledgements This research is funded by Vietnam National Foundation for Science and Technology Development (NAFOSTED) under grant number 102.03-2014.40.

Appendix

Below is source code snippets of the sample program.

```
public class User  {...}
public class Patient extends User { ...
    public String PatientReadMR(MedicalRecord mr){
        if (this.getID() == mr.getPID())
            return mr.getContent();
        return "";
    }
    public MedicalRecord PatientUpdateMR(MedicalRecord mr,String content){
        if (this.getID() == mr.getPID())
```

```
                mr.setContent(content);
            return mr;
        }
    }
    public class Doctor extends User { ...
        public String DoctorReadMR(MedicalRecord mr){
            if (this.getID() == mr.getPID())
                return mr.getContent();
            return "";
        }
        public MedicalRecord DoctorUpdateMR(MedicalRecord mr,String content){
            mr.setContent(content);
            return mr;
        }
    }
    public class AdministrativeStaff extends User{ ...
        public String AdministrativeStaffReadMR(MedicalRecord mr){
            return mr.getContent();
        }
    }
    public class MedicalRecord {//source code}

    public class Main {
        public static void main(String[] args) \{ ...

        //Create users Mary, John, David, Bob, Ann,.. from Java GUI input
        ...

        //Assign created users to their roles
        Mary = new Patient (Mary.getID(), Mary.getName(), "Myocarditis");
        John = new Patient(John.getID(), John.getName(), "Headache");
        David = new Doctor(David.getID(), David.getName(), "Heart");
        Bob = new Doctor(Bob.getID(), Bob.getName(), "Nerve");
        Ann = new AdministrativeStaff(Ann.getID(), Ann.getName());

        MedicalRecord mr1 = new MedicalRecord (Mary.getID(), David.getID());
        MedicalRecord mr2 = new MedicalRecord (John.getID(), Bob.getID());
        mr1=((Patient)Mary).PatientUpdateMR(mr1, content); //violation
        ((Patient)John).PatientReadMR(mr2);
        if (David.getID() == mr1.getDID())
            mr1=((Doctor)David).DoctorUpdateMR(mr1, content);
        mr1=((Doctor)Bob).DoctorUpdateMR(mr1, content); //violation
        ((AdministrativeStaff)Ann).AdministrativeStaffReadMR(mr1);
        }
    }
```

References

1. Mead, N.R., Allen, J.H., Barnum, S., Ellison, R.J., McGraw, G.: Software Security Engineering: A Guide for Project Managers. Addison-Wesley Professional (2004)
2. Ferraiolo, D., Kuhn, D.R., Chandramouli, R.: Role-Based Access Control. Artech House (2003)
3. Lodderstedt, T., Basin, D., Doser, J.: SecureUML: A UML-based modeling language for model-driven security. In: International Conference on the Unified Modeling Language, pp. 426–441. Springer (2002)

4. Ferraiolo, D.F., Sandhu, R., Gavrila, S., Kuhn, D.R., Chandramouli, R.: Proposed NIST standard for role-based access control. ACM Trans. Inf. Syst. Secur. (TISSEC) **4**(3), 224–274 (2001)
5. Sandhu, R.S., Coynek, E.J., Feinsteink, H.L., Youmank, C.E.: Role-based access control models. IEEE Comput. **29**(2), 38–47 (1996)
6. Basin, D., Doser, J., Lodderstedt, T.: Model driven security: from UML models to access control infrastructures. ACM Trans. Softw. Eng. Methodol. (TOSEM) **15**(1), 39–91 (2006)
7. Matulevičius, R., Dumas, M.: A comparison of secureUML and UMLsec for rolebased access control. In: Proceedings of the 9th Conference on Databases and Information Systems, pp. 171–185 (2010)
8. Boadu, E.O., Armah, G.K.: Role-based access control (RBAC) based in hospital management. Int. J. Softw. Eng. Knowl. Eng. **3**, 53–67 (2014)
9. Fisler, K., Krishnamurthi, S., Meyerovich, L.A., Tschantz, M.C.: Verification and change-impact analysis of access-control policies. In: Proceedings of the 27th International Conference on Software Engineering, pp. 196–205. ACM (2005)
10. Clarke, E., Kroening, D., Lerda, F.: A tool for checking ANSI-C programs. In: International Conference on Tools and Algorithms for the Construction and Analysis of Systems, pp. 168–176. Springer (2004)
11. Qamar, N., Ledru, Y., Idani, A.: Validation of security-design models using z. In: International Conference on Formal Engineering Methods, pp. 259–274. Springer (2011)
12. Qamar, N., Faber, J., Ledru, Y., Liu, Z.: Automated reviewing of healthcare security policies. In: International Symposium on Foundations of Health Informatics Engineering and Systems, pp. 176–193. Springer (2012)
13. Gunter, T.D., Terry, N.P.: The emergence of national electronic health record architectures in the United States and Australia: models, costs, and questions. J. Med. Internet Res. **7**(1), e3 (2005)
14. Kierkegaard, P.: Electronic health record: wiring Europes healthcare. Comput. Law Secur. Rev. **27**(5), 503–515 (2011)
15. M. University: Soot: a Java optimization framework. https://www.sable.mcgill.ca/soot/index.html (2012)

Generation of Test Data Using Genetic Algorithm and Constraint Solver

Ngoc-Thi Dinh, Hieu-Dinh Vo, Thi-Dao Vu and Viet-Ha Nguyen

Abstract Search-based testing techniques using genetic algorithm (GA) can automatically generate test data that achieves high coverage on almost any given program under test. GA casts the path coverage test data generation as an optimization problem and applies efficient search-based algorithms to find suitable test cases. GA approaches scale well and can handle any source code and test criteria, but it still has some degrades when program under test has critical path clusters. This paper presents a method for improving GA efficiency by integrating a constraint solver to solve path conditions in which regular GA cannot generate test data for coverage. The proposed approach is also applied to some programs under test. Experimental results demonstrate that improved GA can generate suitable test data has higher path coverage than the regular one.

Keywords Genetic algorithm · Path coverage testing · Automatic test data generation

N.-T. Dinh (✉) · H.-D. Vo · V.-H. Nguyen
VNU University of Engineering and Technology,
E3 Building, 144 Xuan Thuy Street, Cau Giay, Hanoi, Vietnam
e-mail: thidn.di13@vnu.edu.vn; dinhngocthi@gmail.com

H.-D. Vo
e-mail: hieuvd@vnu.edu.vn

V.-H. Nguyen
e-mail: hanv@vnu.edu.vn

T.-D. Vu
Academy of Cryptography Techniques, 141 Chien Thang Street, Thanh Tri,
Hanoi, Vietnam
e-mail: daovt.di10@vnu.edu.vn

© Springer International Publishing AG 2017
D. Król et al. (eds.), *Advanced Topics in Intelligent Information
and Database Systems*, Studies in Computational Intelligence 710,
DOI 10.1007/978-3-319-56660-3_43

499

1 Introduction

Software products are becoming more and more important in our daily life, therefore the appliance of software testing to ensure those product quality has become a vital requirement. However, as most of the software testing is being done manually, the workforce and cost required are accordingly high [1]. In general, about 50% of workforce and cost in the software development process is spent on software testing [2]. Considering those reasons, automated software testing has been evaluated as an efficient and necessary method in order to reduce those effort and costs.

Automated structural test data generation is being the research topic attracting much interest in automated software testing, for it not only enhances the efficiency but also can reduce considerably costs of software testing. In our paper, we will focus on path coverage test data generation, considering that almost all structural test data generation problems can be transformed to the path coverage test data generation one. Moreover, Kernighan and Plauger [3] also pointed out that path coverage test data generation can find out more than 65% of bugs of the given program under test.

Although path coverage test data generation is the major unsolved problem [4], various approaches have been proposed by researchers. These approaches can be classified into two types: static analysis and dynamic analysis.

Symbolic execution (SE) is the state-of-the-art of static analysis approaches [5]. Even though there have been significant achievements, SE still faces difficulties in handling infinite loops, array, procedure calls and pointer references in each program under test [6].

There are also random testing, local search [7], and evolutionary methods [8–10] in dynamic analysis approaches. As the value of input variables is assigned when program executes, problems encountered in static analysis approaches can be avoided in dynamic analysis approaches.

Being an automated searching method in a predefine space, genetic algorithm (GA) was applied to test data generation since 1992 [11]. Micheal et al. [6], Levin and Yehudai [10], Joachim et al. [12] indicated that GA outperforms other dynamic analysis methods e.g. local search or random testing.

Even though GA-based test data generation has already proved its efficiency in generating test data for dynamic approaches, it still faces difficulties when the programs under test have test paths with low generation probability for coverable test data. For example, consider program under test Sample() as below [13]:

```
1   void Sample(double x, double y, double z) {
2      if (Math.cos(z)- 0.95 < Math.exp(z)) {
3         if ((x + y == 1024) && (y > 1000))
4            // path 1
5         else
6            // path 2
7      }
8      else
9         // path 3
10 }
```

By using constraint solvers, even though symbolic execution (SE) can solve the second condition, it cannot for the first because of it having the Math library functions of the Java language. In the contrary, GA can generate test data for the first condition but it has degraded with the second one. This means that if using only SE or GA, we cannot have test data for the path 1.

In order to solve the problem, this paper presents a proposal to improve GA in generating test data to cover all paths in the above program under test. Our approach combines constraint solvers into GA. The static program analysis phase is applied to find out paths of the program under test which are difficult to be covered. In this paper, the difficult path means the path contains if-else statement which is difficult to generate test data for coverage. For these difficult paths, the constraint solver Z3 [14] is used to solve the path conditions. Basing on the result, the constraint satisfaction is used as mutated chromosome in the procedure of generating new populations in GA.

The rest of this paper is organized as follows: Sect. 2 gives some theoretical backgrounds including path coverage test data generation as an optimization problem, and conditional statements in Java. Section 3 summarizes some related works, and Sect. 4 presents the proposed approach in detail. Section 5 shows the experimental results and discussions. Section 6 concludes the paper.

2 Background

This section describes the theoretical background being used in our proposed approach.

2.1 Transform Path Coverage Testing into an Optimization Problem

When using GA, a path coverage test data generation is transformed into an optimization problem. To cover a test path during execution, we must find appropriate values for the input variables which satisfy related branch predicates. The usual way is to use Korel's branch distance function [7]. For example, if a branch predicating B is $(x \leq y - 5)$, then apply the Korel function $f(B) = x - (y - 5)$. As a result, generating test data for a desired branch is transformed into searching input values which minimizes the return value of its Korel function. Table 1 gives some common formulas which are used in branch distance functions. To generate test data for a desired path P, we define a fitness function $F(P)$ as the sum of all related branch distance functions. For these reasons, generating path coverage test data can be converted into searching input values which can minimize the return value of function $F(P)$.

In Korel function, k is the smallest step between 2 operands in the condition. In this paper, all variable types are double, therefore as to simplify we assume $k = 0$.

Table 1 Korel's branch distance function [7]

No	Predicate	Branch distance function
1	x = y	f(B) = abs(x − y)
2	x ≠ y	f(B) = k
3	x < y	f(B) = (x − y) + k
4	x ≤ y	f(B) = (x − y)
5	x > y	f(B) = (y − x) + k
6	x ≥ y	f(B) = (y − x)
7	B1 ∧ B2	f(B) = f(B1) + f(B2)
8	B1 ∨ B2	f(B) = min(f(B1), f(B2))

2.2 The if-else Statement

The if-else statement allows choosing between two alternatives. The syntax of the if-else statement is as below:

```
if (condition)
    then-statement
else
    else-statement
```

The condition is an arbitrary expression of boolean type. There are 4 types of if-else statement as below [15].

(1) A variable of boolean type

```
boolean bDone;
// ...
if (bDone)
```

(2) Compare between variables (or expressions) of a primitive type by using predicate operators (==, !=, >, <, >=, or <=)

```
double x, y, z;
// ...
if (x == y - z)
```

(3) A call to the method which returns a value of type boolean

```
String password;
// ...
if (password.compareTo("Tomcat"))
```

(4) A complex boolean expression contains some simpler expressions which use the boolean operators !, &&, and ||

```
int x, y, z, t;
double e, f;
// ...
if ((x > (y + z)) || (x == t) && !(Math.abs(e - f) > 10))
```

In this paper, we will focus on 2 types of if-else statement which are (2) (only one comparison operator) and (4) (a complex boolean expression).

3   Related Work

The path coverage literature using GA started with Lin and Yeh [16] in 2000. They extended Jones et al. work [17] from branch coverage to path coverage. The ordinary (weighted) Hamming distance was extended to handle different orders of target paths having same branches. The fitness function is called SIMILARITY, which computes similar items with respect to their ordering between two different paths: actual executed path and target path. Only one program was used to test the approach, i.e. simple triangle classifier. It was reported that the approach outperformed random search. However, in this method, test data generation must be called many times to generate the test data for the most difficult path to be covered. In addition, because their work only used GA, the program under test `Sample()` in the Sect. 1 cannot cover all test paths.

Chen and Zhong [18] developed a multi-population genetic algorithm for path testing. This work has improved the GA-based path testing as described in Sect. 2.2. The work revealed that the proposed approach outperformed the traditional genetic algorithm based one by using the triangle classifier as the program under test. Like our approach, Chen and Zhong also targeted finding the test data to cover path conditions of the most difficult path to be covered in the program under test. As it approached the parallel processing, test data generating time was better than that of the regular GA. But number of test data generation was still high (21073 test data generations were required by average count).

Malburg and Fraser [13] introduced a novel mutation operator for evolutionary search which was based on dynamic SE. Path constraints were collected during execution, and negation of one of the path constraints yielded a new individual which followed a different execution path. This approach is different from ours, as it does not use direct constraint solver to solve the path constraints, therefore for some programs in their benchmark test, e.g. ASW or WBS, there are still test paths with low coverage.

Thi et al. [19] also proposed one method of combining static analysis program with GA. This method also identified test paths which were difficult to generate test data in each program under test, then basing on the path conditions of these difficult paths to perform adjustment in the procedure in order to generate new population of GA. This method was also proven to be effective with the program under test triangle classifier, however there was still a limitation on manual adjustment process. In this paper our proposed approach has a greater improvement. The constraint solver Z3 is used to solve path conditions, then save the constrain satisfaction into one structure to perform automatic adjustment in GA.

4 Proposed Approach

This section describes details of our proposed approach for automatic test data generation using GA. We propose a 2-phase approach as in Fig. 1.

4.1 Perform Static Program Analysis

The purpose of this phase is to create a list of input variables and their setting values by performing static analysis program and using constraint solver. This list is used as the conditions for procedure adjustment of GA in the next phase. To create this list, we take the following steps:

4.1.1 Solve Path Conditions

In this paper, we analyze two types of if-else statement Java language as discussed in Sect. 2.2, which are "only one comparison operator" and "complex boolean expression". We also store test path ID corresponding to each path condition. To solve these path conditions, the open source C++ constraint solver Z3 [14] is applied.

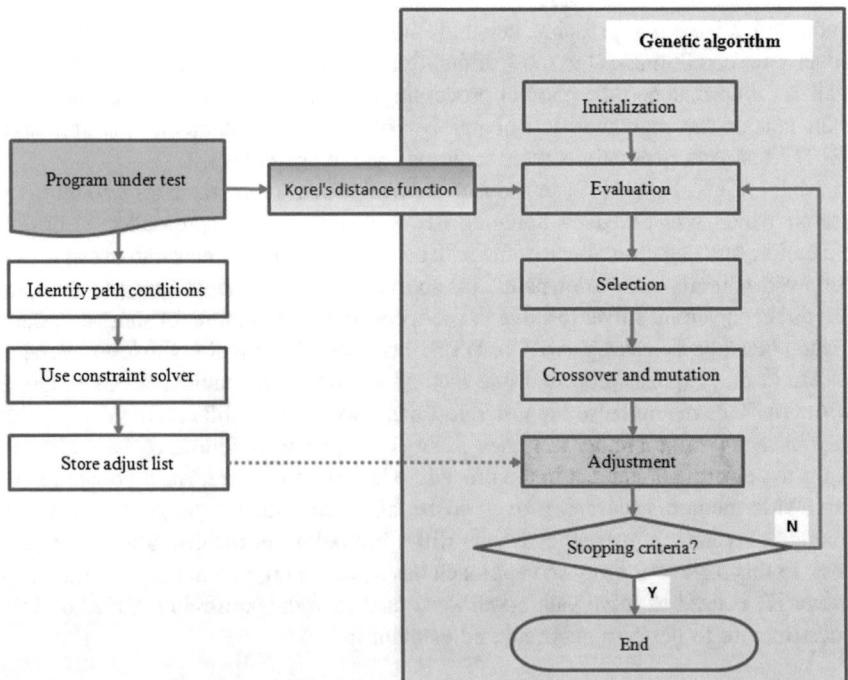

Fig. 1 Main steps of our proposed approach

From our experiments, we find out that, with the equal condition in the if-else statement and without any adjustment in GA, we cannot generate test data to satisfy these condition statements. Therefore, we extract equal condition from the if-else statement and solve them by using widely known constraint solver Z3, then store the constraint satisfaction (also called mutated individual) in a list to adjust in GA.

Algorithm 1 Solve list of path conditions

Input: L_{in} list of of path conditions
Output: L_{out} list of constraint satisfactions
1: **for each** path condition in L_{in}
2:      use Z3 to solve constrains;
3:      store constraint satisfaction also test path ID into L_{out}
4: **end for**
5: **return** L_{out}

4.1.2   Store Constraint Satisfaction

This paper uses the Adjust class below to hold each constraint satisfaction (output from constraint solver Z3) of a program under test.

```
class Adjust {
    public int index;        // order of input variable
    public double value;     // assigned value of parameter
    public int testpathID;   // corresponding test path ID
}
```

Value for each Adjust object will be set by using the following algorithm.

Algorithm 2 Store list constraint satisfactions

Input: L_{in} list of constraint satisfactions
Output: L_{out} list of Adjust objects
1: **for each** constraint satisfaction in L_{in}
2:      create a new Adjust object;
3:      Adjust.index = order of variable in constraint satisfaction;
4:      Adjust.value = value of variable in constraint satisfaction;
5:      Adjust.testpathID = corresponding test path ID;
6:      add Adjust object to L_{out};
7: **end for**
8: **return** L_{out}

To illustrate this idea, back to the `Sample()` program under test mentioned in Sect. 1, the second condition statement `((x + y == 1024) && (y > 1000))` (line 2) will be solved by constraint solver Z3. We will get constraint satisfaction {x = 23, y = 1001} and store in the list of Adjust classes as below:

```
adjust[0].index = 0;       // input variable x (order = 0)
adjust[0].value = 23;      // assigned value of x
adjust[1].index = 1;       // input variable y (order = 1)
adjust[1].value = 1001;    // assigned value of y
```

4.2   Execute GA

To automatically generate test cases, we use GA with the below procedures:

4.2.1   Population Representation

Depend on the input variables of the program under test, GA uses a double or an integer array as a chromosome $chrom = (x_1, x_2 \ldots x_n)$ to encode a test case. The array size is same as the number of input variables. The array type depends on the type of each input variable. For example, we use an array `double x [3]` to encode a test case of the program under test `Sample()` in Sect. 1.

4.2.2   Initial Population

This procedure set all parameters for GA. At first, it needs to identify a fixed number of chromosomes in a population and a maximum population generation for each time of GA run. After that, it initializes the random values for all chromosomes in the first population.

4.2.3   Fitness Function

Korel's branch distance function (mentioned in Sect. 2.1) is used as the fitness function in improved GA. To apply the Korel's branch distance function, which is similar to the previous approaches [14], we manually insert instrumented code into program under test and use it as the fitness function of GA. For example, with program under test `Sample()` in Sect. 1, instrumented code will be manually inserted into the original code at line 2 and 11 as below:

```
1  double Sample(double x, double y, double z) {
2     double ret = (Math.cos(z) - 0.95) - Math.exp(z);
3     if (Math.cos(z)- 0.95 < Math.exp(z)) {
4       if ((x + y == 1024) && (y > 1000))
5         // path 1
6       else
7         // path 2
8     }
9     else
10       // path 3
11    return ret;
12 }
```

4.2.4 Selection

The selection uses Korel function as fitness function to evaluate fitness value for each chromosome which encodes a test case. A chromosome with higher fitness value is considered as the better chromosome from the current population to the next generation. Selected chromosomes are also called parent chromosomes.

4.2.5 Crossover and Mutation

After selection, the crossover creates new chromosomes (child chromosomes) from the selected chromosomes (parent chromosomes) by swapping the values of vector $x = (x_1, x_2 ... x_n)$ between two selected chromosomes. This process is repeated for different parent chromosomes until the next generation has enough chromosomes. Thereafter, the mutation operator is applied to randomly alter some values in child chromosomes in order to introduce a better solution for optimization.

4.2.6 Constraint-based Adjustment

The purpose of the constraint-based adjustment procedure is to help GA to generate test data which can cover the entire test paths of the given program under test. Therefore, after executing the mutation of regular GA, based on the list of constraint satisfactions which are contained in the list of Adjust classes, we need to adjust the values of each chromosome in the population. The adjustment will be executed as follows:

Algorithm 3   Constraint-based adjustment
Input: List of Adjust objects
Output: Adjusted chromosome
1: **for each** Adjust object in the list
2:       if the Adjust.testpathID is not coverage
3:               chromosome.x[Adjust.index] = Adjust.value
4:       end if
5: **end for**
6: **return** the adjusted chromosome

5   Experimental Result

In this section, we present the experimental results of test data generation of improved GA for 3 given programs under test, and then compare the results with those of regular GA.

5.1   Programs Under Test

Besides the program under test Sample() presented in Sect. 1, to demonstrate the effectiveness of our proposed approach, 2 other programs under test Tritype() and QuadEq2() are executed as follows:

5.1.1   Tritype Program Under Test

This program has 3 input variables which are the angle and two sides of a triangle. This program also uses the Math library functions of the Java language to determine if 3 input variables show an equilateral, isosceles, scalene triangle or not. SE-based testing cannot solve the condition statement in line 2 and 4, while regular GA also faces problems with condition statement in line 3.

```
 1  void Tritype(double corn,double edge1,double edge2) {
 2    if (corn > 0 && corn < Math.PI) {
 3      if (edge1 == edge2) {
 4        if(Math.abs(Math.toDegrees(corn) - 60) < 0.01) {
 5          // path 1: Equilateral
 6        }
 7        else {
 8          // path 2: Isosceles
 9        }
10      }
11      else {
12        // path 3: Scalene
13      }
14    }
15    else {
16      // path 4: Not a triangle
17    }
18  }
```

At first we perform static program analysis to obtain equal condition statement which can be solved by constraint solver Z3 (edge1 == edge2). Constraint solver Z3 returns a constrain satisfaction {edge1 = 1, edge2 = 1} which is transferred to GA.

5.1.2 QuadEq 2 Program Under Test

This program finds all roots of a quadratic equation with 3 coefficients a, b and c being the input variables.

```
 1  void QuadEq2(double a, double b, double c) {
 2    double root1, root2;
 3    if (a == 0) {
 4      if (b != 0) {
 5        root1 = (-c)/b; // path 1
 6      }
 7      else {
 8        // path 2
 9      }
10    }
11    else if (((b*b) - (4*a*c)) < 0) {
12        // path 3
13    }
14    else {
15      if (((b*b) - (4*a*c)) == 0) {
16        root1 = (-b)/(a*2); // path 4
17      }
18      else {
23        // path 5
19        root1 = (-b + Math.sqrt(((b*b) - (4*a*c))))/(2*a);
19        root2 = (-b - Math.sqrt(((b*b) - (4*a*c))))/(2*a);
24      }
25    }
26  }
```

The QuadEq 2() has 3 equal condition statements which can be solved by constraint solver Z3 { (a == 0 && b != 0), (a == 0 && b == 0), ((b*b) - (4*a*c)) == 0) }. Constraint solver Z3 returns 3 constrain satisfactions {(a = 0, b = 1), (a = 0, b = 0), (a = 1, b = 2, c = 1)} which are transferred to GA.

5.2 GA Parameters Setting

Parameter settings of both regular GA and improved GA are as follows:

- Length of the chromosome: 3 (=3 input variables)
- Selection method: based on fitness value
- Two-point crossover probability: 0.5
- Mutation probability: 0.1
- Stopping criteria: all test target paths are covered.

Also, each program under test still requires other parameters as below (Table 2):

- Type: type of input variables
- Range: range of input variables
- Maxgen: maximum population generation for each time to run GA
- Popsize: number of chromosome for each population.

5.3 Results

The result test data generations of improved GA and regular GA are shown in the following tables. We will evaluate basing on two criteria: the number of covered test paths and the number of times to perform test data generation.

5.3.1 Test Path Coverage

This criterion is evaluated basing on the capacity to generate test data which can cover test paths of given program under test. The Table 3 proves that for all 3 programs under test, improved GA can generate test data with higher test path coverage than the regular one.

Table 2 GA parameter setting for each program

Program	Type	Range	Maxgen	Popsize
Sample	Double	[−10000, 10000]	150	250
Tritype	Double	[0, 10000]	150	250
QuadEq 2	Double	[−10000, 10000]	150	250

Table 3 Comparison on test path coverage between regular and improved GA

Program	Feasible path	Regular GA	Improved GA
Sample	3	2(67%)	3(100%)
Tritype	4	2(50%)	4(100%)
QuadEq 2	5	2(40%)	5(100%)

Table 4 Comparison on test path coverage between improved GA and Malburg and Fraser's approach [13]

Program	Lines of code	Feasible path	[13]	Improved ga
ASW	308	98	83(85%)	90(91%)
WBS	170	90	59(66%)	68(76%)

In addition, we also compare our proposed approach with the method given by Malburg and Fraser in their paper [13]. We did the experiments with 2 programs of benchmark test mentioned in this paper, the Altitude Switch (ASW) and the Wheel Brake System (WBS) (Table 4).

Both ASW and WBS are written by Java language. ASW has 308 lines of code with 98 feasible paths. WBS has 107 lines of code with 90 feasible paths. J. Malburg and G. Fraser's approach [13] can cover 83 paths (85%) for ASW and 59 paths (66%) for WBS, while ours can cover 90 paths (91%) and 68 paths (76%) respectively, proving that our proposed GA is more effective.

5.3.2 Test Data Generation Counts

This evaluation criterion is based on the number of times to perform test data generation which can cover the entire feasible paths in the given program under test.

From Table 5, we can see that improved GA just uses limited test data generation to cover the entire data paths of the given program under test, while the regular GA cannot do this.

Table 5 Test data generation counts

Program	Feasible path	Regular GA	Improved GA
Sample	3	Cannot cover all paths	252
Tritype	4	Cannot cover all paths	1098
QuadEq 2	5	Cannot cover all paths	756

6 Conclusion

In software development life cycle, even though software testing is one of the critical phases, it takes a lot of manual labor effort and cost. Therefore, much attention and interest have been raised about automated test data generation in both industry and academia. In this paper, we have applied the GA to generate test data automatically for feasible execution paths.

Our proposed approach is from a given program under test, we find out test paths which are difficult or impossible for GA to generate coverage test data, and then use the widely known constraint solver Z3 tool to solve these path conditions. The constraint satisfaction obtained from the Z3 will be used again in generating population of GA. The experimental results of some programs under test demonstrate that improved GA generated test data can cover all feasible paths having path conditions which cannot be covered by test data generated from regular GA.

In the future, we will investigate how to automatically insert instrument code into the given program under test. Moreover, we are going to extend our proposed method to apply to more complex programs under test.

References

1. Antonia, B.: Software testing research: achievements, challenges, dreams. In: Future of Software Engineering, pp. 85–103. IEEE Computer Society, Washington (2007)
2. Myers, G.J.: The Art of Software Testing, 2nd edn. Wiley (2004)
3. Kernighan, B.W., Plauger, P.J.: The Elements of Programming Style. McGraw-Hill Inc, New York (1982)
4. Weyuker, E.J.: The applicability of program schema results to programs. Int. J. Parallel Prog. **8**, 387–403 (1979)
5. Pasareanu, C.S., Visser, W., Bushnell, D., Geldenhuys, J., Mehlitz, P., Rungta, N.: Symbolic PathFinder: integrating symbolic execution with model checking for java bytecode analysis. Autom. Softw. Eng. J. (2013)
6. Michael, G.M., Schatz, M.: Generating software test data by evolution. IEEE Trans. Softw. Eng. **27**, 1085–1110 (2001)
7. Korel, B.: Automated software test data generation. IEEE Trans. Softw. Eng. **16**, 870–879 (1990)
8. Wegener, J., Baresel, A., Sthamer, H.: Evolutionary test environment for automatic structural testing. Inf. Softw. Technol. **43**, 841–854 (2001)
9. Wegener, J., Kerstin, B., Hartmut, P.: Automatic test data generation for structural testing of embedded software systems by evolutionary testing. In: Genetic and Evolutionary Computation Conference. Morgan Kaufmann Publishers Inc. (2002)
10. Levin, S., Yehudai, A.: Evolutionary testing: a case study. In: Hardware and Software, Verification and Testing, pp. 155–165 (2007)
11. Xanthakis, S., Ellis, C., Skourlas, C., Le Gall, A., Katsikas, S., Karapoulios, K.: Application of genetic algorithms to software testing (Application des algorithmes genetiques au test des logiciels). In: 5th International Conference on Software Engineering and its Applications, Toulouse, France, pp. 625–636 (1992)
12. Joachim, W., Baresel, A., Harmen, S.: Suitability of evolutionary algorithms for evolutionary testing. In: 26th International Computer Software and Applications Conference on Prolonging

Software Life: Development and Redevelopment. IEEE Computer Society, Washington (2002)

13. Malburg, J., Fraser, G.: Search-based testing using constraint-based mutation. J. Softw. Test. Verif. Reliab. **24**(6), 472–495 (2014)

14. Moura, L.D., Bjørner, N.: Z3: An efficient SMT solver. In: 14th International Conference on Tools and Algorithms for the Construction and Analysis of Systems, pp. 337–340. Springer Press, Berlin (2008)

15. Java programming lectures. https://www.inf.unibz.it/~calvanese/teaching/04-05-ip/lecture-notes/uni05.pdf

16. Lin, J.C., Yeh, P.L.: Using genetic algorithms for test case generation in path testing. In: 9th Asian Test Symposium, pp. 241–246. IEEE Computer Society, Washington (2000)

17. Jones, B.F., Sthamer, H.H., Eyres, D.E.: Automatic structural testing using genetic algorithms. Softw. Eng. **11**(5), 299–306 (1996)

18. Chen, Y., Zhong, Y.: Automatic path-oriented test data generation using a multi-population genetic algorithm. In: 4th International Conference on Natural Computation, vol 1, pp. 566–570 (2008)

19. Thi, D.N., Hieu, V.D., Ha, N.V.: A technique for generating test data using genetic algorithms. In: International Conference on Advanced Computing and Applications (accepted). IEEE Press, Can Tho (2016)

Investigating the Issues of Using Agile Methods in Offshore Software Development in Sri Lanka

V.N. Vithana, D. Asirvatham and M.G.M. Johar

Abstract Offshore software development (OSD) is the process of developing software by globally distributed teams and it has become increasingly popular due to the many benefits it offers. Reduced costs, access to wide range of skilled people and reduction in application development time are the major attractions to OSD. Agile methods are popular software development methodologies that help companies deliver high quality products rapidly. In order to stay competitive in today's dynamic business environment many software companies are embracing Agile methods. Agile methods require frequent informal communication between all stakeholders of the project. OSD introduces time, cultural and geographical barriers to this requirement. Previous studies in other countries have identified some of the challenges to implement Agile methods in an offshore environment. However, the impact of the challenges on the project success has not been studied by many. This study uses empirical and theoretical evidences in order to develop a model of challenges and their impact on project success. OSD as the fifth largest export earner of Sri Lanka, plays an important role in the Sri Lankan economy. Results of this study will help researchers as well as the ICT industry to obtain an awareness and to alleviate the challenges of OSD.

Keywords Offshore software development · Agile software development methods · Project success · Issues · Challenges · Global software development

V.N. Vithana (✉) · M.G.M. Johar
Management and Science University, Malaysia, Malaysia
e-mail: nipunika.vithana@gmail.com

D. Asirvatham
Taylor's University, Malaysia, Malaysia
e-mail: david.asirvatham@taylors.edu.my

© Springer International Publishing AG 2017
D. Król et al. (eds.), *Advanced Topics in Intelligent Information and Database Systems*, Studies in Computational Intelligence 710,
DOI 10.1007/978-3-319-56660-3_44

1 Introduction

Offshore software development (OSD) is the process of developing software by globally distributed teams. OSD also known as Global software Development (GSD) has become increasingly popular and thus the norm of developing software worldwide [2, 9].

The major benefits of OSD are lower cost, access to abundance of skilled people willing to work for a lower labour cost, improved quality and to reduce development time by developing the product for nearly 24 h [1, 9, 23].

Thus, many developing countries such as India, Sri Lanka, Philippines, and Bangladesh have become offshore outsourcing destinations. OSD has many barriers due to spatial distance, temporal distance and cultural differences [1, 14].

Software development methods constantly evolve due to new technologies and demands of the users. Organizations need to continuously adapt new structures, strategies and policies to stay competitive in today's dynamic business environment. Thus, agile software development methods are becoming increasingly popular with the great flexibility they provide organizations to adapt to changing requirements and to market products rapidly.

Agile methods were originated by a group of experienced professionals in 2001 based on their years of experience in the industry. According to the agile manifesto there is more value on, Individuals and interactions over processes and tools; Working software over comprehensive documentation; Customer collaboration over contract negotiation; Responding to change over following a plan.

'Agile' means being able to deliver quickly. Change quickly [13]. Agile methods are iterative and incremental methods. Agile development methods have been designed to solve the problem of delivering high quality software on time under constantly and rapidly changing requirements.

Agile methods were originally developed for projects where all team members are in the same physical proximity where face to face communication is possible. With OSD, close proximity with all team members at all times is impossible. Several past research studies have identified challenges of implementing Agile methods in an offshore environment.

There are several research studies conducted on success and failure stories of implementing agile in an offshore setting. The researcher could not find any studies which have identified all the challenges and their impact on the project success. Further, there are very few studies conducted in Sri Lanka on this area. Sri Lanka is recognized as a top ranked outsourcing destination and currently, OSD is the fifth largest export earner for Sri Lanka. Hence, it would be very beneficial for the industry as well as the researchers to obtain a holistic view of the challenges and their impact on the project success.

Thus, this research attempts to answer the following research question: What are the challenges of using Agile methods in offshore software development according to the peer reviewed articles in the literature?

The rest of this paper is organized as follows. Literature review section discusses the identified challenges and issues. In the next section, the theoretical model of challenges is presented.

2 Literature Review

This section discusses the challenges in implementing agile methods in an offshore setting. Many of the previous studies are case studies. According to the literature there are 9 identified challenges that affect the project success. They are communication, personal selection, work culture, different time zones, trust, tools and infrastructure, knowledge management, following agile methods and testing. Following section discusses these challenges in detail.

2.1 Communication

Throughout the literature the importance of communication in Agile methods is highly emphasized [2–5, 17, 24]. As agile methods were originally designed for small co-located teams, they heavily rely on face-to-face communication.

This is not possible in global teams where the members are geographically distributed and the communication between on-site and offshore teams is possible only through technology mediation. Another limiting factor may be the time difference due to different time zones. This hinders synchronous communication even with the usage of telephones or video conferencing systems. The geographical and temporal distance prohibits informal communication [5, 10, 17]. According to Fowler [12], Offshore development brings these two issues which conflict with the principles of agile development.

David Parnas states that poor communication among the team members and team members and the customer is the main reason for project failure [1]. Communication with geographically distributed team members often happen via asynchronous communication methods. Effective communication might be difficult to achieve with asynchronous methods. Email, chat, teleconferencing, video conferencing are the commonly used tools for communication [5]. In some instances video conferencing was used only at critical stages of the project as it was the most expensive. Many companies encouraged daily meetings as a successful strategy. Though there are well developed tools used for communication, it is still a challenge to achieve 100% effective communication as in face to face communication.

2.2 Personal Selection

Personal selection in a team is identified as another challenge in GSD [5]. Agile development requires cross functional teams with high communication skills [18]. Members with a mind set for plan driven approach will not be suitable in this environment. Communication would be more effective between members with similar cultural backgrounds and similar mind set. Team members need to be autonomous rather than following the orders of the project manager. Hence selecting the right team members with the correct attitude is a challenge [8].

According to the literature different companies used different strategies in selecting the personnel for projects. Many successful scenarios emphasized the importance of having a good team leader who can coordinate the various activities. According to the literature some companies chose to have well experienced, creative people who have worked with each other before to be in the team. Some companies chose to have self-organizing teams who could manage their work responsibilities with minimum supervision [5].

2.3 Work Culture

As Fowler [12] states another weak point of offshore development is cultural differences between offshore team members. Agile development works best with close communication and an open culture. Cultural differences might act as a barrier to develop trust between team members. This might lead to less effective communication. Important information might not get communicated. As Agile is a light weight method, less documentation is implemented. Hence most of the information is communicated informally [5, 12]. The characteristics of the people involved with the project and the culture plays an important role in the success of an agile project. It is said that offshore teams prefer plan driven development. Hence cross functional teams in an agile environment might not be effective at first [15, 24, 25].

2.4 Different Time Zones

Having to work in different time zones is identified as another challenge in GSD [5, 8]. A globally distributed team has to work in time zones that vary vastly. Agile methods rely on several face to face meetings even within a single iteration. Heavy documentation is not practiced as a way of communication. Thus distributed environment impose barriers for conducting regular meetings that are informed in short notice [12].

2.5 Trust

Achieving trust between team members is identified as another significant challenge in GSD. Since it is difficult to conduct regular meetings with all the team members as in co-located teams, it demotes trust between team members [5, 19, 25]. Cultural differences are another barrier that impedes trust. It takes longer time for the off-shore members to feel close than co-located members. Team spirit where the whole team holds responsibility and works towards a common goal is essential in agile development. It is a challenge in offshore environment where Sri Lankan and USA teams might have an 'Us versus Them' attitude [17].

2.6 Knowledge Management

Documentation plays an important role in distributed agile more than in the co-located teams. Agile philosophy promotes 'just enough documentation'. But with the absence of proper documentation certain knowledge might not be efficiently conveyed among all stakeholders [4, 9, 16]. Various tools such as wikis, code repositories are used for knowledge management. It is important that proper strategies and tools are used for efficient knowledge management in order to effectively utilize tacit knowledge of experienced programmers [5].

2.7 Following the Agile Development Process

The agile practices are often modified in the distributed environment in order to meet the organizations requirements. Agile philosophy advocates a framework to software development. Details of practices to be used are not elaborated. For an example how to elicit requirements or what kind of documents to produce are not stated. Thus practices are adjusted according to the company's requirements. As Marambe et al. [17] states following the proper techniques are challenged due to lack of knowledge, tight schedules, and distributed nature. These modified techniques might provide short term benefits while introducing unforeseen long term problems. According to Marambe's study, following the agile development process has the most significant impact on project success [8, 17, 24].

2.8 Tools and Infrastructure

Usage of correct tools and infrastructure has a huge impact on the project success especially in GSD [3]. Unavailability of the correct tools and infrastructure or improper handling of the tools will impede the project success. Large companies will have all the facilities where as small and medium organizations might not be able to afford the cost [2, 22]. Schwaber [21] states the following "Prior to scaling any project, an appropriate infrastructure must be put in place. For instance, if a project will employ multiple collocated teams, a mechanism for frequently synchronizing their work must be devised and implemented." Smite, Moe and Agerfalk [1] states the following "Although no single tool is strictly mandatory for a successful agile project, the right set of tools greatly facilitate realizing the various agile processes".

2.9   Testing

Agile philosophy emphasizes early and continuous testing of the product. Unit testing, integration testing, test driven development and test automation are some of the techniques used. Due to time constraints sometimes integration testing might not be possible for all features. This could result in decreased quality of the product. An industry expert opinion was to use test automation as a solution. Then the cost factor would be high. The distributed nature imposes challenges to continuous testing [12, 17]. Bavani (2009) [5] states that testing is a critical aspect when rolling out agile projects on a distributed scale.

Table 1 lists the identified challenges and the number of times they were stated in the literature.

According to the literature review communication and trust are the main listed challenges which have a major impact on the project success.

Using the identified challenges, the Theoretical model depicted in Fig. 1 is developed.

Table 1 Frequency of identified challenges

Challenge	Number of times	References
Communication	8	[2, 3, 5, 6, 10, 12, 17, 20]
Personal selection	2	[6, 20]
Following the agile development method	3	[6, 17]
Testing	4	[5–7, 17]
Tools and infrastructure	4	[5, 6, 17, 21]
Trust	7	[5–7, 17, 20, 24, 25]
Differences in time zone	3	[2, 6, 12]
Differences in work culture	4	[2, 6, 12, 25]
Knowledge management	2	[6, 12]

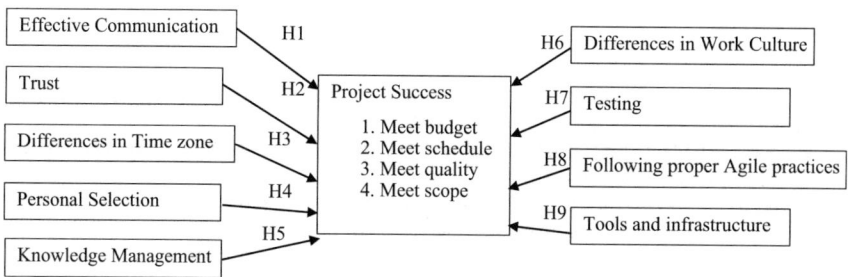

Fig. 1 Theoretical framework

3 Theoretical Framework

A successful project should meet the scope, time, cost and quality constraints [11, 22]. According to the literature 9 challenges that affect the project success were identified as depicted in Fig. 1. Following hypothesis were developed accordingly.

H1—Effective communication has a positive impact on project success
H2—Trust between team members has a positive impact on project success
H3—Differences in time zone have a negative impact on project success
H4—Personnel selection has a positive impact on project success
H5—Knowledge management has a positive impact on project success
H6—Differences in work culture has a negative impact on project success
H7—Following Testing methods has a positive impact on project success
H8—Following Proper agile practices has a positive impact on project success
H9—Having Proper tools and infrastructure has a positive impact on project success

4 Future Research

In the future we will be testing the proposed theoretical model using quantitative methods. A random sample of software professionals who practice agile methods in an offshore setting in Sri Lanka will be selected. The population will consists of software professionals such as project managers, technical leaders, business analysts, software engineers and quality assurance engineers. It is planned to collect data using a questionnaire. The questionnaire would be personally given and data would be collected. The sampling method would be stratified random sampling and unit of analysis would be a software organization.

Using quantitative methods it is planned to identify the most important factors that have an impact on the project success. OSD is currently the fifth large export earner for Sri Lanka. Many software companies are engaged in OSD using agile methods. Hence ICT industry as well as Sri Lanka as a country would greatly benefit from the findings of this study.

References

1. Agerfalk, P.J., Fitzgerald, B.: Flexible and distributed software processes: old petunias in new balls? Commun. ACM **49**(10), 27–34 (2006)
2. Alzoubi, Y.I., Gill, A.Q.: Agile global software development communication challenges: a systematic review. Pac. Asia Conf. Inf. Syst. (2014)
3. Ambler, S.W.: Communication on Agile Software Teams. Retrieved from Agile Modeling. http://www.agilemodeling.com/essays/communication.htm (2014, July)
4. Bjarnason, E., Wnuk, K., Regn, B.: Requirements are slipping through the gaps—A case study on causes and effects of communication gaps in large-scale software development. In: IEEE 19th International Requirements Engineering Conference, pp. 37–46 (2011)
5. Bavani, R. Critical success factors in distributed agile for outsourced product development. In: International Conference on Software Engineering, Chennai, India: Computer Society of India (2009)
6. Bose, I.: Lessons learned from distributed agile software projects: a case-based analysis. Commun. Assoc. Inf. Syst. **23**(1), 619–632 (2008)
7. Cockburn, A.: Agile Software Development, USA (2000)
8. Cohn, M.: Succeeding with agile: software development using Scrum. Addison-Wesley, Upper Saddle River, NJ. http://www.slideshare.net/rodrigorac2/succeeding-with-agile-software-development-using-scrum-addisonwesley-2010 (2010)
9. Cunningham, W.: Retrieved from Manifesto for Agile Software Development. http://www.agilemanifesto.org/ (2001)
10. Dingsøyr, T., Smite, D.: Managing Knowledge in global software development projects. IT Prof. IEEE Comput. Soc. 22–29 (2014)
11. Dorairaj, S., Noble, J., Malik, P.: Effective Communication in Distributed Agile Software Development Teams. Springer, Heidelberg (2011)
12. Dyba, T., Dinsoyr, T.: What do we know about Agile software development. IEEE Softw. 6–9 (2009)
13. Fowler, M.: Writing The Agile Manifesto. http://martinfowler.com/articles/agileStory.html (2006)
14. Highsmith, J.: Agile Software Development Ecosystems. Addison-Wesley, Boston, MA (2002)
15. Jaanu, T., Paasivaara, M., Lasseni, C.: Effects of four distances on communication processes in global software projects. In: ESEM '12: Proceedings of the ACM-IEEE International Symposium on Empirical Software Engineering and Measurement. ACM, Lund, Sweden (2012)
16. Lalsing, V., Kishnah, S., Sameerchand, : People factors in agile software development and project management. Int. J. Softw. Eng. Appl. (IJSEA) **3**(1), 117–137 (2012)
17. Lindvall, M., Basili, V., Boehm, B., Costa, P., Dangle, K., Shull, F., Zelkowitz, M.: Empirical findings in agile methods. In: Proceedings of Extreme Programming and Agile Methods—XP/Agile Universe 2002, pp. 197–207 (2002)
18. Marambe, A., Jayasundara, C.: The challenges of offshore agile software development in Sri Lanka and effects on the project outcome. Int. Sci. J. Manag. Inf. Syst. **9**(3), 10–20 (2014)
19. Misra, S.C., Kumar, V., Kumar, U.: Identifying some important success factors in adopting agile software development practices. J. Syst. Softw. (2009)
20. Paasivaara, M., Durasiewicz, S., Lassenius, C.: Using Scrum in Distributed Agile Development: A Multiple Case Study. In: Proceedings of the 2009 Fourth IEEE International Conference on Global Software Engineering. ACM, pp. 195–204 (2009)
21. Ramesh, B., Baskerville, R., Cao, L.: Agile requirements engineering practices and challenges: an empirical study. Inf. Syst. J. **20**(5), 449–480 (2010)
22. Schwaber, K., Sutherland, J.: The Scrum Guide. http://www.scrumguides.org/docs/scrumguide/v1/Scrum-Guide-US.pdf#zoom=100 (2013)

23. Shrivastava, S.V., Rathod, U.: Risks in distributed agile development: a review. Sci. Direct **133**, 417–424 (2014)
24. Vogel, D., Connolly, J.: Best practices for dealing with offshore software development. In: Hand Book of Business Strategy (2005)
25. Yaggahavita, H.D.: Challenges in Applying Scrum Methodology on Culturally Distributed Teams. Sheffield Hallam University (SHU), UK (2011)

A Method for Automated Test Cases Generation from UML Models with String Constraints

Thi Dao Vu, Pham Ngoc Hung and Viet Ha Nguyen

Abstract This paper proposes an automated test cases generation method from sequence diagrams and class diagrams with string constraints. The method supports UML 2.0 sequence diagrams including twelve combined fragments. An algorithm for generating test scenarios are developed to avoid test paths explosion without having data sharing points of threads in parallel fragments or weak sequencing fragments. Test data are also generated with solving constraints of string variables. We standardize string constraints and equations at the boundary of variables that are input formula of Z3-str solver. Comparing with the current approach of the solver, some preprocessing rules are extended for other operations such as charAt, lastindexOf, trim, startsWith and endsWith. If a result of the Z3-str is SAT, test data of each test scenario are generated to satisfy the constraints with boundary coverage. A tool is implemented to support the proposed method, and some experiments are also presented to illustrate the effectiveness of the tool.

1 Introduction

Model-based testing plays a significant role in research and practice due to great benefits. There are some approaches for model-based testing: test data generation, test cases generation from behavior models and test scripts generation from abstract tests [6]. Test data generation from solving constraints has focused on primitive data type of variables. However, there are many applications being faulty in doing strings

T.D. Vu (✉)
Academy of Cryptography Techniques, 141 Chien Thang Str.,
Thanh Tri District, Hanoi, Vietnam
e-mail: vtdao@bcy.gov.vn; vuthidao@gmail.com

P.N. Hung · V.H. Nguyen
VNU-University of Engineering and Technology, Cau Giay, Vietnam
e-mail: hungpn@vnu.edu.vn

V.H. Nguyen
e-mail: havn@vnu.edu.vn

© Springer International Publishing AG 2017
D. Król et al. (eds.), *Advanced Topics in Intelligent Information
and Database Systems*, Studies in Computational Intelligence 710,
DOI 10.1007/978-3-319-56660-3_45

processing. In addition, one of major approaches is generation of test cases from Unified Modeling Language (UML) models. In this approach, an intermediate model helps to generate the control flow sequences. There are three options to choose the intermediate models [5] such as activity diagram, control-flow graph (CFG) [4, 8] and Colored Petri Nets. The test scenarios which are abstract test cases help to find errors during implementation of software systems.

Many works have been proposed in order to show the approach. However, an approach [1] did not address different types of combined fragments, especially in case of nested combined fragments. And the method in [1] did not also generate test data. A method [4] dealt with five interaction fragments such as loop, alt, opt, break and parallel fragments in UML 2.0 sequence diagrams. Moreover, in [8] a method was developed for eight kinds of combined fragments describing control flow of systems, and the method only solved test data generation with numeric data type.

There are many string solvers such as REX [7], DPRLE [2] and HAMPI [3] that only use string operations [9], but many non-string operations in applications are also popular. Moreover, the string operations interact with the non-string operations that cause errors. An analyzing of string-only will be the shortage of pure integer constraints. There are many solvers converting string-to-integer constraints that are not precise enough. Therefore, a Z3-str solver [9] is used by supporting of a combined logic both strings and non-string operations.

The paper proposes a method in order to generate automatically test cases from sequence diagrams and class diagrams with string constraints. This method is to solve all twelve kinds of fragments in UML 2.0. An algorithm for generating test scenarios is developed to avoid test paths explosion without having data sharing points of threads in parallel (par) or weak sequencing fragments (seq). String constraints of each test scenario and equations at the boundary of variable are converted into input of Z3-str solver. If output of the solver is SAT, a possible model is given. Test data are given with satisfying the constraints and boundary coverage from the possible model. Comparing with Z3-str solver, some preprocessing rules are extended for other operations such as charAt, lastindexOf, trim, startsWith and endsWith. A tool is implemented to support the proposed method. Some experiments are illustrated the effectiveness of the tool.

The paper is organized as follows: Sect. 2 mentions transforming UML sequence diagrams and class diagrams into CFG, Sect. 3 describes the algorithm of test scenarios generation, Sect. 4 presents solving string constraints. A tool to implement the proposed method and some experiments to validate its feasibility and effectiveness are shown in Sect. 5. We conclude the paper and discuss future works in Sect. 6.

2   Control Flow Graph Generation

Test sequences generation from UML 2.0 models needs an intermediate model. Elements of the model will be processed easily. A CFG is chosen in our approach. The proposed technique of CFG generation requires UML diagrams in xmi file. To solve

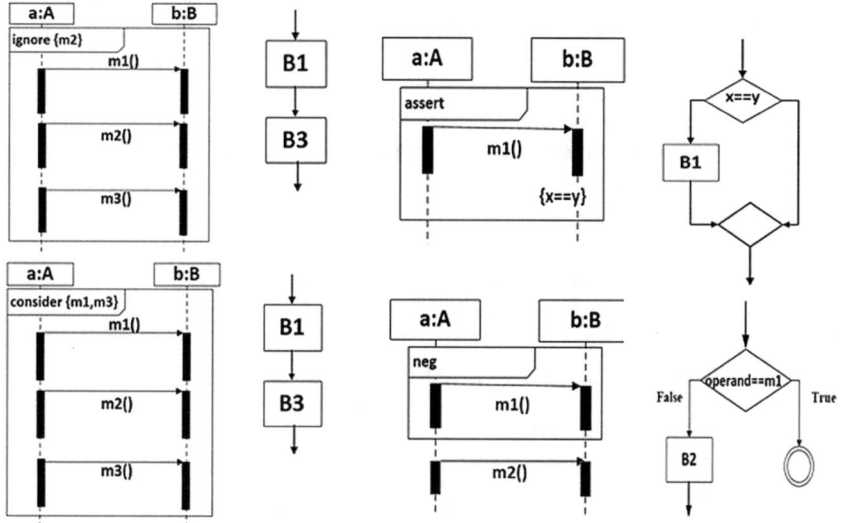

Fig. 1 General structure of CFG (for ignore, consider, assert and neg fragments)

all twelve fragments in UML 2.0 sequence diagrams, CFG generation is extended [8] for the remaining four fragments: ignore, consider, negative and assertion. String constraints of variables are derived from class diagrams and conditions of sequence diagrams which are solved to generate test data.

The definitions of CFG, a block node (BN), a decision node (DN), a merge node (MN), a fork node (FN) and a join node (JN) are mentioned in [8]. We use again Algorithm 1 in [8] for generating CFG from analyzing the queue, and develop Algorithm 2 for ignore, consider, negative and assertion fragments (shown in Fig. 1 and detailed description in below Algorithm 2 in Appendix).

3 Test Scenarios Generation

Input to the test scenarios generation is a CFG. After constructing the CFG, the CFG is traversed automatically to generate the test scenarios which satisfy coverage criteria. The test scenarios denote abstract test cases which are paths starting from the initial node to a final node. Each given test scenario begins with the initial node (*in*) of the CFG.

Algorithm 1 Generating the test scenarios

Input: Control-flow Graph G with initial node *in* and final nodes are fn_i
Output: T is a collection of test scenarios, t is a test path

1: $T = \emptyset$; $t = \emptyset$; *queue* $= \emptyset$;
2: *curNode* $= in$; //current node starts from *in*
3: **repeat**
4:      t.append(curNode);
5:      move to next node
6:      **if** (*curNode* == *DN* and decision==TRUE) **then**
7:          Append true part of BN up to MN in t
8:      **else**
9:          Append false part of BN up to MN in t
10:     **end if**
11:     **if** (*curNode* == *FN*) **then**
12:         active all nodes of threads; nodes = ready;
13:         put a beginning node x of queue; x= waiting;
14:         **repeat**
15:             remove front node y of queue; y=processed;
16:             t.append(y);
17:             The neighbour(z) of y having ready status add to the end of queue;
18:             z = waiting;
19:         **until**(*queue* is empty)
20:     **end if**
21:     **if** (*curNode* == fn_i) **then**
22:         $T = T + \{t\}$;
23:     **end if**
24: **until** Graph end

The selection coverage criterion of sequence diagrams is used to cover the diagrams during testing (the test scenarios ensure that each branch of selector modeled is traversed at least once). Depth-first search (DFS) algorithm was used in [4] to generate test scenarios, but the method did not address the issues of the synchronization and data safety. A method in [8] solved this issue to avoid test explosion by selecting switch points of threads in par or seq fragments. However, there are many applications without having sharing data points of threads in these fragments. In this case, an Algorithm 1 is developed (GenerateTestScenarios) to traverse CFG using both DFS and breadth-first search (BFS) algorithm. The BFS is useful for traversing CFG if current node is a fork node. Besides, the proposed method traverses those nodes of threads in case of parallel by BFS to avoid finding the sharing data points of threads in par or seq fragments. The remaining of CFG is traversed by using DFS. Therefore, it avoids wasting time of finding the data sharing points and test paths explosion.

4 Solving String Constraints

The test scenarios obtained denote the sequences of messages. The sequence is a
feasible sequence of messages if we find test data (test input) to satisfy all constraints
along the scenario. Many current researches [4, 8] solve the equations to find values
that satisfy these constraints. However, it generates test data in case of numeric data
type and rarely considers string constraints. Using a Z3-str solver generates test data
if constraints are satisfiable. The input of the solver are all the constraints along the
scenario and equations at the boundaries of the domains of variables. If output of the
solver is SAT that means all constraints are satisfiable, a possible model is given. We
take examples from the possible model, and test data are satisfiable with boundary
coverage.

4.1 Input Formula of Z3-str Solver

Z3-str can handle a boolean combination of atomic formulas, it is converted into
conjunction of literals. We will use an example of string solving for Z3-str. Consider
string constraints: $c1, c2, x$: *String*; $vi1$: *Integer*; $c1 = c1.concat("te")$;
$c2 = "aaaa\_efg\_bbbb\_efg"$; $x = c1.concat(c2)$; $vi1 = x.indexOf("efg")$; $vi1 \geq 4$;
The core treats the string operations as five independent Boolean variables ($e1, e2,$
$e3, e4$ and $e5$) and tries to assign values to them.
$e1 : c1 = c1.concat("te")$; $e2 : c2 = "aaaa\_efg\_bbbb\_efg"$;
$e3 : x = c1.concat(c2)$; $e4 : vi1 = x.indexOf("efg")$; $e5 : vi1 \geq 4$;
Consider the string constraints above, the input formula of Z3-str is converted as
follows assert ($e1 \bigwedge e2 \bigwedge e3 \bigwedge e4 \bigwedge e5$). Each operation above is transformed into:
(assert (concat c1 "te")); (assert (= c2 "aaaa_efg_bbbb_efg")); (assert (= x (concat
c1 c2)); (assert (= vi1 (indexOf x "efg"))); (assert (≥ $vi1$ 4))
If the output of Z3-str of their respective input is satisfiable, values of variables are
given by using get-model. A few good data values are chosen as test inputs when
there are a number of possible input values using boundary coverage. There are a
lot of faults in the system under testing that are located at the frontier between two
functional behaviors. In our approach, constraints added at a boundary point of predicates are input of Z3-str (input of Z3-str is added by $vi1 == 4$). Therefore, test data
are generated and satisfiable boundary coverage.

4.2 Improving Preprocess Rules for Other Operations

In [9] plug-in of Z3-str supports the string operations: string equation, concatenation, length, substring, contains, indexof, replace and split. They have three primitive
operations: string equation, concatenation and string length. That method reduces

Table 1 Pre-processing rules for other operations

Expression	Rule New Formula
c = x.charAt(i)	$charAt(x, i) = c \rightarrow x = x_1.t.x_2 \bigwedge t = c \bigwedge length(x_1) = i$
$i = x_1.lastIndexOf(x_2)$	$lastIndexOf(x_1, x_2) = i \rightarrow (x_1 = x_{s1}.x_{s2}.x_{s3}) \bigwedge (i = -1 \bigvee i \geq 0) \bigwedge ((i = -1) \leftrightarrow (\rightarrow contains(x_1, x_2)) \bigwedge ((i \geq 0) \leftrightarrow (i = length(x_{s1}) \bigwedge x_{s2} = x_2 \bigwedge (\rightarrow contains(x_{s3}, x_2)))$
$x_2 = x_1.trim$	$trim(x_1, x_2) \rightarrow (x_1 = x_{s1}.x_{s2}.x_{s3}) \bigwedge (x_{s2} = x_2) \bigwedge ((x_{s1}="") \bigvee (x_{s3}=""))$
$j = x.startsWith(x_t, i)$	$startsWith(x, x_t, i, j) \rightarrow (x = x_1.x_2.x_3) \bigwedge (j = 1 \bigvee j = 0) \bigwedge ((j = 1) \bigwedge x_2 = x_t \bigwedge length(x_1) = i) \bigwedge ((j = 0) \bigwedge (\rightarrow contains(x, x_t))$
$i = x.endsWith(x_t)$	$endsWith(x, x_t, i) \rightarrow (x = x_1.x_2.x_3) \bigwedge (i = 1 \bigvee i = 0) \bigwedge ((i = 1) \bigwedge x_2 = x_t \bigwedge \rightarrow contains(x_3, x_t)) \bigwedge ((i = 0) \bigwedge \rightarrow contains(x, x_t))$

other string operations to an equivalent formula based on above primitives. They performed pre-processing to translate substring, contains, indexOf, replace and split operations into formulas using concatenation and length operations. Extension of some rules of the preprocessing is presented for other operations such as charAt, lastindexOf, trim, startsWith and endsWith when comparing to [9]. The rules are converted into the primitive operations which are as follows (Table 1),

charAt. Takes two arguments x, i and the result is c. This operator returns the character c located at the specifying i of string x. The indexing of the string x starts from zero. A formula of charAt can be converted with concatenation and string length operations. Particularly, we break the argument x into three pieces x_1, t, x_2, and assert the middle piece t which equals to the return character c. We assert the lengths of x_1 to respect the position of constraints.

lastIndexOf. If the string x_2 argument occurs one or more times as a substring within string x_1, then it returns the index of the first character of the last substring x_2. If it does not occur as a substring, -1 is returned. We break x_1 into three pieces x_{s1}, x_{s2} and x_{s3}. The result value i options: if and only if string x_1 does not involve x_2, i is -1. Otherwise, if and only if x_{s2} equals to x_2, its predecessor x_{s3} does not contain x_2, and i equals to the length of x_{s1}.

trim. Method returns x_2 that is a copy of the string x_1 and omits leading and trailing whitespace. We break x_1 into three pieces x_{s1}, x_{s2} and x_{s3}, and assert the middle piece x_{s2} which equals to the return string (x_2). We assert x_{s1} or x_{s3} to respect whitespace.

startsWith. It tests whether the string x_t is a substring of string x and x_t starts with the specified prefix beginning (i). It returns true ($j = 1$) if the character sequence represented by the argument (x_t) is a prefix of the character sequence represented by this string x; false ($j = 0$) otherwise. The first argument x is broken into three pieces x_1, x_2 and x_3. The result value j options: if and only if string x does not contain x_t, j is 0. Otherwise (j is 1), we assert the string x_2 and length of x_1 to respect the string x_t and the specified index (i).

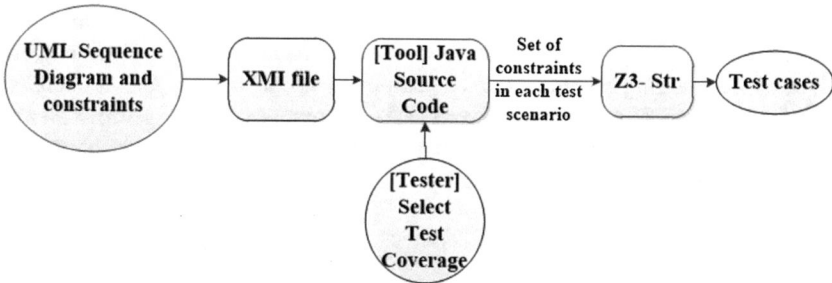

Fig. 2 Figure showing architecture of SequenceString

endsWith. This method returns true ($i = 1$) if the character sequence represented by the argument (x_t) is a suffix of the string (x), else false ($i = 0$). We break x into three pieces x_1, x_2 and x_3. The result value i options: if and only if string x does not contain x_t, i is 0. Otherwise, if and only if x_2 equals to x_t, its predecessor x_3 does not contain x_t.

5 Experiments

This section proposes a tool, SequenceString which is developed to support the proposed method. A case study is conducted to examine the method. And then some experiments are analyzed about performance and errors found in test scenarios that is used to evaluate its effectiveness.

5.1 Tool Support

In this section we discuss the results by implementing the proposed method. The method is implemented using JAVA and JDK version 1.8. Our method is developed for generating test cases automatically from UML sequence diagrams and string constraints, data type of variables from class diagram. The architecture of the tool is shown in Fig. 2. The implemented tool is available at the site.[1]

The Tool consists of 2216 lines of code and has the following functionality:

1. Preprocessing: Enterprise Architect ver.11 is used to produce the UML design artefact. The tool imports the UML sequence diagrams and constraints, data type of variables from class diagram (in XMI file).
2. Generating test scenarios: for each coverage criterion, the tool generates test scenarios from CFG and gives constraints along with each test scenario.

---

[1]http://www.uet.vnu.edu.vn/~hungpn/SequenceString/.

3. Generating test data: with set of constraints along with the test scenario, it is converted into input formula of Z3-str solver. The constraints include numeric constraints, string constraints and equations of variables satisfying at boundary value. If the output of Z3-str solver is SAT, a possible model is given. Test data of each test scenario are generated from taking examples in the possible model, and it is satisfiable boundary coverage.

5.2 Case Study

In this section, the test cases generation is illustrated from UML 2.0 sequence diagram and string constraints, data type of variables from class diagram. Figure 3 shows an example that has input string s, and data types of variables are given that are: String s, s1; int i = s.lastIndexOf(','); int f = Long.parseLong(s); BigDecimal d1 = BigDecimal(-1); s1 = s.substring(i+1);int x = Integer.parseInt(s1); There are six test scenarios in accordance with selection coverage criterion in the example. Test scenario passing $m5()$ is considered. Firstly, it checks whether input string s starts with '-' (that can represent a negative number). Then if s is of a format (the string involves beginning with '-', followed by at least one digit, and comma, lastly 3 digits). After that, $(i = s.lastIndexOf(','))! = -1$ means that it checks whether comma appears in s. With String s1 = s.substring(i+1) if string s1 which is string after comma is substring of string s, and then s1 is converted into the integer x. Finally, it continues by checking whether x is greater than or equals to 100. If the condition is satisfiable, test scenario passes $m5()$.

When sequence diagram (XMI file) and data type of variables are input into our tool, CFG is generated. When we click one test scenario, the details of test path show a set of constraints along with the test scenario. For example, this path (passing $m5()$) has conjunction of some constraints which are as follows,

s.charAt(0) == '-';s.notMaches("-\d+,\d{3}"); i = s.lastIndexOf(',');

i != -1; s1 = s.substring(i+1); x = parseInt(s1); $x \geq 100$

Equation at boundary value is x == 100. It is converted into input formula of Z3-str. If output of Z3-str is SAT, variable s of this test scenario is "−1,000,100".

5.3 Evaluation

Our evaluation consists of two experiments. In the first experiment, three applications have some functions with string constraints and relational operators. We compare the percentage of errors found in applications which are used by our proposed method and random test data generation (Table 2). In the second experiment, we compare the performance of tool with test data generation in the same test scenario when to use or not to use the preprocessing rules for some operations such as charAt, lastIndexOf, trim, startsWith and endsWith. All experiments are run on an Intel Core i3-6100U CPU 2.30 GHz with Ram 4 GB.

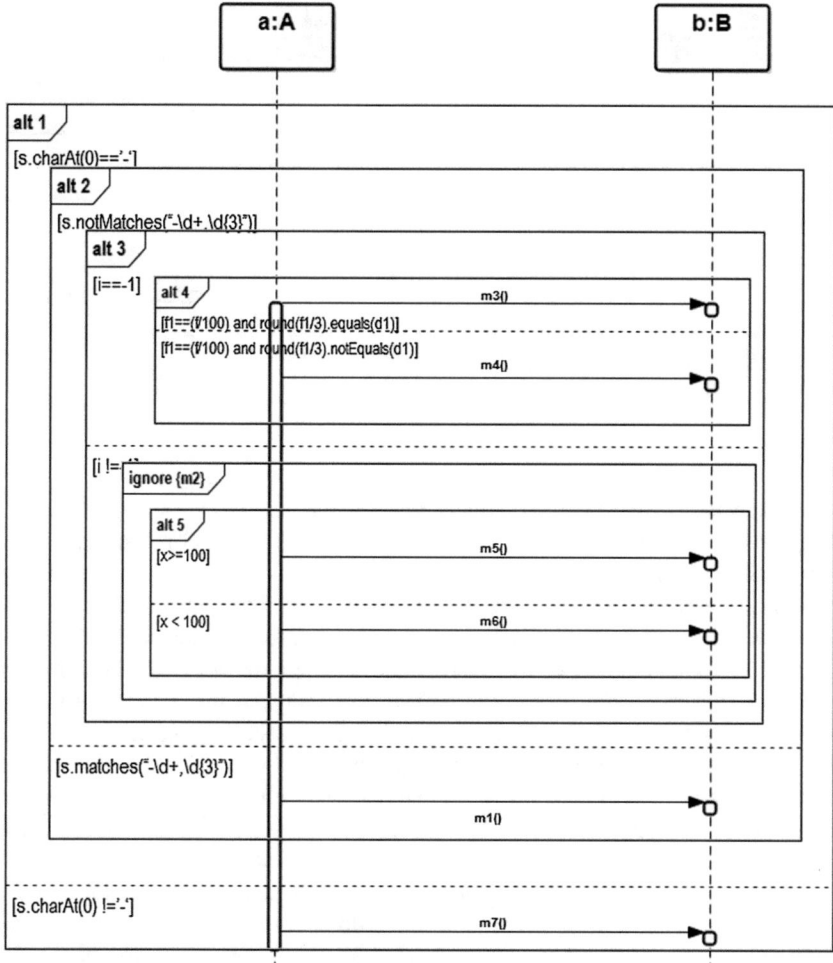

Fig. 3 The sequence diagram for a checking information of web application

Table 2 Comparison errors found in applications

Application	Description	Inputs	Errors found of our method (%)	Errors found of random test data (%)
A	Checking information of user registration	3(strings)	100	42.5
B	Business ordering	5(strings)	100	36.5
C	Insurance registration	4(strings)	90	35.6

Table 3 Comparison performance with applying preprocessing rules

Fragment	Inputs	Time of Z3-str (s)	Time of Z3-str with preprocessing rules (s)
concat	8	0.035	0.035
indexOf	12	0.055	0.055
charAt	12	0.048	0.041
match	13	0.036	0.036
replace	15	0.045	0.045
substring-charAt	10	0.056	0.045
split-startsWith	9	0.066	0.061
lastIndexOf	12	0.036	0.031
lastIndexOf-replace	10	0.042	0.038

Comparison errors: in the functions of three applications, errors are injected at the points of boundary. We perform checking information of user registration function, business ordering and insurance registration. The user registration function has three string variables and two relational operators. The business ordering has five string variables and three relational operators. The last function has four string variables and zero relational operator. From sequence diagrams of the applications, our tool is used to generate test cases. We compare errors found in the same test scenarios but test data are generated by our method and random test data generation. With the proposed method, the fourth column is the percentage of errors found in test scenarios in total errors which inserted into functions. The fifth column is the percentage of errors in the same test scenarios with random test data generation. Therefore, the ability of our method in terms of finding errors is better than that of random test data.

Comparison performance: because Z3-str is open-sourced. We are looking for the communication between Z3-str and the string theory to improve performance. Some rules of the preprocessing are extended for other operations such as charAt, lastindexOf, trim, startsWith and endsWith when comparing with [9]. We use 9 test cases of the checking information of user registration. Then, we run both 30 times for each test scenarios and take average of the execution time in Z3-solver and Z3-solver with precessing rules of some operations. We can see Z3-str with precessing rules of some operations is faster than Z3-str in case of charAt, lastindexOf, trim, startsWith and endsWith operations (Table 3).

6   Conclusion

The paper presents the automated test data generation method based UML sequence diagrams, class diagrams. The method supports UML 2.0 sequence diagrams including all twelve kinds of combined fragments. From CFG, the algorithm for generating test scenarios is developed to avoid test paths explosion without having the points of shared data of threads in parallel or weak sequencing fragments. The constraints of each test scenario and equations at the boundary of variable are converted into input formula of Z3-str solver. Test data are given by a possible model of the Z3-str. Moreover, these test data are satisfiable the constraints with boundary coverage. In test data generation, some preprocessing rules are extended for other operations such as charAt, lastindexOf, trim, startsWith and endsWith. Our tool is implemented to support the proposed method. Some experiments are shown the effectiveness of the tool.

We are also going to develop completely automated test case generation that is automatic standardization the inputs of Z3-str. The proposed method is extended for other UML diagrams (e.g., state-chart diagrams, activity diagrams). Moreover, we would like to investigate, evaluate further the fault-detection effectiveness, costs, and the coverage criteria.

Acknowledgements This work is supported by the project no. QG.16.31 granted by Vietnam National University, Hanoi (VNU).

Appendix

When analyzing xmi file, parameters of ignore fragment are named parFrag. If fragment is ignore and message m is considered insignificant (line 3), the algorithm makes each message of operand corresponding to BN except for message m. In line 6, if a fragment is consider and a parameter of the fragment is message m, the method only creates a BN corresponding to message m. If assert fragment shows message m1() occur at this point, following by state invariant $\{x == y\}$ (line 9, the algorithm only creates a BN corresponding to m1() if $\{x == y\}$ is true). In line 18, if a fragment is neg and a operand of the fragment is m1 then the algorithm returns exitNode and goes back the Algorithm 1 in [8].

Algorithm 2 Analyzing queue for ignore, consider, neg, assert fragments

Input: Class diagram CD, queue, *curNode* ∈ A
Output: *exitNode* ∈ A

---

function processElement(queue, CD:class diagram, curNode:A):A
1: **while** queue != empty **do**
2:     x= queue.pop();
3:     **if**((x==frag)&(x.type=="ignore")&(parFrag==m)& *(operand ≠ m)*) **then**
4:         Create a BN ;
5:         ConnectEdge(curNode,BN);
6:     **else if**(x==frag)&(x.type=="consider")&(parFrag==m) &(operand==m)**then**
7:         Create a BN ;
8:         ConnectEdge(curNode,BN);

```
9:    else if ((x==frag)&(x.type=="assert")) then
10:       Create a DN;
11:       attachGuard_DN; curNode = DN;
12:           if (guard==true) then
13:               Create a BN ;
14:               ConnectEdge(curNode,BN);
15:           end if
16:           Create a MN;
17:           ConnectEdge(curNode,MN);curNode = MN;
18:    else if ((x==frag)&(x.type=="neg")) then
19:       if (operand==m1) then
20:           return exitNode
21:       else
22:           Create a BN;
23:       end if
24:       curNode=BN;
25:    end if
26:    return exitNode;
27: end while;
```

References

1. Dhineshkumar, M., Galeebathullah: An approach to generate test cases from sequence diagrams. In: Proceedings of the 2014 International Conference on Intelligent Computing Applications, ICICA '14, pp. 345–349. IEEE Computer Society, Washington, DC, USA (2014)
2. Hooimeijer, P., Weimer, W.: A decision procedure for subset constraints over regular languages. In: Proceedings of the 30th ACM SIGPLAN Conference on Programming Language Design and Implementation, PLDI '09, pp. 188–198. ACM, New York, USA (2009)
3. Kiezun, A., Ganesh, V., Guo, P.J., Hooimeijer, P., Ernst, M.D.: Hampi: A solver for string constraints. In: Proceedings of the Eighteenth International Symposium on Software Testing and Analysis, ISSTA '09, pp. 105–116. ACM, New York, NY, USA (2009)
4. Nayak, A., Samanta, D.: Automatic test data synthesis using UML sequence diagrams. J. Object Technol. 9(2), 115–144 (2010)
5. Shirole, M., Kumar, R.: Testing for concurrency in UML diagrams. SIGSOFT Softw. Eng. Notes 37(5), 18 (2012)
6. Utting, M., Legeard, B.: Practical Model-Based Testing: A Tools Approach. Morgan Kaufmann Publishers Inc., San Francisco, CA, USA (2006)
7. Veanes, M., de Halleux, P., Tillmann, N.: Rex: Symbolic regular expression explorer. In: Proceedings of the 2010 Third International Conference on Software Testing, Verification and Validation, ICST '10, pp. 498–507. IEEE Computer Society, Washington, DC, USA (2010)
8. Vu, T.-D., Hung, P.N., Nguyen, V.-H.: A method for automated test data generation from sequence diagrams and object constraint language. In: Proceedings of the Sixth International Symposium on Information and Communication Technology, SoICT 2015, pp. 335–341. ACM, New York, NY, USA (2015)
9. Zheng, Y., Zhang, X., Ganesh, V.: Z3-str: a z3-based string solver for web application analysis. In: Proceedings of the 2013 9th Joint Meeting on Foundations of Software Engineering, pp. 114–124. ACM New York, NY, USA (2013)

Mobile Application for Calculation of Optimal Route Between Searched Points of Interest

Veronika Nemeckova, Jan Dvorak and Ondrej Krejcar

Abstract Searching for gifts' inspirations for relatives is becoming increasingly difficult. A person can collect ideas on the internet, ask the relatives or get inspiration from gifts, which the person received in the past. This study focuses on the suggestion and development of an algorithm for the calculation of the optimal route between searched shops, which is a part of the mobile application called "Tip na dárek/Gift idea" for Android platform. The application contains two basic functions: search in the "gift ideas" database and searching for shops with the option of planning the route between chosen shops. The introduction of the article concentrates on a more detailed description of the developed application with the calculation of functionalities that this application offers.

Keywords Google Maps and Places API · Minimum spanning tree · Database

1 Introduction

The shopping for gifts is a task which is relevant during the whole year—for occasions such as birthday, name day and other. The search for the most suitable gift becomes more difficult each year. With the upcoming Christmas, more users search online for the gifts' inspirations for relatives. This is shown in the numbers of search results for the terms "tipy na dárky" [1] or "gift ideas" [2], which are

V. Nemeckova · J. Dvorak · O. Krejcar (✉)
Faculty of Informatics and Management, Center for Basic and Applied Research,
University of Hradec Kralove, Rokitanskeho 62,
500 03 Hradec Kralove, Czech Republic
e-mail: ondrej@krejcar.org

V. Nemeckova
e-mail: veronika.nemeckova@uhk.cz

J. Dvorak
e-mail: jan.dvorak@uhk.cz

© Springer International Publishing AG 2017   537
D. Król et al. (eds.), *Advanced Topics in Intelligent Information
and Database Systems*, Studies in Computational Intelligence 710,
DOI 10.1007/978-3-319-56660-3_46

Graph 1 Search for term "Tipy na dárky" in years 2011–2015 [1]

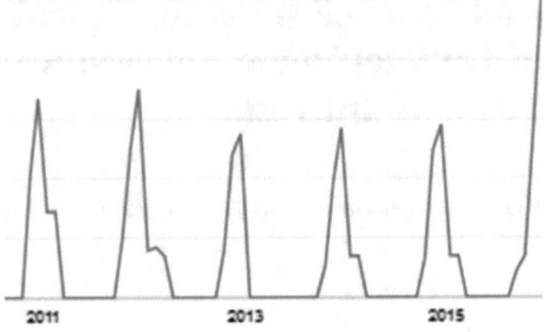

Graph 2 Search for term "Gift ideas" in years 2011–2015 [2]

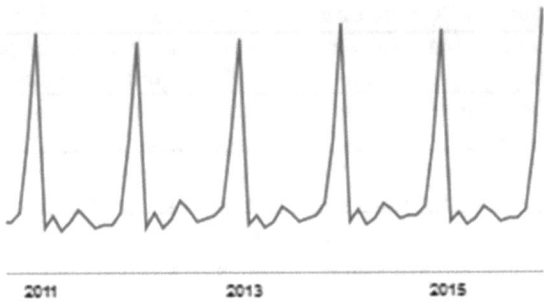

available on Google Trends. The graph of interest for such related topics (gifts, gifts for women, gifts for men, etc.) increases in values.

By using the Google Trends [3] it is possible to study the trends for search for various topics on Google. The graphs (Graphs 1 and 2) show the popularity of the terms' search "tipy na dárky" (Graph 1) and "gift ideas" [2] in the course of time. The values in the graph show the number of queries which relate to these terms. However, these are no absolute numbers of search queries, but the normalised values on a scale from 0 to 100. The "0" value does not have to necessarily mean that in the given time period there was no interest in the topic, rather that there was not enough values for evaluation [4].

Nowadays, it is very common to use the mobile phones for resolving various matters. Therefore, the rich usability of this mobile application for gift ideas search is undisputable. This application could simplify the selection of gifts for relatives and reduce possible stress caused by searching for the most suitable presents for the given occasion.

The aim of this study is to create a mobile application for the Android platform which would contain two basic functions: search for gift ideas on the basis of the given parameters and shop search according to the chosen place or on the basis of the current location of the user.

The search for gift ideas will be implemented using a series of queries, by which the user can narrow the search radius—for example the person's gender for whom is the gift intended, the type of the gift, price range, etc. On the basis of the given values, the application will run the query into the database and will show the user the list of gift ideas matching the user's preferences.

The second functionality of this application—shop search—is described in the rest of this article. The user will be able to find the shops of the chosen type (bookstores, jewellery stores, liqueur stores, etc.) according to the place or location, as well as the option to show the chosen shops on the map and plan the optimal route between the found shops.

There are many ways, how can the user plan the route between stores. In order to solve this issue, two solutions arise: use the principle of Hamiltonian path search in the graph or to use one of the algorithms for search of minimum spanning tree. These algorithms are applied even by authors in [5–7] for the problems' solutions based on the route search, radius, or minimum spanning tree in the evaluated graph.

2 Problem Definition and Related Works

The aim of this study is to suggest a suitable algorithm for the optimal route search between found shops that will be based on the principle of searching for the minimum spanning tree in the evaluated graph.

A. Search for the minimum spanning tree in the graph

The algorithms for minimum spanning tree in the evaluated graph are used for the connection of all vertices with the "lowest" spending's. There are multiple algorithms that focus on this problem. The most common ones will be introduced in this section.

(1) Boruvka's algorithm

Boruvka's algorithm for search of minimum spanning tree in the graph gradually connects the components into larger units, until a continuous subgraph (spanning tree of a graph), containing all vertices of the previous graph is created [8].

At the beginning each vertex of the graph is a component. In the next step, each component receives an edge with the lowest price and this edge is added into the given component. This step is repeated so many times until it obtains the searched minimum spanning tree [8].

Pseudo-code of Boruvka's algorithm, taken from [9]

- Begin with a connected graph G containing edges of distinct weight, and an empty set of edges T
- While the vertices of G connected by T are disjoint
    - Begin with an empty set of edges E
    - For each component
        - Begins with an empty set of edges S
        - For each vertex in the component
            - Add the edge of minimum weight from the vertex in the component to another vertex in a disjoint component to S
        - Add the minimum weight edge in S to E
    - Add the resulting set of edges E to T
- The resulting set of edges T is the minimum spanning tree of G

(2) Jarník-Prim algorithm

Jarník's, in the international literature also known as Prim's algorithm starts to search for minimum spanning tree graph from one (randomly chosen) vertex from which it then searches for edges with the lowest price. These are progressively added into the component of the future minimum spanning tree [10].

At the beginning, a random vertex is chosen, from which the search is started. From this vertex, all of the edges to other vertices are found. The cheapest option is then chosen. The step with search and selection of the edge with the lowest price is repeated until all vertices of the previous graph are contained in the component [10].

Pseudo-code of Jarník-Prim algorithm, taken from [11]

- Make a queue (Q) with all the vertices of G (V);
- For each member of Q set the priority to INFINITY
- At each step
    - We add to the cloud the vertex u outside the cloud with the smallest distance label
    - We update the labels of the vertices adjacent to u.
- A primary queue stores the vertices outside the cloud (key - distance, element - vertex)
- Result is the minimum spanning tree of G

(3) Kruskal's algorithm

Kruskal's algorithm for searching of the minimum spanning tree in the evaluated graph gradually choses the edges from the list of ordered edges in order to not create a circle [12].

Before the beginning of the search, all the edges of the graph are first ordered according to their price. Furthermore, the algorithm gradually studies the list and chooses the edges which after being added into the component do not create a circle. The algorithm comes to an end when all the vertices of the previous graph are added.

Pseudo-code of Kruskal's algorithm, taken from [13]

- `T (the final spanning tree) is defined to be the empty set`
- `For each vertex v of G, make the empty set out of v`
- `Sort the edges of G in ascending (non-decreasing) order`
- `For each edge (u, v) from the sored list of step 3`
  - `If u and v belong to different sets`
    - `Add (u,v) to T`
    - `Get together u and v in one single set`
- `Return T`

B. Search for Hamilton's path

Hamilon's path contains all vertices of the graph just once and each of these vertices has a degree of maximum two. The search for minimum Hamilton's path or circle, in practice known as the Problem of the business traveller, is a difficult task which goes beyond the scope of this study [14].

Therefore, the focus of the following sections will concentrate on the algorithm which will be based on the algorithm of searching for the minimum spanning tree of the evaluated graph. However, the resulting path between the found shops will be the Hamilton's path.

3 New Solution

The calculation of the optimal path between found shops will be based on the distance between the individual shops.

The user will have the option to choose, where and which shops should be found. It will be possible to search the shops according to the current position, or on the basis of the entered location. The search can be also narrowed according to the type of a store to shops with jewellery, liqueur, books, clothes, household goods or electro.

The application will show the user a list of stores according to the entered parameters which correspond to the user's requirements. These found stores are then shown on the map with the possibility of also displaying the calculated path between these shops.

The store search according to the given parameters will be implemented using the Google Places API service from Google. In order to obtain information about the points of interest, it is necessary to create a query which would contain the coordinates of the place, radius in which the stores should be searched and the types of stores that interest the user. This query is then executed by Google Places API service that returns the data in the JSON format from which the information that is needed can be extracted. In this case, the name of the store, address and the coordinates will be sufficient [15].

In order to display the found shops on the map, the Google Maps API service, also from Google, will be implemented. This service enables the display of the found places on the map—for each store it creates a Marker that is then placed on the position on the basis of previously obtained coordinates of this store [16].

The main part of the application is the route calculation between the found shops. In order to build the optimal route, it is necessary at first to obtain the distances between the individual shops. The distance between the stores can be found on the basis of a query to the Google Maps Directions API that must contain the coordinates of the starting and target destination. The service, similarly as in the previous case, returns data in the JSON format that also contain the needed distance. These data will be used for the calculation of the optimal route [17].

The developed algorithm for the purpose of this survey will be based on the principle of Kruskal's algorithm for minimum spanning tree search in the evaluated graph. It is not wanted for the result path to contain points, to which the user would have to come back in order to continue to the next stores. Therefore, it will be controlled that the resulting path is a path between stores. This means that within one calculated route, there will be only one path to the store, as well as one path from the store (except border points). This will be achieved by the addition of a condition that will control the number of paths to/from the shop.

At the beginning there will be a list of routes between individual stores, where each path will contain starting and ending shop and the distances between them. This list will be ordered according to the distance—from shortest route to the longest—and only then the resulting route will be build.

Further, for each route between the shop A and shop B it will be monitored, if a circle is created by adding a new route to the resulting path. This would mean that the created route already contains both of the shops and therefore, this part of the path would be then discontinued. At the same time, it will be controlled if the shop A or B already contains two paths (from shop and to the shop), because in this way the unwanted branching would be created. If the processed path fulfils both requirements, it is added into the final route and the stores A and B are also included to the already processed stores. This process continues until the number of processed stores is not equal to the number of found stores.

At the end, the calculated optimal route between the searched shops will be displayed for the user on a map. Suggested algorithm in pseudo-code:

4   Implementation

A. Store search and their display on the map

The following form was created for the input of the parameters, according to which the stores will be searched for (Fig. 1).

Fig. 1 Screenshot from the
application—search form

The found stores can be viewed by the user on a map as points or they can be directly displayed as a route between these stores.

The Google Places API from Google Play services was used for store search. This service processes the query for points of interest according to the user's parameters and returns the result in JSON format.

For the purposes of this study, it is sufficient to select from the search results for each store its name, address and coordinates. The parsing of the data for each individual store appears flowingly:

```
String name = placeObj.getString("name");
String address = placeObj.getString("vicinity");
JSONObject locObj = placeObj
        .getJSONObject("geometry")
        .getJSONObject("location");
double lat =
        Double.valueOf(loc.getString("lat"));
double lng =
        Double.valueOf(loc.getString("lng"));
```

If the user selects the option to have the results shown on a map, the map will be displayed with the marked stores on it. The Google Maps API technology was used for the displaying of the map, marking of the found stores and plotting of the optimal route. This topic will be covered later.

For adding of the stores into the map, the addMarker() method was used, which is the parameter of the MarkerOptions method which contains the information about the particular store. The implementation of the marker creation for shops and their addition into the map can be done in the following way:

```
// creation of MarkerOptions
stores[p]
        = new MarkerOptions()
    .position(new
    LatLng(store.getLatitude(),
    store.getLongitude()))
    .title(store.getName())
    .snippet(store.getVicinity());
// adding into the map
mMap.addMarker(stores[p]);
```

B. Calculation of the optimal route between stores

In the first step, it is necessary for each store to obtain the distances to the other stores. For this task the Google Maps Directions API technology is used. The query with the coordinates of the starting and target destinations and the type of the transportation used (walking, car) is sent to the service.

```
String distanceUrl = GoogleApiUrlBuilder.getDirectionUrl(
        selectedStores.get(i).getLatitude(),
        selectedStores.get(i).getLongitude(),
        selectedStores.get(j).getLatitude(),
        selectedStores.get(j).getLongitude());
new GetDistanceTask().execute(distanceUrl);
```

The result of the query is, identically as in the previous case, in the JSON format. The distance between the stores is obtained in the completely same way as for store search.

The data which is prepared in this way is then used for the actual route calculation.

(1) Route calculation algorithm

At the input, there is the list with all routes between individual store which are represented by the LocationDistance class (Fig. 2).

LocationDistance represents the edge of the graph, the attributes locationA and locationB are then its vertices.

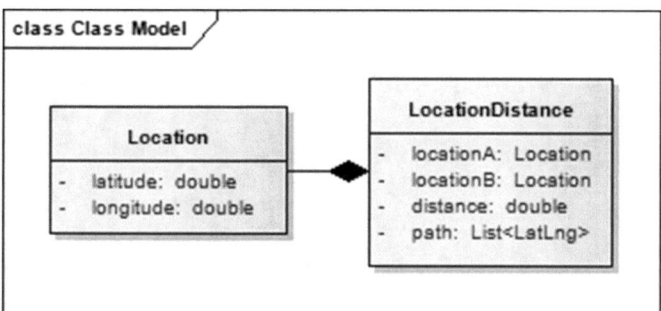

Fig. 2 Class diagram

At the beginning the empty container is prepared, into which the locations (stores) that were already added into the planned route and list which contains the future optimal route are stored.

In this first step, it is necessary to order the list of all routes between individual stores according to the distance from shortest paths to the longest.

Consequently, each route from the list is checked if a circle is created by adding a new path or if one store would contain a third path. If the route complies with both of the requirements, it is added to the optimal route and the new added locations are saved.

5 Testing of Developed Application

The testing is focused on the optimisation of the process for obtaining the distance between individual stores, calculation of the optimal route and acquiring the list of coordinates for depicting the route between the stores. This part of the algorithm takes the longest.

In the first test, the calculation time and the route depiction is to be tested. Here it is necessary to first obtain the distances between individual stores and the list of the coordinates for the route depiction is only acquired for those routes that were returned by the algorithm for the optimal route calculation.

The second test focuses on the improved algorithm with which it is possible to save the coordinates at the time of distance determination between the stores. However, this is done in the form of symbols chain which is decoded into coordinates only until those paths that were included into the optimal route between stores.

The algorithm in the last test is able to decode the text chain, where the coordinates for the future route are saved, immediately during the recognition of the distance between the stores for all routes. This list is then saved.

The measurement was in all cases applied into the list of ten stores. The results of the testing are contained in the following subsections (Fig. 3).

Fig. 3 Depicted route on the
map

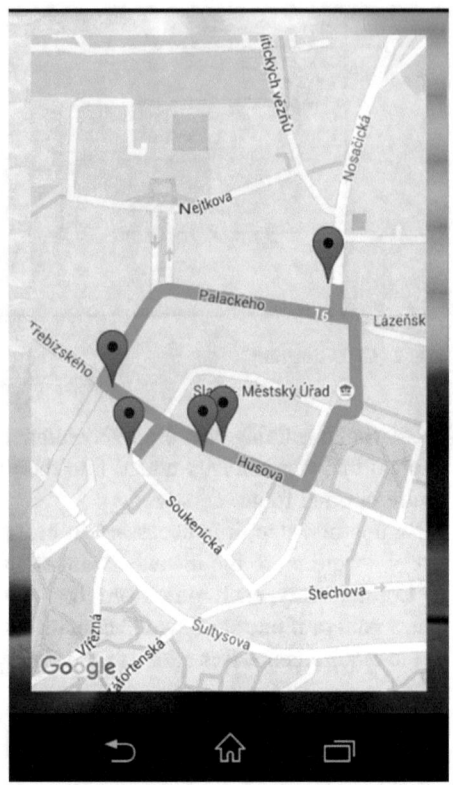

The tests proved that the most suitable option is to directly save the list of
coordinates that create the path between the two stores, and that applies as well for
the paths that in the end will be included in the final optimal route between stores.
At the same time, it was shown that there is not such a large difference between the
access, during which the coordinates' list is directly decoded or the one during
which the data are left in the text form and then transferred into the coordinates list
only for the paths that are depicted in the result.

The aim of this testing was to optimise the suggested algorithm in order to
achieve calculation and depiction of the obtained route in the shortest possible time.
This was successfully achieved.

6 Conclusions

The aim of this study was to suggest and implement an algorithm for optimal route
search between searched stores. However, the requirement was to achieve calculation and depiction of the obtained route in the shortest possible time. This was

successfully achieved. The advantage of this new algorithm is that it is possible to use it for the search of any points of interest, as far as the coordinates of this place are available. By using a simple modification, it can be achieved that the resulting route must not be the path between the chosen points, but the real minimum spanning tree of the evaluated graph.

Acknowledgements This work and the contribution were supported by project "SP-2017 - Smart Solutions for Ubiquitous Computing Environments" Faculty of Informatics and Management, University of Hradec Kralove, Czech Republic.

References

1. Google Trends—Tipy na dárky: Google Trends. https://www.google.com/trends/explore#q= tipy%20na%20d%C3%A1rky. Accessed 28 Dec 2015
2. Google Trends—Gift Ideas: Google Trends. https://www.google.com/trends/explore#q=gift% 20ideas. Accessed 28 Dec 2015
3. Google Trends: https://www.google.cz/trends/ (2016). Accessed 28 Dec 2015
4. Grafy v Trendech: Google Trends. https://support.google.com/trends/answer/4355164?hl= cs&ref_topic=4365530 (2016). Accessed 28 Dec 2015
5. Contreras-Bolton, C., Gatica, G., Barra, C.R., Parada, V.: A multi-operator genetic algorithm for the generalized minimum spanning tree problem. Expert Syst. Appl. **50**, 1–8 (2016). doi:10.1016/j.eswa.2015.12.014
6. Moncla, L., Gaio, M., Nogueras-Iso, J., Mustiére, S.: Reconstruction of itineraries from annotated text with an informed spanning tree algorithm. Int. J. Geogr. Inf. Sci. **30**(6) (2016)
7. Shangin, R., Pardalos, P.: Heuristics for the network design problem with connectivity requirements. J. Comb. Optim. **31**(4), 1461–1478 (2016)
8. Boruvkuv algoritmus: Algoritmy.net. https://www.algoritmy.net/article/1396/Boruvkuv-algoritmus. Accessed 30 Dec 2015
9. Sanjay Kumar, P.A.L.: Renovation of minimum spanning tree algorithms of weighted graph. http://ubiquity.acm.org/article.cfm?id=1353566. Accessed 30 Dec 2015
10. Jarnik - Primuv algoritmus: Algoritmy.net. https://www.algoritmy.net/article/1393/Jarnik-Primuv-algoritmus. Accessed 30 Dec 2015
11. Computer Algorithms: Prim's Minimum Spanning Tree. Stoimen.com. http://www.stoimen. com/blog/2012/11/19/computer-algorithms-prims-minimum-spanning-tree/. Accessed 30 Dec 2015
12. Novotny, J., Dvorak, J., Krejcar, O.: User based intelligent adaptation of five in a row game for android based on the data from the front camera. Lect. Notes Comput. Sci. LNCS **9768**, 133–149 (2016)
13. Computer Algorithms: Kruskal's Minimum Spanning Tree. Stoimen.com. http:// www.stoimen.com/blog/2012/11/12/computer-algorithms-kruskals-minimum-spanning-tree/. Accessed 30 Dec 2015
14. Novotny, J., Dvorak, J., Krejcar, O.: Face-based difficulty adjustment for the game five in a row. Lect. Notes Comput. Sci. LNCS **9847**, 121–134 (2016)
15. Behan, M., Krejcar, O.: Adaptive graphical user interface solution for modern user devices. Lect. Notes Comput. Sci. LNCS **6592**, 411–420 (2012)
16. Benikovsky, J., Brida, P., Machaj, J.: Proposal of user adaptive modular localization system for ubiquitous positioning. LNCS **7197**, 391–400 (2012)
17. Behan, M., Krejcar, O.: Modern smart device-based concept of sensoric networks. EURASIP J. Wirel. Commun. Netw. **2013**(155) (2013)

On Implementation of the Assumption Generation Method for Component-Based Software Verification

Chi-Luan Le, Hoang-Viet Tran and Pham Ngoc Hung

Abstract The assume-guarantee verification has been recognized as a promising method for solving the *state space explosion* in modular model checking of component-based software. However, the counterexample analysis technique used in this method has huge complexity and the computational cost for generating assumptions is very high. As a result, the method is difficult to be applied in practice. Therefore, this paper presents two improvements of the assume-guarantee verification method in order to solve the above problems. The first one is a counterexample analysis method that is simple to implement but effective enough to prevent the verification process from infinite loops when considering the last action of *counterexample* as suffix in implementation. This is done by finding a suffix that can make the observation table not closed when being added to the suffix set of the table and use that suffix for the learning process. The second one is a reduction of the number of membership queries to be asked to *teacher* when learning assumptions. This results in a significantly faster speed in generating assumption than that of the original algorithm. An implemented tool and experimental results are also described to show the effectiveness of the improvements.

1 Introduction

Software quality nowadays plays an important role in our society because software has helped us to improve all matters of our life such as our homes, schools, jobs, etc. With almost all of software in practice, testing has been considered as a major

C.-L. Le
University of Transport Technology, Thanh Xuan, Vietnam
e-mail: luanlc@utt.edu.vn

C.-L. Le · H.-V. Tran · P.N. Hung (✉)
VNU University of Engineering and Technology, Cau Giay, Vietnam
e-mail: hungpn@vnu.edu.vn

H.-V. Tran
e-mail: 15028003@vnu.edu.vn

© Springer International Publishing AG 2017
D. Król et al. (eds.), *Advanced Topics in Intelligent Information and Database Systems*, Studies in Computational Intelligence 710,
DOI 10.1007/978-3-319-56660-3_47

549

solution for guaranteeing software quality. However, testing is not enough for high quality software that requires no error such as plane, train controller systems, etc. With such systems, we will need formal methods to ensure the correctness of systems in both of design and implementation phases. Therefore, many researches have been carried out to improve software quality while keeping software development fast and effective. Two well known approaches that address this problem are theorem proving and model checking [4]. In regards to model checking, assume-guarantee verification has been used as one of the most important method to verify component-based software (CBS) with model checking. This is because it helps us to do verification in a full automatic manner. Moreover, the method is not only suitable for CBS but also for solving the *state space explosion* problem in model checking. It does this by allowing us to verify a target system composed from components by model checking each of them separately.

In assume-guarantee verification, the major problem is how to generate assumptions that satisfy the rules of assume-guarantee. The problem can be solved by using the proposed framework in [5]. The idea of the framework is to generate an assumption A as a contextual assumption about system environment using the L^* algorithm [1, 12]. In this framework, the learning process is performed by the interaction between L^* (from now on called *learner*) and *teacher*. During the learning process, the *teacher* must be able to answer correctly two kinds of queries from *learner* to learn the unknown regular language of A, denoted by $L(A)$. The first one is membership query which is to ask whether a trace σ belongs to $L(A)$. The second one is equivalence query which is to ask if the language of a conjecture C (denoted by $L(C)$) is equivalent to $L(A)$. If $L(C)$ is equivalent to $L(A)$, then C is the needed assumption and *teacher* answers *yes* to *learner*. If $L(C)$ is not equivalent to $L(A)$ but the system does not really violate the given property, *teacher* will return a counterexample *cex* that witnesses the difference between $L(C)$ and $L(A)$. Otherwise, *teacher* will return *no and cex*, where *cex* is corresponding to the actual violation. For the purpose of generating conjectures, *learner* maintains an observation table in form of (S, E, T) and updates it frequently by using membership queries. Whenever this table is closed, *learner* will create a conjecture from the table and submit it to *teacher* as an equivalence query. When *teacher* returns *cex*, *learner* analyzes it to find out a suffix that should be added to E for generating a better conjecture. The algorithm proposed in [5], instead of providing a detailed method to retrieve the suffix to be added to E, refers to the method proposed in [12] for retrieving that suffix. However, the method proposed in [12] has huge complexity. Implementing this method will not always suitable for large scale systems in practice. On the other hand, if we simply add the last action of *cex* to E as suffix, it will lead to a case where the learning process comes into an infinite loop. Therefore, the process fail to generate assumption even though the system does not violate the given property. Moreover, although the assume-guarantee verification method have been well known for a long time, its application in practice is very limited due to the high computational cost in generating assumptions. This is due to the reason that there are many duplicate membership queries which have been asked to *teacher*. Therefore, the method needs improvements so that it can run correctly with lower cost to generate assumptions.

This paper proposes two improvements of the assumption generation method. The first one is an algorithm that simplifies the counterexample analysis process so that it can run without infinite loop in most of the cases in a reasonable time cost. The key idea of this algorithm is to try to add each of the suffixes with the length from one to the length of *cex* to E. After that, the table is updated to see if the updated one is closed. If a suffix can make the observation table not closed, we can add it to the suffix list E. The table is then used to generate a new conjecture to submit to *teacher* as an equivalence query. The learning process continues until *teacher* answers *yes* or *no* with *cex*. The second one is an algorithm to reduce the number of membership queries when updating observation tables. The key idea of the algorithm is that we should only ask membership query for a specific trace σ only once and store the result in a dictionary for later using in the learning process. As a result, the number of membership queries to be submitted to *teacher* will be minimal. This results in the reduction of the computational cost for generating assumptions.

The rest of this paper is organized as follows. At first, we review the original assumption generation method in Sect. 2. An improved counterexample analysis algorithm will be presented in Sect. 3. This section will also describe the algorithm to reduce the number of membership queries when learning assumptions. A support tool and experimental results will be shown in Sect. 4. Section 5 presents an overview about the researches that are related to the topic. Finally, we conclude the paper in Sect. 6.

2 The Original Assumption Learning Algorithm

2.1 Generating Assumption Using L* Algorithm

Let M be a system that consists of two components M_1 and M_2 and a property p. The original assumption learning algorithm proposed in [5] generates a contextual assumption using the L^* algorithm [1]. The details of this algorithm are shown in Algorithm 1. In order to learn assumption A, Algorithm 1 maintains an observation table (S, E, T). The algorithm starts by initializing S and E with λ (i.e., an empty string) (line 2). After that, the algorithm updates the observation table (S, E, T) by using membership queries (line 4). While (S, E, T) is not closed, the algorithm continues adding sa to S and updating the observation table to make it closed (from line 5 to line 8). When the observation table is closed, the algorithm creates a conjecture C from (S, E, T) and asks *equivalence query* to *teacher* (from line 9 to line 11). If C is the needed assumption, the algorithm stops and returns C (line 13). Otherwise, it analyzes the returned counterexample *cex* to add the suffix e that witnesses the counterexample to E (line 15) and continues the learning process again from line 4.

Algorithm 1: Learning Assumptions for Compositional Verification

---

```
 1 begin
 2  │  Let S = E = {λ}
 3  │  while true do
 4  │  │    Update T using membership queries
 5  │  │    while (S, E, T) is not closed do
 6  │  │    │    Add sa to S to make (S, E, T) closed where s ∈ S and a ∈ Σ
 7  │  │    │    Update T using membership queries
 8  │  │    end
 9  │  │    Construct candidate DFA M from (S, E, T)
10  │  │    Make the conjecture C from M
11  │  │    Ask equivalence query for the conjecture C
12  │  │    if C is correct then
13  │  │    │    return C
14  │  │    else
15  │  │    │    Add e ∈ Σ* that witnesses the counterexample to E
16  │  │    end
17  │  end
18 end
```

2.2 Updating Observation Table

While Algorithm 1 is learning assumption, a very important step is to update the observation table. The details of this step are presented in Algorithm 2. For every $s \in S$ or $sa \in S.\Sigma$ (line 2), the algorithm concatenates this with each of e in E (line 3). Then, it asks *teacher* a *membership query* for $s.e$ or $sa.e$ (line 4) (where "." is the concatenation operator). After that, it updates the corresponding T in (S, E, T) with the result of this membership query (line 5). When finishing this process, it returns the updated observation table (S, E, T) (line 8).

Algorithm 2: Updating Observation Table

---

input : An observation table (S, E, T)
output: The updated observation table (S, E, T)

```
 1 begin
 2  │  forall s ∈ S or sa ∈ S do
 3  │  │    forall e ∈ E do
 4  │  │    │    t ← ask membership query for s.e or sa.e
 5  │  │    │    Update the corresponding T in (S, E, T) with t
 6  │  │    end
 7  │  end
 8  │  return (S, E, T)
 9 end
```

---

3 Two Improvements for the Assumption Generation Method

From our observation, we have seen that if we simply apply Algorithm 2 to update observation tables, there will be a lot of duplicated membership queries. This will dramatically affect the assumption generation process when dealing with large scale systems. Following sub-sections propose algorithms to choose suffix from counterexample and to reduce the number of membership queries to be asked to *teacher*.

3.1 An Improvement on Counterexample Analysis

Although the idea of trying all of the possible suffixes with the length increased one by one from the counterexample is not new and can be found in other works such as in [10], no one has ever applied the idea in assume-guarantee reasoning. When applying the idea in assume-guarantee reasoning, we have an effective method to analyze the counterexample as shown in Algorithm 3. When *teacher* processes an equivalence query with a conjecture argument C, if C is not the satisfied assumption, but M does not violate the property p, *teacher* will return a counterexample *cex*. Algorithm 3 analyzes *cex* to choose an appropriate suffix to add to E. The idea of this algorithm is to try to add each of the suffix which has length from one to *cex*'s length to E to find out which suffix will make the observation table not closed. For this purpose, the algorithm uses a loop for all of the possible suffixes of *cex* from line 2 to line 10. For each of the suffix e, the algorithm clones the observation table (S, E, T) and stores in OT (line 3). This is because processing with OT will not affect the current (S, E, T). After that, the algorithm adds e to E of OT, updates OT, and checks if OT is closed (from line 5 to line 7). If the updated OT is not closed, then adding e to E of (S, E, T) will make (S, E, T) not closed. In order to make it closed, an *sa* will need to be added to S. This will make the next conjecture C' different from the previous conjecture C. As a result, Algorithm 1 can continue learning assumption. Therefore, Algorithm 3 returns *yes* and e as the needed suffix for adding to the input observation table (line 8). If there is no e that can make the observation table not closed, then the algorithm returns *no*. This means that Algorithm 1 will come into an infinite loop. We will need to find another solution so that the algorithm can continue running correctly. This kind of solution will be one of our future work and not be mentioned in this paper.

3.2 Reducing the Number of Membership Queries

Although the method shown in Algorithm 3 can prevent Algorithm 1 from running infinitely in some special cases, it costs us more time to do that than Algorithm 1. We propose an algorithm to improve the performance of the whole learning process

---

Algorithm 3: Choosing suffix from counterexample

---

    input : The current observation table (S, E, T), the counterexample *cex*
    output: *yes* + *e* **or** *no*

1 **begin**
2   **foreach** *counter* = *1* **to** *cex's length* **do**
3          $OT \leftarrow$ *the cloned table of* (S, E, T)
4          $e \leftarrow$ *suffix* with length is counter
5   Try to add e to $OT's\ E$
6   Update OT with membership queries
7   **if** OT *is not closed* **then**
8   **return** *yes* + *e*
9   **end**
10   **end**
11   **return** *no*
12 **end**

---

by reducing the number of membership queries. Although this improvement seems to be trivial and obvious when implementing the assume-guarantee reasoning, but with the large number of membership queries can be reduced and in the context of software evolution where the software needs to be rechecked whenever there is any changes, the improvement can play an important role in reducing the cost of software verification in practice. Details of the algorithm is shown in Algorithm 4. For this purpose, we use a dictionary *dict* to store list of query results in form of couple $\langle str, t \rangle$, where *str* is the trace that is passed to *teacher* as a membership query and *t* is the corresponding result. For each *str* to be passed to *teacher* as a membership query, if it exists in *dict*, then its value will be used to update the observation table (line 5 to line 7) without asking a new membership query result to *teacher*. Otherwise, it will ask a new membership query to *teacher*, store the result to *dict*, and update the observation table (line 8 to line 11). *dict* will be used throughout the assumption learning process to improve the learning performance.

4 Experiments

We have implemented the two improvements in Algorithm 3 and Algorithm 4 into an application called IAGTool[1] in order to compare assumption generation performance of the original in [5] and improved algorithms. The algorithm is developed using Microsoft Visual Studio 2015 Community [11]. The test is carried on a machine with the following system information: Processor: Intel(R) Core(TM) i5-3230M; CPU: @2.60 GHz, 2601 MHz, 2 Core(s), 4 Logical Processor(s); OS Name: Microsoft Windows 10 Home; IDE: Visual Studio Community 2015. The experimental results are shown in Table 1. In this table, there are three kinds of results of the original

---

[1] http://www.coltech.vnu.edu.vn/~hungpn/IAGTool/.

Algorithm 4: Improved Observation Table Update

---

input : An observation table (S, E, T), the Membership queries result dictionary *dict*
output: The updated observation table (S, E, T)

```
1  begin
2  |   forall s ∈ S or sa ∈ S do
3  |   |   forall e ∈ E do
4  |   |   |   str ← s.e or str ← sa.e
5  |   |   |   if dict contains str then
6  |   |   |   |   t ← get value of str from dict
7  |   |   |   |   Update the corresponding T in (S, E, T) with t
8  |   |   |   else
9  |   |   |   |   t ← ask membership query for str
10 |   |   |   |   Store ⟨str, t⟩ to dict
11 |   |   |   |   Update the corresponding T in (S, E, T) with t
12 |   |   |   end
13 |   |   end
14 |   end
15 end
```

---

learning algorithm without Algorithm 3 (denoted by *Original*), original learning algorithm with Algorithm 3 (denoted by *Original+*), and improved learning algorithm with Algorithms 3 and 4 (denoted by *Improved*) to compare for each of test cases (with safety property p). The number of membership queries shown in column "Queries Number". That allow us to calculate how many queries are saved using the improved algorithm. The "Time (ms)" columns show us how much time the improved algorithm and the original one take to generate assumptions. "–" values indicate cases where the learning algorithm failed to generate required assumptions. Among the test cases shown in Table 1, "TestCase1" is the example presented in [5]. From the experimental results shown in Table 1, we have the following observations:

- The *Original* algorithm can only generate assumptions for TestCase1 and Test-Case4. In most of other test cases, the algorithm failed to generate assumption

Table 1 Experimental results

No.	M1	M2	p	Original		Original+		Improved	
				Queries number	Time (ms)	Queries number	Time (ms)	Queries number	Time (ms)
TestCase1	3	3	2	52	4	52	3	17	1
TestCase2	42	5	3	–	–	875	10,278	161	4032
TestCase3	23	11	6	–	–	465	8889	82	3380
TestCase4	78	22	7	78	13,495	78	7361	29	3981
TestCase5	30	12	3	–	–	2440	11,064	339	3982
TestCase6	65	32	3	–	–	1421	651,131	229	276,253

correctly due to the infinite loop during the learning process. In the meantime, when using the improved counterexample analysis algorithm, both of the *Original+* and *Improved* algorithms successfully pass the loop and are able to generate the required assumptions for these test cases.

- The number of membership queries is dramatically reduced by using Algorithm 4. There is no test case where the number of membership queries in *Original+* is the same or less than that of the *Improved* algorithms because even with the smallest test case (TestCase1), the learning process needs two closed observation tables during assumption generation. Therefore, several membership queries have been saved thanks to Algorithm 4.
- The *Improved* algorithm runs much faster than the *Original+* algorithm. This is because Algorithm 4 implemented in *Improved* algorithm has saved several membership queries during the learning process. As a result, generating assumptions of the *Improved* algorithm has faster speed than the original algorithm in [5].
- When running with such small test cases as shown in Table 1, the *Improved* algorithm runs much faster than the *Original+* one. Therefore, in practice, it could improve the assumption generation speed dramatically.
- In order to implement Algorithm 4, we need to create a dictionary to store membership query results. That costs us some more memory. However, with the current hardware technology, this will be a cost-effective method to improve the whole speed of verification process.

5 Related Works

There are a lot of researches related to optimizing the L^* based assume-guarantee verification. Consider only the most current works, we can refer to [2, 3, 6–9].

Chaki and Strichman proposed three optimizations in [2] to the L^* based automated Assume-Guarantee reasoning algorithm for the compositional verification of concurrent systems. The paper suggested an optimization that uses some informations that are already available to *teacher* in order to avoid many unnecessary membership and candidate queries. Sharing concern about improving the assumption generation speed, our researches proposed two improvements on this. The first one is to improve suffix choosing process that will prevent the original assumption generation method from coming into an infinite loop if choosing the last action *cex* as suffix. The second one is to reduce the number of membership queries by another kind of observation that traces submitted to *teacher* for membership queries are duplicate many times.

In a series of papers of [7–9], Hung et al. proposed a method for generating minimal assumptions, improving, and optimizing that method to generate those assumptions for compositional verification. However, that is for the result of the verification, not improve the method to generate the assumption itself. This paper shares the interest of improving the compositional verification, but we focus on improving the method itself so that it has faster speed than the original one.

In 2010, Chen et al. proposed a pure method for learning assumption through implicit learning in [3]. This has a great result on having faster speed than the original assumption generation method. Nevertheless, it focuses on a brand new approach that uses a specification method with Boolean functions. We share the interest about compositional verification, but we focus on the original assumption generation method to improve it in order to prevent it from running infinitely if choosing the last action *cex* as suffix and to have the faster speed.

In [6], Gupta et al. proposed a method to compute an exact minimal automaton to act as an intermediate assertion in assume-guarantee reasoning, using a sampling approach and a Boolean satisfiability solver. This is an approach which is suitable to compute minimal separating assumption for assume-guarantee reasoning for hardware verification. Our approach focuses on the original assumption generation method to improve it by reducing the number of membership queries and improving the counterexample analyzing algorithm to choose correct suffix that prevent the algorithm from running infinitely if choosing the last action *cex* as suffix.

In [10], Maler and Pnueli have mentioned the idea of analyzing the counterexample when learning infinitary Regular Sets. We share the idea of analyzing counterexample when implementing the L^* algorithm, but we apply it for the context of assume-guarantee reasoning. With this small finding, our proposed algorithm makes it easier for the implementation of software verification. Although it seems to be small change, but it can prevent the original algorithm from running infinitely if choosing the last action *cex* as suffix.

6 Conclusion

In order for the assume-guarantee paradigm to be used effectively in practice with large-scale systems, its assumption generation time must be improved as much as possible. We have presented a method to do this by preventing *learner* from asking membership queries for traces that are already been asked. This is done by creating a dictionary to store membership queries results and use it whenever *learner* wants to ask membership queries. *Learner* can only ask membership query for a trace that has never been asked. We have also applied the idea for choosing suffixes from counterexamples in [10] in the implementation of assume-guarantee reasoning so that the learning process will not run endlessly if we simply consider the last action of *cex* as suffix. This is done by trying to add suffixes with length from one to the length of *cex* to E of the observation table. If a suffix can make the observation not closed, that is the one to be added to E. The experimental results included in this paper also show that the proposed method have improved the assumption generation time significantly and there is no infinite loop when learning assumptions for the presented test cases.

Although the presented methods have a very positive effect on the assumption learning process, there are a lot of things need to be done. The first one is already described in Sect. 3.1. The solution in Algorithm 1 will also not be able to run correctly when Algorithm 3 returns *no*. We will need another research to analyze such cases to find out a solution so that we can generate the required assumption if it exists. The second one is that we are analyzing if there is any case where the learning process faces infinite loop. The third one is that the proposed framework in [5] is not for evolving systems. What needs to be done in order to generate the best assumption for evolving systems. Additionally, we are also in process of applying the proposed algorithms to larger systems in practice in order to verify their correctness and usefulness when doing verification.

Acknowledgements This work is supported by the project no. QG.16.31 granted by Vietnam National University, Hanoi (VNU).

References

1. Angluin, D.: Learning regular sets from queries and counterexamples. Inf. Comput. **75**(2), 87–106 (1987)
2. Chaki, S., Strichman, O.: Optimized L*-based assume-guarantee reasoning. In: Tools and Algorithms for the Construction and Analysis of Systems: 13th International Conference, TACAS'07. Proceedings, pp. 276–291. Springer, Berlin, Heidelberg (2007)
3. Chen, Y.-F., Clarke, E., Farzan, A., Tsai, M.-H., Tsay, Y.-K., Wang, B.-Y.: Automated assume-guarantee reasoning through implicit learning. In: Touili, T., Cook, B., Jackson, P. (eds.) Computer Aided Verification. Lecture Notes in Computer Science, vol. 6174, pp. 511–526. Springer, Berlin, Heidelberg (2010)
4. Clarke Jr., E.M., Grumberg, O., Peled, D.A.: Model Checking. MIT Press, Cambridge, MA, USA (1999)
5. Cobleigh, J.M., Giannakopoulou, D., Păsăreanu, C.S.: Learning assumptions for compositional verification. In: Proceedings of the 9th International Conference on Tools and Algorithms for the Construction and Analysis of Systems, TACAS'03, pp. 331–346. Springer, Berlin, Heidelberg (2003)
6. Gupta, A., Mcmillan, K.L., Fu, Z.: Automated assumption generation for compositional verification. Form. Methods Syst. Des. **32**(3), 285–301 (2008)
7. Hung, P.N., Nguyen, V.H., Aoki, T., Katayama, T.: A Minimized Assumption Generation Method for Component-Based Software Verification. In: Theoretical Aspects of Computing—ICTAC'09: 6th International Colloquium. Proceedings, pp. 277–291. Springer, Berlin, Heidelberg (2009)
8. Hung, P.N., Nguyen, V.H., Aoki, T., Katayama, T.: An improvement of minimized assumption generation method for component-based software verification. In: Computing and Communication Technologies, Research, Innovation, and Vision for the Future (RIVF), pp. 1–6, Feb 2012
9. Hung, P.N., Nguyen, V.H., Aoki, T., Katayama, T.: On optimization of minimized assumption generation method for component-based software verification. IEICE Trans. **95-A**(9), 1451–1460 (2012)
10. Maler, O., Pnueli, A.: On the learnability of infinitary regular sets. Inf. Comput. **118**(2), 316–326 (1995)
11. Microsoft: Visual studio community. https://www.visualstudio.com/en-us/products/visual-studio-community-vs.aspx (2015)
12. Rivest, R.L., Schapire, R.E.: Inference of finite automata using homing sequences. In: Proceedings of the Twenty-first Annual ACM Symposium on Theory of Computing, STOC '89, pp. 411–420. ACM, New York, NY, USA (1989)

Author Index

© Springer International Publishing AG 2017
D. Król et al. (eds.), *Advanced Topics in Intelligent Information
and Database Systems*, Studies in Computational Intelligence 710,
DOI 10.1007/978-3-319-56660-3

Printed in the United States
By Bookmasters